# Biomedical Engineering Principles in Sports

**BIOENGINEERING, MECHANICS, AND MATERIALS: PRINCIPLES AND APPLICATIONS IN SPORTS**

**Series Editors: George K. Hung**
*Rutgers University*
*Piscataway, New Jersey, USA*

**Jani Macari Pallis**
*Cislunar Aerospace, Inc.*
*San Francisco, California, USA*

---

BIOMEDICAL ENGINEERING PRINCIPLES IN SPORTS
Edited by George K. Hung and Jani Macari Pallis

# Biomedical Engineering Principles in Sports

Edited by

George K. Hung
*Rutgers University*
*Piscataway, New Jersey*

and

Jani Macari Pallis
*Cislunar Aerospace, Inc.*
*San Francisco, California*

**Kluwer Academic / Plenum Publishers**
New York, Boston, Dordrecht, London, Moscow

Library of Congress Cataloging-in-Publication Data

Biomedical engineering principles in sports/edited by George K. Hung and Jani Macari Pallis.
p. cm. — (Bioengineering, mechanics, and materials: principles and applications in sports)
ISBN 0-306-48477-3
1. Sports—Physiological aspects. 2. Human mechanics. 3. Biomedical engineering. I. Hung, George K. II. Pallis, Jani Macari, III. Series.

RC1235.B56 2004
612′.044—dc22

2004043510

ISSN 1573-2428

ISBN 0-306-48477-3


233 Spring Street, New York, New York 10013

http://www.wkap.nl/

10 9 8 7 6 5 4 3 2 1

A C.I.P. record for this book is available from the Library of Congress



Printed in the United States of America

To my mother, Dorothy Loh Hung, who nurtured me through illness and health; to the memory of my father, Ernest K. Hung, who brought his five children to the United States with $40 in his pockets and a desire to improve our lives; to my sisters Evelyn and Mary and brothers Anthony and Henry for their unbridled support; and to my wife Xiaosi Yang, for her love and devotion:

- G.K.H.

To my husband, Jim, for his constant patience and love; and to George Hung and Aaron Johnson (publishing editor at Kluwer) for the great opportunity they have provided.

- J.M.P.

## FOREWORD

First there were coaches, then came specialty coaches, followed by fitness experts, rehabilitations therapists, sports nutritionists, strength and conditioning specialists, sport psychologists, orthopedic specialists, movement specialists, message therapists and biomechanical experts. All of this to support athletes in their quest to find the outer limits of what's possible for them and for their teams. Science and technology have transformed nearly every dimension of our lives and sport is no exception. Sport science technology is taking sport to unprecedented levels of safety, health and performance. It's also challenged traditional beliefs about what's possible and not possible in sport. Just as Tiger Woods has redefined the boundaries of golf, the science of sport is helping to redefine the boundaries of sport for nearly every participant. And at the world class level, the science of sport has opened new doors for unprecedented, awe-inspiring athletic achievements.

And now there is a new kid on the block-sport engineers! They specialize in applying biomedical engineering to the world of athletic performance. Their goals are fewer injuries, faster learning, better execution, more enjoyment, better equipment and breakthrough performances. Sports engineers are all part of what I refer to as the "total solution for athletes." The achievement of maximum performance is a multi-dimensional quest and real solutions must also be multi-dimensional. Sport engineers bring fresh new interdisciplinary, multi-dimensional solutions to the performance challenges faced by today's Athletes. From world class sports to little league pitching, sports engineers add exciting new tools for analysis and measurement and bring new thinking and technology to the living laboratory of athletic performance. The future of all science is in connecting the big ideas, in sharing findings, methodology and perspective. Science's greatest breakthroughs invariably come from interdisciplinary, holistic thinking. This is precisely why this book offers so much promise. It truly represents frontier, interdisciplinary science applied to the exciting arena of sport.

The thinking and science contained in this book push the envelope in our understanding of an appreciation for the world of athletic performance.

Included are chapters on the physics, mechanics and engineering methodology applied to golf as well as tennis biomechanics, shoe-surface interactions in tennis and the aerodynamics of the tennis ball. Also included are chapters in the biomechanics of pitching and batting in baseball, the mechanics of the forward pass and tackling in football and biomechanical considerations for soccer. Critical issues in basketball and selected topics in women's and children's sports participants are also reviewed. The information contained in this book will clearly change the way athletes, coaches and sport science practitioners function in the world of sport.

Dr. Jim Loehr

Chairman and CEO
LGE PERFORMANCE SYSTEMS, INC.
9757 Lake Nona Road
Orlando, Florida 32827
www.fullengagement.com

## PREFACE

There are hundreds of books on physics, biomechanics, and physiology of sports. They all provide important contributions to the understanding of the science of sports. However, most concentrate on one aspect, such as physics of sports, that deal in great depth but also into areas that may be of relatively little concern to the intelligent layperson. Moreover, some of the books are at least a decade old, and could not have incorporated some of the significant recent advances in imaging, modeling, and signal processing techniques. This edited book attempts to bridge this gap by providing the biomedical engineering student, the dedicated athlete, and the intelligent layperson, a book with the latest advances in biomechanics and sports medicine in the sports that they either actively participate in or enjoy watching. This includes golf, tennis, baseball, football, soccer, and basketball.

Within each sport, there are many facets associated with good performance. For example, in golf, club selection based on wind, terrain, and green hardness are important considerations. In tennis, court coverage and angles of shots are important aspects of defensive and offensive strategies. In baseball, the batting history of a batter is important to the pitcher in determining what type of pitch to deliver. Also, in football, the defensive formation must be changed in accordance with the offensive set. Moreover, in soccer as well as basketball, dribbling and passing are vital fundamentals. However, these topics are not covered in this book, because incorporating them would make the book too broad and rather unwieldy. Instead, we are primarily concerned with the simple acts of throwing, shooting, hitting, or kicking a ball, as well as tackling an opponent. We believe these are the areas that the sports-minded person enjoys most.

The book is divided into six sections. The first five sections correspond to the sports areas discussed above (with American football grouped with soccer). The sixth section is concerned with training normal vision for improving performance, maturation of skills in children, and medical advances in the treatment of sports injuries. In two of the chapters (Chapters 10 and 15), biographical sketches of the authors are provided to highlight their unique contributions to the science and practice of sports. The Foreword is written by Dr. Jim Loehr, a pioneer in the field of performance psychology and the recipient of numerous awards related to sports and health.

The book is aimed first towards college seniors and beginning graduate students in biomedical engineering, mechanical engineering, sports medicine, physiology, optometry, and psychology, who will gain a broad understanding of quantitative analysis of sports science. In addition, it has considerable depth and important new information in each area to be useful as an updated reference and tutorial for graduate and post-doctoral students, as well as established sports scientists.

George K. Hung, Ph. D.
Piscataway, New Jersey

Jani Macari Pallis, Ph.D.
San Francisco, California

## CONTENTS

# I

# GOLF

# Chapter 1

# Aerodynamics of the Golf Ball

Alexander J. Smits[1], Steven Ogg[2]

[1]*Department of Mechanical and Aerospace Engineering, Princeton University, Princeton, NJ 08544, PH (609) 258 5117, FX (609) 258 1918, EM asmits@princeton.edu*

[2]*Callaway Golf, 2180 Rutherford Road, Carlsbad, CA 92008-7328, PH: (760) 804-4061, FX: (760) 804-4003, EM: SteveO@callawaygolf.com*

## 1.1 INTRODUCTION

The aerodynamics of golf balls is still not well understood, primarily because the aerodynamic performance of the golf ball depends crucially on the details of the airflow over the ball, and those details are controlled by many factors. As the ball flies through the air, it develops lift and drag forces that depend on its velocity, spin rate, the atmospheric conditions, and, most importantly, the shapes of the individual dimples, and their arrangement over the surface of the ball. It has been said that it is more difficult to predict the performance of a golf ball than it is to predict the performance of a full-scale aircraft (see, for example, Moin & Kim, 1997). Nevertheless, a great deal of experimental work has been done, and our basic understanding is reasonably complete. In this chapter, we describe the factors that govern the aerodynamic performance of a golf ball. We will discuss the complicated nature of the flow over a spinning golf ball, and why the dimples, and their arrangement on the surface, play such a crucial role. We will describe the design methods used by the golf ball companies in their search for better performance, and the testing procedures used by the regulating bodies, notably the USGA, in enforcing the Rules.

We can identify three separate phases in the motion of a golf ball. Consider a typical tee shot by a low-handicap golfer (Fig. 1.1). First, there is the impact of the club with the ball. The contact time is extremely short, typically about 400 to 500 μs, but during this time the ball accelerates from rest to a velocity of about 230 ft/s and a spin rate between 2000 and 3000 rpm. The ball deforms extensively during contact with the club, compressing to about 80% of its diameter (Fig. 1.2), but regains its original shape within a couple of diameters along its trajectory.

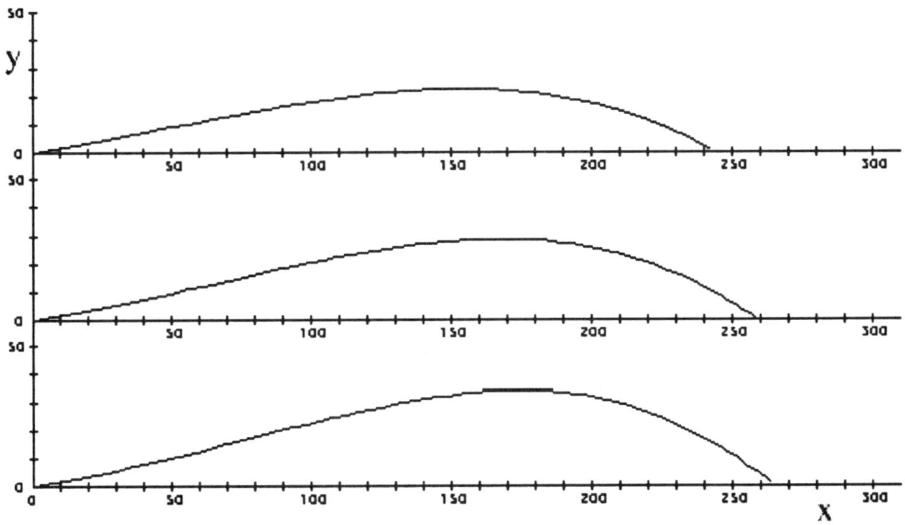


*Figure 1.1.* Typical trajectories of a driver shot, initial velocity 240 ft/s, launch angle 10°. Top: 1800 rpm; Middle: 3000 rpm; Bottom: 4200 rpm.

The second phase is the actual flight of the golf ball, which is about 250 to 300 yards long and about 35 yards high at its maximum point. Aerodynamics plays a crucial role in this phase, and it will dictate where and when the ball lands. The third and final phase of the motion begins when the ball impacts the ground. It may bounce a number of times before rolling some distance and finally coming to rest. Aerodynamics does not play a significant role in this phase.

The study of golf ball aerodynamics may have a history as long as the game itself (Thomson, 1910). The most significant event in golf ball aerodynamics, however, took place soon after the introduction of the gutta-percha ball in 1848, invented by Rev Adam of St. Andrews (before that time, golf balls were made of leather covers stuffed with feathers, called the "feathery" golf ball). Gutta percha is a natural rubber, and the ball was made by filling a mold with latex, which was then pressed and cured at high temperature. Golfers discovered, probably by accident, that the ball flew further and better when scored or marked. Chase (1981) credits this discovery to a professor at St. Andrew's University, but it clearly became common knowledge very rapidly, and it initiated a wave of innovation in cover design (Fig. 1.3).

*Figure 1.2.* Deformation of a golf ball on impact with a driver (photo courtesy of Callaway Golf Company).

*Figure 1.3.* Some early gutta percha balls showing different surface treatments. From left to right: "Smooth gutty," circa 1850's; "Knife cut gutty," circa 1860 to 1870's; "Ocobo molded gutty," circa 1890's; "Finch bramble gutty" circa 1890-1900. Images courtesy of Leo M. Kelly, Jr.: http://www.oldgolf.com.

By the turn of the century, the solid gutta-percha ball had given way to the wound ball, where rubber thread was wound under high tension around a small rubber core, and the wound core was encased in a gutta-percha cover. This was the Haskell ball, designed and patented by Coburn Haskell in 1898. Balata (a thermoset plastic) then replaced gutta percha for the cover, since it had a better adhesion to the rubber windings, and a better color and toughness. It is the first rubber-cored ball. The modern golf ball is generally of a "two-piece" design, first introduced by Spalding through the Executive ball in 1972, where the rubber windings were replaced by a solid polybutadiene core and the cover was made of blends of synthetic polymers that gave greater consistency and better durability (Sullivan and Melvin,

1994), although some manufacturers, notably Maxfli and Titleist, still manufacture a wound ball.

Dimples were first patented by William Taylor in 1901 (Martin, 1968) and by 1909 had begun to replace score marks. By 1930, the round dimple had become accepted as the standard design for golf balls, with 330 or 336 round dimples in regular rows (Bearman and Harvey, 1976). The last twenty years have seen great innovation in dimple design. By one count over 2907 patents on golf ball design were issued between 1976 and 2003, with about 163 focusing exclusively on dimple design. The number of dimples has ranged from as low as 252 up to a high of 812 (Thomas, 2002), and their shape has gone from circular to hexagonal, with many variations in between. Their arrangement on the surface has also seen many variations with those based on the icosahedral design, where the ball surface is divided into segments defined by 20 equilateral triangular patches, becoming very popular (see, for example, the 1932 British patent by Pugh, and the 1984 US patent by Ayoma, #4,560,168). Today's ball is likely to have about 384 dimples, of 5 to 6 different sizes, covering about 80% of the ball surface (Fig. 1.4).

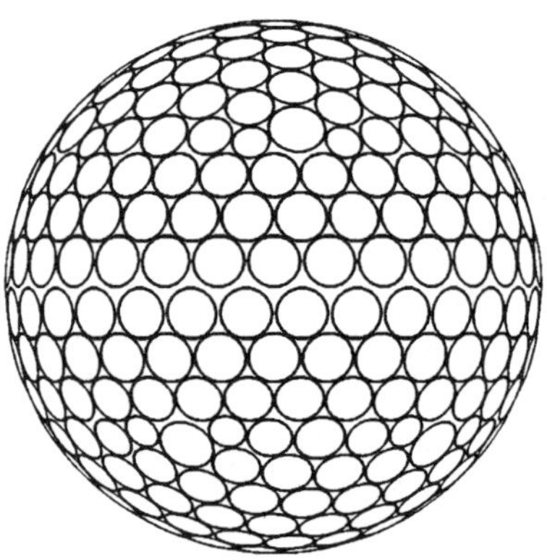

*Figure 1.4.* Callaway CTU 30 ball, featuring a solid polybutadiene core and a cast thermoset urethane cover. Image courtesy of Callaway Golf Company.

Once the ball leaves the club, its motion is governed by the laws of aerodynamics and the force of gravity. The trajectory of the golf ball is independent of its construction, and the only important characteristics are its geometrical shape, size, mass, and its moment of inertia. That is, the shape and distribution of the dimples on a golf ball, together with its size and inertial characteristics, completely determine its aerodynamic performance. It is for this reason that there is currently such intense competition among golf ball manufacturers to produce innovative dimple shapes and patterns. Very small changes in dimple design can have important consequences for the ball trajectory, particularly its "carry" distance (the distance between tee and impact with the ground), the maximum trajectory height, and the angle of incidence at the point where it hits the ground. These quantities are important to a golfer, and there is continuous innovation required by the manufacturers to satisfy the demand from golfers for a better, longer, more accurate ball.

The desire for increased distance comes squarely up against the Rules of Golf, as published jointly by the United States Golf Association (USGA) and the Royal and Ancient Golf Club of St. Andrews. In particular, the Overall Distance Standard requires that the ball, when struck by a mechanical golfer using a driver under specified conditions, not travel more than a total distance (carry and roll combined) of 296.8 yards, which includes a +6% "testing tolerance" that could be reduced to 4% as testing methodology improves. The golf ball manufacturers, therefore, have invested heavily in developing sophisticated tools and techniques for measuring and predicting the performance of golf balls, both to design better balls, and, most importantly, not to infringe the Rules.

## 1.2 BASIC AERODYNAMIC PRINCIPLES

Once the ball is launched, the only forces acting on the ball are those due to gravity and the air acting on the ball. The aerodynamic force has a lift component (due to the spin on the ball) and a drag component (due to the friction between the air and the surface of the ball).

### 1.2.1 Scaling Laws

In principle, the forces acting on the ball can be found by solving the equations governing the flow of fluids, known as the Navier-Stokes equations, together with the appropriate boundary conditions. For most practical flows, this represents a very challenging problem, and it is almost always necessary to resort to a computer. However, for the flight of a golf ball under realistic conditions, it is extremely difficult to obtain computer-generated solutions, even when using the most powerful computers available today. We resort, by necessity, to experimental work. Experiments are also challenging, and it can be labor-intensive to obtain accurate and complete results. Fortunately, we can use a powerful tool called dimensional analysis to significantly reduce the amount of work involved. Dimensional analysis is a common tool in aerodynamics, as well as in other fields where the complexities of the physical problem do not allow analytical or computational solutions. In particular, it allows the organization and presentation of experimental data in a very efficient and elegant manner (see, for example, Kline, 1965).

We begin by noting that the aerodynamic force $F$ acting on the ball depends on a number of important parameters, including the velocity of the body $V$, its size, the spin rate $\omega$, the fluid density $\rho$ and viscosity $\mu$, and the nature of the surface. Note that the spin rate at any point in the trajectory depends on the initial spin imparted to the ball by the club, and the rate of spin decay, which depends on the moment of inertia, among other parameters (see Section 1.3.3). The viscosity of a fluid is a material property that measures the rate of deformation of the fluid under the application of a stress. Honey, for example, has a higher viscosity than water, which in turn has a higher viscosity than air. The size of the ball is described by its diameter, $D$, or radius $R$ and its associated cross-sectional area, $A = \pi R^2$. The nature of the surface is determined by the dimple pattern, and each particular dimple pattern will have its unique influence on the ball performance. To start with, we will assume that the effects of the dimple pattern can be described by a single length scale, such as, for example, the rms amplitude of the dimple depth, $k$. In functional form, we have

$$F = f\left(V, D, \omega, \rho, \mu, k\right) \tag{1.1}$$

where f indicates a functional relationship. Dimensional analysis allows us to write this relationship as

$$\frac{F}{\frac{1}{2}\rho V^2 A} = g\left(\frac{\rho V D}{\mu}, \frac{\omega R}{V}, \frac{k}{D}\right) \tag{1.2}$$

Non-dimensionalizing the problem in this way has an important consequence: instead of describing the relationship between the force acting on the ball and the other parameters in terms of 7 dimensional variables, as in Eq. 1.1, we can describe it in terms of 4 non-dimensional variables, as in Eq. 1.2.

In terms of the lift and drag forces $F_L$ and $F_D$ (defined, respectively, as the components of the total force taken in the direction normal to and along the direction of motion of the ball; see Fig. 1.11 below), we can write this relationship as

$$C_L = g_1(\mathrm{Re}, W, \varepsilon) \tag{1.3}$$

and

$$C_D = g_2(\mathrm{Re}, W, \varepsilon) \tag{1.4}$$

where $C_L$ and $C_D$ are the lift and drag coefficients, respectively, defined by

$$C_L = \frac{F_L}{\frac{1}{2}\rho V^2 A} \qquad \text{and} \qquad C_D = \frac{F_D}{\frac{1}{2}\rho V^2 A}. \tag{1.5}$$

The other non-dimensional groups are the Reynolds number Re, the spin rate parameter $W$, and the relative roughness, $\varepsilon$, are given by

$$\mathrm{Re} = \frac{\rho V D}{\mu}, \qquad W = \frac{\omega R}{V}, \qquad \varepsilon = \frac{k}{D}. \tag{1.6}$$

To understand the role played by the dimples and the spin rate, it is useful to consider first the case of a non-spinning ball. Here, there is no lift, only drag (we shall see why later), and Eq. 1.4 reduces to

$$C_D = g_3(\mathrm{Re},\ \varepsilon) \tag{1.7}$$

That is, the drag coefficient is only a function of Reynolds number and relative roughness. Figure 1.6 shows this behavior, obtained as the result of many experiments on smooth and rough spheres, including a representative golf ball (the range of Reynolds numbers experienced by a golf ball during a typical driver shot is approximately 50,000 to 200,000, corresponding to balls traveling at 60 to 240 ft/s in air at 70°F and atmospheric pressure). There are different curves depending on the value of the relative roughness, but all curves show a similar behavior: at low Reynolds numbers the drag coefficient is approximately constant at a value of about 0.5, while at higher Reynolds numbers the drag coefficient suddenly reduces to a much lower value of about 0.1 or 0.25, before slowly increasing.

The value of the Reynolds number where the drag coefficient takes a sudden dip is called the "critical" Reynolds number. For a golf ball its value lies somewhere between 40,000 and 60,000, whereas for a smooth ball it is between 350,000 and 400,000 (Fig. 1.5). For a Reynolds number above the critical value, a golf ball has about half the drag of a smooth ball. That is, for velocities greater than about 70 ft/s, a golf ball experiences less than half the air resistance experienced by a smooth ball of the same size.

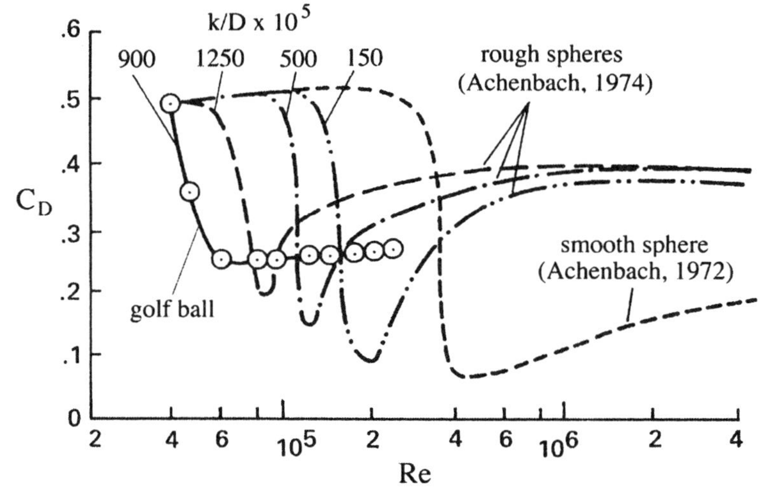


*Figure 1.5.* Drag coefficients as a function of Reynolds number for spheres with different degrees of roughness. Adapted from Bearman and Harvey (1976). With permission of the Royal Aeronautical Society

## 1.2.2 Laminar and Turbulent Flow

To understand the physical mechanisms that cause this sudden change in drag coefficient, we need to understand more about the flow over a sphere in the region of the critical Reynolds number. In the frame of reference of a ball moving at a velocity V, the air appears to be moving toward the ball at a constant velocity V. However, in the vicinity of the ball, the air velocity changes significantly. Near the front of the ball, it first decreases to a value less than V because of the blocking effect of the ball, and then increases to a value greater than V as the flow reaches the point of maximum area. According to Bernoulli's principle, the region of accelerating flow is also a region of falling pressure. Beyond the point of maximum area, the flow slows down again, and so the pressure starts to rise.

These observations apply to the flow behavior in what is called the "freestream" region, but there is another very important region found near the surface where a thin "boundary layer" is found. Boundary layers are produced whenever a fluid flows over a solid surface, and they represent a region where the velocity varies from zero at the surface (relative to the ball), to the full freestream value a short distance away. The air in contact with the ball surface is forced to travel at the same speed as the ball by the strength of the attractive forces between the molecules of air and the molecules of the solid surface (this is called the "no-slip" condition). The boundary layer, therefore, is a region where strong velocity gradients exist that lead to significant viscous forces. The thickness of the boundary layer decreases with Reynolds number. At a value of about 100,000, the boundary layer thickness on a golf ball is less than 1% of its diameter.

The forces acting on the ball critically depend on the state of flow inside the boundary layer. Below the critical Reynolds number, the flow in the boundary layer over the entire front face of the ball is laminar. Laminar flow is smoothly-ordered flow, with no fluctuations in velocity or pressure, so that the flow is steady. Laminar flow may be observed by looking at water issuing at low speed from a faucet, and it has a smooth appearance free from disturbances. As the flow passes over the "crest," where the cross-sectional area of the golf ball is a maximum, the freestream velocity (the velocity just outside the boundary layer) reaches its maximum value, and then begins to decrease as the flow diverges over the rear of the ball. According to Bernoulli's principle, the pressure increases in this diverging flow, so that the resultant force on the flow in the boundary layer due to pressure differences acts in the upstream direction (since the boundary layer is very thin, the pressure change across the layer is very small and may be neglected). At the same time, the fluid momentum in the lower part of the boundary layer, that is, near the surface of the ball, is low because the velocities are low. The force due to pressure differences builds up very quickly with downstream distance to a level where it pushes the low-momentum fluid in the lower part of the boundary layer in the upstream direction, so that the a region of reversed flow forms. We say that the flow is "separated." Instead of following the surface of the ball, the freestream velocity vectors are now directed away from the surface, and as a consequence a large region of low velocity and low pressure forms over the rear of the ball (Fig. 1.6). Downstream of the ball, in its wake, velocities are significantly below the freestream value, and a simple momentum balance would show that a large drag force acts on the ball. This drag force actually has two main contributions: (1) the viscous drag due to the boundary layer on the front face of the ball, and (2) the force due to the pressure differences between the front and rear faces of the ball. The second part is 50 to 100 times bigger than the first part, so that the drag due to separation (also called "form drag") dominates the drag force. The form drag is the dominant contribution to the overall drag for a golf ball in flight under all conditions.

So far we have described the flow state that exists below the critical Reynolds number. Above the critical Reynolds number, an important change takes place. At these Reynolds numbers, the boundary layer over the front face of the ball becomes turbulent. Instead of orderly, steady flow, large, random fluctuations in velocity are observed in the boundary layer, of the order of 10% of the freestream velocity. Turbulent flow may be seen in the flow from a faucet at high flow rates, where the jet has a ragged and choppy appearance. Turbulent flows are very common in nature, and the fluctuating velocity is very effective at mixing and transporting pollutants in streams and in the atmosphere.

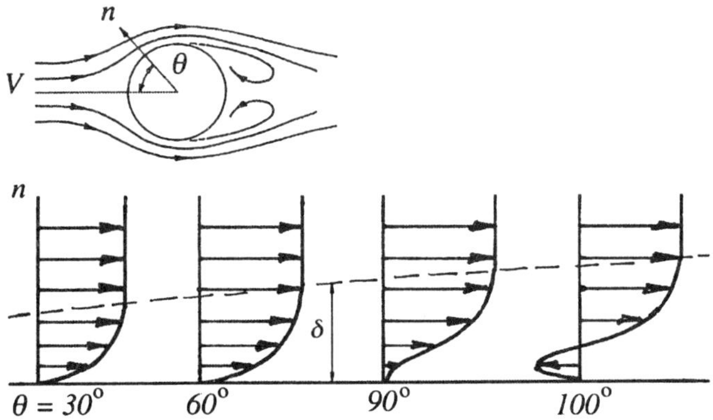


*Figure 1.6.* Flow over the ball at Reynolds numbers below the critical value. A laminar boundary layer is found on the upstream face of the ball, and the flow separates near the point of maximum cross-section. A large wake region is formed.

Although the boundary layer is now turbulent, the flow just outside the boundary layer is still very much the same as that seen at lower Reynolds numbers: that is, the freestream velocity reaches a maximum value near the crest, and then begins the decrease. Again, a pressure difference is generated that leads to a force on the fluid in the boundary layer that acts in the upstream direction. However, a turbulent boundary layer responds quite differently to a laminar boundary layer when acted upon by this "adverse" pressure gradient, primarily because of the turbulent mixing that occurs inside the boundary layer. First, the turbulent mixing tends to even out the velocity gradients in the boundary layer, so that the velocity profile is "fuller" (see Fig. 1.7), and the flow enters the region of adverse pressure gradient a with relatively high momentum near the wall. A turbulent boundary layer, therefore, will be able to resist the decelerating force due to pressure differences more effectively than a laminar one. Second, even as the boundary layer enters the adverse pressure gradient region, turbulent mixing continues to take place, helping to replenish the momentum loss near the wall by bringing higher momentum fluid from the outer part of the boundary layer closer to the wall. As a result of these two phenomena, both related to turbulent mixing, a turbulent boundary layer is able penetrate further into the adverse pressure gradient before separating. The width of the wake will be reduced, and the total momentum deficit in the wake will also be reduced, leading to a lower drag force.

We see that the change from laminar to turbulent flow in the boundary layer on the front face of the ball has a profound effect on the drag force. The transition to turbulence reduces the size of the wake, which results in a considerably lower total drag force. Although the friction drag in a turbulent boundary layer is higher than in a laminar boundary layer at the same Reynolds number, the total drag is dominated by the momentum loss in the wake, and since the turbulent boundary layer leads to a smaller wake, the overall drag is considerably lower.

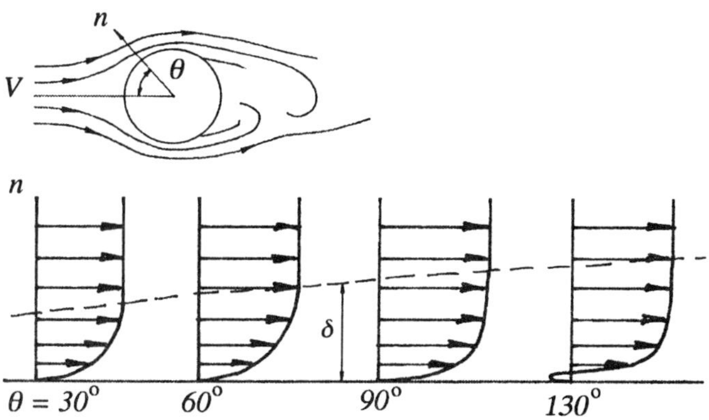


*Figure 1.7.* Flow over the ball at Reynolds numbers above the critical value. The boundary layer becomes turbulent on the upstream face of the ball, and the flow separates well downstream of the point of maximum cross-section. A smaller wake region is formed than in the case with fully laminar flow.

### 1.2.3 Effect of Dimples

Figure 1.5 showed how the critical Reynolds number depends on the surface roughness. At Reynolds numbers below the critical value, the boundary layer on the front face of the ball is laminar, separation occurs near the crest, and the ball has a high drag coefficient. Above the critical value, the boundary layer on the front face of the ball is turbulent, separation occurs well past the crest, and the ball has a low drag coefficient. When the ball is made rough, as by scoring or marking it in some way, the roughness elements introduce disturbances in the boundary layer, and this causes the boundary layer to become turbulent at a Reynolds number lower than in the smooth case. The critical Reynolds number is lowered, and if the roughness is of the right type and amplitude, the entire flight envelope will occur above the critical Reynolds number. Consequently, the drag force acting on the ball will be lower and the ball will fly further.

Dimples are a particularly effective form of roughness. For the Reynolds number range experienced by golf balls during a driver shot (50,000 to 200,000) the ball operates almost entirely above the critical Reynolds number. To compare the difference in overall carry between a smooth ball and a dimpled ball, we note that a professional golfer will drive a golf ball about 265 yards on a tee shot. The same golfer, hitting a smooth ball the same size and weight of a regular golf ball, will only drive it about 140 yards. Roughness, and dimple design, are critical factors in influencing golf ball performance.

However, rougher is not necessarily better. Each individual roughness element, that is, each dimple, makes a contribution to the overall drag, and when the roughness becomes too large, the gains achieved by manipulating the critical Reynolds number are offset somewhat by the drag due to the roughness elements. This effect is seen in the slow rise of the drag coefficient at Reynolds numbers beyond the critical value (see Fig. 1.5). Therefore, the design of the dimple pattern, and the choice of dimple sizes and shapes, is a balance between achieving the lowest critical Reynolds number and the lowest form drag contribution at the same time.

One of the most significant differences among golf balls relates to their performance near the end of the trajectory. An effective dimple pattern will promote transition to turbulence over the entire Reynolds number range experienced in flight, so that the flight envelope is always above the critical Reynolds number. If the dimples are not effective at low Reynolds numbers, transition to laminar flow may occur, and the drag on the ball will suddenly rise near the end of its trajectory. It will literally seem to fall out of the sky, reducing the total carry. In addition, because its velocity at impact will be reduced, and its angle of impact with the ground is increased, the roll distance will also be reduced.

Therefore, the dimple pattern should be designed to promote turbulence at the lowest possible Reynolds number, for the range of spin rates encountered in flight, without creating an undesirable amount of extra drag due to roughness. For example, it may be possible to promote turbulence at very low Reynolds numbers using highly exaggerated dimple shapes, but it is likely that the total drag at higher Reynolds numbers would not be optimal because of the relatively high skin friction drag due to the added roughness (even if the skin friction drag is small compared to the form drag, it may still make an important contribution to the total drag).

### 1.2.4 Effect of Spin

Spin is responsible for generating a lift force on the ball: without spin, there would be no lift. The sign of the lift force depends on the direction of spin, and we know that backspin produces a positive lift force. Similarly, sidespin produces a lift force in the horizontal plane, which results in a curved trajectory for hooks and slices.

To gain a qualitative understanding of how spin is related to lift, consider an observer moving with the ball. In this reference frame, the backspin reduces the relative velocity between the surface and the freestream at the top of the ball and increases it at the bottom of the ball. The lift due to spin is a result of the increased momentum in the boundary layer on the upper surface and the decrease on the lower surface. As a result, separation on the upper surface is delayed, and the new pressure distribution about the golf ball produces a net lift force.

The spin rate can also change the drag since it affects the local Reynolds number. Since the transition to turbulence has such a significant effect on the drag, spin has a particularly significant effect near the critical Reynolds number. For example, with W = 0.08, there could be as much as an 8% difference in the Reynolds number between the top and bottom of the ball. Thus, the flow over the top of the ball with backspin may be above the critical Reynolds number, while over the bottom it may be below the critical Reynolds number. This would result in a turbulent flow separation on the top, and a laminar separation on the bottom, with sharp changes in the lift and drag forces. For a typical driver shot, the critical Reynolds number is approached only near the end of the trajectory, so these effects may not have much influence on the total carry distance, although the velocity and angle of impact with the ground may be significantly affected.

As the ball moves along its trajectory, the spin rate decreases slowly. The rate of spin rate decay, $\dot{\omega}$, depends on the spin rate, the velocity, the density and viscosity of the air, and the radius of gyration $r$ ($r^2$ = moment of inertia/$m$). In functional form,

$$\dot{\omega} = h(\omega, V, \rho, \mu, r) \tag{1.8}$$

Non-dimensionalizing, we have

$$SRD = h_1(W, \mathrm{Re}, M), \tag{1.9}$$

where the non-dimensional spin rate decay is given by $SRD = \dot{\omega}\,(R/V)^2$, and $M = r/R$ is the non-dimensional radius of gyration.

Typical spin rate decay measurements are given in Fig. 1.8. By using the non-dimensional representation shown, the results for this particular ball collapse onto a single curve for all Reynolds numbers, which is close to a straight line, and suggests that at constant velocity the spin rate decreases exponentially.

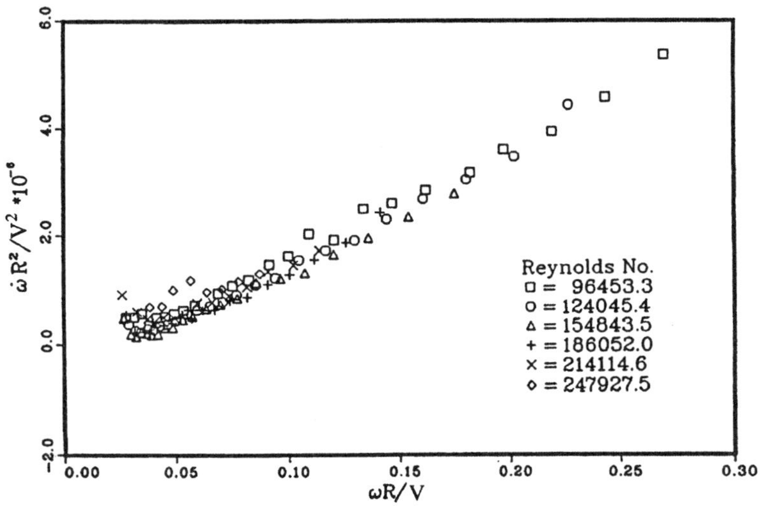


*Figure 1.8.* Spin rate decay for different Reynolds numbers. Reprinted from Smits and Smith (1994), with permission of E & FN Spon.

## 1.3 TRAJECTORY ANALYSIS

To properly balance the conflicting criteria of a low critical Reynolds number and minimum drag at high Reynolds numbers it important to clearly define the operating envelope and hence design conditions for golf balls. The highest Reynolds number will be achieved with high ball speed, low temperature, and high pressure. A Reynolds number between 65,000 and 240,000 will account for the vast majority of driver shots. Launch spins typically vary from 1500 up to 4000 revolutions per minutes.

As we indicated earlier, the requirement for lift at the low end of the operating envelope is of particular importance to the aerodynamic design of the golf ball. This aspect of the design tends to drive the robustness of the design and the trajectory.

### 1.3.1 Equations of Motion

Consider the free body diagrams shown in Fig. 1.9a,b. Figure 1.9b depicts the forces acting on the ball as it ascends. The lift force is defined to act normal to the velocity vector. The drag force is defined to act in a direction opposite the velocity vector. Note that there is a component of the lift vector, Vh, that acts in a horizontal direction opposing the travel of the ball. There is a component lift force opposing the gravitational force, Lv, but there is also a horizontal component, Lh, that is resisting the down-range travel of the ball. Also note that the drag force has a component that acts in a vertical direction opposing the vertical component of lift. The drag force is not only causing the ball to decelerate, but is pulling the ball down! Given that with the creation of additional lift comes increased drag, it becomes clear that for ascending flight minimization of lift is critical.

This picture becomes quite different after the golf ball has passed the apex and is in descending flight (Fig. 1.9a). Here the lift force is not only opposing the force of gravity, via component Lv, but as indicated by the horizontal component, Lh, is pulling the ball forward! The lift force is therefore contributing to increased carry distance and, by reducing the incoming angle to the ground, increased roll distance. But what about the increase in drag associated with maximizing lift at this low speed condition? While the drag force certainly has a negative decelerating impact on the ball, observe that it now has a vertical component, Dv, that is opposing the force of gravity and hence helping to keep the golf ball in the air.

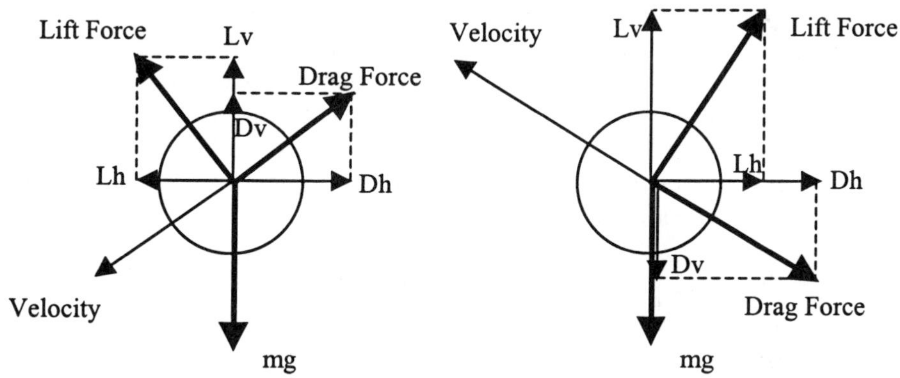


*Figure 1.9.* Free body diagrams showing the forces acting on a golf ball in flight: (a) Descending ball; (b) Ascending ball.

With reference to Fig. 1.9, for a ball in flight without crosswind, hook or slice, we have:

$$m\ddot{x} = -F_D \cos\alpha - F_L \sin\alpha \qquad (1.10a)$$

$$m\ddot{y} = -F_D \sin\alpha + F_L \cos\alpha - mg \qquad (1.10b)$$

where m is the mass of the ball, $F_D$ is the total drag force, $F_L$ is the total lift force, and $\alpha$ is the angle the ball trajectory makes with the horizontal direction, x, and g is the acceleration due to gravity in the vertical direction y. The notation $\ddot{x}$ denotes the second derivative of x with respect to time t.

If we define a parameter $K = \rho A/2m$, we have:

$$\ddot{x} = KV^2(-C_D \cos\alpha - C_L \sin\alpha) \tag{1.11a}$$

$$\ddot{y} = KV^2(-C_D \sin\alpha + C_L \cos\alpha) - g \tag{1.11b}$$

Since $\dot{x}^2 + \dot{y}^2 = V^2$, and $\dot{x} = V\cos\alpha$ and $\dot{y} = V\sin\alpha$, we obtain

$$\ddot{x} = K\left(\frac{\dot{x}}{\cos\alpha}\right)^2 (-C_D \cos\alpha - C_L \sin\alpha) \tag{1.12a}$$

$$\ddot{y} = K\left(\frac{\ddot{x}}{\cos\alpha}\right)^2 (-C_D \sin\alpha + C_L \cos\alpha) - g \tag{1.12b}$$

These equations can be solved numerically to find the ball trajectory and its point of impact once the initial conditions are specified. The trajectories shown in Fig. 1.1 were obtained in this manner. However, the equations cannot be solved without accurate measurements of the lift and drag characteristics of golf balls since the variation of the lift and drag coefficients as functions of Reynolds number and spin rate is a necessary input to the trajectory calculation.

### 1.3.2 Results From Wind Tunnel Testing

Outdoor testing, using, for example, a hitting machine such as the USGA's "Iron Byron," is not suitable for measuring accurate lift and drag coefficients. Even if all the atmospheric and wind conditions are carefully controlled, it is usually only possible to determine average values, rather than the instantaneous lift and drag coefficients that apply along the trajectory. Recently, some attempts to measure instantaneous values have been made using stereoscopic viewing of the trajectory using multiple cameras (Chikaraishi *et al.* 1990, Aoyama, 1990), but the inherent difficulties of outdoor testing (for example, the wind varies with time, with distance, and with height) make it very difficult to obtain accurate values. Consequently, most fundamental investigations use wind tunnels or indoor ballistic ranges.

The first comprehensive measurements of the lift and drag of a spinning golf ball were not made until after World War II by Davies (1949). Davies measured the aerodynamic forces acting on the ball by dropping it through the horizontal air stream of a wind tunnel. By measuring the point of impact of the ball on the tunnel floor, and by repeating this measurement for balls with clockwise and counterclockwise spin, the lift and drag could be determined. The spin rate was varied up to 8000 rpm (a spin rate parameter $W$ of 0.089), but only one tunnel speed $V$ of 105 ft/s was used (corresponding to a Reynolds number of about 94,000), whereas typical driver shots have a maximum velocity of approximately 235 ft/s. Davies also presented some trajectory calculations using the measured lift and drag information, but, as he stated: "Neither the manner in which $L$ and $D$ vary with translational speed nor the time rate of change of rotational speed are known."

The next major contribution was made by Bearman and Harvey (1976), who suspended a large model of a golf ball in a wind tunnel using fine wire supports. The ball could be spun in either direction using a motor mounted inside the ball. The interference due to the support wires was determined by reference to previous results for spinning smooth spheres, and it was found that when the ratio of the support wire diameter to the ball diameter was less than about 0.005 the interference effects were negligible. Results were obtained for Reynolds numbers between $1.26 \times 10^5$ and $2.38 \times 10^5$, and spin rate parameters of 0.02 to 0.3, and they demonstrated a marked difference between hexagonal dimples and conventional dimples by trajectory calculations. It should be pointed out that in these calculations the time rate of change of rotational speed, that is, the spin-rate decay parameter *SRD* was assumed to be zero.

More recently, Aoyama (1990) adapted Davies' method for finding lift and drag from drop tests by recording the trajectory of the ball as it fell using short exposure video techniques. This approach can give more accurate results for lift and drag since there is no need for simplifying assumptions regarding the trajectory shape, the equations of motion, or the relative velocity differences between the ball and the airstream. Lift and drag coefficients were presented for ball velocities of 100 to 250 ft/s, and spin rates from 1000 to 3500 rpm. They showed that the lift and drag coefficients for the balls using the 384 icosahedron dimple pattern were consistently lower than the 336 Atti patterned balls (an octahedron pattern, split into eight concentric straight-line rows, named after the main producer of molds for golf balls).

Smits and Smith (1994) used wind tunnel testing to develop an aerodynamic model of the golf ball. The balls were mounted in a drag balance using thin metal spindles for support. The spindles allowed the ball to rotate freely, and one side could be connected to a motor to spin the ball to high rotation rates. When the desired rotation rate was reached, the spindle was released from the driving motor and the motor was withdrawn from the tunnel. As the spin on the ball decreased, lift and drag measurements were taken simultaneously. The measurements were corrected for the interference effects of the spindles, one of the problems in using a drag balance. Spin rate decay measurements over the entire Reynolds number range were taken in a separate series of tests. The results were obtained using only one brand of golf ball, the Slazenger two-piece ball with a Surlyn cover. This brand was chosen because previous tests had indicated the balls are highly symmetrical, thereby avoiding many of the mechanical resonance problems experienced at high spin rates with some other brands.

Typical results for the lift coefficient as a function of spin rate are given in Fig. 1.10. The data at spin rates less than about 0.04 is not reliable and it should be discounted. Clearly, the Reynolds number dependence is weak, at least for Reynolds numbers greater than approximately 90,000. Below this value of 90,000, however, there is a decrease in lift coefficient, followed by a sudden increase at a Reynolds number of about 60,000.

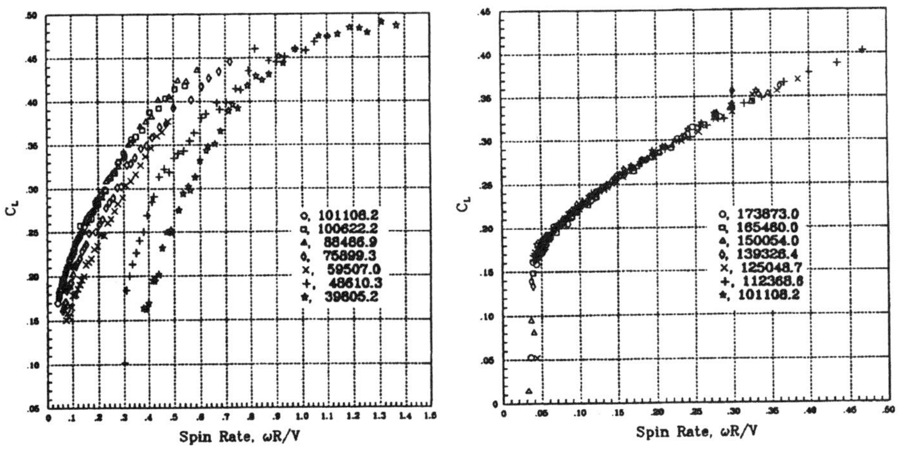


*Figure 1.10.* Lift coefficient as a function of spin rate parameter for different Reynolds numbers. Reprinted from Smits and Smith (1994), with permission of E & FN Spon.

Typical results for the drag coefficient as a function of spin rate are shown in Fig. 1.11. Three observations can be made. First, all the results indicate a strong dependence on spin rate. Second, the low Reynolds number results (<50,000) show a minimum in the drag coefficient at a certain non-dimensional spin rate, which corresponds to a particular rate of rotation (about 43 revolutions/s). Third, for Reynolds numbers greater than about 50,000, the drag coefficient increases with Reynolds number. Smits and Smith (1994) noted that when the Reynolds number exceeds approximately 200,000, there appeared to be a relatively sudden decrease in drag coefficient, which they ascribed to compressibility effects. These trends are more evident in Fig. 1.12, where the data of Fig. 1.11 are re-plotted as a function of Reynolds number. This figure also demonstrates the drag rise below a Reynolds number of about 60,000, the "critical" Reynolds number. The evidence from Fig. 1.11 indicates that the critical Reynolds number is a function of spin rate.

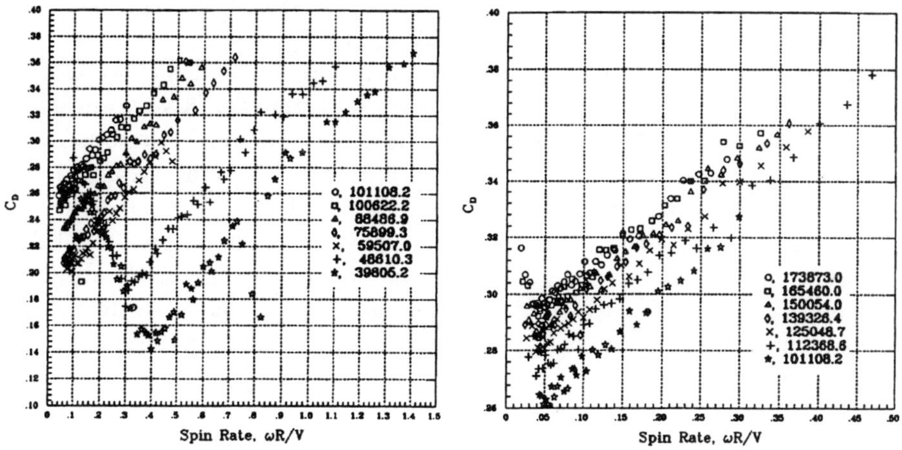


*Figure 1.11.* Drag coefficient as a function of spin rate parameter for different Reynolds numbers. Reprinted from Smits and Smith (1994), with permission of E & FN Spon.

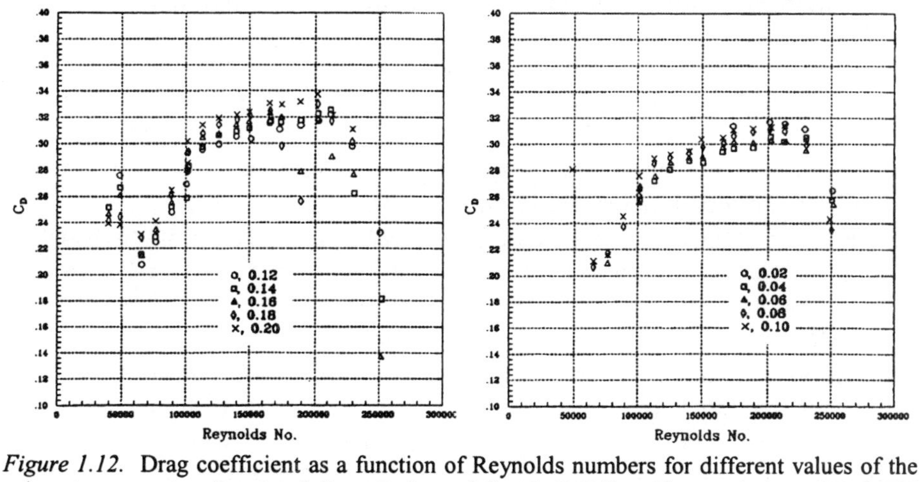


*Figure 1.12.* Drag coefficient as a function of Reynolds numbers for different values of the spin rate parameter. Reprinted from Smits and Smith (1994), with permission of E & FN Spon.

## 1.3.3 Results From Indoor Testing

Indoor testing seems to be the best approach for obtaining high quality data. The first successful Indoor Testing Range (ITR) was designed and constructed at the USGA Research and Test Center in Far Hills, New Jersey (Zagarola *et al.* 1994). The test range uses a calibrated launching machine to provide precisely known initial velocity, launch angle and spin rate. The velocity of the ball is then measured at three down range stations, along with the vertical and horizontal position, all with a high degree of precision. This facility can be used to obtain fundamental aerodynamic data and complement outdoor test functions for conformance tests. By changing the launch conditions, the entire flight regime may be studied using a relatively short range (60 ft in the case of the USGA range).

An Indoor Testing Range has also been developed by Callaway Golf in Carlsbad, CA. The launching machine used by Callaway is very similar to that used by the USGA, but the ball position as a function of time is found from image data acquired by pairs of stereoscopic cameras arranged at a large number of stations along its length. Particular care is taken in calibrating the sensors in order to obtain the most accurate positional data possible. The spin rate of the golf ball at the time of measurement of the lift, drag, and spin decay is measured by marking three orthogonal stripes on the ball and use image processing to determine the rotational orientation at numerous locations.

The accuracy of the system depends on the spatial extent over which the trajectory and ball orientation are measured. An accurate calibration is crucial. The three-dimensional trajectory and precise spin axis orientation needs to be determined to accurately determine the lift and drag coefficient.

Having measured the trajectory of the golf ball flight, Newtonian physics allows one to calculate the lift and drag forces acting on the ball along with speed. These forces are non-dimensionalized using the speed, air density, and the cross-sectional area of the golf ball as previously described in section 9.2.1. The speed is also used in the calculation of the independent variables of spin parameter and Reynolds number.

The lift coefficient, drag coefficient, and spin decay are measured at specific values of spin parameter and Reynolds number. A suitable surface fitting program may then be used to allow interpolation between the acquired conditions. This information is provided to the trajectory calculation to make accurate predictions of golf ball trajectories. Typical results for lift and drag coefficient are shown in Figs. 1.13 and 1.14. Note how different the drag results look when plotted for a constant spin rate (as in Fig. 1.13), compared to results plotted for a constant value of the non-dimensional spin rate parameter (as in Fig. 1.11).

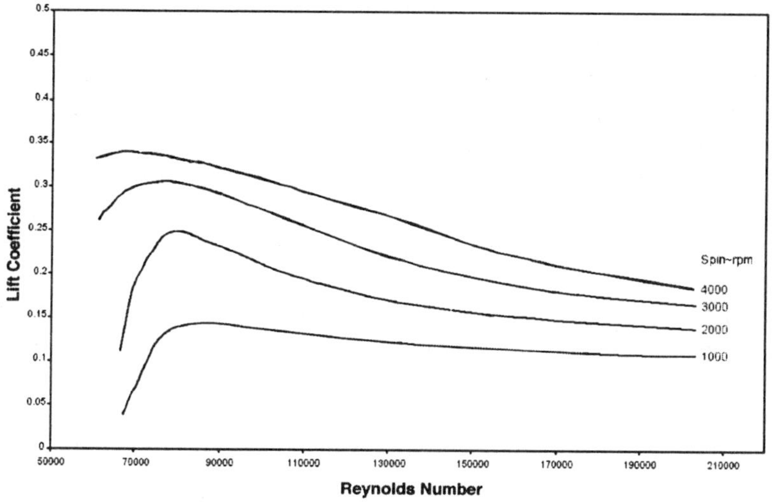


*Figure 1.13.* Lift coefficient as a function of Reynolds numbers and spin. Courtesy of Callaway Golf Company.

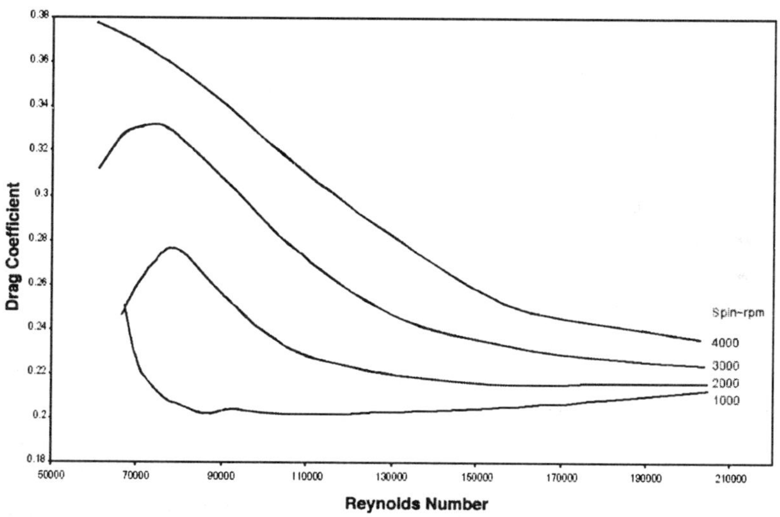


*Figure 1.14.* Drag coefficients as a function of Reynolds numbers at 3000 rpm. From Ogg Patent 6,290,615 (1999).

## 1.4 DESIGN PROCEDURES

At present, golf balls are designed almost entirely by empirical means. A dimple pattern is proposed, imprinted on a test ball, and then indoor or outdoor testing is used to determine its performance. Many iterations are usually required before a ball is brought to market. The effective aerodynamic design of golf balls requires the development of several systems. The first of these is a geometry creation system. A computer aided design (CAD) software package is the preferred means for creating three dimensional solid golf ball models. The CAD system is most highly leveraged when an associative model is created that is based on a few key parametric geometrical features representing the cross sectional shape of the dimples and/or tubes, and a definition of the pattern that dictates their distribution on the surface of the sphere. It is essential that the key geometrical features selected be at the characteristic length scales that influence the transition and separation of the boundary layer. Parameters that describe a particular dimple geometry include: the radius of the dimple, its chord depth, the entry angle, and the edge radius.

The second is a tooling fabrication system necessary to produce golf balls for testing purposes. The repeatability and reproducibility of this tooling fabrication system is of great import to the accurate representation of the CAD geometry in the test articles. If the cavities do not accurately represent the CAD geometry they must in the least provide a consistent transformation. It should be kept in mind that, depending on the nominal value, a change in dimple depth of as little as 0.0002" can make a significant change in aerodynamic characteristics. There are numerous methods for fabricating golf ball cavities; including hydroforming, hobbing, electrical discharge machining, machining, and metallic deposition. The fabrication of tooling can be out-sourced to a tool shop, but for the quickest development cycle and greatest ability to protect proprietary information, in-house fabrication can be beneficial.

The third is a prototype system for the fabrication of test articles. This is preferable to using a production system as long as all the key production process variables that influence the surface geometry are accurately represented. The flexibility and responsiveness of a prototype lab shortens the development cycle, prevents loss of production throughput, and avoids possible "escapes" of prototype balls into commercial packaging. It is important not to overlook the influence of process variables on the aerodynamics. The surface tension of the cover material, the wear or surface tension imparted by surface preparation, and the modification of the surface shape by the finishing system, all can have a significant impact of the aerodynamic characteristics.

In addition, measurement systems for size, weight, and moment of inertia are required in order to determine the aerodynamic coefficients. For each system, a thorough gage repeatability and reproducibility study should be performed. The acceptance criteria for physical properties will depend on the aerodynamic testing methodology. Specifically, if the prototype golf balls are to be treated as a single homogeneous group, upper and lower specification limits for size and weight must be determined.

Measurement systems for temperature, pressure, and humidity are necessary to calculate the Reynolds number and the aerodynamic coefficients. It is paramount that the temperature and pressure are known precisely so that the density and kinematic viscosity of air may be found accurately. According to Smits and Zagarola (2000), the density and kinematic viscosity of air at normal temperatures and pressures are known to within 0.4% and 0.9%, respectively. Most importantly, an aerodynamic force measurement system that is valid, repeatable, and reproducible is the cornerstone of a comprehensive aerodynamic development system. These days, this means an indoor test range, such as those described in Section 1.3.3.

The selection of design goals is important to a successful aerodynamic design. There is a strong interdependence between the construction and aerodynamic design of the golf ball. For a golf ball with relatively high driver spin, the severity of the surface must be greater to avoid excessive lift, which causes high drag and results in an undesirably high trajectory. When designing golf balls for the general golfing public, a range of launch conditions should be evaluated. Initial golf ball speeds of 105 to 160 mph, launch spins of 2000 to 3500 rpm, and launch angles of 8 to 13 degrees cover a broad spectrum of golfers.

Variations in atmospheric conditions should also be considered. Temperature and pressure variations representative of the intended market should be evaluated. In particular it is important to evaluate distance at conditions of high temperature and low pressure, which result in low Reynolds numbers at low ball speeds. As we noted earlier, there is a sudden drop in lift coefficient that occurs at Reynolds numbers between 60,000 and 80,000, depending on spin parameter and surface geometry. If this drop in lift occurs at a Reynolds number above the lowest level achieved during the ball flight, the golf ball will descend rapidly and lose both carry and roll distance.

The fundamental design challenge in optimizing golf ball aerodynamics is achieving the lowest possible drag level at high Reynolds number while ensuring a high lift coefficient to the lowest Reynolds number in the design space. These two conditions are in direct conflict because the drag coefficient is largely influenced by the degree of separation of the boundary layer, and hence dictates that the surface roughness be just severe enough to cause boundary layer transition at this high Reynolds number. Yet maintaining a high lift coefficient at low Reynolds number requires that the boundary layer be at the transition point over at least part of the golf ball throughout this condition, which dictates a significantly more severe roughness than that necessary at the high Reynolds number condition.

## 1.5 FUTURE DIRECTIONS

Many questions remain regarding golf ball aerodynamics. What is the effect of a side wind on lift and drag coefficients? What is the precise nature of boundary layer transition on the golf ball? What causes some surface geometries to have a much higher maximum lift coefficient than others at a given spin parameter? At this point, we lack any detailed understanding of the flow in and around dimpled surfaces, even without spin, so that there is still much to be learned about the science of golf ball aerodynamics. It seems highly likely that we will begin to see full-scale direct numerical simulations of the flow around golf balls, and we may then learn the answers to many of our questions. Given that these computations can yield valuable fundamental insights, they may give us better future guidance in choosing dimple geometries and patterns. What of the limits on golf ball performance? Some people have claimed that advances in golf ball aerodynamics are near their end; that it is a mature technology. We believe the contrary is true, and we expect that as we gain deeper quantitative insights into golf ball aerodynamics we will continue to bring new value to the average golfer.

## 1.6 REFERENCES

Achenbach, E., 1972, Experiments on the flow past spheres at very high Reynolds numbers. *J. Fluid Mechan.* **54**: 565.

Aoyama, S., 1990, A modern method for the measurement of aerodynamic lift and drag on golf balls. *Science and Golf,* A. J. Cochran, ed., E. & F. N. Spon, London, pp. 199-204.

Bearman, P. W., and Harvey, J. K. 1976, Golf ball aerodynamics, *Aeronautical Quart.* **27**: 112-122.

Chikaraishi, T., Alaki, Y., Maehara, K., Shimosaka, H., and Fukazawa, F., 1990, A new method on measurement of trajectories of a golf ball., *Science and Golf,* A. J. Cochran, ed., E. & F. N. Spon, London, pp. 193-198.

Davies, J. M., 1949, The aerodynamics of golf balls, *J. Applied Physics.* **20**: 821-828.

Kline, S. J., 1965, *Similitude and Approximation Theory,* McGraw-Hill, New York.

Maccoll, J. , 1928, Aerodynamics of a spinning sphere. *J. Royal Aeronautical Soc.* **32**: 777.

Martin, J., 1968, *The Curious History of the Golf Ball*, Horizon Press, New York .

Mehta, R., 1985, Aerodynamics of sports balls, *Ann. Rev.Fluid Mechan.* **17**: 151-189.

Moin, P. and Kim, J., 1997, Tackling turbulence with supercomputers, *Sci. Amer.* **276**: 62-76.

Smits, A. J., and Smith, D. R., 1994, A new aerodynamic model of a golf ball in flight, *Proc. Second World Scientific Congress of Golf,* E. & F. N. Spon, London, pp. 340-347.

Smits, A. J., and Zagarola, M.V., 2000, Applications of Dense Gases to Model Testing for Aeronautical and Hydrodynamic Applications, *Proceedings International Workshop on Dense Gas Dynamics*, Institute for Advanced Physics, Boulder, CO.

Sullivan, M. J. and Melvin, T., 1994, The relationship between golf ball construction and peformance, *Science and Golf II,* A. J. Cochran and M. R. Farrally, eds., E. & F. N. Spon, London, pp. 334-339.

Thomas, F., 2002, The aerodynamics of golf, *The Falconer*, No. 22, pp. 2-5.
Thomson, J. J., 1910, The dynamics of a golf ball, *Nature*, **85**: 2151-2157.
Zagarola, M. V., Lieberman, B., and Smits, A. J., 1994, An indoor testing range to measure the aerodynamic performance of golf balls, *Proc. Second World Scientific Congress of Golf,* E. & F. N. Spon, London, pp. 348-354.

# Chapter 2

# Engineering Methodology in Golf Studies

Arthur C.P. Chou

*Chou Golf Design Labs, 3512 Deer Park Drive, Santa Rosa, CA, PH: (650) 996-9640, FX: (707) 528-0450, EM: chou.golf@pipeline.com*

## 2.1 INTRODUCTION

Golf is a difficult game at best. The concept of hitting a ball that is 1.6" in diameter into a hole that is 4" in diameter from a distance of over 400 yards in four swings seems a challenge to some, impossible to many. The primary factor in determining golf performance is the swing; learning a powerful, efficient, repeatable swing is the fastest path to golfing prowess. However, recent advances in equipment technology have made it easier for those even with the most flawed swing to experience improved performance on the golf course, and in turn, increased enjoyment of the game. In the following pages, we examine some of the equipment designs and technologies that make it easier for players of all abilities to get that ball in the hole.

## 2.2 OVERVIEW OF CURRENT TECHNOLOGY

### 2.2.1 Materials

One of the main areas of golf technology growth has been new materials. Persimmon wood and carbon steel are out; titanium and composite fibers are in. With the economic growth of the golf industry, materials that maximized performance became the top priority; cost efficiency became an important but secondary concern.

The main factor in material selection for club design is the strength-to-weight ratio (Fig. 2.1). A strong but lightweight material allows the designer to use less material to accomplish the basic strength requirements of the clubhead or shaft. A shaft made of carbon fiber instead of carbon steel has a lighter overall weight, allowing the average person to swing it faster. Lighter materials in the clubhead means that there is more weight left over to be selectively placed in the head to achieve certain moment of inertia or center of gravity goals. For example, a driver made of titanium instead of steel can either be made larger (to create better performance on mis-hits) or with a lower center of gravity (by adding heavyweight materials in the sole) to create a higher ball trajectory.

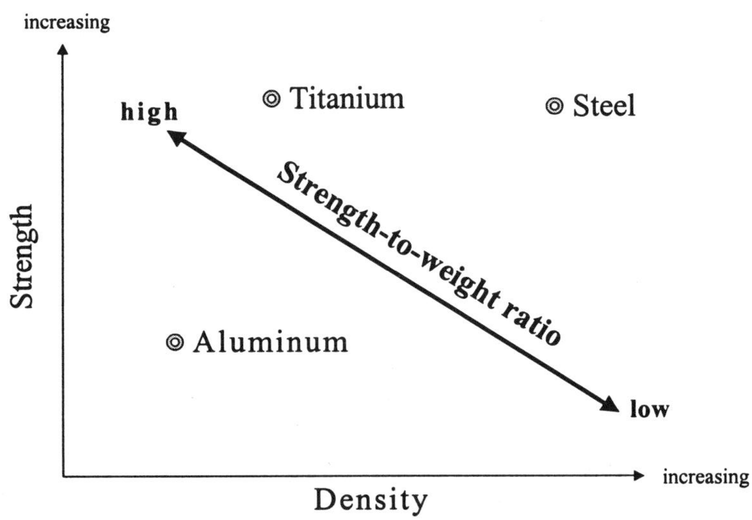


*Figure 2.1.* Strength-to weight ratios of various clubhead materials.

Durability and ease of manufacturing are additional issues to consider in choosing new materials for your club. Not only do clubheads have to survive different dirt and sand conditions (and the occasional cart path), but your entire bag of clubs typically experiences temperature ranges of 40-110 deg F while constantly being bounced around in your car trunk or golf cart. Ease of manufacturing directly affects the cost of the club. After all, cost may be a secondary consideration to performance, but it still plays a main role in the design process.

Example: Iron heads were traditionally forged from carbon steel and then chrome plated to protect against rusting. While some irons continue to be made this way, the large majority of irons are cast from stainless steel. The casting process is much less expensive than forging. Moreover, the stainless steel doesn't require any plating for maximum durability.

One of the latest material issues in club design has been ductility or elongation. In laymen's terms, this refers to the "springiness" of the material. Modern drivers aim to maximize the rebound, or restitution, of the clubhead (Fig. 2.2). The goal is to choose the material and face thickness combination that creates the optimum amount of face deflection and recovery time for the split second (actually, .5 milliseconds) that the ball is deformed against the clubface. If the face is too stiff, the ball deforms too much and loses energy during impact. If the face is too soft, the recovery time is too slow and is not timed with the ball deflection. Thin-faced (.070"-.090") drivers made of beta titanium alloys are the current state of the art in maximizing rebound.

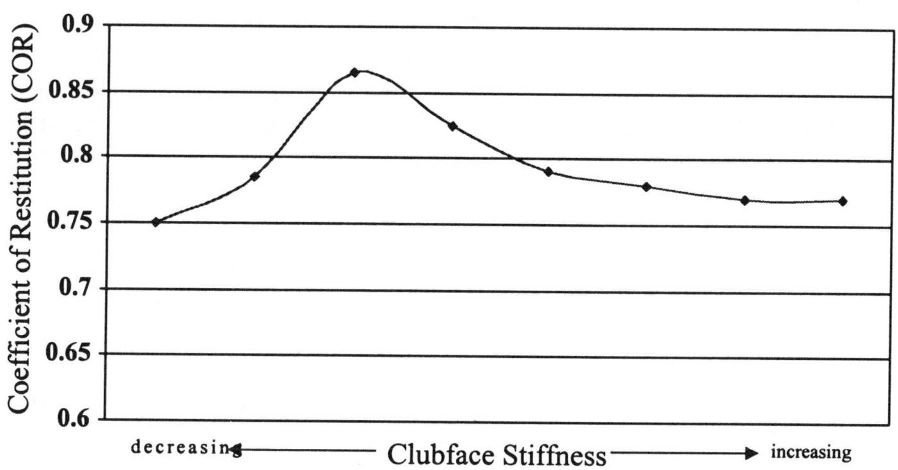


*Figure 2.2.* Clubface restitution, or rebound, as a function of face stiffness (calculated). An increase in COR of 0.05 results in approximately a 3% increase in ball speed. Adapted from Cochran (1994) with permission of E and FN Pons.

## 2.2.2 Design & Analysis

Design methodology for golf equipment has not changed appreciably since the game was invented. Whether it's persimmon wood hitting feather-stuffed leather or beta titanium striking multi-core urethane construction, the goal is still to maximize performance in the key areas of distance, control, forgiveness, and feel.

For all of the sameness in design methodology, recent technology has had a profound effect on the analytical methods used in the design process. Whereas there is always a certain amount of trial and error when designing products for human use, advanced analytical methods have taken much of the guesswork out of the process.

Computer-aided design (CAD) and solid modelling software allows designers to identify the mechanical properties of the clubhead, such as moment of inertia and center of gravity location, for a variety of head designs. Finite element analysis can take any head design and computer simulate impact with the ball; this not only yields critical information regarding the structure of the head, but also can predict the launch characteristics of the ball

Once a computer design is finalized, rapid prototyping methods can create a 3-dimensional model within hours. Methods such as stereolithography not only can produce wax prototypes for review; these wax models can actually be used in the investment casting process to make working metal heads. This combination of analysis and prototyping software can result in multiple iterations of a certain head design in days or weeks, in contrast to previous methods that required months.

Computer developed heads and composite shafts are then tested by mechanical hitting robots. Designed to simulate a human golf swing, the most advanced robots reproduce most of the range of motion of the upper body during the swing, from shoulder rotation to wrist release. They can produce a variety of swing types and speeds including a high draw from Tiger Woods to a typical weekend golfer's banana slice.

One of the most recent advances in analyzing golf equipment performance is the use of launch monitors (Fig. 2.3). Used for both robot and real golfers, the typical monitor is placed alongside the ball and uses high-speed photography methods to capture the clubhead and ball motions during the impact process. Critical data such as clubhead loft at impact and ball velocity are instantaneously calculated for each shot, allowing real-time cause and effect analysis for each piece of the club-ball equation.

*Figure 2.3.* Mechanical swing robot and launch monitor (photo courtesy of Golflabs, San Diego, CA).

Launch monitors create the ultimate custom fitting for the serious golfer; nowhere is this more evident than on the PGA Tour. In their quest for maximum distance off the tee, most Tour pros now understand that distance is not just a function of how "hard" they hit the ball, but how they launch the ball. Launch monitors tell each player their ball velocity, launch angle, and backspin for each shot. If a player is losing distance due to excessive backspin, a modification in shaft, clubhead, or ball type is made until the same swing produces more optimum launch conditions. This approach has played more than a minor role in the increase in driving distance on the PGA Tour over the past ten years.

## 2.3 THE FIVE DISCIPLINES OF GOLF EQUIPMENT DESIGN

Golf equipment design can be divided into five main disciplines that cover from start to finish every aspect of hitting a golf shot. While several of these areas are necessarily intertwined with the act of the golf swing, this chapter will focus on how they affect equipment design. In sequential order, they are:

1. Aim and Alignment
2. Clubhead Delivery
3. Club/Ball Impact
4. Ball Flight
5. Ball/Turf Interaction

### 2.3.1 Aim and Alignment

Aim and alignment is defined as the part of the pre-swing preparation in which the golfer orients his/her body and swing towards the intended target line.

The shot process begins with the player identifying the intended shot direction, or target line. The clubhead is then aimed at the target line and the golfer's stance and body position are oriented relative to the clubhead.

Although this discipline is an underappreciated part of the design process due to its lack of technology "sizzle", it still remains a critical piece of the swing puzzle. Club alignment leads to body alignment, which governs swing plane alignment. If you are not aimed correctly it is almost impossible to execute a successful shot.

Ball markings have always been used to aid alignment. Logos or stampings on the ball in the shape of a line or arrow allow the golfer to aim their ball at the target line. This is most commonly used during putting, where the rules of golf allow you to pick up your ball and then replace it when it is on the green. Some players will even draw a line on their ball specifically for this purpose.

Clubhead shapes and markings are also designed to help the aiming process. Logos on the crown of a driver and paint filled in an iron's scorelines are both ways to help orient the club at the target line. The bottom corner of the clubface, or leading edge, is often designed to be straight for alignment purposes.

The largest use of clubhead shape and marking for alignment is in putter design. Putting has always been a more creative stroke compared to a full swing, allowing for a wide variety of successful methods. This variety, combined with greater flexibility in the Rules of Golf regarding putter design guidelines, has resulted in a plethora of alignment-aiding shapes and markings.

The most common approach is to utilize a combination of lines and surfaces that are either parallel or perpendicular to the clubface. Long, straight toplines and paintfilled sightlines in the putter cavity are typical. Circular markings are sometime used to simulate the outline of one or more golf balls, creating the illusion of orienting the real ball along a visual "dotted" line. One novel alignment design featured a crosshair arrangement on the putter to not only assist in club aim, but also orient the golfer's eyes directly over the clubhead.

## 2.3.2 Clubhead Delivery

Clubhead delivery occurs over the time interval from the moment the swing begins until the moment just prior to club/ball impact

The main purpose of the golf swing is to deliver the clubhead to the ball in the most consistent and efficient manner. Just prior to impact, the key issues are the clubhead velocity, orientation, and angle of attack. Top professionals may all have different swings depending upon their build and personal style, but their clubhead properties at impact are strikingly similar.

Clubhead velocity has a dominant effect on shot performance. In simple terms, the faster the clubhead velocity at impact, the faster the resulting ball velocity. The easiest way to increase shot distance is to increase clubhead velocity. Professional golfers routinely generate 10-20% greater clubhead velocity than average players due to efficiencies in both swing and equipment (Fig. 2.4).

| Player Type | Clubhead Velocity (driver, mph) |
|---|---|
| Professional | 105 + |
| Avg. Male Amateur | 85-95 |
| Avg. Female Amateur | 70-80 |

*Figure 2.4.* Typical driver clubhead velocities for various golfer types

Clubface orientation and angle of attack determine how efficiently energy is transferred from club to ball. The ideal attack angle (in the horizontal plane or when viewed from above) is to have the clubhead approach the ball on the same line as the intended target line. The ideal clubface orientation is exactly perpendicular to the attack angle, so that the clubface points directly at the target at impact. This combination results in zero ball sidespin, maximizing the amount of energy devoted to ball speed.

If the clubface orientation is not perpendicular to the attack angle, the result is ball sidespin that creates a hooking (curving to the left for a right-handed swing) or slicing (curving to the right) shot. If the attack angle is not directly on line with the target line, the result is a shot that either starts to the left (pulled) or the right (pushed). In general, the clubhead attack angle determines the initial ball flight direction; the relationship between clubface orientation and clubhead angle of attack determines the curvature of the shot.

Clubhead angle of attack in the vertical plane (or when viewed from the side) affects the amount of ball backspin (Fig. 2.5). Generally speaking, a more descending angle of attack generates increased backspin. Increased backspin energy means less energy devoted to ball speed; the result is the trade off of increased control (ball stops more quickly) and decreased distance. This is why short shots that demand control (maximize backspin) are struck with a more descending blow that creates a large divot while long tee shots that demand distance (minimize backspin) are hit off a tee that allows a flat to ascending attack angle.

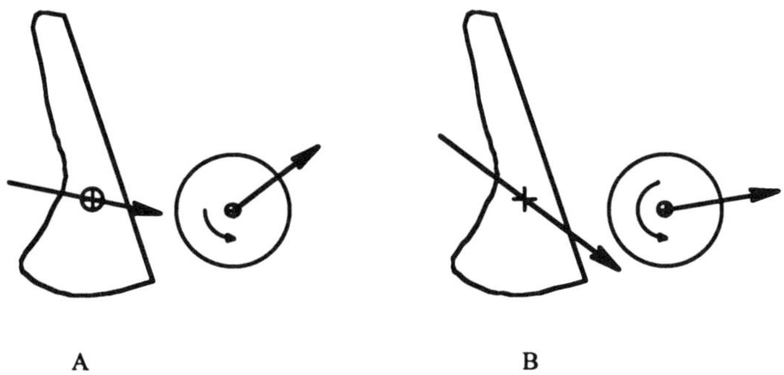


*Figure 2.5.* (A) A shallow angle of attack results in a steeper initial launch angle but less spin. This type of "sweeping" swing creates a flatter trajectory and more roll. (B) A steeper angle of attack results in a lower launch angle but more backspin. This type of downward swing creates a higher spinning trajectory that starts low, flies high, and stops quickly.

How can equipment affect clubhead delivery?

The shaft has a significant effect on all three areas of clubhead delivery. A shaft that flexes too much or too little for a given swing results in decreased clubhead velocity; this is a result of a mismatch in the timing of the swing and shaft bending and release. Shaft flex also affects clubface orientation; too much or too little flex can cause the clubhead to over- or under-rotate during the downswing, resulting in a closed (pointing to the left) or open (to the right) clubface at impact. This also can affect the clubhead angle of attack by altering the timing of the shaft flex during the swing arc. A shaft that flexes too late in the swing will still be on the downward portion of the swing arc at impact, resulting in a steeper angle of attack and a lower shot trajectory.

The location of the clubhead center of gravity (Cg) not only has a major effect on the club/ball impact dynamics, but it also plays a role during the clubhead delivery process (Fig. 2.6). During the downswing, the club is swung in an arc, where the clubhead traces a circular path with the shaft as a radius. The centrifugal forces on the clubhead pull in a radially outward direction during the swing; these forces tend to make the Cg of the clubhead try to align directly along the same radial line as the shaft. Imagine swinging a rock on a string: the centrifugal forces would tend to make the rock align itself so that the string pointed directly at the rock's Cg.

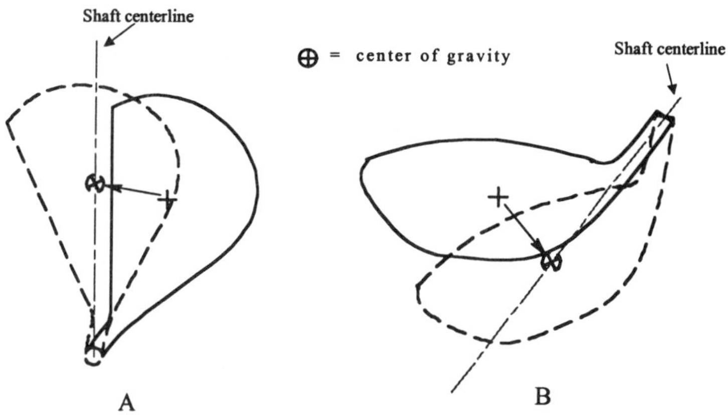


*Figure 2.6.* The effect of center of gravity (Cg) location on head rotation during the swing. Swing forces tend to make the Cg rotate towards the shaft centerline. The top view in A shows an offset Cg helping the clubhead rotate more closed. The side view in B shows the Cg rotation causing the club to flex in the 'toe down" direction during the swing.

This affects clubhead delivery in two main areas. In the plane of the lie angle, the centrifugal forces bend the club in a "toe down" direction, effectively flattening the lie angle. This is why a club that has a proper dynamically fit lie angle appears to be "toe up" in a resting position.

In the horizontal plane, this effect is why clubhead offset is used as a design tool. The term "offset" describes shaping the clubhead so that the hitting area is moved back from the plane of the shaft. This is generally used to reduce sliced shots by helping the clubface rotate more "closed" during the downswing. The offset head impacts the ball slightly later in the swing arc, allowing more time for the clubhead to "close". Also, the centrifugal forces aligning the Cg along the swing radius tend to make the clubface angle more "closed" prior to impact.

The term "inset" describes shaping the clubhead so that the shaft centerline projects more towards the center of the club in the heel-to-toe direction. This affects the player's ability to control the clubface angle at impact. Think of the clubhead as a door that is swung open during the backswing and rotated closed during the downswing. Ideally the door swings perfectly shut at impact where the clubface points directly at the target line.

Inset moves the shaft closer to the clubhead Cg, effectively making the "door" smaller. A smaller door rotates easier, making it easier for the skilled player to control the face angle. A less skilled player commonly has difficulty completely closing the "door" during the downswing; a slightly open "door" at impact is a typical trait for slicers. An inset club helps reduce slicing by allowing the door to "close" easier.

### 2.3.3 Club/Ball Impact

Club/ball impact occurs during the very brief time interval from the moment the club touches the ball to the moment the ball leaves the clubface.

During the impact process, the clubhead can be considered to behave like a free body, essentially "detaching" from the shaft for dynamic analysis purposes. The goal is efficient transfer of energy from club to ball. There are several major factors that affect club/ball interaction at impact: Cg location, moment of inertia, and clubface deformation.

Cg Location – The relative positions of the ball and clubhead Cg to each other at impact affect shot direction and spin. Because most shots occur with a round ball of standardized size resting on the ground, the main Cg effects of equipment take place in the clubhead.

The lower the clubhead Cg, the higher the resulting ball launch angle and spin rate after impact. This is why many "game improvement" clubs designed to help average golfers get the ball more airborne feature large, wide soles sometimes with extra weight added. This is also why more accomplished players who tend to hit the ball too high off the tee prefer a deep-faced (taller in the face) driver with a higher Cg.

Cg location in the horizontal plane also affects trajectory and shot shape. If impact occurs where the club and ball Cg are not directly aligned along the target line, the clubhead will rotate during impact. The amount of clubhead rotation is a function of the distance and angle between the impact location and the clubhead Cg.

This is the source of the "gear effect" that occurs in metal-wood type clubs, where the Cg is back away from the clubface (Fig. 2.7). A shot hit on the toe causes the club to rotate clockwise (as seen from above). The shared point of contact between club and ball means that this creates a corresponding counter-clockwise rotation in the ball, resulting in a "hooking", or right-to-left, type of spin. Clubs that create this "gear effect" usually have faces that are convex. The face curvature compensates for the gear effect by sending this toe shot off to the right of the target line, where the associated hooking spin brings the ball back on line.

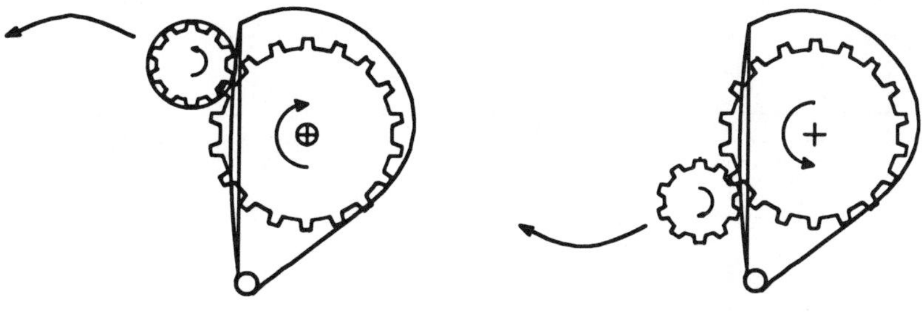


*Figure 2.7.* Clubhead rotation on off-center hits creates a "gear effect" as the clubhead and ball move together at impact. The toe impact in A causes the head to rotate clockwise, which in turn causes the ball to rotate counter-clockwise. The face curvature starts this toe shot off to the right to compensate for the hook spin that brings the ball back to the target line. Similarly, gear effect and face curvature in B make the heel shot start to the left and curve to the right.

Moment of Inertia - Clubhead rotation at impact is also affected by the moment of inertia (MOI) of the head. A clubhead where the weight is distributed far away from the Cg has a higher MOI and resistance to twisting. This generally creates a more efficient energy transfer on off-center impacts; less energy is wasted in clubhead twisting and more energy can be transferred to the ball. This is why larger clubheads are generally more forgiving on mishits.

Clubface Deformation - During impact the ball experiences significant deformation and then recovers its shape as it leaves the clubface. Ball compression, or hardness, contributes to the efficiency of this process. A ball that is too soft for the impact velocity may deform excessively, resulting in lost energy and a soft feel. A ball that is too hard has less resiliency and a harsher feel.

Recent equipment technology has focused on clubface rebound, sometimes referred as "trampoline" effect (Fig. 2.8). The basic principle is to reduce the amount of energy that is lost during ball deformation by allowing the clubface to deform slightly during impact. If the timing is such that the clubface expends less energy deforming than is correspondingly saved in the reduced ball deformation, the overall energy transfer is more efficient. This rebound effect is a function of the size, thickness, and material of the trampoline, or clubface. In general this requires a larger, thinner clubface made of a material that doesn't experience permanent deformation at the proper impact conditions. When properly designed, this effect can create an increase in ball speed for a given clubhead speed.

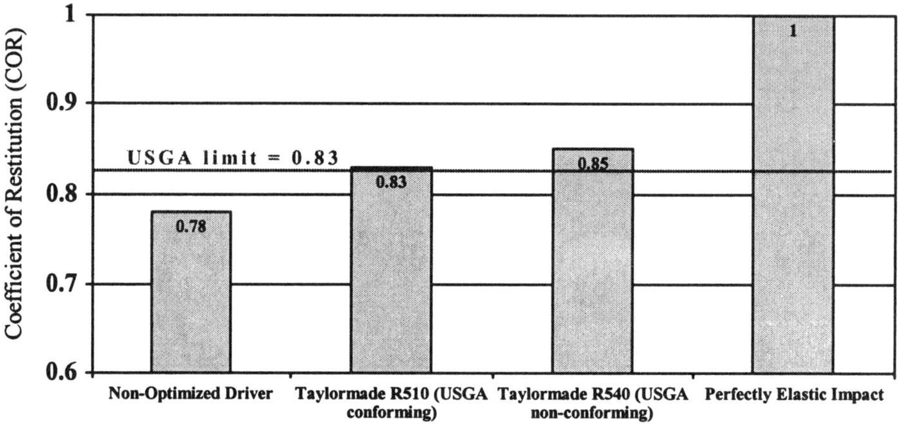


*Figure 2.8.* Rebound effect for various drivers. Coefficient of restitution (COR) measures the ratio between the incoming and outgoing velocities of a ball fired into the clubface. A COR of 1.0 would indicate no energy lost by ball or club during impact, or perfectly elastic impact. Golf's rulemaking body, the United States Golf Association (USGA), has established a maximum allowable COR of 0.83.

## 2.3.4 Ball Flight

Ball flight occurs during the time interval from just after a struck ball is airborne to when it lands. A struck ball is propelled into the air with a specific set of launch conditions that describe its initial velocity, angle, and spin. The resulting shot shape is a function of these launch conditions and the design characteristics of the ball (Fig. 2.9). The goal is to design the ball to create the desired trajectory and shot shape for a given set of launch conditions.

The main design factor in ball performance is construction. Modern golf balls are made of several layers including a core, a cover, and a range of mid-layers in between. The hardness, resiliency, and weight of each of these layers determine the launch conditions and ensuing flight characteristics for each impact scenario.

Traditionally, there has always been a trade-off in ball performance. Tee shots demand maximum distance, requiring the high flying but low spinning flight that is produced by a harder ball. These balls are designed to retain maximum energy after impact and produce additional rolling distance after landing. Shorter shots demanding maximum accuracy and distance control require the higher spinning performance of a softer cover ball. The increased spin helps the better player control the shot shape while also helping the ball to stop immediately upon landing.

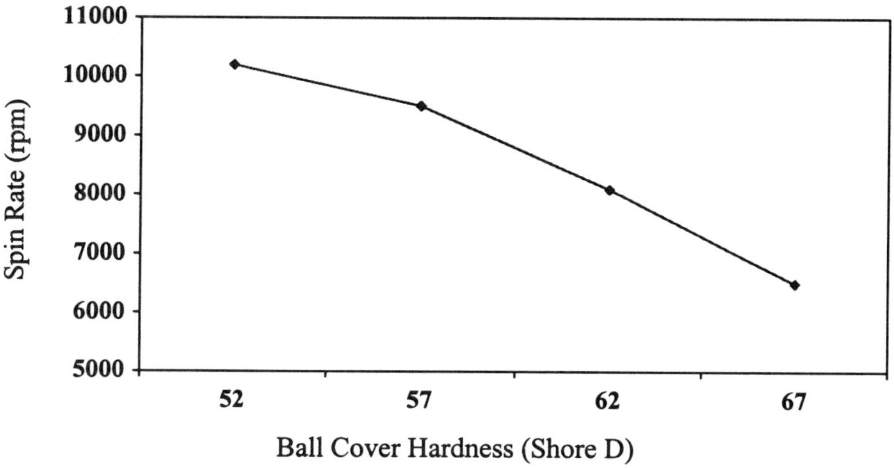


*Figure 2.9.* Ball spin rate as a function of cover hardness for a 9-iron. Adapted from Sullivan and Melvin (1994) with permission of E and FN Spon.

A ball's core plays the major role in dictating the performance on longer shots. A full swing with a low-lofted club (such as a driver or long iron) creates the maximum deformation of the ball at impact. As the core constitutes a large percentage of the total ball volume, its hardness and resiliency dictate how the ball deforms and recovers at impact. The proper combination of deformation and recovery for a given impact speed results in the most efficient transfer of energy from club to ball.

The ball's cover or exterior layer takes the leading role in shorter shots because the increased club loft and slower swing speed result in minimal core deformation. Under these conditions, a softer cover material is generally used to create the optimum combination of spin and feel that are required for maximum accuracy.

Recent ball technology has reduced the trade-off in distance versus control by combining the best of both worlds. These advanced designs essentially use the harder core required for distance and the softer cover required for control to create a multi-layer construction for all shot conditions.

It is important to remember that a higher spinning ball not only produces more backspin, but also results in an increase in the sidespin that creates the left or right curvature in the shot. Better players who prefer to control the shot shape on demand will get better results from a higher spinning ball. However, less accomplished players who want to reduce involuntary sliced or hooked shots would benefit from a lower spinning ball (Fig. 2.10a,b).

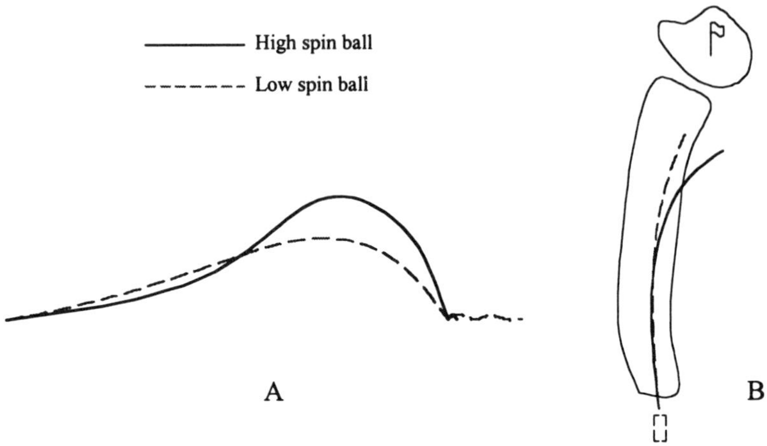


*Figure 2.10.* (A) This side view shows that a higher spinning ball rises higher and stops shorter than a lower spin ball. (B) This aerial view illustrates that a high spin ball also creates more sidespin, potentially turning a slight fade into a large slice.

The dimples on the outside cover help create the aerodynamic lift that allow a spinning ball to remain in the air longer. These markings result in the turbulent airflow and corresponding pressure differential to produce lift in the same way that the cross sectional shape of a wing generates lift in an airplane. Dimple patterns are matched with the spin and velocity characteristics of each ball construction type to produce the optimum lift conditions. One interesting note is that a smooth ball without any dimples produces minimal lift and falls to the ground quite quickly after being hit.

## 2.3.5 Ball/Turf Interactions

This final discipline involves how the ball reacts after it has landed on the ground.

Every golf shot ends with the ball bouncing and then rolling on the ground. How much it bounces and rolls is a function of its landing conditions: velocity, angle, and spin.

The shot trajectory and the ball construction determine the landing conditions of the ball. A higher spinning trajectory tends to climb and drop at a steeper angle, resulting in an increased first bounce but reduced roll. This creates a shot that stops close to where it lands, which is preferred for maximum accuracy situations like approach shots into the green.

A lower spinning trajectory is "flatter" and tends to climb and drop at a much shallower angle, resulting in minimal bounce but increased roll. This is preferred for maximum distance situations like shots off the tee.

A ball's landing spin is not only a function of how much spin it has when launched but also how much spin it retains during its flight. All balls experience some spin decay during flight; this is primarily a function of its moment of inertia. As different materials are used for the different ball layers, the variance in density of these materials results in a change in moment of inertia depending on the construction.

The most common ball/turf interaction occurs during putting. Once struck with a putter, a ball bounces and then skids before beginning its characteristic end-over-end rolling action. It is generally most preferable to minimize the amount of bouncing and skidding because the distance and direction during the rolling phase is the most predictable.

The relationship between rolling and skidding is a function of the putter's loft at impact. All putters have a few degrees of loft; this is designed to lift the ball in the air slightly at impact away from the bumpy imperfections of the green. Too much loft at impact and the ball is launched too high and will bounce too much. Increased loft also creates increased backspin, causing increased skid. Too little loft and the ball is immediately subject to any green imperfections that can deflect it from its intended path.

A common misconception is that a ball can be struck with topspin that improves its rolling characteristics. What is perceived as topspin is actually a decrease in backspin. Every ball hit with a lofted club from a resting position on the ground has backspin; a reduced amount of backspin allows the ball to skid less and roll sooner in an end-over-end manner.

A main equipment-related factor on ball roll is the loft and offset of the putter. Combined with the player's putting style and ball position, they help to determine the effective loft at impact. Players who tend to forward press their hands in front of the ball before impact are effectively de-lofting the putter; these players need a putter with increased offset and/or increased loft to achieve the best roll.

The other piece of equipment that has the greatest influence on rolling performance is the lawn mower. Advances in lawn equipment and turf science have allowed greens and fairways to be cut at very low heights. This results in increased roll distance off the tee and faster greens that putt more "true".

## 2.4 CASE STUDIES

### 2.4.1 The Chronic Slicer

The majority of golfers are slicers, meaning that their typical bad shot is a ball that curves excessively left-to-right (for a right-handed golfer). This is caused by a clubhead path that cuts across the ball from outside to in and a clubface that is open (pointed to the right) in relation to the target line. The result is a shot that starts to the left but quickly curves to the right.

Correcting an out-to-in swing path is mostly a function of swing technique (read: lessons) but clubs that aid a player's alignment may also help. Many slicers unconsciously compensate for their ball flight by aiming their bodies to the left of their target. This "open" body position increases the out-to-in club path, which only exacerbates the original problem. A club that helps this player to align the body in a position that is parallel, or "square", to the target line should reduce the problem.

Many of the equipment-effects highlighted in the section on "Clubhead Delivery" can be used to minimize the open clubface aspect of the sliced shot. The most direct approach is to use a club that has a face angle that is intentionally closed; without any other changes, a club that is more closed at address will be more closed at impact. Increased offset and inset both involve hosel or shaft locations that are designed to create a more closed clubface at impact. A softer flexing shaft, especially in the tip area, can also result in a closed clubface that reduces slicing. Finally, increased loft reduces the sideways curvature of any shot; the increase in loft generates more backspin and leaves less energy for the sidespin that creates curvature.

### 2.4.2 The Slower Swinger

Many golfers, especially seniors and women, experience shorter shots due to limited clubhead speed. Simply put, they swing the club slower.

Aside from the obvious point of a slower clubhead speed produces a slower ball velocity, these players generally also lose distance because the ball doesn't stay in the air long enough. Shots that hit the ground sooner lose their velocity (and distance) through ground contact.

Slower swingers benefit from equipment that helps them hit the ball higher. Increased loft, especially in the longer hitting clubs (driver, long irons), is the most direct way to increase shot height. Weight distribution can also play a role; weight that is located low in the head and further back in the head (away from the target) increases ball flight height. It is also important to remember that increased backspin can help raise the trajectory; a softer, higher spinning ball may actually increase distance for a slow swinger.

Finally, a softer shaft flex benefits the slow swinger. The slower flexing nature of these shafts is better timed to maximize clubhead speed while the increased flex in the tip region (i.e., region closer to the clubhead) helps to increase dynamic loft at impact.

### 2.4.3 The Hooker

A ball flight that draws (right-to-left curvature for a right hander) is the result of a club path that is slightly in-to-out and a clubface angle that is square to the target or slightly closed. When this mis-match in club path and face angle increases, the result becomes a shot that hooks, or curves sharply from right-to-left.

Hookers are generally harder swingers who naturally draw the ball. To combat this problem, hookers should look for equipment in the opposite direction from the slicer. Clubface angles that are square or slightly open reduce the chances of a closed face at impact. Reduced offset and decreased back weighting also should help the hooker. Stiffer shafts, especially those with stiff tip sections, make it more difficult to close the clubface during the swing.

## 2.5 CONCLUDING REMARKS

Like much of our modern culture, golf continues to walk the fine line between tradition and technology. While the basic nature of the game and its implements remain similar to the game invented by Scottish shepherds, its growing commercial popularity over the past 15 years has attracted the full interest of the latest technology and innovations. Interestingly, the average golfer's handicap as tracked by the USGA has not changed dramatically during this period. The equipment is better, but the scores are the same.

From the equipment perspective, this overall stagnation in handicap may be explained by looking at the recent focus of technological innovation. Most advances over the past 15 years have involved improving the efficiency of the club-ball impact and the resulting ball flight. However, the main deficiency in the average golfer's performance lies in their inability to deliver the clubhead to the ball in the preferred orientation. This would point to the discipline of clubhead delivery as the most likely area for significant advancement in the future.

## 2.6 REFERENCES

Chou, A., and Chou, P. C., 1996, Can a Smooth Clubface Increase Spin? in: *Proceedings of The First International "Golf - Theory In Practice" Conference*, Coleraine, Northern Ireland, pp. 70-73.

Chou, A., Gilbert, P., and Olsavsky, T., 1995, Clubhead Designs: How They Affect Ball Flight, in: *Golf The Scientific Way*, A. J. Cochran, ed., Aston Publishing Group, Hertfordshire, United Kingdom, pp. 15-24.

Chou, P. C., Liang, D., and Yang, J., 1994, Contact Forces, coefficient of restitution, and spin rate of golf ball impact, in: *Science and Golf II: Proceedings of the World Scientific Congress of Golf,* A. J. Cochran & M. R. Farrally, eds., E & FN Spon, London, United Kingdom, pp. 296-301.

Cochran, A. J., 1994, Club Face Flexibility and Coefficient of Restitution, in: *Science and Golf III: Proceedings of the World Scientific Congress of Golf,* A. J. Cochran & M. R. Farrally, eds., Human Kinetics, Leeds, United Kingdom, pp. 486-492.

Sullivan, M. J., and Melvin, T., 1994, The relationship between golf ball construction and performance, in: *Science and Golf II: Proceedings of the World Scientific Congress of Golf,* A. J. Cochran & M. R. Farrally, eds., E & FN Spon, London, United Kingdom, pp. 334-339.

# Chapter 3

# Physics and Mechanics of the Golf Swing

Stephen J. B. Mather

*Dept. of Mechanical Engineering, University of Nottingham, University Park, Nottingham, NG7 2RD, United Kingdom, PH: +44-115-951-3773, FX: +44-115-951-3800, EM: stephen.mather@nottingham.ac.uk*

## 3.1 INTRODUCTION

The golf swing is one of the most complex biomechanical motions and human can make in sport. Larson's Encyclopaedia of Sports Sciences and Medicine (1970) quotes *a good golf swing requires a solid stance, a firm grip, good balance, excellent timing, rotation and flexion extension of wrists and elbows, rotation of both arms in shoulder joints with the right shoulder kept mostly in adduction and the left going from adduction to full abduction, twist of the trunk and lateral thrust action through the pelvis, none of which should cause of the neck and head to shift from its central position*

To add to the problem of achieving this motion is the fact that most golfers are over 40 years old when starting the game, usually caused by financial and social conditions. Their biological state and, in many cases, their prior muscular development, preclude them ever being able to accomplish other than a very rudimentary swing pattern, sufficient to make contact with that the ball, but insufficient to become proficient at the game.

The game has no barriers or restrictions for play. There are many golf courses all over the world, and, whereas access to the top courses can be limited, most welcome all types of golfer. There is no age or ability limit, which means that the golfing fraternity comes is all shapes and sizes, with good and poor co-ordination and fitness, and with all manner of physical ability and disability. The equipment must therefore satisfy the needs of all potential players.

The manufacturers of golf equipment, in general, seem to have little understanding of the biomechanics of the swing and rely for sales mainly on subtle marketing. Those who have studied the swing quickly realise the magnitude of the problem of designing equipment for such a wide variety of ability. At best therefore, they accept that they will produce a design for the average golfer and one for a good golfer, which is precisely what the rest of the market does.

In this chapter, we shall first describe the equipment that is currently on the market and what it is supposed to do. At the same time we will add information, from the latest research, which will point the way to how equipment is to be made in the future. The ultimate situation must be that of matching the clubs to individuals, and even allowing that matching to alter as the player either improves or grows older.

### 3.1.1 The Essence of Golf

The aim of the golfer ought to be to propel the ball in the desired direction for a given distance. The manufacturers of golf clubs, more specifically their marketing sectors, have engendered the golfing public with the idea that distance alone is the prime element to measure performance on the golf course. Quite obviously the game is to hole out the ball within the very small cup on the green. This involves a number of shots, typically drive, iron shot, and a few putts. In a round of, say, 90 shots, an average round for many golfers, a rough breakdown would be:

| | |
|---|---|
| Shots from the tee | 18 |
| Shots from the fairway | 23 |
| Shots from the rough/hazard | 13 |
| Putts | 36 |

It would appear that reducing the number of putts would be the simplest way of improving the round of golf. Any golfer would note, however, that this was far too simplistic a concept, bearing in mind that greens are 50 feet (16m) in diameter and the hole is only 4¼ inches (108 mm) in diameter. The chances of reducing the figure of 36 are slim. It would appear from the statistics from the tour that professional golfers average considerably less than 36 putts per round. However the tour has a unique way of assessing what is a putt, a system which always reduces the number from those actually taken.

Returning to what is a sensible performance indicator for club design, Riccio (1990) shows clearly that hitting the ball straight and onto the green is the one clear way of improving. For each additional green hit in regulation figures (1 for a par 3 hole; 2 for a par 4; and 3 for a par 5) the average score over the year will reduce by 2 shots! Reducing the number of times the shot is played from the rough rather than the fairway, will reduce the score only by 1 shot for every additional fairway hit. From this, we conclude that ensuring better play with the irons and woods from the fairway is the most effective way of reducing the score on the round.

To counter the market's argument over hitting the ball long distances, Riccio shows that the reduction in score is only 1 shot for every increase in driving distance of 10 yards!

Propelling a golf ball requires the expenditure of energy. The ball is accelerated from rest to speeds of 70 m/s or more in a time of 500 μsecs and in this time it acquires up to 113 J of kinetic energy in translational movement and up to 4J in rotation. To achieve this the golf club needs to be accelerated from rest (at the top of the swing) to 40 m/s at impact. The time for this acceleration is about 0.3 secs, though it is not linear over this period. The elements of the swing involved in achieving this acceleration are:

the club head
the shaft
the hands
the wrists
the arms
the shoulders
the upper trunk
the lower trunk
and the legs

Daish (1972) estimates that about 2.5 kW on energy is expended on moving these elements in such a co-ordinated way as to create a club head speed of 40 m/s. Much of this energy is consumed in accelerating the club and therefore one aspect of design ought to be to temper the energy required so that it matches that of the golfer. A young athletic player can withstand much more force and generate much more energy than a retired and unfit septuagenarian.

We should not lose sight of the fact that efficient conversion requires excellent co-ordination of the body elements involved in the swing. The equipment is currently designed to aid the competent golfer with this conversion but few golfers have this level of skill. From this we conclude that much of the equipment on the market is quite unsuited to the requirements of the average golfer.

Finally, in this introduction, anyone who has played in competition and followed the professional game will know that an understanding of the mental or psychological aspect of the game is vital. This topic is not dealt with in this chapter, because there are some excellent books available which deal with the subject at the right level.

## 3.2 THE SWING

The aim of the golfer must be to make a consistent strike of the ball. If that consistent strike is one where the face at impact is perpendicular to the intended line of flight of the ball and the head of the club is moving along that intended line, then the strike will be pure and the flight of the ball will be straight. To improve on that, all that remains is to increase the speed of the head at impact, and the ball will gain more energy. By club design, this can be converted to linear velocity to make the ball fly further or rotational velocity to increase the spin rate

### 3.2.1 Description of the Swing

A general description of the golf swing would be as follows, concentrating mainly on the downswing:

a) At the top of the swing, the left arm is generally straight the right arm bent to accommodate the position of the club, and the wrists held at an angle so that the shaft of the club is horizontal and pointing towards the intended direction of flight,

b) much of the weight of the body is carried on the right side,

c) the downswing is initiated by movements of the body which result in the arms rotating approximately in an arc about the shoulders, with the angle between the shaft and the wrists maintained at a constant value, the rotation being similar to that of a solid body,

d) since the majority of the mass of the golf club lies in the head, the angular acceleration results in a couple on the wrists, and the wrists have to apply the opposite couple to maintain the angle,

e) the body continues to accelerate the arms until a point where they are at an angle of approximately 60 degrees before the impact with the ball,

f) the couple on the wrists is now released, and the club is allowed to rotate. It should be noted that there is some dispute as to whether the wrists are released or whether a torque is applied by the golfer to accelerate the club in this second phase of the swing. What is clear is that the action can only occur at the right point in the swing as a result of automating the swing after years of practice.

g) the subsequent motion of the arms and the club are similar to that of a double pendulum, with the hands travelling in an arc at a constant velocity,

h) the club rotates about the axis of the shaft, this rotation being brought about by the wrists. Again there is some doubt whether this is a positive action· or as a result of the biomechanical geometry of the swing.

i) if the timing of the swing is correct, the acceleration of the club about the wrists will bring the head of the club into impact with the ball at the same time as the hands reached the bottom of the arc.

High-speed photographic images of many elite players confirm this analysis of the swing. However full compliance with this procedure is not always essential and some elite players, notably Lee Trevino, have developed their personal system of arriving at the condition marked (i) above.

### 3.2.2 Amateur Swings

Only the lower handicap players achieve the swing as defined above. So what do the rest do? Expert teachers of the game point to one major fault which creates a divergence from the required pattern. Usually the golfer does not comply with the requirements to initiate the downswing with a downward pull of the arms. Instead, the shoulders are rotated about the spine in almost a horizontal plane and the arms and the club are now forward of the plane of the body. From this point, the golfer goes on to perform the correct swing with a downward pull of the arms. The release of the wrists takes place as required and the head is rotated into impact with ball. Because of the initial rotation of the shoulders the club head is now made to travel along a plane angled to the required line of flight. If the head of the club has been rotated so that the face is travelling perpendicular to the arc of the swing then the ball will be hit to the left (for right-handed players). In golfing jargon this is called a pull. In most cases, however, the rotation of the head is incomplete and the impact takes place with the face angled to the plane of the swing. This creates side spin on the ball and results in the common slice. Such a swing is often called the out-to-in swing, reflecting the geometry of the golfer's action.

Paradoxically, such a motion of the body is easily achievable, and may even be said to be more consistently achievable than that of the correct swing. Golfers with such slicing swings often play consistent golf, the only error occurring when the rotation of the wrists does not produce the required slice or the pull.

At some stage, manufacturers may consider changing the current designs either to prevent the initial out-of-plane movement, or to accept it but ensure that the pull shot always occurs. This would be of great benefit to the 90%+ golfers who have the out-to-in swing.

The following diagrams, taken from Cooper & Mather (1994), show the measured angular velocity of the club head for golfers with a range of ability (Fig. 3.1a-d). The plots show the average velocity (which has been scaled to the maximum average velocity) together with vertical bars showing the range of deviation from this average. From these results, the authors postulated that those golfers who regularly play can be placed in one of three groups categorized by the velocity pattern. The general trend of groups D is one of continuously increasing velocity up to the impact point. The deviations for this group are too small to plot. For Group C, the maximum velocity is achieved close to impact but the deviations are larger than for Group D. For group B, the maximum velocity is achieved about 40 degrees before impact takes place, but the deviations are still small. It can be said for Groups B and C, that the pattern of the golfers' swings in each group is consistent. On the other hand the patterns for group A show very large deviations and, in almost all of the swings, the maximum velocity is achieved well before impact. Group D are professional golfers. Groups B and C are regular players with moderate handicaps, from 5 to 15 and, of these, Group B mainly demonstrate a slice. Group A are golfers who either have little co-ordination, or play infrequently, or who have only recently taken up the game.

Repeated tests by the authors suggest that regular players can always be placed in one of the first three categories.

The similarity of the results for groups C and D suggests suggest that golfers in group C have learnt the rudiments of the correct swing and are consistently applying it. Group B are also consistent but have the signs of an out-to-in swing. Group A need to be converted to the swings of groups C and not those of group B.

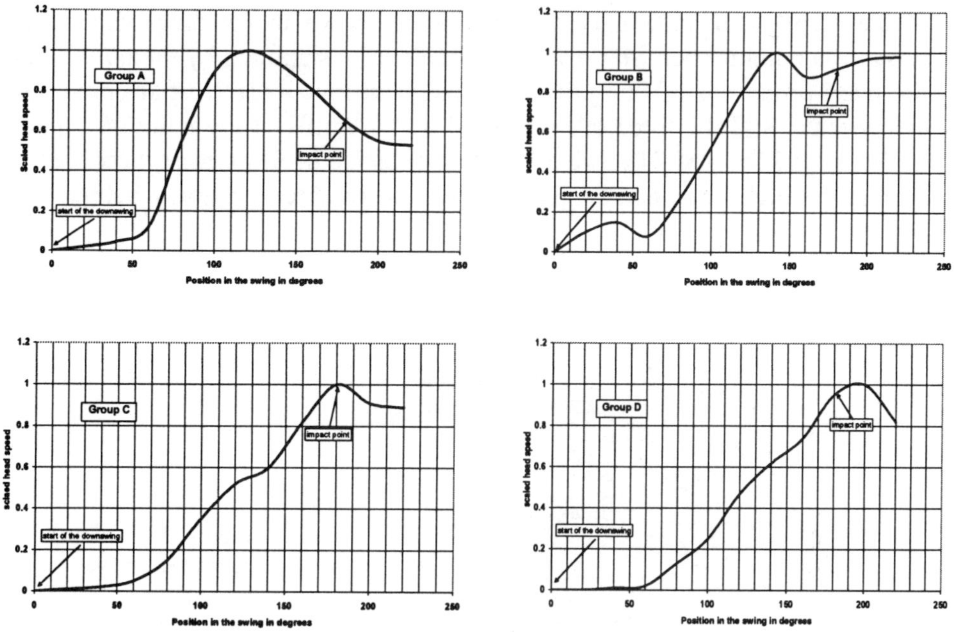


*Figure 3.1.* Variation of clubhead speed during the swing for: (a) new players, high handicappers; (b) regular players, mid-handicappers; (c) regular players, low handicappers; and (d) professionals.

## 3.2.3 Mathematical Models of the Swing

Williams (1968) was one of the first to model the swing as a double pendulum. He compared his model with that of one of the best ever golfers, Bobby Jones, and was able to show a good correlation. He concluded that the good swing was as described in (a) to (i) in the previous section. The concept of the double pendulum was taken further by Jorgensen (1970) who posed the question as to the effect of releasing the wrists at different times in the downswing, by introducing a "Zeit" or delay factor. His conclusion was that the head velocity varied with the delay but the spread did not match the variation found in amateur golfers.

Since then, many efforts have been made to model the biomechanics of the body in addition to elasticity of the club. Miura and Naruo (1998) applied various torque patterns to the two-link model. A three-link model was developed by Turner and Hills (1999) and modeling the shoulder rotation has been done by Rees-Jones (2002).

In all of these, the muscular biomechanics of the body are ignored. Figure 3.2, taken from Williams, shows the forces generated by the motion of the club for an elite player. The forces shown in the diagram are the result of a combination of the angular accelerations applied by the golfer and the mass properties of the golf club. The centripetal acceleration increases as expected, and the force to be resisted by the golfer at the point of impact is almost 107 lbf (480 N). The body must organise itself to resist these forces so that the club head can be placed with the required accuracy behind the ball.

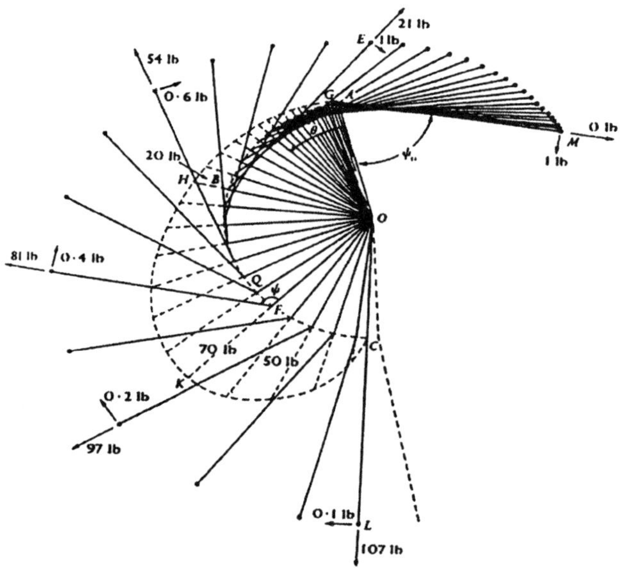


***Figure 3.2.*** Forces generated at various time instances of the swing. Adapted from Williams (1968).

Of equal importance is the figure of over 81 lbf (360 N) which the body must oppose at that part of the downswing where the club is approximately horizontal. In this region the body is ill positioned to resist the force, the golfer can become imbalanced and body movement destroys the swing. In the downswing of elite golfers, the legs are angled, particularly in the region of the knees. Without this lower body movement the correct swing cannot be achieved. It is clear from all of the modelling and from the investigations of the biomechanics of the golfer that Williams suggestion from some 40 years ago is still correct. He postulated that golfers of varying ability should attempt to achieve the same swing pattern as elite golfers, but in order to withstand the large forces and to achieve the correct timing, the pattern should simply be repeated at a lower tempo.

## 3.2.4 Effect of Golf Club Design on the Golf Swing

What of the mass properties of the golf club? The mathematical models clearly show the dependence of the swing on the first and second moments of inertia of the golf club about the wrists. Of these, the first is in common use by manufacturers in "matching" golf clubs and is given the title "swing weight". This is defined, somewhat empirically, as the product of the mass of the golf club with a moment arm which is the distance between the centre of gravity of the club and a point 14 inches (35.6 cm) from the butt end of the club. The mass of the head is a dominant element in this calculation.

The second moment of inertia of the club is not a quantity in common use in either the design or the production of clubs, although it is clearly a major factor in the force generation in the swing, and is contained explicitly in most mathematical models. Figure 3.3 shows the values of second moment of inertia and swing weight for a typical commercial set of clubs made to modern standards.

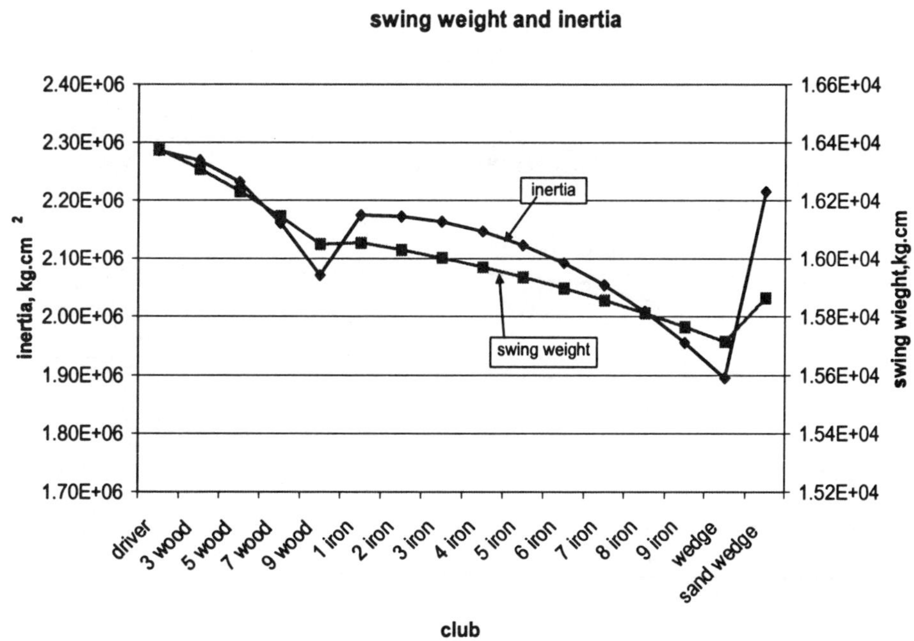


*Figure 3.3.* Values of swing weight and inertia for different golf clubs.

Contrary to popular belief, the swing weight is strictly not constant across the range of clubs, but there is only a 2 percent variation from a 1 iron to a wedge, and the driver is only 3 percent more difficult to swing than a six iron. The variation in inertia is, however, significantly more. Across the range of irons the variation is 15 percent, and the driver is almost 10 percent more difficult to swing than a six iron. Since golfers find the long clubs (driver etc.) significantly more difficult to swing than the shorter clubs (7, 8 iron etc.) this suggests that neither the swing weight nor the inertia are correct parameters for design. Over the years, manufacturers have attempted to manufacture golf clubs which have truly constant swing weight from the driver to the wedge, and at least one company produced in clubs with constant inertia. None of these were very successful.

So the current consideration of these mechanical parameters do not lead to good club design. What is clearly needed are golf clubs which match the muscular strength, co-ordination and flexibility of the golfer. To think that one set of golf clubs is sufficient for the whole golfing population is clearly ridiculous.

Still on the matter of club design, the Rules of Golf place no restriction on the length of the clubs or on the mass of the clubs. Daish (1972) performed an interesting experiment some 40 years ago. He asked players with reasonable ability (including himself) to use drivers whose lengths were

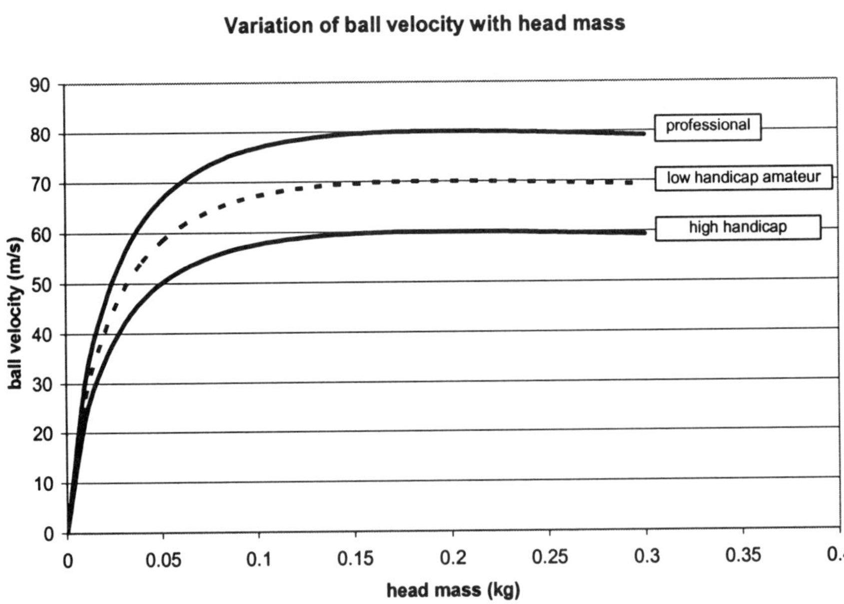


*Figure 3.4.* Effect of clubhead mass on the initial velocity of the ball. Adapted from Daish (1972).

constant but whose head masses varied from 100 gms to 300 gms. Remembering that the norm is 190-200 gms, some of Daish's test clubs were well outside the normal range. The initial ball velocity achieved by these clubs are shown in Fig. 3.4. It is clear that manufacturers did not take account of what Daish showed which implies that golfers with low muscle mass, and perhaps also those with little co-ordination, could use clubs with light head masses without significant reductions in performance. Indeed, it could be argued that such golfers would actually show a benefit since the control they would then attain could result in a better strike of the ball and in consequence more accurate shots.

## 3.3 IMPACT PROCESS

We now turn our attention to the impact process. Figure 3.5 shows the result of a typical impact of an iron club with a normal ball. The ball has begun to be compressed against the face of the club and it is obvious from the location of the ball on the tee that the ball has yet to move forward. This is typical and of all regular impact collisions in golf.

*Figure 3.5.* Initial impact of the club and the ball.

Figure 3.6 shows the full process of impact, together with a plot of the force between the ball and the club face. The forward motion of the head compresses the ball and the force increases swiftly. This force accelerates the ball and decelerates the head until at maximum compression the ball and the head are travelling at the same speed. This is the moment in the impact when the force is greatest. The strain energy built up in the ball is now released, the ball expands and the force reduces to zero when the ball leaves the club. Because stress waves travel through the ball as it is compressed, the impact is not perfect and energy is lost. With modern balls the loss of energy can be as low as 15 percent for a two-piece ball with a hardcover and as high as 20 percent for the softball of the type used by elite and professional golfers.

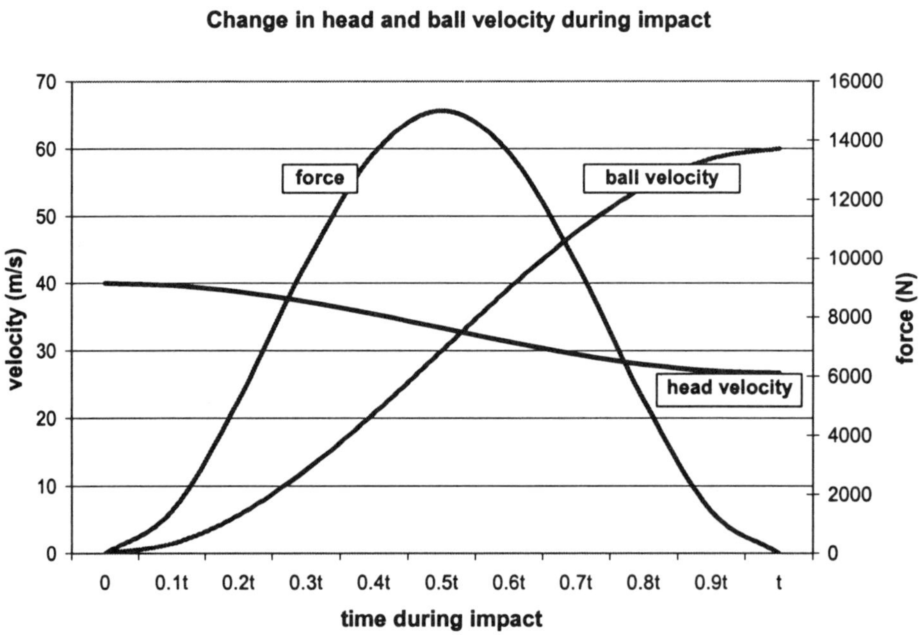


*Figure 3.6.* Velocity and force variations during the impact of the club and the ball.

The impact process is further complicated when loft is placed on the club. Now, the force is at an angle to the face. Simplistically, this force can be resolved into a normal force which induces the velocity of the ball and a tangential force which creates spin (see Fig. 3.7). The situation has been carefully studied by many researchers, notably Gobush (1996), Johnson (1994), and Lieberman (1990). The aim of their research is twofold. First, by conducting suitable experiments, to understand the mechanisms involved in the impact process, and secondly to produce a physical model of these processes which can be used as input to the collision calculation for both the ball and the club head. In this way, clubs can be matched to the characteristics of the ball to achieve the desired velocity and spin values.

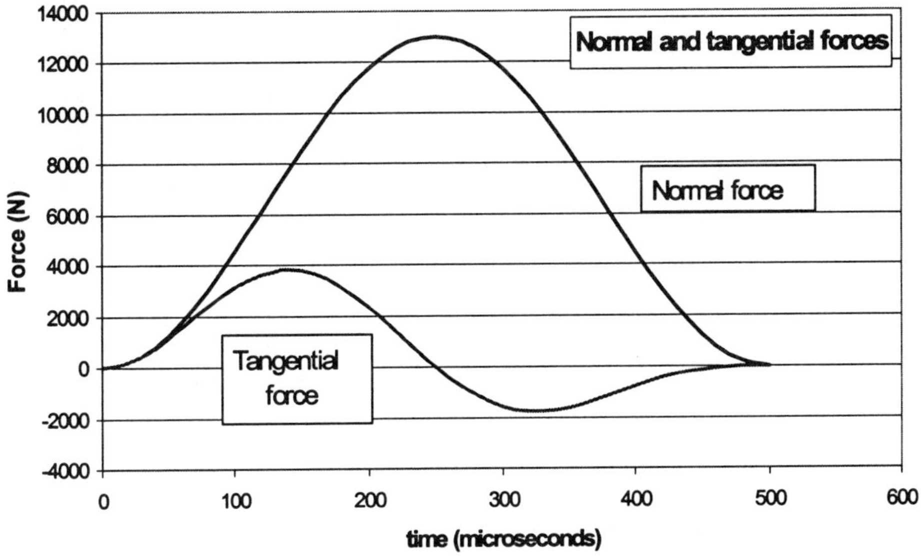


*Figure 3.7.* The variation of normal and tangential forces during the impact of the ball and an angled plate. Adapted from Gobush (1996).

The measured results of such an impact are not easily interpreted. Figure 3.8 shows a typical result produced by Gobush for an impact at an angle of 40 degrees and that the velocity of 40 m/s. He used an angled flat plate mounted by force transducers onto a massive block. The normal force is as we would expect, as is the initial part of the tangential force. Towards the end of the impact, however, when the ball is expanding, the tangential force reverses implying that the backspin generated during compression is now reduced. The reason for this is not fully understood but a possible mechanism is shown in the next diagram (Gobush (1996)). During the compression phase, it is suggested that the ball, with it is many layers, winds up in the same way as a clockwork spring. As the ball expands away from the face of the club, the clockwork spring unwinds not only reducing the tangential force but, if the spring force is large enough and the ball is still in contact with the face, reversing its direction. A mathematical model was attempted by Maw, Fawcett, and Barber (1975), and modified recently by Serandour (2001) which can be used, qualitatively, to understand the relevance and importance of the mechanical parameters of the ball. The degree of reverse spin will depend upon the construction of the ball as well as the head velocity and loft angle.

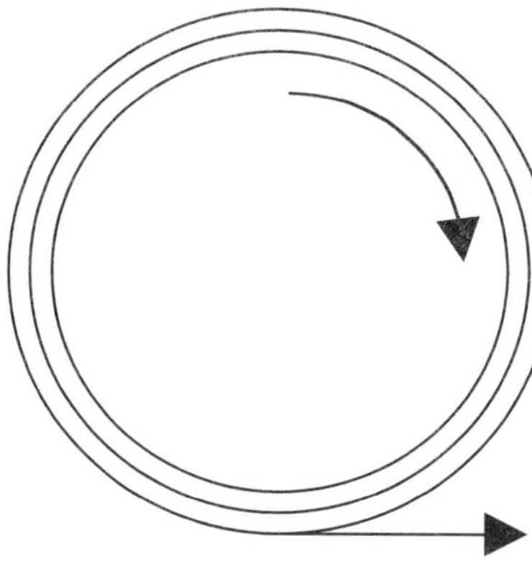

*Figure 3.8.* Possible explanation of how the layers of the golf ball wind up during the compression phase. Adapted from Gobush (1996).

Figure 3.9a shows the phase of the impact where the ball is still in contact with the club and rotation has just started. In Fig. 3.9b, the ball has left the face of the club, and the spin is clearly visible.

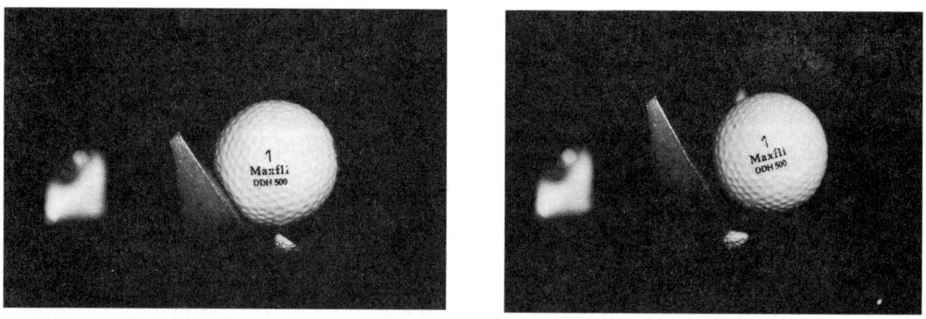

*Figure 3.9.* (a) Left. Expansion phase of the impact. Initial period of rotation of the ball. (b) Right. Eccentric deformation of the ball during the compression phase.

It remains to explain why some balls spin more than others. For instance a hard and durable ball, when struck by a lofted club will spin up to 40 percent less than a soft ball as used by elite golfers. One possible explanation for this is shown in Fig. 3.10. Because of the angled nature of the impact, the pressure distribution on the face of the club is not uniform. The centre of pressure of this distribution may not pass through the centre of gravity of the ball, thereby creating a torque on the ball. The softer the ball in the larger its deformation and the bigger the moment arm of the talk about the centre of gravity. If this explanation is correct, then spin comes partly from deformation and therefore the available energy for propelling the ball forward is reduced.

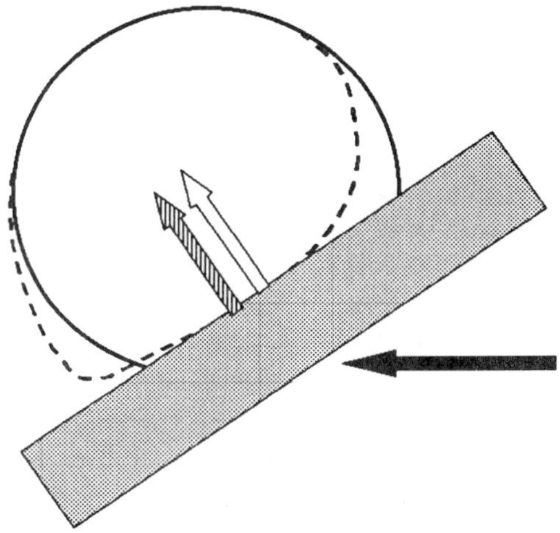

*Figure 3.10.* Eccentric deformation of the ball during the compression phase. Adapted from Gobush (1996).

The manufacturers of modern golf balls have three aims:

to reduce the negative compartment of the tangential force
to create high spin without reducing forward velocity
to maintain a durable product.

Some excellent papers have been published, and the reader is encouraged to look at *Golf, the Scientific Way,* edited by Alastair Cochran.

## 3.4 SHAFT DESIGN

The swing model described in Section 3.2 shows the forces on the body of the golfer. These forces, transmitted through the shaft and the grip, are three-dimensional, and the shaft will bend and twist in response to these accelerations and a Coriolis force. The significant centripetal acceleration is very important, introducing a tensile component which changes the properties of the shaft considerably.

A brief history of the development of the shaft is important to the understanding of its role in golf. Early shafts were made of hickory and their dimensions were governed almost entirely by the size of the head which could be cast and the size of the grip which could be comfortably held by the golfer. The result was a gentle taper from head to butt, where the golfers hands are placed. In the early part of the 20th-century wood was replaced by steel, and later by aluminum, titanium, and plastic composite. For some reason, all of these continued to replicate the linear taper of the wooden shaft, although whether this is correct is very debatable. In the case of the steel shaft the taper was usually achieved by a series of steps of different length and diameter. For other materials the taper is often continuous.

It has frequently been stated that the shaft is simply a device to connect the head of the golf club to the golfer. It is clearly more than that. The way that the shaft flexes and twists during the downswing will determine critically the attitude with which the head is presented to the ball. Realising that an error of ±1 degree in placing the face of the head on the ball will result in the ball flying off the fairway, a knowledge of the detailed motion of the shaft is very important. In addition, as we have seen in the previous section, the force of impact is enormous. The shaft must therefore minimise deflection during impact as much as the designer wishes. Finally the shaft must continue to operate consistently over a sensible period of time.

The force pattern, reported in an earlier section, causes the motion shown in Fig. 3.11.

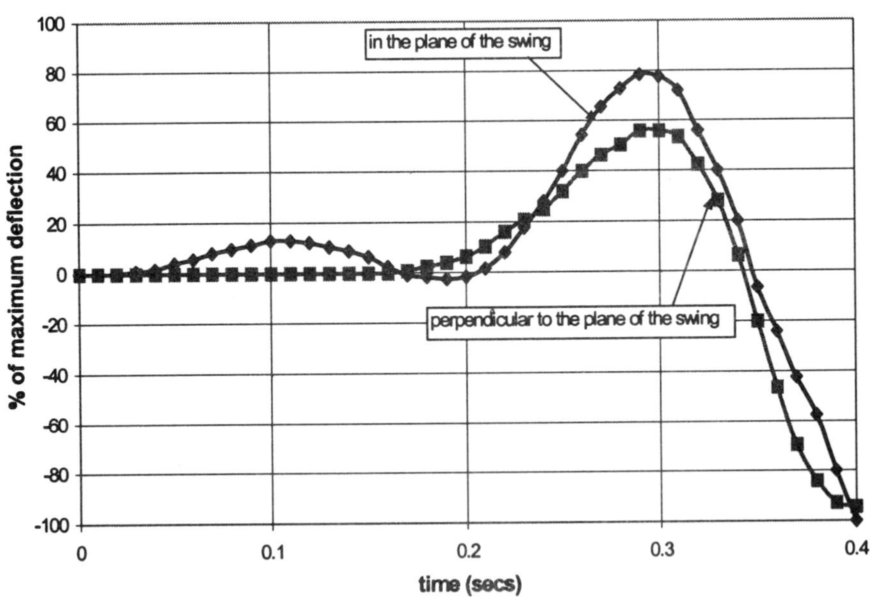


*Figure 3.11.* Deflection of the shaft and club head during the downswing.

Making the assumption that the swing is roughly planar the measured deflection in plane and out of plane is plotted for the swing of a good golfer using a standard driver. At the start of the downswing, which is initiated by the hands, the inertia of the head bends the shaft downwards. In the next phase of the downswing the acceleration is small and the deflected shaft returns to its neutral position. The wrists now release the club which accelerates significantly. The shaft and bends backwards and at the same time the Coriolis acceleration deflects the shaft rearwards. Towards the end of the swing the hands rotate the head and the centrifugal acceleration acting through the centre of gravity, which is offset form the axis of the shaft, bends the shaft forwards and downwards. The latter is shown in Fig. 3.12, where it should be understood that the CG of the head remains in the same position.

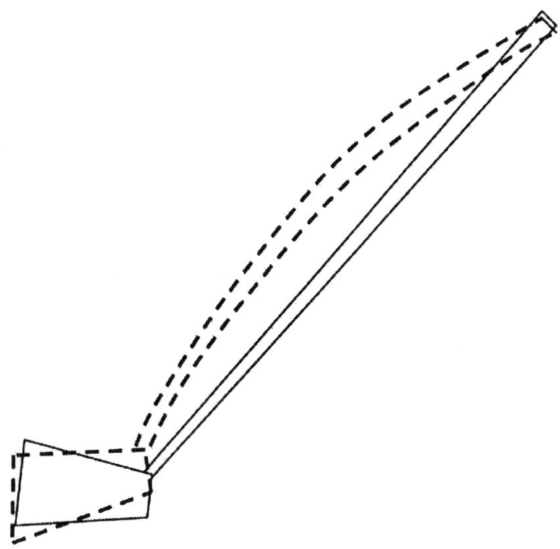

*Figure 3.12.* Bending of the shaft in droop.

The forward bend of the shaft increases the loft which the club face presents to the ball. The exact attitude will depend upon the flexibility of the shaft, and the acceleration pattern applied by the golfer.

In many cases the manufacturers of golf clubs use shafts which are more flexible in order to increase the loft of the club for less expert golfers. It is often suggested that a better arrangement would be to increase the loft of the club and to use a stiffer shaft thereby transmitting more energy to the ball.

The set of photographs in Fig. 3.13a-d show the flex of the shaft during the impact period: (a) the shaft is bent slightly forwards as stated above; (b) as impact takes place, the head is decelerated by the force of impact, and the bottommost part of the shaft is bent rearward; (c) this bending action continues whilst the force of impact is applied; and (d) even when the ball loses contact with the face of the club, the bending action continues until the whole of the shaft is bent rearwards.

The construction of the shafts determines their performance. In most cases the diameter of the shaft at the head (tip) and at the hands (butt) is fixed, surprisingly with few exceptions. For a shaft in a driver or a fairway wood, the diameter at the tip will be 8.5 mm, whereas in an iron it will be 9 mm. For all clubs the diameter at the hands is 15 mm. The stepped steel shaft is produced by passing a block of steel through dies of various diameter. By varying the distribution of the length and diameter of steps, the shaft can be made stiffer or more flexible and the distribution of the stiffness can be varied. The wall thickness has little effect on the stiffness and is mainly chosen for strength and reduction of fatigue. The mass of a finished shaft ready for installation in a club is about 100 to 120 g. In recent times the manufacturers have tried to reduce this mass in order to ease the playability of the club. Some steel shafts have been made with masses of less than 70 g, but the durability is suspect. Metals of lighter density such as aluminum and titanium can help to solve the problem. Titanium, in particular, has a specific strength similar to that of steel.

From the description of the attitude of the shaft during the swing, the designer would wish to have the ability to vary the bending stiffness and torsional resistance separately. With metals, increasing the bending strength will automatically increase the torsional resistance, thereby reducing the ability of the designer to match the shaft to the player performance.

This problem can be solved with the use of plastic composites. These shafts are made of a number of layers of resin impregnated carbon fibres. Laying the fibres in the direction parallel to the axis of the shaft will increase the bending resistance whereas laying the fibres at 90 degrees to this axis will affect the torsional resistance. Careful selection of a number of layers with varying angles of fibre enables the designer to produce the correct balance between bending stiffness and torsional stiffness. The designer is constrained, however, by the restrictions on the diameter of the shaft as it enters the head of the club. Such a restriction is unnecessary, and the manufacturers should be encouraged to allow this diameter to vary to accommodate shafts more suiting to the range of ability of golfers.

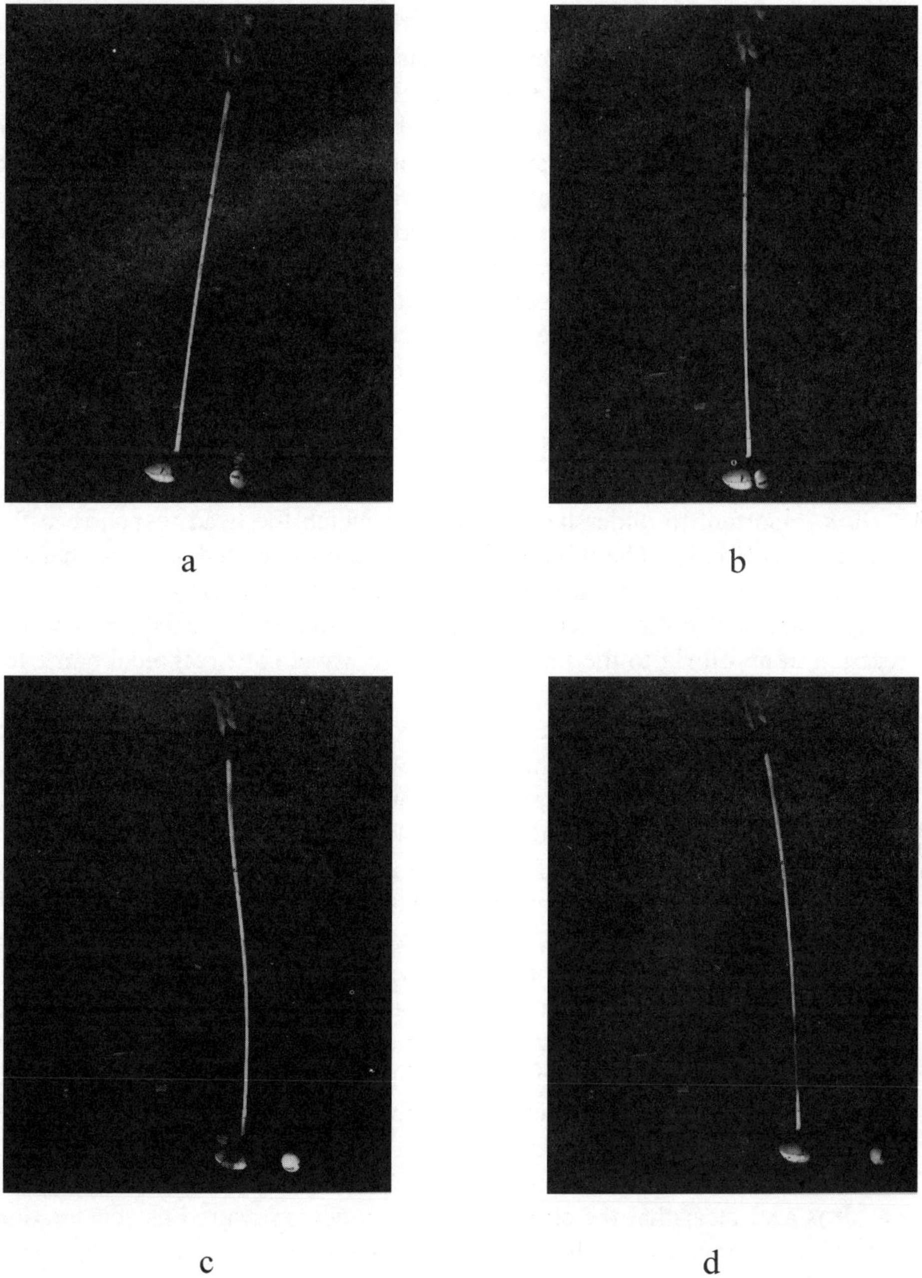

*Figure 3.13.* Attitude of the shaft: (a) before impact, (b) during impact, (c) immediately after impact, and (d) well after impact.

Whereas metal fatigue and durability have to be considered in the design of steel shafts, the problem with composite shafts lies in the quality of their construction. A cross section of a reasonably carefully wound shaft is shown in Fig. 3.14a together with Figs. 3.14b and c, where the quality is suspect. The irregular nature of this shaft and the voids in the material are obvious. Not only will these shafts produce inconsistent performances, and perhaps failure, but they do not comply with the rules of golf which require the shaft to have the same bending properties in any direction and the same torsional properties in both directions.

Much needs to be done on the design of shafts. This cannot be done in isolation from the design of the head or a consideration of the requirement of the golfer.

## 3.5 MODAL ANALYSIS

It is important to understand the way in which the head responds to the impact of golf ball. The whole of the club can be treated as a mechanical structure, subjected to a large force for a very short period of time. The analysis applied to the system is exactly the same as that used on, say, the reaction of an oil rig to the impact of a large wave. The technical name for such a process is modal analysis.

Every structure vibrates at its natural frequencies. Each frequency is associated with a displacement pattern of the structure which is called a mode. Theoretically there are an infinite number of modes of each structure, but in practice only a few modes at the lowest frequencies are important in the vibration of the structure. Further, the number of modes excited depends upon the nature of the impact process. A low-frequency process will excite only the lowest of frequencies. If the duration of the impact is short and the amplitude high a large number of modes will be excited.

There is a well-tried-and-trusted method of finding the displacement of a structure for a known input, by deriving the frequency response function. A typical frequency response function (FRF) is shown in Fig. 3.15 for a five iron golf club when the impact takes place at the centre of the club. The frequencies range up to 700 Hz and in this range, 6 modes are visible. The first mode is at 7 Hz, the second mode at 60 Hz, the third at 100 Hz and so on. It is also clear that the strength of the modes, measured as acceleration in this case, varies across the frequency range. Only the zero axis has been included.

a

b

c

*Figure 3.14.* Layering of: (a) regular composite shaft; (b) a poorly made composite shaft; and (c) a used shaft.

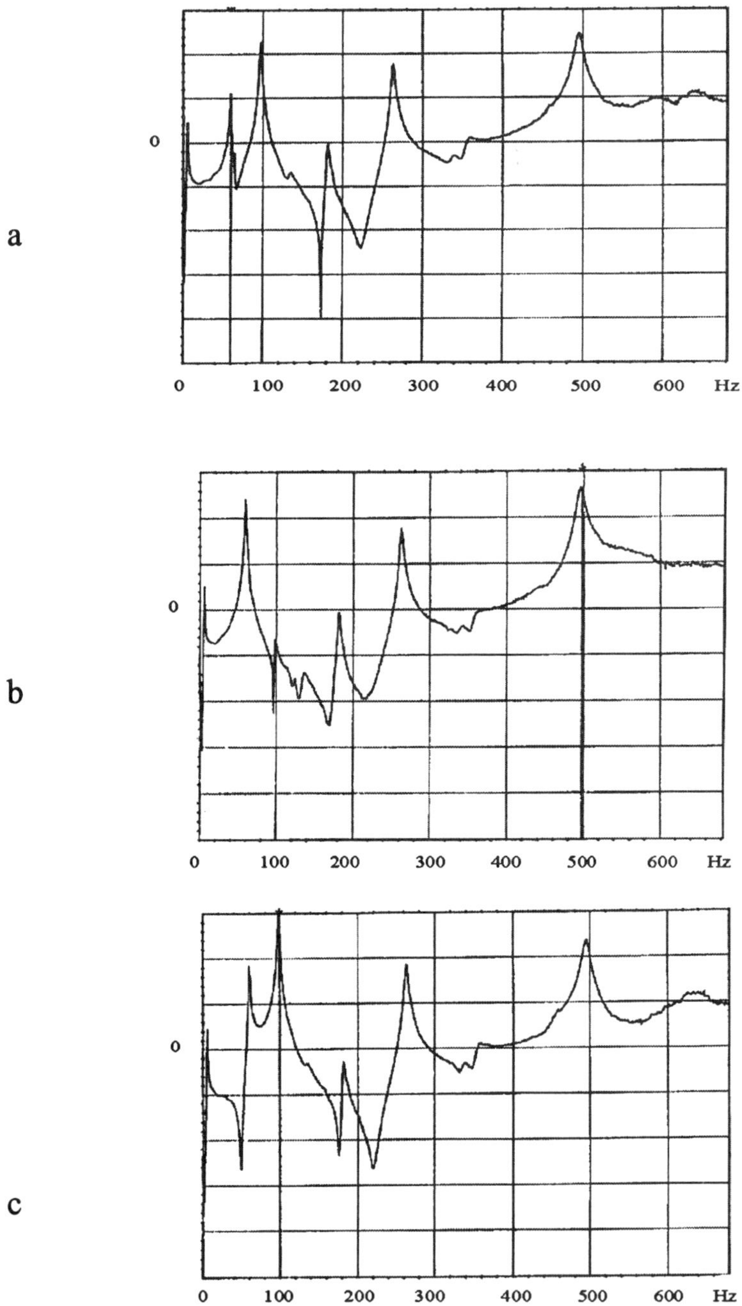


*Figure 3.15.* Frequency response function for a 5 iron with: (A) centre, (B) heel, and (C) toe impact.

Figure 3.16 shows the shapes of the displacement patterns associated with these modes. The lowest mode at 7Hz is a cantilever about the butt end of the golf club. There is an identical cantilever mode at the same frequency but vibrating at 90 degrees to the first mode. The mode at 60 Hz is torsional and the diagram shows the head rotating about an axis which passes through the head and, in this case, roughly parallel to the axis of the shaft. The mode at 100 Hz is also torsional, but the axis about which the vibration takes place is now at a different angle. Because no displacement occurs along these axes at the relevant frequencies, they are referred to as nodal axes.

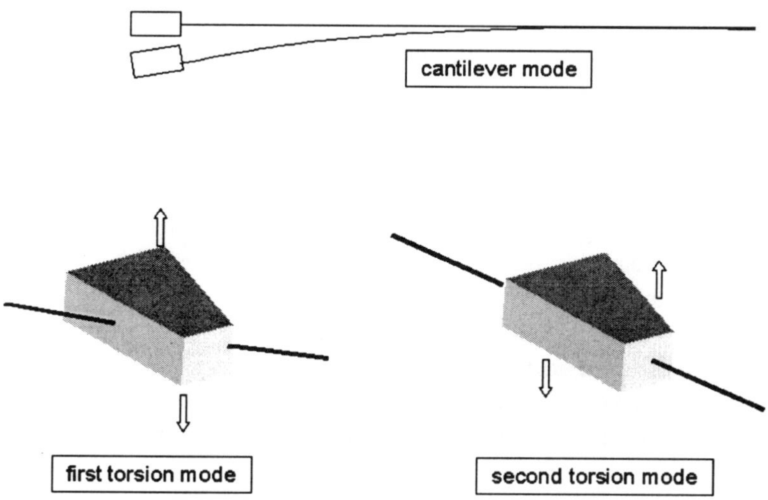


*Figure 3.16.* Modal shapes for cantilever and torsional modes.

Tests have been done on a very large number of club's of different types from which it is concluded that modes at higher frequencies than these four contributing very little to the movement of the head impacted by golf ball. Therefore for the time being, we will concentrate on these four modes, remembering that every movement of the head whether in bending or in twist removes energy from the impact, energy which is therefore lost to propelling of the ball.

Figure 3.15b shows the same club with the impact is taking place in the heel of the golf club towards the shaft. Modes at the same frequencies appear but the amplitudes at each are not the same. The cantilever mode at 7 Hz is almost identical but the amplitude of the mode at 60 Hz is more than doubled and that at 100 Hz is reduced to almost nothing. Figure 3.15c shows the FRF when the impact is on the toe. The amplitude distribution is again different, with a large contribution from the 100 Hz mode and more form the 60 Hz mode than if the impact was on the centre of the face (see Fig. 3.15a). This begs the question whether it is possible to design a golf club where the response at these four contributing modes is always low?

First let us ask why there is a variation? Referring back to the modal response, it is known that the impact on or close to the axis of a mode will produce no response in that mode. It would be useful, therefore, for the impact to take place along or close to the nodal axes of the various modes of the structure. Figure 3.17 shows the nodal axes of the torsional modes superimposed on the outline of the five iron golf club. In order not to excite the third and fourth modes, the impact should take place at the point marked A where the nodal axes cross. The figure clearly shows that, for this five iron, the impact point in the centre of the club face is a well away from point A. In golfing terms point A could be referred to as the "sweet spot" of the club, which is therefore defined as that the point which transmits the most energy to the ball. When the impact is at an adjacent point and then the response of the head will be determined by its mechanical properties, such as inertia, and we can therefore define an area where the energy transfer is still good. It is perhaps logical to design the club to have the sweet spot at or near to the impact area of the golf ball. This would be true for a "forgiving" golf club to be used by an amateur. But this situation is not always desirable since an elite golfer will want to be able to move the ball left and right and hit the ball high or low and to do this will require a small area of high energy transfer, and a rapid response outside of this area.

In line with the above, some manufacturers measure the performance of their products using a process called "face mapping". Figure 3.17 also shows the result of one of these tests, where the iron has been swung by a robot, and the ball has been placed to impact the face at different points. The terms R and L stand for right and left of the desired line of hit, and the

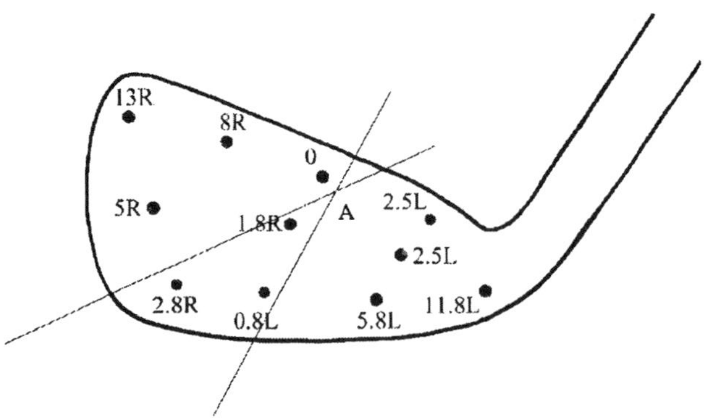


*Figure 3.17.* Face mapping of a 5 iron. The nodal axes are also shown. The sweet spot is at A.

numbers give the value in yards of the dispersion. Superimposed on the map are the measured nodal axes of this particular club. The lowest dispersion occurs when the impact takes place where the axes cross. High in the toe, well away from the nodal axes, the dispersion is high. The same is also true of the impact point low in the heel. For points close to or on the nodal axes the dispersion is much smaller.

Figure 3.18a,b shows some of the nodal axes on golf clubs which are commercially available. For most, the sweet spot is high in the heel of the club (Fig. 3.18a). For some, the sweet spot is not even on the face at all (Fig. 3.18b)!

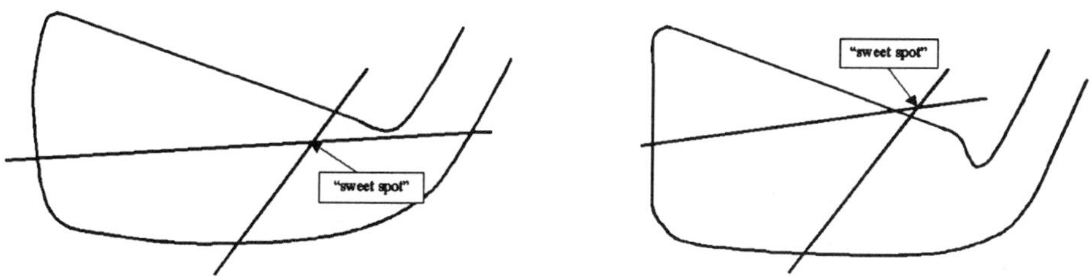


a b

*Figure 3.18.* Commercial club sweet spot located: (a) high on the heel, and (b) outside of clubface.

Turning now to the situation with the woods, consisting of the driver and fairway clubs, the same of four modes occur. In this case, however, the nodal axes are displaced rearwards towards the centre of the club because the centre of gravity is also displaced rearwards, and therefore it is impossible to produce a club with a sweet spot at the point where the impact takes place. Designers have to minimise the rotation of the head, which will inevitably occur, by increasing the rotational inertia of the club and putting curvature onto the face.

Finally, in all of the above, for both the irons and woods, the cantilever modes will always be present. The amount of energy removed during the impact in the energy transfer to the ball will therefore depend upon the stiffness of the shaft both in bending and in torsion.

Figure 3.19 shows the result of modeling the driver as an ellipsoidal mass. Modes 3 and 4 are the same as the torsional modes on the iron. In this case, it is found that a further mode, labeled 5, also contributes to the deflection of the head. By varying the mass distribution around the head, the location of the nodal axes changes. Configuration (b) would be unacceptable to all golfers, whereas configurations (a) and (d) would be suitable for good golfers. Configuration (c) should produce more backspin than the others, and therefore would be useful for new players.

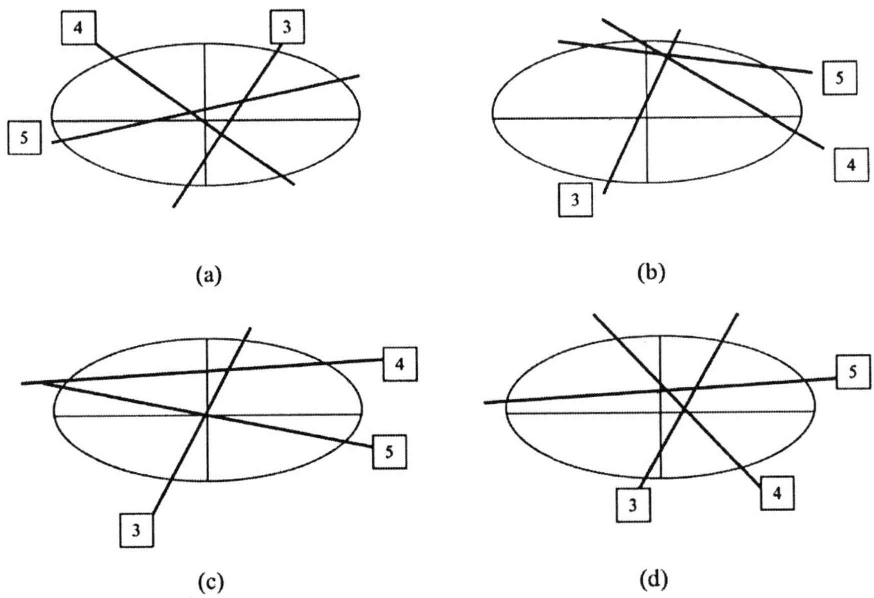


*Figure 3.19.* The location of nodal axes and sweet spots on ellipsoids using finite element modeling.

## 3.6 REFERENCES


Daish, C., B., 1972, *The Physics of Ball Games,* Hodder and Stoughton, London.

Gobush, W., 1990, Impact force measurements on golf balls, in: *Science and Golf, Proceedings of the First World Scientific Congress of Golf,* A. J. Cochran, ed., E and FN Spon, London, pp. 219-224.

Gobush, W., 1996, Spin and the inner workings of the golf ball, in: *Golf the Scientific Way,* A. J. Cochran, ed., Aston Publishing, pp. 141- 145.

Johnson, S. H., and Lieberman, B. B., 1994, An analytical model for ball-barrier impact, part 1, models for normal impact, in: *Science and Golf II, Proceedings of the Second World Scientific Congress of Golf,* A. J. Cochran, and M. R. Farrally, eds., E and FN Spon, London, pp. 309-314.

Johnson, S. H., and Lieberman, B. B., 1994, An analytical model for ball-barrier impact, part 2, a model for oblique impact, in: *Science and Golf, II, Proceedings of the Second World Scientific Congress of Golf,* A. J. Cochran and M. R. Farrally, eds., E & F.N. Spon, London, pp. 315-320.

Jorgensen, T., 1970, On the dynamics of the swing of a golf club, *American.J. Physics*, **38**: 644-651.

Larson, L. A., ed., 1970, *Encyclopedia of Sports Sciences and Medicine*, Macmillan, NY.

Lieberman, B. B., 1990, The effect of impact conditions on golf ball spin rate, in: *Science and Golf, Proceedings of the First World Scientific Congress of Golf,* A. J. Cochran, E and FN Spon, London, pp. 225-230.

Mather, J. S. B., and Cooper, M. A., 1994, Categorisation of golf swings, in: *Science and Golf II, Proceedings of the Second World Scientific Congress of Golf,* A. J. Cochran and M. R. Farrally, eds., E and FN Spon, pp. 65-70.

Maw, N., Barber, J. R., Fawcett, J. N., 1975, The oblique impact of elastic spheres, *Wear*, **38**: 101-114.

Miura, K., and Naruo, T., 1998, Accelerating and decelerating phases of the wrist motion of the golf swing, in: *The Engineering of Sport,* S. J. Haake, ed., Blackwell Science, Oxford, United Kingdom, pp. 455-463

Rees-Jones, J., 2002, A spatial model of the rigid body club swing, in: *Science and Golf IV, Proceedings of the Fourth World Scientific Congress of Golf,* E. Thain, ed., Routledge, pp. 3-12.

Riccio, L. J., 1990, Statistical Analysis of the average golfer, in: *Science and Golf, Proceedings of the First World Scientific Congress of Golf,* A. J. Cochran, ed., E and FN Spon, London, pp. 153-158.

Serandour, 2001, The application of the oblique impact of an elastic sphere theory to the study of the golf ball behaviour, *Project report for Diplôme d'Ingénieur de l'Institut National des Sciencess Appliquées de Toulouse.*

Turner, A. B., Hills, N. J., 1999, A three-link mathematical model of the golf swing, in: *Science and Golf III, Proceedings of the Third World Scientific Congress of Golf,* M. R. Farrally, and A. .J. Cochran, eds., Human Kinetics, pp. 3-12.

Williams, D., 1968, The dynamics of the golf swing, *Quart..J..Applied Mathematics*, **20**: 247-255.

## Chapter 4

# Eye and Head Movements During the Golf Putting Stroke

George K. Hung

Dept. of Biomedical Engineering, Rutgers University, 617 Bowser Rd., Piscataway, NJ 08854-8014, PH: (732) 445-4137, FX: (732) 445-3753, EM: shoane@rci.rutgers.edu

## 4.1 INTRODUCTION

Sports medicine is becoming a very popular means of addressing specific questions posed by athletes concerning the body's forces and actions during athletic motions. It has, for example, provided valuable information about the golf swing and physical forces impacting on the golf ball (Hay, 1978; Nicklaus and Bowden, 2002). Much of this information has been obtained using high-speed photography (Barnes, 1919; Bobby Jones' swing photograph, 1939; golfswingphotos.com). Indeed, the components of the golf swing have been studied in great detail over past 50 years. However, there is a surprising lack of objective simultaneous measurements of eye and head motion during the golf swing, especially in putting.

Putting is a crucial element in golf. This is demonstrated by statistics compiled by the Professional Golfers Association, which showed that the best players in the world expend approximately 40% of their total strokes in a round on putting (partners.golfserv.com). Golf teaching professionals and sports psychologists specific to golf have long taught the importance of minimal or no eye and head movements throughout the putting stroke. The eyes are important because they provide accurate perception of ball location to result in successful execution of a putting stroke. If the eyes are fixated elsewhere at a position other than the ball, this can lead to an improper stroke and a missed putt. Head position is also important because it allows for maintenance of stability of the visual environment. If the head moves during the stroke, this can lead to misalignment and a missed putt.

Different putter grip styles can be seen in the golf course. Through empirical experimentation or having been taught by particular school of thought, golfers have adopted different putting grips. Indeed, many golfers try different grips over the years, and sometimes even in the course of a round However, there has not been any quantitative measures of physiological and physical parameters related to the putting stroke to determine why a certain grip works for a particular individual or under a particular circumstance (e.g., a short versus a long putt). . Previous quantitative studies related to golf grips were concerned not with their effect on putter performance, but rather on driving distance and accuracy. For example, Alderman (196) tested 33 subjects and found no significant difference between the overlapping (Vardon) and interlocking grip with respect to distance, accuracy, and velocity and angle at impact. Also, Walker (1964) compared driving performance using the overlapping, interlocking, and baseball grips in 24 male golfers, and found no significant difference with respect to distance and accuracy.

In modern golf, there are basically three types of putting grips: conventional, cross-hand, and one-handed (Pelz and Mastroni, 1991; Degunther, 1996; Farnsworth, 1997). One-handed putting represents the grip used for the long-shafted putter, where the left hand hold the putter near the chin or chest and is anchored there. A detailed description of these grips is given in Section 4.2.2. Also, for this chapter, only the right-handed putting style will be considered, although the results can be generalized for the left-handed player.

To quantitatively assess the effect of grip style on eye and head movements, a study was undertaken to record the eye, head, and putter positions simultaneously during a putting stroke using the three putting-grip styles (Hung, 2003). Seven novice players served as subjects. These players were chosen in part because they possessed less pre-conceived bias in their putting-grip style than more experienced players. Nevertheless, the findings should be applicable to golfers regardless of their level of expertise. The results show differences in eye and head movements for the different putting-grip styles, which may provide important insights into the underlying mechanisms behind the selection of particular putting-grip styles by individual golfers.

## 4.2 METHODS OF MEASUREMENT

### 4.2.1 Apparatus

The laboratory environment was designed to present reasonably closely a real-life putting green environment while allowing for the recording of data. It consists of a putting platform, putter position sensor, helmet mounted eye movement sensors, and head movement measurement device (Patent Pending; see Figs. 4.1a,b). The putting platform is covered by green artificial turf that mimics the texture of the putting green. The platform is constructed of standard $\frac{5}{8}$ in. thick particle boards and are stacked up three-boards thick to provide the full thickness of the platform. The platform is comprised of three sections. The first section consists of embedded phototransistors and is where the subject stands and executes the putts. The other two sections can be adjusted so that putting distances of either 3 or 9 ft. are possible.

The putting position sensor consists of 16 infrared photo-transistors place in the middle-level particle board in the first section (Figs. 4.2a,b). They are inserted into different pin positions in three electronic breadboards which are sandwiched and secured in place by particle boards. The resolution of the putting sensor is provided by the minimum spacing between phototransistors, which is 1.2 cm. The breadboards, and in turn the photo-transistors, are aligned in the direction of the putting stroke. A ribbon cable connected to the photo-transistors transfers the electronic signals to a circuit board outside of the platform. The circuit board processes the electronic signals to provide a digital signal representing the instantaneous position of the putter face. The signal is input to a PC.

The binocular horizontal eye movement monitor is a Skalar-Iris (Model 6500) helmet-mounted infrared reflection device (Fig. 4.3). It has a linear range of ± 25 deg., a resolution of 5 min. of arc, and a bandwidth of 200 Hz. The voltage output signals representing eye rotations in the two eyes are input to a PC.

The head rotation movement sensor consists of a potentiometer that is placed in an adjustable mount above the subject's head (Figs. 4.4a,b). A small rectangular magnet is glued to the end of the potentiometer and is bounded by two plastic rectangular pieces to form the female coupling member (Fig. 4.4a). Similarly, a small rectangular magnet is screwed into a male coupling member (Fig. 4.4b). At the beginning of an experimental session, while the subject takes his/her normal putting stance, the position of cylindrical unit, which itself is screwed onto the top the helmet, to form the head movement sensor unit is adjusted until the male and female components are joined and the axis of the potentiometer is aligned

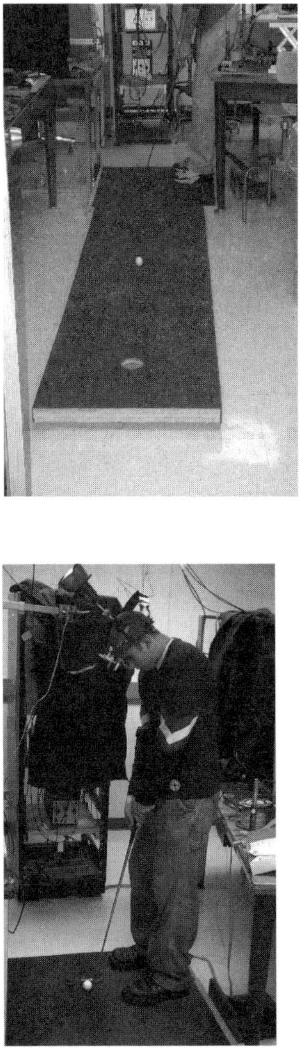

(A)

(B)

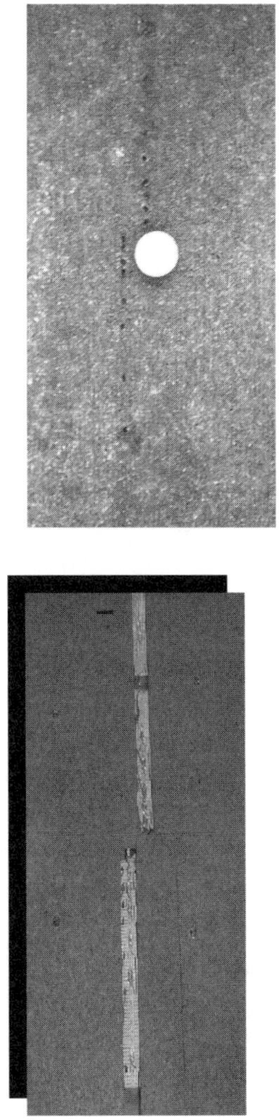

*Figure 4.1.* Subject standing on putting platform during an experimental trial. (a) Subject wears a helmet (for eye and head movement measurements) and prepares to putt on the first section of the platform. Note that under this section are the embedded photo-detector arrays (see Fig. 4.2). (b) Two other sections are combined with the first section to provide the setup for the 9-ft putts. Turning around the third section (where the hole is) and removing the middle section allows for 3-ft putts.

*Figure 4.2.* (a) Top view of first section of platform where the subject stands (see Fig. 4.1a) showing ball and photo-detector locations. (b) Arrays of photo-detectors embedded under the artificial turf in the platform. Note the front array is offset so that the ball movement will not trigger the detectors. The relative position of the golf hole is in front of the photo-detectors (i.e., in the direction below the figure).

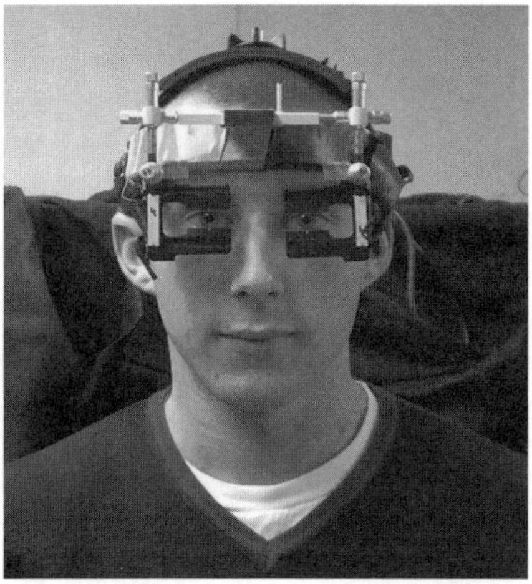

*Figure 4.3.* Subject wearing helmet containing the binocular eye movement sensor unit. The eye movement sensor unit in each eye contains arrays of infrared emitters (upper black rectangular piece) and detectors (lower black rectangular piece). As the eye rotates, the amount of infrared reflection from the white part of the eye changes, which can be recorded as a voltage change.

*Figure 4.4a.* The female component of the head movement sensor consists of a potentiometer cemented to a swivel unit (upper left). A small square aluminum platform is cemented to the end of the potentiometer shaft. The rectangular magnet is cemented on top of the platform, and is flanked by 2 rectangular plastic members.

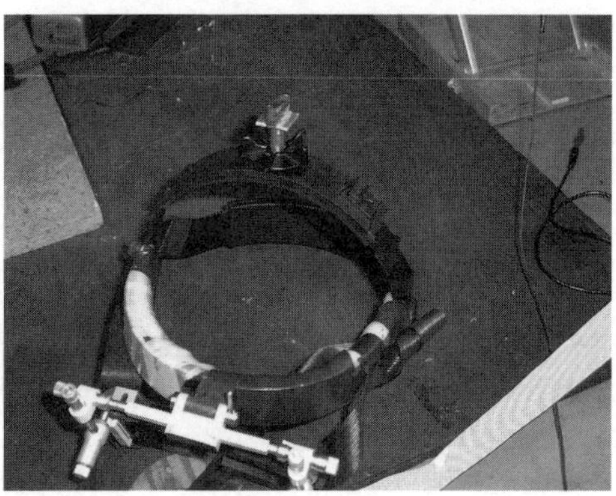

*Figure 4.4b.* The male component of the head movement sensor consists of a rectangular magnet mounted on top of a threaded cylindrical unit, which in turn is screwed onto the top portion of the eye movement helmet.

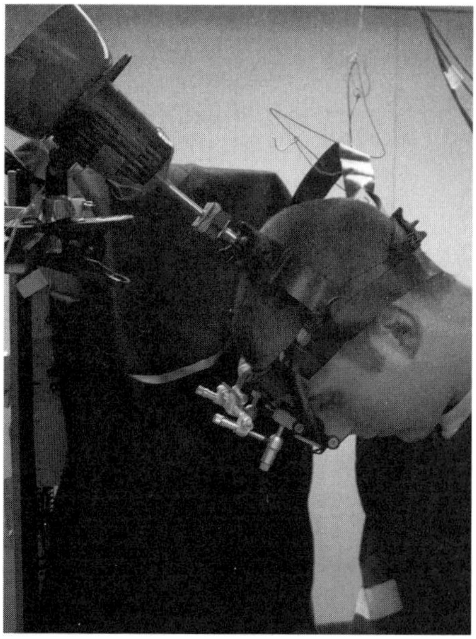

*Figure 4.5.* The head movement sensor is assembled by attaching the female component in the head sensor swivel unit (see Fig. 4.4a) to the male component in the helmet (see Fig. 4.4b). The swivel unit is clamped onto a frame with three directions of freedom of movement. The swivel unit itself can be pivoted and rotated about its stem. This allows for different subject heights and head angles during the putting stroke. During the experimental setup, the shaft of the potentiometer that is attached to the swivel unit (see Fig. 4.4a) is aligned with the head in such a way that only rotation about an imaginary axis through the top of the head and its center-of-rotation will turn the potentiometer, which in turn provides the head rotation - signal.

approximately with an imaginary axis through the top of the head and its center-of-rotation (Fig. 4.5). Rotation of the head about this axis will result in rotation of the potentiometer, which in turn causes a change in voltage in the head sensor output signal. The head movement sensor has a resolution of 0.5 cm and is linear throughout the ± 25.0 cm display range (in units of equivalent platform displacement in cm; see below for conversion from head angular rotation in degrees to equivalent platform displacement in cm). The voltage signal, which represents the head position, is then input to the PC. The head movement sensor can be easily disengaged from the helmet by simply having the subject take a step back away from the device.

A more recent innovation in our laboratory is the use of miniature eye movement and head acceleration sensors, along with a remote putter motion sensor, to wirelessly transmit the signals to a personal computer for display and recording (Patent Pending).

## 4.2.2 Putting Grip Styles

Three putting-grip styles were used. In the conventional grip (Figs. 4.6a,b), the left hand holds the top of the grip portion of the putter shaft while the right hand is placed below the left. In most players, the little finger of the right hand is placed in a position just overlapping the gap between index and middle fingers of the left hand. In the cross-hand grip (Figs. 4.7a,b), the positions of the hands are reversed. The right hand holds the top of the grip portion of the putter shaft while the left hand is placed below the right. In this position, the tip of the index finger of the right hand may be in contact with the wrist of the left hand. In the one-handed grip (Figs. 4.8a,b), which represents the action used for the long-shafted putter, the right hand holds approximately the middle portion of the putter grip. For consistency in interpreting the results, the same putter (37 in. Acushnet Bulls Eye FLU5S) was used for all three putting-grip styles in all subjects.

# Conventional Grip

(A)

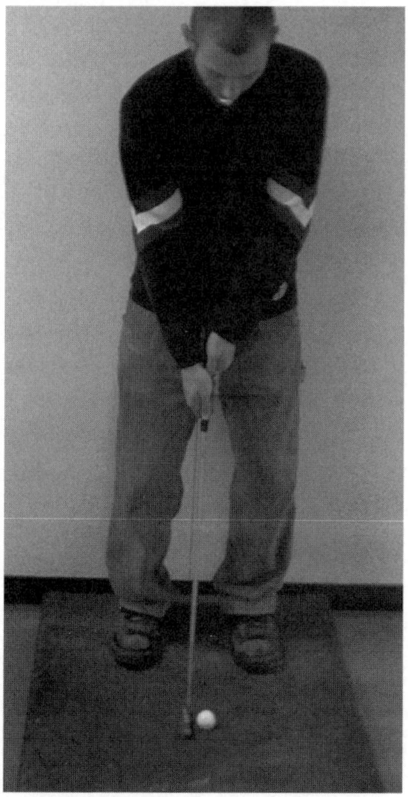

(B)

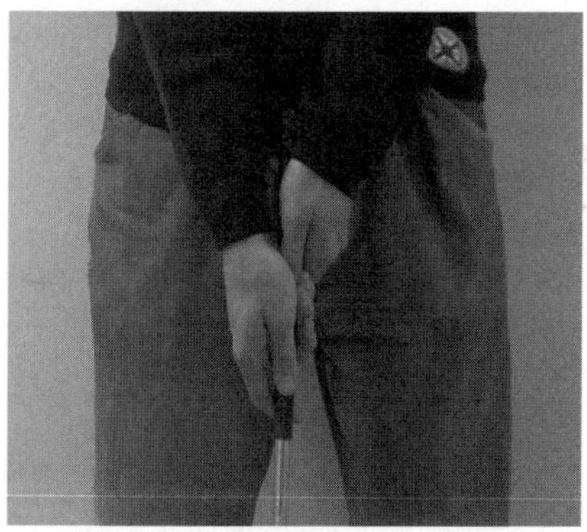

*Figure 4.6.* (a) Overall view of conventional grip for putting in a novice player (GH), where the right hand is *below* the left hand. (b) Close-up view of the conventional grip.

# Cross-Hand Grip

(A)

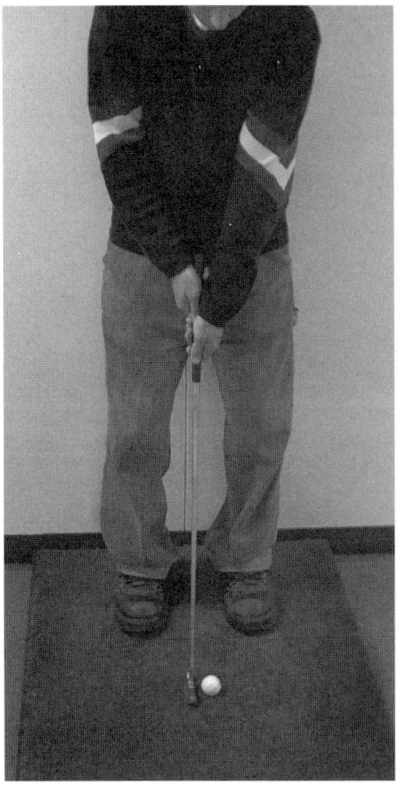

(B)

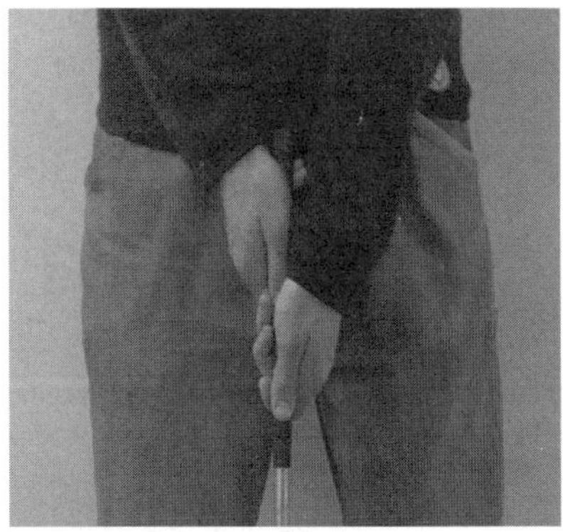

*Figure 4.7.* (a) Overall view of cross-hand grip for putting in a novice player (GH), where the right hand is *above* the left hand. (b) Close-up view of the cross-hand grip.

# One-Handed Grip

(A) (B)

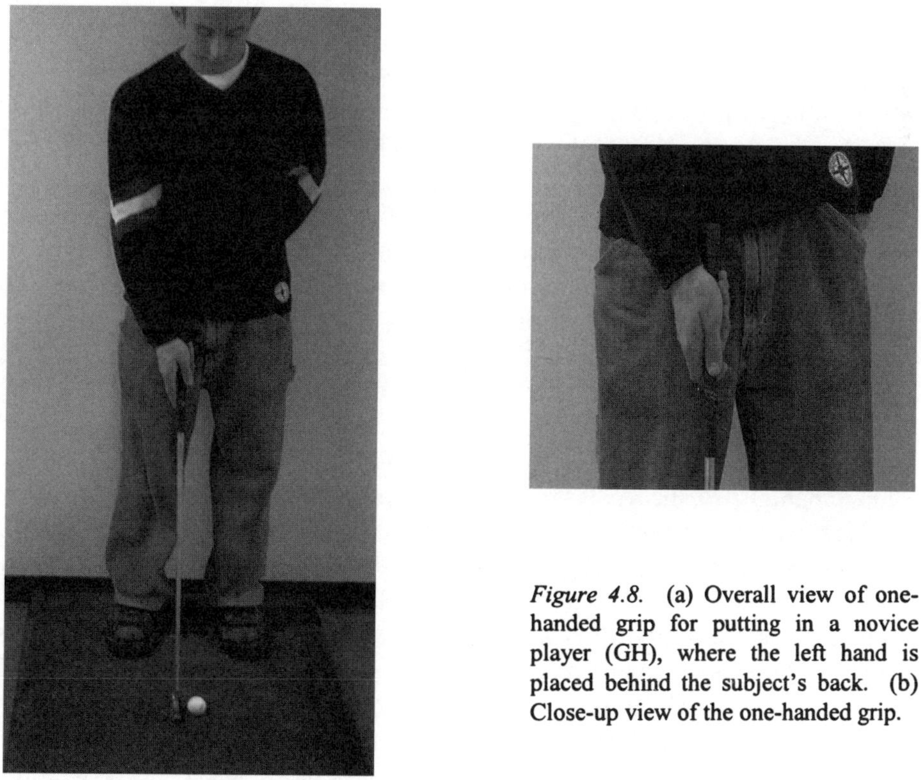

*Figure 4.8.* (a) Overall view of one-handed grip for putting in a novice player (GH), where the left hand is placed behind the subject's back. (b) Close-up view of the one-handed grip.

### 4.2.3 Experimental Procedure

Seven unpaid volunteer subjects were recruited from the Rutgers University student community via word-of-mouth and advertisement. They were all novice right-handed golfers ranging in age from 21 to 22 years, with corrected vision of 20/20 or better. All were male except for one female.

The experimental paradigm for each subject was as follows: the subject first practiced 5 putts without the helmet to become familiar with the conditions. Also, for each of the three putting-grip styles, as the subject took a putting stance, the distance from the ball on the platform to the subject's eyes was measured. Just prior to the beginning of the experiment, the helmet was placed on the subject's head, and the positions of the infrared detectors for the two eyes were properly adjusted. At the beginning of the trials, an eye movement calibration trial was performed, where the subject fixated sequentially on three known locations (left end, middle, and right end of the photo-transistor array) on the platform. This provided accurate recording of eye fixations during a putting stroke, measured in units of cm of displacement on the platform. For the putting trials, the head movement sensor was attached to the head. The potentiometer was pre-calibrated to provide the conversion from a measured voltage change to the corresponding angular rotation of the potentiometer shaft, which in turn corresponded to the angular rotation of the head. This angle was converted to the displacement of a hypothetical beam emanating from the center-of-rotation of the head (approximated by the position of the center between the two eyes) on the platform, in cm. In this way, all measurements (i.e., putter, eye, and head movements) had the consistent unit of cm of platform displacement. The subject was instructed to attempt to putt the ball into the hole using as natural a stroke as possible. Each putt constituted a trial lasting 3 sec, and the result of the putt (i.e., where the putt ended up relative to the hole) was tabulated in a data sheet. Twenty trials were performed for each of two distances: 3 and 9 ft. This was repeated for each of the three putting-grip styles. The entire experiment for each subject required about 90 min.

For each 3-sec trial, the putter, left eye, right eye, and head position time courses were recorded on a PC at a sampling rate of 100 Hz. The resulting data were analyzed using programs written in C++ and MATLAB codes. The results were displayed in 4 channels as position time courses for putter, left and right eye movements, and head movement. Also displayed were the corresponding velocity traces. For each record, the beginning of the putt as well as the end of the putt (i.e., point of impact, or at the point corresponding to the return to the initial putter position) were marked visually. The data were analyzed to provide: the duration of the putt and the standard deviations (STD; i.e., relative to the initial position) of combined left and right eye movements as well as head movements during the putt (i.e., from the beginning of the backstroke to the point of impact).

The data from each subject was averaged, and the averaged data for the seven subjects were used in the statistical analysis. One-tailed t-tests (MATLAB Statistical Analysis Toolbox) were performed to assess the statistical significance of differences between the different putting-grip styles for the various parameters: height to eyes, putt amplitude, duration of putt, percentage of putts made, STD of combined left and right eye movements during the putt, and STD of head movements during the putt.

## 4.3 EXPERIMENTAL RESULTS

### 4.3.1 Typical Movements

Typical 9-ft putt time courses for the three putter-grip styles are shown in Figs. 4.9a-c to illustrate the general pattern of responses. The sign convention is as follows: the forward stroke of the putt, leftward eye movement, and leftward head rotation (i.e., nose points leftward towards the hole) are assigned positive values. For the conventional-grip putt (Fig. 4.9a), note the slight rightward rotation of the head (i.e., negative value, or away from the hole direction), probably due to shoulder rotation, during the backstroke, and simultaneous automatic compensatory smooth eye movements (called vestibulo-ocular response (VOR), or compensatory rotation of the eyes in a direction opposite to head rotation to maintain stable fixation in space) (Ciuffreda and Tannen, 1995) in the positive (or towards the hole) direction. These two movements generally tend to cancel each other to provide relative steady eye fixation on the ball. However, incomplete compensation or disruptions due to other eye or body motions will result in inappropriate eye fixation in space. Note also that at about the moment of impact, there is a saccadic eye movement (or rapid rotation of the eyes to acquire a new target, such as that seen during reading) (Bahill and Stark, 1979; Hung, 2001; Hung and Ciuffreda, 2002; Pola, 2002; Stark, 1968) in the direction of the hole, which may possibly affect the release of the putter following impact. The putting stroke shows relatively good positive acceleration in the forward stroke, reaching approximately peak velocity at impact (0 cm putter position).

# Conventional Grip Response Characteristics

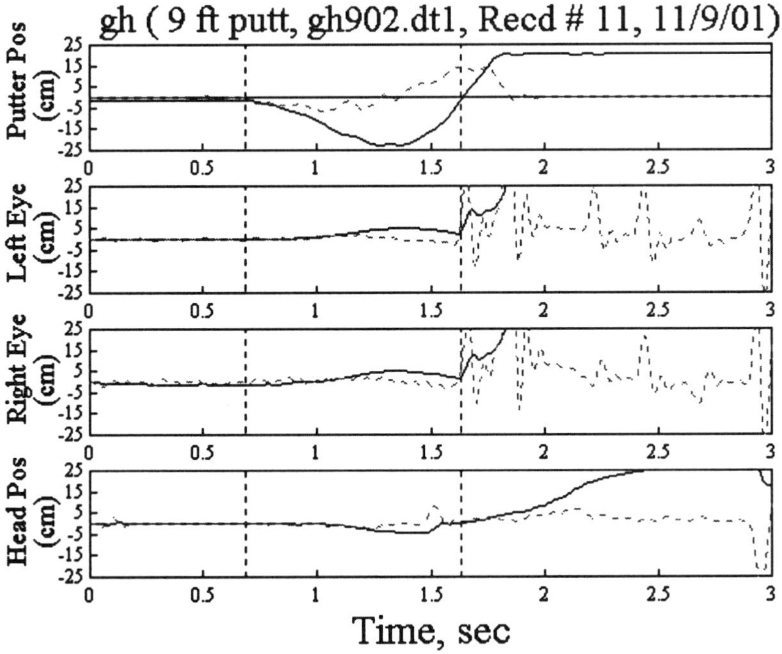


*Figure 4.9a.* Time traces of putter, eye, and head movements for a typical 9-ft putt using the conventional grip. The hole location is in the direction of the top of the page. Positive and negative numbers represent movements towards or away from the hole, respectively. Top trace - Putter position. 2nd and 3rd traces - Left and right eye positions, respectively. The eye movements, via the calibration procedure, are converted to units of cm displacement on the platform. Bottom trace - Head rotation. The angle of rotation about an imaginary axis through the top of the head and its center-of-rotation has been converted to equivalent movement of an imaginary beam projected from the center-of-rotation of the head (approximated by the center position between the two eyes) onto the platform, in cm. For all traces, the position and velocity are designated by solid and dashed lines, respectively. Vertical dashed lines were visually selected to encompass the duration of the putt from the beginning of the backstroke to the point of impact. Sub. GH.

## Cross and One-Handed Grip Response Characteristics

b

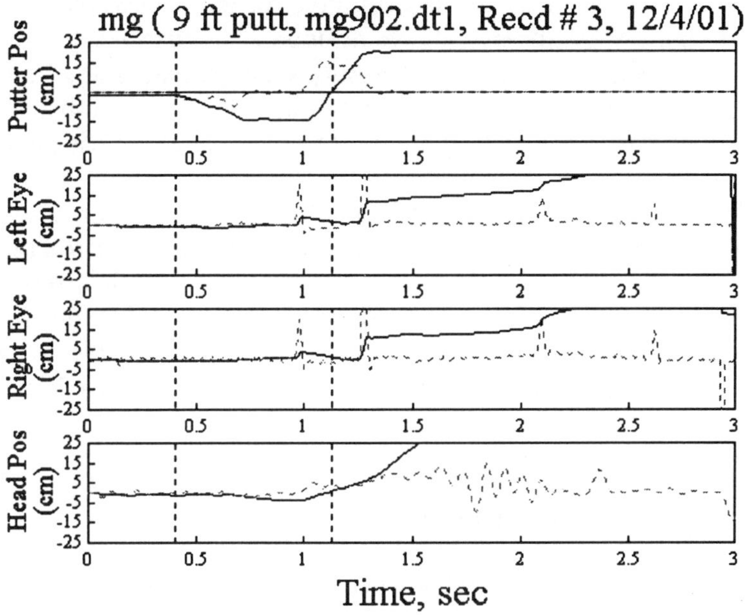


c

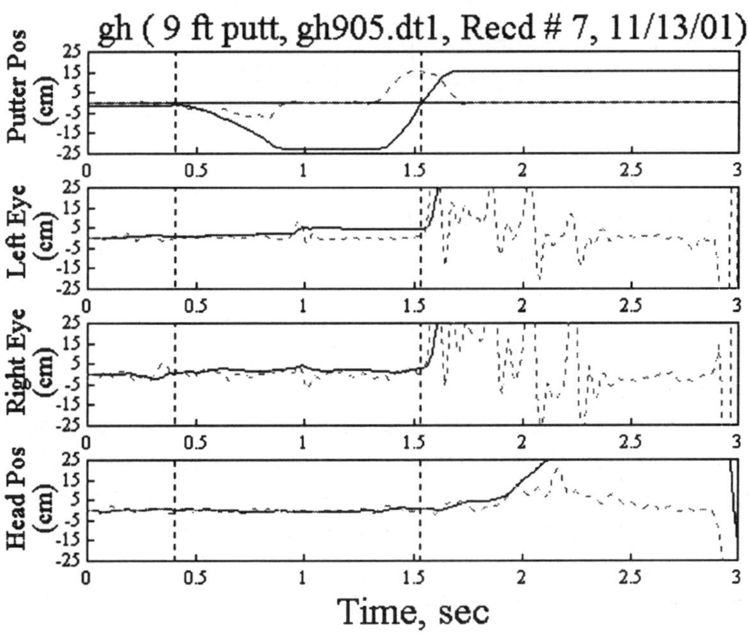


*Figures 4.9b,c.* Time traces for a typical 9-ft putt using : (b) the cross-hand grip (Sub. MG) and (c) one-handed grip (Sub. GH).

For the cross-hand grip putt (Fig. 4.9b), note the slight rightward head rotation away from the hole, along with a slight compensatory VOR during the putt. There is also a saccade in the direction of the hole, which occurs prior to impact, but the overall eye position is approximately directly on the ball at the moment of impact. The subsequent saccades and head rotations are simply tracking movement in following the path of the ball towards the hole. Moreover, the putting stroke is near peak velocity at the moment of impact.

For the one-handed grip putt (Fig. 4.9c), note the near absence of head movements during the stroke, and the eyes are fixated approximately on the ball throughout the stroke. Also, the putting stroke is near peak velocity at the moment of impact. Subsequent saccadic and head rotations are associated with viewing the path of the ball towards the hole. Also note that the putter is held for a relatively prolonged period during the backstroke.

## 4.3.2 Across All Subjects

The results show differences between putting-grip styles [conventional (Conv); cross-hand (Cross); one-handed (One)] for various parameters: height to eyes, putt amplitude, putt duration, percentage of putts made, STD left and right eye movements during the putt, STD of head movements during the 3 ft (Table 4.1a) and 9 ft (Table 4.1b).

The data in Table 4.1a,b can be interpreted using an example. For the first two columns under "Height to Eyes" for the 3-ft putt, the mean for the conventional and cross-hand grips are 142.6 and 139.3 cm, respectively. To determine whether the difference between the means is significant (i.e., is conv > cross), a one-tailed t-test was applied (first column). The result is that $p = .031$, indicating the difference is significant (i.e., $p < 0.05$, the criterion level). A similar interpretation is used for all the other columns.

"Height to Eyes" is highest for one-handed, intermediate for conventional, and lowest for cross-handed. This is because for the one-handed putt, the subjects tend to stand more erect. On the other hand, for the cross-hand grip, the left hand is positioned lower on the putter grip, thus lowering the left shoulder. This tends to bring the head down as well, thus lowering the height of the head relative to the platform. The "Height to Eyes" findings are obviously the same for 3- and 9-ft putts. They are shown for both distances for consistency in appearance of Tables 4.1a,b.

| | Height to Eyes | | | Putt Amplitude | | | Putt Duration (sec) | | |
|---|---|---|---|---|---|---|---|---|---|
| Mean | Conv 142.6 | Cross 139.3 | One 145.1 | Conv 16.80 | Cross 16.64 | One 17.04 | Conv 0.79 | Cross 0.75 | One 0.87 |
| Hypothesis | Conv > Cross | Cross < One | One > Conv | Conv > Cross | Cross < One | One > Conv | Conv > Cross | Cross < One | One > Conv |
| t - test | P =.031 SG | P =.002 SG | P =.032 SG | P =.417 NS | P =.334 NS | P =.362 NS | P =.213 NS | P =.005 SG | P =.004 SG |

| | Percent Made (%) | | | STD LE + RE Mov't | | | STD Head Mov't | | |
|---|---|---|---|---|---|---|---|---|---|
| Mean | Conv 71.20 | Cross 78.14 | One 71.09 | Conv 4.41 | Cross 4.15 | One 3.30 | Conv 2.75 | Cross 2.09 | One 1.65 |
| Hypothesis | Conv < Cross | Cross > One | One < Conv | Conv > Cross | Cross > One | One < Conv | Conv > Cross | Cross > One | One < Conv |
| t - test | P =.121 NS | P =.121 NS | P =.493 NS | P =.312 NS | P =.066 TN | P =.025 SG | P =.076 TN | P =.151 NS | P =.018 SG |

*Table 4.1a* - (3 ft putt) Comparisons among conventional (Conv), cross-hand (Cross), and one-handed (One) grips for the averaged data in seven subjects using the t -test (one tail, significance level set at $p = 0.05$) for 3 ft putts. STD = standard deviation. SG = significant, $P < 0.05$. TN = trend, $0.05 \leq P \leq 0.10$. NS = not significant. All values in cm of platform displacement except where indicated.

"Putt Amplitude" is not significantly different for the three putting-grips style for the 3-ft putt. However, for the 9-ft putt, amplitude for the cross-hand grip is statistically significantly smaller than either conventional or one-handed grip. This may be due to a restriction of the right elbow motion in the backstroke as a result of the increased bend of the right elbow to compensate for the higher right hand position on the putter for the cross-hand grip.

"Putt Duration" is longer for the one-handed than either conventional or cross-hand grip, for both 3- and 9-ft putts. This is due to the increased length of the swing arm (putter plus hand and arm) in the one-handed putt. Hence, if one considers this as a pendulum motion, the increased length corresponds to an increase in the period of motion, and in turn an increase in the duration of the putt.

| | Height to Eyes | | | Putt Amplitude | | | Putt Duration (sec) | | |
|---|---|---|---|---|---|---|---|---|---|
| Mean | Conv 142.6 | Cross 139.3 | One 145.1 | Conv 22.13 | Cross 20.38 | One 22.55 | Conv 0.77 | Cross 0.80 | One 0.91 |
| Hypo-thesis | Conv > Cross | Cross < One | One > Conv | Conv > Cross | Cross < One | One > Conv | Conv > Cross | Cross < One | One > Conv |
| t - test | P =.031 SG | P =.002 SG | P =.032 SG | P =.052 TN | P =.016 SG | P =.369 NS | P =.300 NS | P =.001 SG | P =.001 SG |

| | Percent Made (%) | | | STD LE + RE Mov't | | | STD Head Mov't | | |
|---|---|---|---|---|---|---|---|---|---|
| Mean | Conv 33.4 | Cross 47.9 | One 41.6 | Conv 5.91 | Cross 5.06 | One 4.23 | Conv 2.79 | Cross 3.04 | One 2.16 |
| Hypoth esis | Conv < Cross | Cross > One | One > Conv | Conv > Cross | Cross > One | One < Conv | Conv < Cross | Cross > One | One < Conv |
| t - test | P =.006 SG | P =.274 NS | P =.242 NS | P =.099 TN | P =.016 SG | P =.038 SG | P =.358 NS | P =.153 NS | P =.127 NS |

*Table 4.1b* - (9 ft putt) Comparisons among conventional (Conv), cross-hand (Cross), and one-handed (One) grips for 9 ft putts.

"Percent Made" appears to be higher for cross-hand than either conventional or one-handed grip, for both 3- (Table 4.1a) and 9-ft putts (Table 4.1b), but these comparisons are not statistically significant except for one condition. For the 9-ft putt, the "Percent Made" is statistically-significantly higher for the cross-hand than conventional grip (p = .006).

The "STD of Combined Right and Left Eye Movements" is lowest for one-handed, intermediate for cross-hand, and highest for conventional, for both 3- and 9-ft putts. This holds except for the comparison between cross-hand and conventional for the 3-ft putts (p = .312).

The "STD of Head Movements" is lowest for one-handed, intermediate for cross-hand, and highest for conventional, for the 3-ft putts. This holds except for the comparison between one-handed and cross-hand putts (p = .151). However, for the 9-ft putts, there are no statistically significant differences among the putting-grip styles. These findings suggest that head motion plays a more important role in shorter (3-ft) than longer (9-ft) putts, and moreover indicates that one-handed grip, and to a lesser extent the cross-hand grip, result in less head motion during the putt.

A simpler-to-visualize illustration of the same information as in Table 4.1a,b is shown in Table 4.2.

## 4.4 INTERPRETATION OF RESULTS

This study has provided some important quantitative insights into the effect of grip style on eye and head movements during the putting stroke. For minimizing head movements, there appears to be a slight advantage of the one-handed grip over the other two grip styles, especially for the shorter putts. This may be because for the conventional and cross-hand grips, the right and left shoulders are linked due the coupling of the two hands. Thus, during the backstroke, the movement of the hands causes the left shoulder to dip slightly and the right shoulder to rise slightly. The natural linkage of the shoulders to the head causes it to rotate slightly clockwise, as seen from above. The opposite occurs during the forward stroke. In contrast, for the one-handed grip, the two hands are not linked, so that movements of the right hand and arm during the putting stroke rotates the right shoulder, but with relatively little dip or rising motion. Moreover, since there is less constrained coupling of the right shoulder to the head, the head motion is relatively small during the putting stroke. Generally, players are reluctant to attempt the one-handed putt perhaps because of a feeling that they may lose control of the stroke. However, some professional players have used the one-handed putting grip, notably Mike Hulbert, who used it in the mid-1990s, and Lanny Watkins, who used it for short putts. Nowadays, the one-handed putting style is manifested in the use of the long-shafted putter, which is quite popular among senior players. An additional reason for its use among senior players is discussed below.

Variation in eye movements were found to be least for one-handed, intermediate for cross-hand, and highest for conventional grip. This is due in part to the VOR. Thus, the smaller eye movements for one-handed grip is partly due to the smaller head movements. However, not all eye movement effects are a reflection of the VOR, since the statistical comparisons for the different grips are not identical for eye and head movements (see Table 4.1a,b). There may be additional factors, such as posture and initial angle of the head relative to the ball on the platform, that can affect the scanning eye movements during the putting stroke.

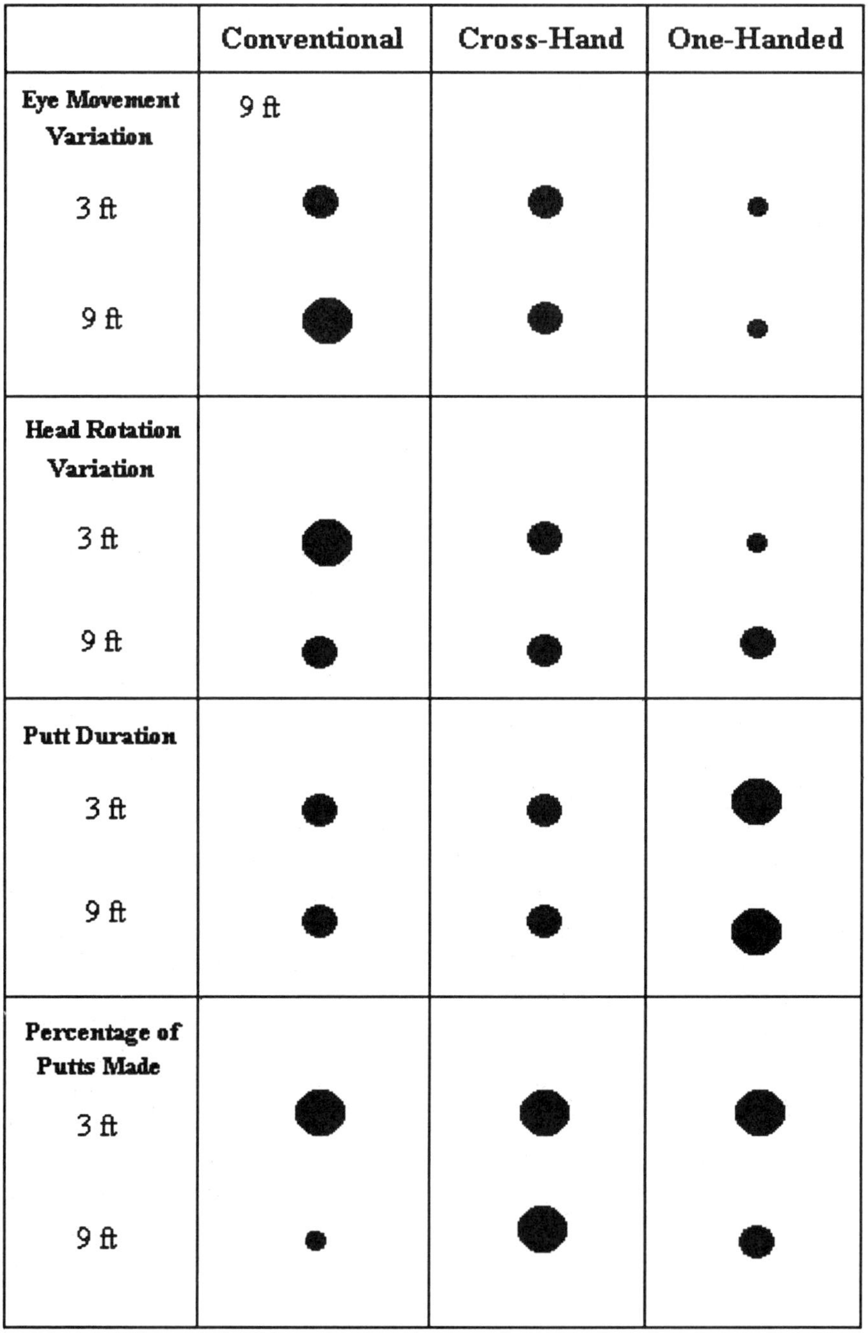


*Table 4.2.* Simpler-to-visualize illustration of same information as in Table 4.1a,b. Note that the circle size represents relative parameter magnitude.

The percentage of putts made was higher for cross-hand than conventional grips for 9-ft putts. It may be possible that the higher percentage made for the cross-hand grip is due a slightly more "solid" impact of the putter with the ball, especially for 9-ft putts, as reported by some of the novice subjects. On the other hand, although the direction of the putt appear to be better for the cross-hand grip, there may be a reduction in control of distance of the putt. This is seen in some of the novice player's putts that "fly past the cup". Thus, there is not a clear-cut advantage of the cross-hand over the conventional grip.

The duration of putts made was longer for one-handed than the other two grip styles. This is due to the greater effective length of the swing arm (putter plus hand and arm) of the putt. If one considers the putting motion as being analogous to that of a pendulum, in which the period is related to its length, then the longer swing arm for the one-hand grip would be associated with a longer duration. A longer duration or slower putt may provide improved timing in the putting stroke, thereby providing greater consistency. This may explain why many senior golfers have, through empirical experimentation with various putters, adopted the long-shafted putter (using effectively the one-handed grip). Perhaps the reduction in timing and control as one ages can be compensated in part by the use of a putting grip style that provide a slower, more controlled, tempo.

The subjects used in this study were all college students between the ages of 21 and 22. Most of them played other sports but all were novice golfers. This pool of novice young golfers has the advantage of not having any preconceived bias in favor of a particular golf grip. Also, they all have relatively good eye-hand coordination, so that this would not be a limiting factor in the study. Moreover, the narrow range of ages ensures a relatively uniform subject pool, without large age disparities. It would, however, be helpful in a follow-up study to investigate this effect in pools of subjects of various age groups. Further, it would be very informative to test this in expert golfers.

## 4.5 CONCLUSIONS

The experimental findings showed smaller variations using the cross-hand and one-handed grips than the conventional grip. This holds both for eye movements during longer putts and head movements during shorter putter. Also, the one-handed grip exhibited longer duration than the two other grip styles. However, when considered in terms of practical applications, each putting grip style has its advantages and disadvantages. The conventional stroke feels more natural and provides a sense of control over the putt. This must be weighed against the perhaps slightly more "solid" impact of the cross-hand grip and the better timing provided by the one-handed grip. The cross-hand grip tends to lock the left hand and arm in a relatively fixed position, and thus reduces any "conflict" between the right and left hands' neuro-motor commands from the left and right cortical hemispheres, respectively, which may cause the "yips" (a golf terminology indicating a psycho-physical condition characterized by jitters during the putting stroke). This, however, must be weighed against the slightly reduced control over the distance of the putt. The one-handed grip provides the least amount of head and eye movements. This occurs both because of the absence of linkage between the shoulders and the hands, and because only the left cortical hemisphere is involved in the neuro-motor command for the right hand, thus eliminating any potential conflict in the command signals. These benefits, however, must be weighed against the feeling of loss of control over the stroke. Yet, actual testing of the one-handed grip on a putting green can reveal that it is surprisingly accurate, and moreover, the sense of loss of control can be offset in part by the use of a long-shafted putter.

What do these finding tell us about the appropriate putting grip? Perhaps a combination of grips should be considered. For example, for the longer putts, use a conventional or cross-hand grip, whereas for the shorter putts, use a one-handed or cross-hand grip. Each individual must decide for oneself, through experimentation on the putting green and the golf course, which grip or combination of grips works best. This study has provided some of the quantitative and scientific rationale for gaining insights into the mechanisms underlying such a decision.

## 4.6 ACKNOWLEDGEMENTS

The author thanks Rutgers University students Gary Horton, Brian Cicalese, Nicholas Frietag, and Shana Groeschler for their assistance in the laboratory.

## 4.7 REFERENCES

Bahill, T., and Stark, L., 1979, The trajectories of saccadic eye movements, *Sci. Am.* **240**: 108-117.

Barnes, J. M., 1919, *Picture Analysis of Golf Strokes. A Complete Book of Instruction.* Lippincott, Philadelphia, PA.

Bobby Jones' Swing Photograph, by Dr. Edgerton of M.I.T. , 1939, *A Peacock? No, It's Bobby Jones in Action!,* The Bodine Motorgram, Bodine Electric company. http://www.bodine-electric.com/Asp/MotorgramViewer.asp?Article=12

Ciuffreda, K. J., and Tannen, B., 1995, *Eye Movement Basics for the Clinician.* Mosby, St. Louis, MO., pp. 112-114, 220-221.

Degunther, R., 1996, *The Art and Science of Putting.* McGraw-Hill, New York, NY.

Farnsworth, C. L., 1997, *See It and Sink It: Mastering Putting Through Peak Visual Performance*. HarperCollins, New York, NY.

golfswingphotos.com - Website for high speed photography of the golf swing; Al Ruscelli Photography.

Hay, J. G., 1978, *The Biomechanics of Sports Techniques*. Prentice-Hall, Englewood Cliffs, NJ, pp. 261-278.

Hung, G. K., 2001, *Models of Oculomotor Control*. World Scientific Publishing Co., Singapore, pp. 3, 102-110.

Hung, G. K. , 2003, Effect of putting grip on eye and head movements during the golf putting stroke. *TheScientificWorld*. **3**: 122-137.

Hung, G. K., and Ciuffreda, K. J., 2002, Models of saccadic-vergence interactions, in: *Models of the Visual System*, G. K. Hung and K. J. Ciuffreda, eds.,. New York: Kluwer Academic/Plenum Publishers, pp. 431-462.

Nicklaus, J., with Bowden, K., 2002, *My Golden Lessons*, New York: Simon & Schuster.

partners.golfserv.com - Web address for Professional Golf Association putting statistics.

Pelz, D., and Mastroni, N., 1991, *Putt Like the Pros: Dave Pelz's Scientific Way to Improve Your Stroke, Reading Greens, and Lowering Your Score*. HarperCollins, New York, NY.

Patent pending: *Non-Contact Embedded Photodetector-Array Platform for Detecting Putter Position*. Serial nos. 60/296,574 and 60/317,944.

Pola. J., 2002, Models of saccadic and smooth pursuit systems, in: *Models of the Visual System*, G. K. Hung and K. J. Ciuffreda, eds., Kluwer Academic/Plenum Publishers, New York, pp. 385-429.

Stark, L., 1968, *Neurological Control Systems, Studies in Bioengineering*. Plenum Press, pp. 250-270.

Walker, A., 1964, *The Relationship of Distance and Accuracy to Three Golf Grips*, M. S. Thesis, Springfield College, Springfield, MA.

# II

# TENNIS

# Chapter 5

# Tennis Ball Aerodynamics and Dynamics

Rabindra D. Mehta[1] and Jani Macari Pallis[2]

*[1]Sports Aerodynamics Consultant, 209 Orchard Glen Court, Mountain View, CA 94043, PH: (650) 960-0587, FX: (650) 903-0746, EM: rabi44@aol.com*

*[2]Cislunar Aerospace, Inc., P. O. Box 320768, San Francisco, CA 94116, PH: (415) 681-9619, FX: (415) 681-9163, EM: deke@cislunar.com*

## 5.1 INTRODUCTION

The detailed aerodynamics and dynamics of a tennis ball have only recently been uncovered with the help of a series of experimental studies (Mehta and Pallis, 2001, Cislunar Aerospace, Inc., 1998). Intuitively, one would expect the (rough) fabric covering on the ball to have an effect on the aerodynamics, but the extent of the effect turned out to be much more significant than had been anticipated. Most of the recent research work on tennis ball aerodynamics was inspired by a decision made by the International Tennis Federation (ITF) to start field testing of a slightly larger "oversized" tennis ball (roughly 6.5% larger diameter). This decision was instigated by a concern that the serving speed in (men's) tennis had increased to the point where the serve started to dominate the game. The fastest recorded serve was produced by Greg Ruzedski in March 1998 and it was measured at 66.6 m/s or 149 mph (Guinness 2000). The main evidence for the domination of the serve in men's tennis has been the increase in the number of sets decided by tie breaks at the major tournaments, in particular on the faster grass courts such as those used at Wimbledon (Haake et al., 2000).

The game of tennis originated in France some time in the 12th century and was referred to as "jes de paume," the game of the palm played with the bare hand. As early as the 12th century, a glove was used to protect the hand. Starting in the 16th century, and continuing until the middle of the 18th century, racquets of various shapes and sizes were introduced. The original game known as Real Tennis, was played on a stone surface surrounded by four high walls and covered by a sloping roof. Only Royalty and the very wealthy played the game and the oldest surviving Real Tennis court, located at Hampton Court Palace, was built by King Henry VIII in about 1530. The present day game of Lawn Tennis was derived from Real Tennis in 1873 by a Welsh army officer, Major Walter Wingfield.

Balls used in the early days of Real Tennis were made of leather stuffed with wool or hair. They were hard enough to cause injury and even death. Starting in the 18th century, strips of wool were wound tightly around a nucleus made by rolling a number of strips into a small ball. Then string was tied in many directions around the ball and a white cloth covering sewn around the ball. The original Lawn Tennis ball was made out of India rubber, made from a vulcanisation process invented by Charles Goodyear in the 1850s. Today's ball design consists of a hollow rubber core (normally with slightly pressurized air within) and covered with a felt cover. The hourglass "seam" on the ball is a result of the glue drying during the curing process. Balls that are approved for play under the Rules of Tennis must comply with the following specifications: "The ball shall have a uniform outer surface consisting of a fabric cover and shall be white or yellow in color. If there are any seams they shall be stitchless."

It was the flight of a tennis ball that first inspired scientists to think and write about sports ball aerodynamics. Newton (1672) had noted how the flight of a tennis ball was affected by spin and he wrote: "I remembered that I had often seen a tennis ball ... describe such a curveline. For, a circular as well as a progressive motion being communicated to it by that stroke, its part on that side, where the motions conspire, must press and beat the contiguous air more violently than on the other, and there excite a reluctancy and reaction of the air proportionably greater." Over two hundred years later, Rayleigh (1877) in a paper entitled: "On the Irregular Flight of a Tennis Ball," commented that: "... a rapidly rotating ball moving through the air will often deviate considerably from the vertical plane." He added the following interesting thoughts: "... if the ball rotate, the friction between the solid surface and the adjacent air will generate a sort of whirlpool of rotating air, whose effect may be to modify the force due to the stream." Despite all this early attention, when the first review article on sports ball aerodynamics was published (Mehta 1985), no scientific studies on tennis balls could be identified and so a discussion of tennis ball aerodynamics was not included.

## 5.2 TENNIS BALL AERODYNAMICS

### 5.2.1 Basic Aerodynamic Principles

Let us first consider the flight of a smooth sphere through an ideal or inviscid fluid. As the flow accelerates around the front of the sphere, the surface pressure decreases (*Bernoulli equation*) until a maximum velocity and minimum pressure are achieved half way around the sphere. The reverse occurs over the back part of the sphere so that the velocity decreases and the pressure increases (adverse pressure gradient). In a real viscous fluid such as air, a boundary layer, defined as a thin region of air near the surface, which the sphere carries with it is formed around the sphere. The boundary layer cannot typically negotiate the adverse pressure gradient over the back part of the sphere and it will tend to peel away or "separate" from the surface. The pressure becomes constant once the boundary layer has separated and the pressure difference between the front and back of the sphere results in a drag force that slows down the sphere. The boundary layer can have two distinct states: "laminar," with smooth tiers of air passing one on top of the other, or "turbulent," with the air moving chaotically throughout the layer. The turbulent boundary layer has higher momentum near the wall, compared to the laminar layer, and it is continually replenished by turbulent mixing and transport. It is therefore better able to withstand the adverse pressure gradient over the back part of the sphere and, as a result, separates relatively late compared to a laminar boundary layer. This results in a smaller separated region or "wake" behind the ball and thus less drag. The "transition" from a laminar to a turbulent boundary layer occurs when a critical sphere Reynolds number is achieved. The Reynolds number is defined as, $Re = Ud/\nu$, where U is the ball velocity, d is the ball diameter and $\nu$ is the air kinematic viscosity. Earlier transition of the boundary layer can be induced by "tripping" the laminar boundary layer using a protuberance (e.g. seam on a baseball) or surface roughness (e.g. dimples on a golf ball).

The flow over a sphere can be divided into four distinct regimes (Achenbach, 1972). These regimes are illustrated in Fig. 5.1, in which the drag coefficient ($C_D$) is plotted against the Reynolds number. The drag coefficient is defined as, $C_D = D/(\frac{1}{2}\rho U^2 A)$, where D is the drag force, $\rho$ is the air density and A is the cross-sectional area of the sphere. In the subcritical regime, laminar boundary layer separation occurs at an angle from the front stagnation point ($\theta_s$) of about 80° and the $C_D$ is nearly independent of Re. In the critical regime, the $C_D$ drops rapidly and reaches a minimum at the critical Re. The initial drop in $C_D$ is due to the laminar boundary layer separation location moving downstream ($\theta_s \cong 95°$). At the critical Re, a separation bubble is established at this location whereby the laminar boundary layer separates, transition occurs in the free-shear layer and the layer reattaches to the sphere surface in a turbulent state. The turbulent boundary layer is better able to withstand the adverse pressure gradient over the back part of the sphere and separation is delayed to $\theta_s \cong 120°$. In the supercritical regime, transition occurs in the attached boundary layer and the $C_D$ increases gradually as the transition and the separation locations creep upstream with increasing Re. A limit is reached in the transcritical regime when the transition location moves all the way upstream, up to the front stagnation point. The turbulent boundary layer development and separation is then determined solely by the sphere surface roughness, and the $C_D$ becomes independent of Re since the transition location cannot be further affected by increasing Re.

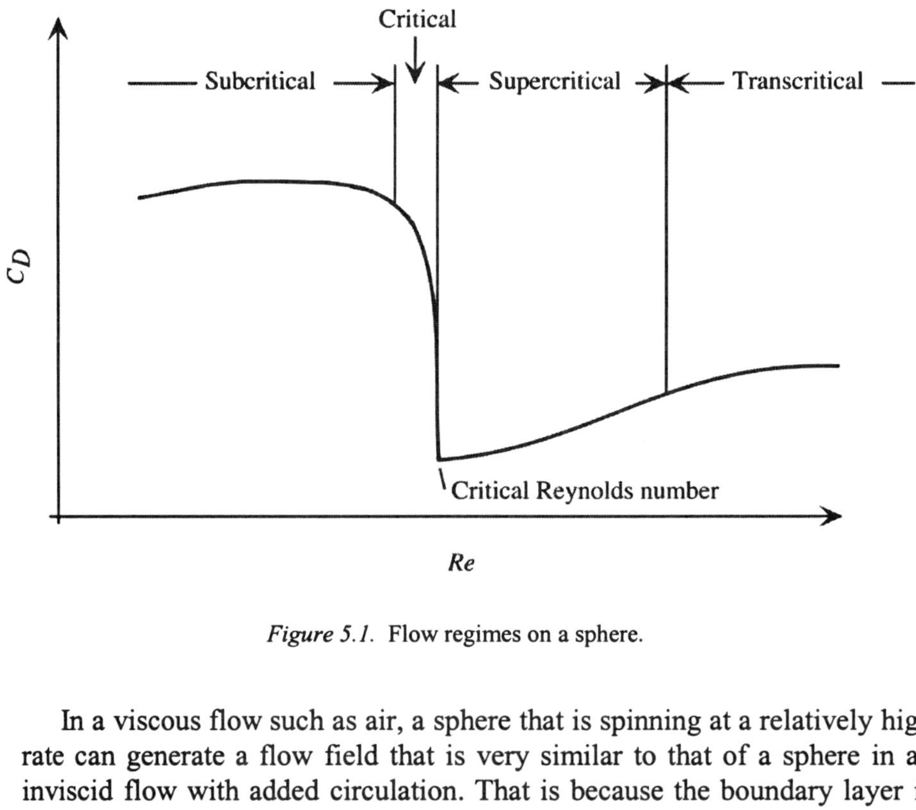


*Figure 5.1.* Flow regimes on a sphere.

In a viscous flow such as air, a sphere that is spinning at a relatively high rate can generate a flow field that is very similar to that of a sphere in an inviscid flow with added circulation. That is because the boundary layer is forced to spin with the ball due to viscous friction, which produces a circulation around the ball, and hence a lift or side force. At more nominal spin rates, such as those encountered on sports balls, the boundary layers cannot negotiate the adverse pressure gradient on the back part of the ball and they tend to separate, somewhere in the vicinity of the sphere apex. The extra momentum applied to the boundary layer on the retreating side of the ball allows it to negotiate a higher pressure rise before separating and so the separation point moves downstream. The reverse occurs on the advancing side and so the separation point moves upstream, thus generating an asymmetric separation and a deflected wake, as shown in Fig. 5.2.

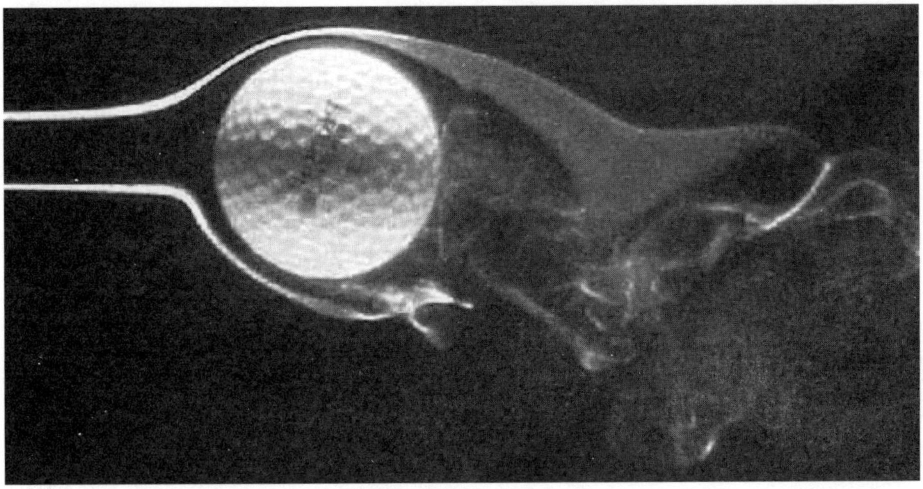

*Figure 5.2.* Flow visualization on a clockwise spinning ball showing asymmetric boundary layer separation and a downward deflected wake. Flow is from left to right. (Photo courtesy of NASA Ames Research Center.)

Following Newton's 3rd Law of Motion, the downward deflected wake implies an upward (Magnus) force acting on the ball. Now the dependence of the boundary layer transition and separation locations on Re can either enhance or oppose (even overwhelm) this effect. Since the effective Re on the advancing side of the ball is higher than that on the retreating side, in the subcritical or (especially) supercritical regimes, the separation location on the advancing side will tend to be more upstream compared to that on the retreating side. This is because the $C_D$ increases with Re in these regions, thus indicating an upstream moving separation location. However, in the region of the critical Re, a situation can develop whereby the advancing side winds up in the supercritical regime with turbulent boundary layer separation, whereas the retreating side is still in the subcritical regime with laminar separation. This would result in a *negative* Magnus force, since the turbulent boundary layer on the advancing side will now separate later compared to the laminar layer on the retreating side. Therefore, a sphere with topspin for example, would experience an *upward* aerodynamic force. So in order to maximize the amount of (positive) Magnus force, it helps to be in the supercritical regime and this can be ensured by lowering the critical Re by adding surface roughness.

## 5.2.2 Aerodynamics of a Nonspinning Tennis Ball

Some flow visualization studies of a 28-cm diameter tennis ball model first revealed the possible flow regimes. For a stationary (nonspinning) case, the first observation at the lower test velocities was that the boundary layer over the top and bottom of the ball separated relatively early, near the apex at about 80 to 90 degrees from the front stagnation point (Fig. 5.3). This normally implies that the boundary layer is laminar (transition to a turbulent state has yet to occur). However, on increasing the wind tunnel velocity, even up to the maximum, resulting in Re = 284,000 and standard-sized ball velocity of 66 m/s (148 mph), no significant changes were observed. This was somewhat surprising since, at some point, the assumed laminar boundary layer was expected to undergo transition, which would be evidenced by a sudden rearward movement of the separation point, thus leading to a smaller wake and less drag. So a new conclusion was reached which presumed that transition had already occurred and that a turbulent boundary layer separation with transcritical flow was obtained over the whole Re range tested. Although the felt was expected to affect the critical Re at which transition occurs, it seemed as though the felt was a more effective boundary layer trip than had been anticipated and that transition occurred at a very low Re. The fact that the boundary layer separation over the top and bottom of the non-spinning ball was symmetric leading to a horizontal wake was, of course, anticipated since a side force (upward or downward) is not expected in this case.

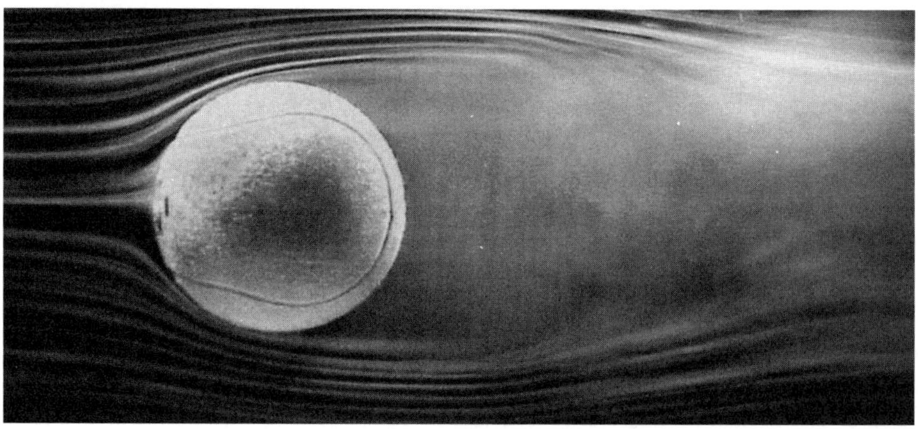

*Figure 5.3.* Flow visualization of 28 cm diameter tennis ball model with no spin (Re = 167,000). Flow is from left to right. (Photo courtesy of NASA Ames Research Center and Cislunar Aerospace, Inc.).

The new (unused) tennis ball $C_D$ data are presented in Fig. 5.4, together with data for a smooth sphere and those for a "bald" ball (the inner rubber core of a tennis ball) over a Re range of about 80,000 to 300,000. In Fig. 5.4, the ball velocity quoted along the upper horizontal axis is that for a standard-sized tennis ball (d = 6.6 cm) moving through air at 20°C and atmospheric pressure. As expected, the smooth sphere has a roughly constant $C_D$ of about 0.5 in this subcritical regime. The bald ball surface was quite smooth except for a small ridge along the equator where the two halves of the ball are joined together. This ball was mounted with the ridge perpendicular to the flow. Transition occurs at a Re of about 140,000, considerably lower than that for a smooth sphere since the ridge acts as a boundary layer trip. The $C_D$ in the supercritical regime remains more or less constant up to about Re = 200,000. In this Re range, transition is still triggered by the ridge on the ball. However, beyond Re = 200,000 the $C_D$ starts to increase again as transition moves upstream of the ridge and the separation also moves towards the front of the ball. The rise in $C_D$ is rather abrupt because the turbulent boundary layer is thickened by the ridge, thus making it more susceptible to separation. Beyond Re ≅ 280,000, the ball would be in the transcritical regime and the $C_D$ would be expected to level off at a constant value, perhaps slightly above 0.4, similar to Achenbach's (1974) observations for rough spheres.

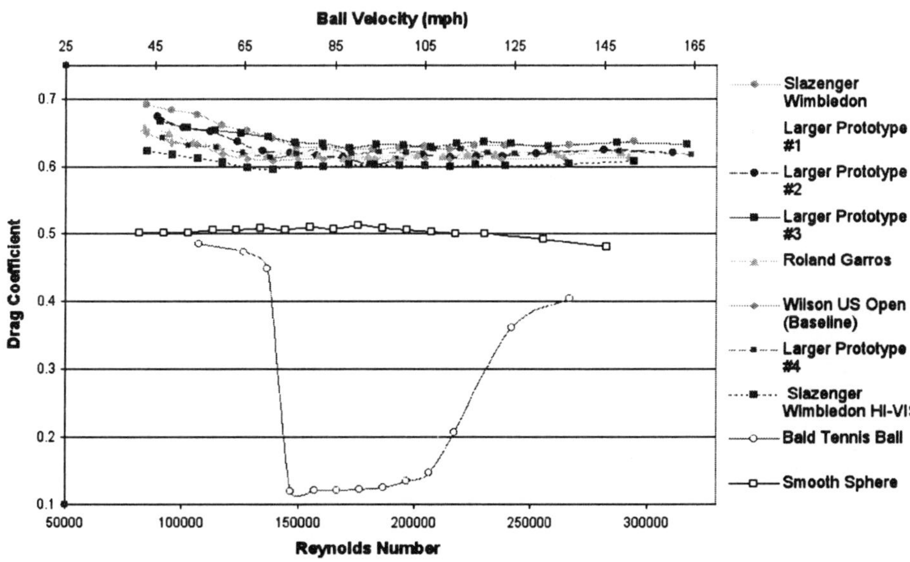


*Figure 5.4.* Drag coefficient versus Reynolds number for new nonspinning tennis balls. From Mehta and Pallis (2001).

The results for the new (unused) tennis balls show an approximately constant $C_D$ of about 0.6 to 0.65 at the higher Re (> 150,000). In this Re range, the tennis ball is clearly in the transcritical regime, only at relatively low Re compared to the smooth and rough sphere data of Achenbach (1972, 1974). So it appears as though the fabric cover on the ball is very effective at causing early transition and also in thickening the turbulent boundary layer.

These data include measurements for some of the new oversized tennis balls. The important point to note is that the $C_D$ values for the larger balls are comparable to those for the regular balls. Of course, this is not all that surprising since a simple scaling of the size should not affect the $C_D$, as long as other parameters, such as the surface characteristics (e.g. the fuzz), are not altered significantly. As a result, the drag on the oversized ball will increase by an amount proportional to the cross-sectional area, and the desired effect of increasing the flight time for a given serve velocity will be attained.

The turbulent boundary layer separation location for the tennis ball appeared to be comparable to that seen for laminar separation at relatively low Re. Now, the total drag on a bluff body, such as a sphere or tennis ball, consists mainly of pressure drag and a very small contribution due to viscous or skin friction drag. The pressure drag is solely determined by the boundary layer separation location on the ball. Since the separation location for the tennis ball is comparable to that for laminar separation at low Re, one would expect the $C_D$ for the tennis ball to be around 0.5. However, all the recent measurements give a $C_D$ of higher than 0.5 (Stepanek, 1988; Chadwick and Haake, 2000; Haake et al., 2000; Mehta and Pallis, 2001). The main task was to determine why the tennis ball $C_D$ was higher than 0.5 and where the additional drag contribution was coming from.

#### 5.2.2.1 "Fuzz" Drag on a Tennis Ball

As discussed above, in the supercritical regime, once transition has occurred, the transition and separation locations start to creep upstream and so the $C_D$ starts to increase. At some point the transition location moves all the way up to the stagnation location and the separation location is then totally determined by the development of the turbulent boundary layer. With increasing roughness, the boundary layer growth rate is increased, thus resulting in earlier separation and higher $C_D$. The constant level achieved by the $C_D$ in the transcritical regime is also expected to increase with increasing roughness, as evidenced in Achenbach's (1974) measurements. However, Achenbach's data show an upper limit of $C_D \cong 0.4$ on spheres with increasing roughness. The measured separation location for this value of $C_D$ was about $\theta_s \cong 100°$. This is still in the region of the adverse pressure gradient and so one would expect the boundary layer separation location to continue moving upstream with increasing surface roughness. However, the point to note is that while the boundary layer growth (rate of thickening) increases with increasing roughness, so does its skin friction coefficient and the behavior of the separation location is then determined by the behavior of these competing effects. The increasing skin friction coefficient makes the boundary layer more resilient to separation, thus opposing the tendency of a boundary layer to separate as it thickens. So it is entirely possible that for certain types of roughness, such as the round glass beads investigated by Achenbach for example, a limit is reached for the $C_D$ level in the transcritical regime because the effects of the boundary layer thickening are offset by those due to the increasing skin friction coefficient.

In principle, though, there is no reason why the separation location cannot continue to creep forward for other types of roughness elements, which may be more effective at thickening the boundary layer than increasing the skin friction coefficient. It is proposed here that the absolute limit for the turbulent boundary layer separation location in the transcritical regime is the same as that for laminar boundary layer separation in the subcritical regime ($\theta_s \cong 80°$). Laminar boundary layer separation occurs upstream of the sphere apex because of the presence of an adverse pressure gradient in this region. The adverse pressure gradient is generated in this region due to an upstream influence of the separated near wake and streamline curvature effects as discussed by Mehta and Pallis (2001). Assuming that a very thick (weak) turbulent boundary layer can become as prone to separation as a laminar layer, then it will separate as soon as it encounters an adverse pressure gradient (at about $\theta_s \cong 80°$), just like the laminar layer. So if the location for turbulent separation in the transcritical regime is similar to that of laminar separation in the subcritical regime, then the pressure drag should also be comparable, thus giving a total drag of $C_D \cong 0.5$.

On examining the tennis ball, the relatively rough surface on the felt is readily apparent. The roughness actually results from the junctions of the fuzz elements, where they are embedded within the fabric covering on the ball. However, in addition, the fuzz elements have a finite thickness and length and this forms an additional porous coating on the ball through which air can still flow. So the tennis ball can be thought of as a very rough sphere with a porous coating. Subsequently, each fuzz element will also experience pressure drag and when this is summed up for all the fuzz elements on the ball's surface, one obtains the additional drag contribution and this has been termed "fuzz" drag (Mehta and Pallis, 2001). The fuzz drag contribution to the total drag on a tennis ball can be as much as 20% to 40%, depending on the Re.

In the lower Re range, ($85{,}000 < Re < 150{,}000$), it was rather surprising to see higher values of $C_D$, achieving values of up to 0.7. A Reynolds number dependence also develops in this range with the $C_D$ decreasing with increasing Re. All past measurements for rough spheres had shown that the approach to a constant $C_D$ level in the transcritical regime was from below, from a lower, critical value ($C_D$ increases with Re in the supercritical and early part of the transcritical regimes until a constant level is attained). Once again the fuzz elements and their behavior with varying Re were identified as the only likely causes for the observed effect.

The first effect, which is perhaps not too surprising, is the change in orientation of some of the filaments. As the flow velocity is increased, many of the filaments that are initially standing almost perpendicular to the ball's surface are forced to lay down due to aerodynamic drag effects. Hence the contribution of the fuzz drag is reduced at the higher flow speeds or Re. Also, the fuzz element Re (based on filament diameter) is estimated to be of order 20, and this puts it in a range where the $C_D$ (for a circular cylinder) is much higher ($C_D \cong 3$) and a strong function of the Re, with the $C_D$ decreasing with increasing Re. So the higher $C_D$ level at the lower ball Re, is attributed to the combined effect of fuzz filament orientation and Re effects on the individual filaments.

### 5.2.2.2 Effects of Surface Wear on Tennis Ball Drag

The critical role of the fuzz in determining the tennis ball drag is borne out most succinctly in the results for some used balls (Fig. 5.5). One of the

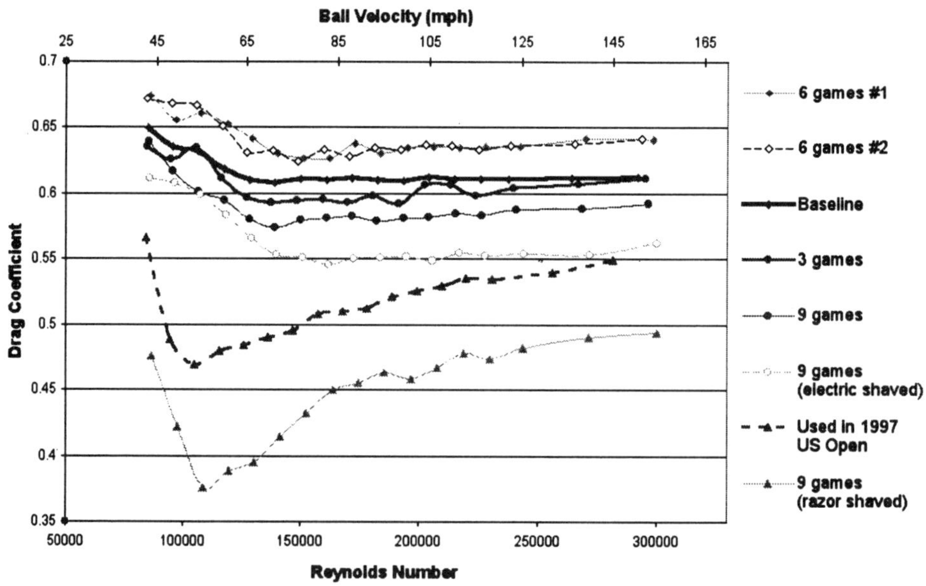


*Figure 5.5.* Drag coefficient versus Reynolds number for used Wilson US Open tennis balls. From Mehta and Pallis (2001).

balls was used in the 1997 US Open for nine games and the other used balls were played with by recreational players for the noted number of games, using only two balls at a time. The baseline data for a new Wilson US Open ball are also shown for reference. The 1997 US Open ball seems to indicate a supercritical behavior with the critical Re of about 100,000 and then a gradual approach towards the transcritical regime. For the balls used by the recreational players, after three games the $C_D$ behavior is comparable to that of the new ball. However, after six games the $C_D$ is clearly higher and to confirm this increase, both the used balls were tested and they show a very consistent trend with excellent repeatability. This initial increase in $C_D$ is supposedly known to tennis players who often refer to the felt as having "fluffed-up" (fuzz is raised), which would obviously account for the higher drag. The nine-game ball clearly exhibited a worn felt and so it was not too surprising to see a lower measured $C_D$.

The $C_D$ for the nine-game ball is still a lot higher and the behavior with Re quite different compared to that of the 1997 US Open ball. This is not too surprising since the recreational players are obviously not expected to hit the ball as hard or have as many rallies as the professionals at the US Open. In order to try and replicate the 1997 US Open ball data, it was decided to accelerate the wear on the nine-game ball by shaving-off the remaining fuzz. An electric razor was used at first, which removed some of the fuzz and the $C_D$ in the transcritical regime was reduced as a result. However, the electric razor also tended to fluff-up the felt, since the filaments often caught in the razor when it was lifted off the surface, and the higher rise in $C_D$ at the lower Re is attributed to this effect. As a final effort, a razor blade was used to shave-off as much of the fuzz as possible. Note that even this level of shaving does not remove the fuzz and associated roughness completely; the surface is similar to that on a man's face with a about a two day stubble. This ball clearly shows a $C_D$ behavior that is comparable to that of the 1997 US Open ball, with about the same critical Re and supercritical behavior. However, the $C_D$ levels are lower throughout the Re range and this is because there is no contribution due to the fuzz drag. The effective surface roughness on this shaved ball and the 1997 US Open ball is probably comparable, which explains the similar value of the critical Re. Also, as is hypothesized above, the maximum $C_D$ in the transcritical regime on a very rough sphere, with no contribution from the element drag (fuzz drag), should be 0.5, and that is exactly the level this shaved ball $C_D$ seems to be approaching asymptotically.

### 5.2.3 Aerodynamics of a Spinning Tennis Ball

In Fig. 5.6a, the ball is spun in a counter-clockwise direction to simulate a ball with topspin while in Fig. 5.6b the ball is spun in a clockwise direction (underspin or backspin). For both cases, the wind tunnel flow speed was 9 m/s and the model was spun at 4 revs/sec, which corresponds to a standard tennis ball velocity of 39 m/s (87 mph) and spin rate of 72 revs/sec (4320 rpm); this would represent a typical second serve in men's professional tennis. In Fig. 5.6a, the boundary layer separates earlier at the top of the ball

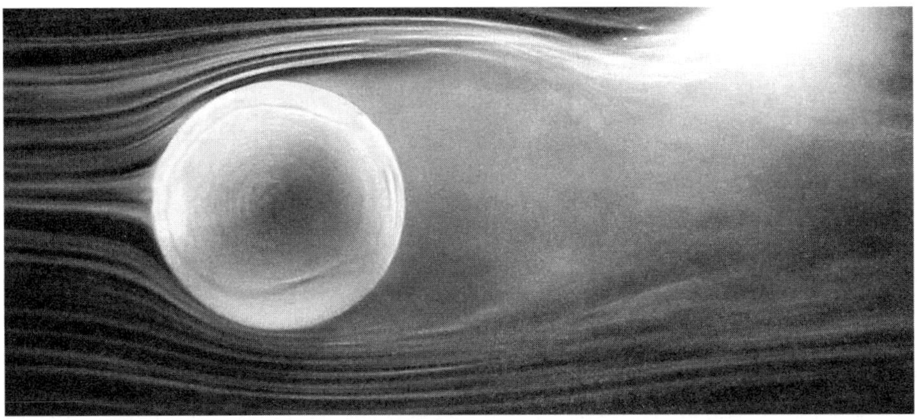

*Figure 5.6a.* Flow visualization on ball with topspin (counter-clockwise rotation at 4 revs/sec, Re = 167,000). Flow is from left to right. (Photo courtesy of NASA Ames Research Center and Cislunar Aerospace, Inc.)

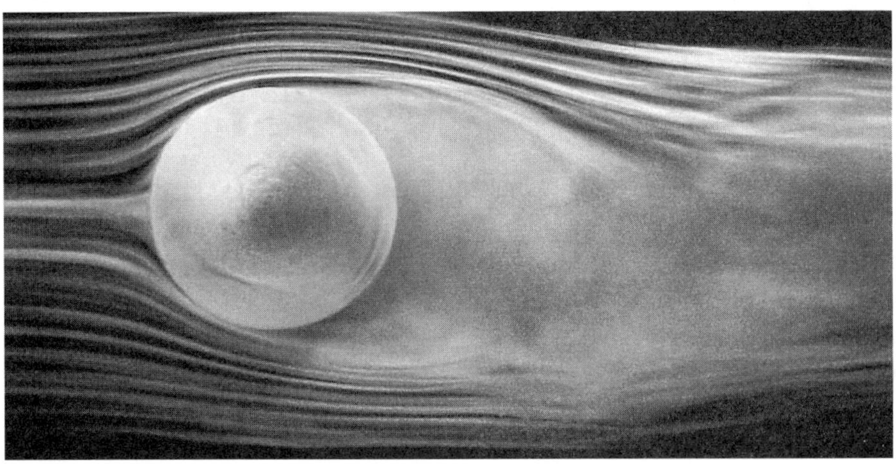

*Figure 5.6b.* Flow visualization on ball with underspin (clockwise rotation at 4 revs/sec, Re = 167,000). Flow is from left to right. (Photo courtesy of NASA Ames Research Center and Cislunar Aerospace, Inc.)

compared to the bottom. This results in an upward deflection of the wake behind the ball, and following Newton's 3rd Law of Motion, implies a downward (Magnus) force acting on it which would make it drop faster than a non-spinning ball. On the other hand, in Fig. 5.6b, the wake is deflected downwards which means that the ball has an upward Magnus force acting on it that would make it travel further than a non-spinning ball. The fact that the Magnus force occurred in a direction that was expected (positive Magnus force) over the whole speed and spin ranges further suggested that transition had occurred and that the boundary layer over both sides (top and bottom) of the ball separated in a turbulent state. A negative Magnus force is generated when transition occurs on the advancing side, but the boundary layer on the retreating side still separates in a laminar state.

The first published study of tennis ball aerodynamics was that due to Stepanek (1988) who measured the lift and drag coefficients on a spinning tennis ball simulating the topspin lob. The aerodynamic forces were determined by projecting spinning tennis balls into a wind tunnel test section. Empirical correlations for the lift and drag coefficients ($C_L$ and $C_D$) were derived in terms of the spin parameter (S) only; it was concluded that a Reynolds number dependence could be neglected. The spin parameter is defined as the ratio of the equatorial velocity at the edge of the ball (V) to its translation velocity (U), hence S = V/U. Stepanek measured values of between 0.55 and 0.75 for $C_D$, and between 0.075 and 0.275 for $C_L$, depending on the spin parameter, which was varied between about 0.05 and 0.6.

In recent years, several experimental studies have been initiated where the aerodynamic properties of spinning tennis balls were to be investigated, but the data have yet to be published.

## 5.3 EFFECTS OF AERODYNAMICS AND DYNAMICS ON THE GAME

### 5.3.1 Application of Aerodynamics

While tennis players and teaching professionals may not be versed in aerodynamics or the mathematics of the Magnus force, they none-the-less utilize the resultant effects to position the ball as part of their game strategy.

In tennis, backspin is called "underspin." In underspin, the top of the ball rotates opposite the direction of translational motion. For topspin, the top of the ball rotates in the direction of the translational motion. A ball with no (or a very small amount of) spin is called "flat."

However, balls are seldom hit with absolutely no spin or the rotational direction exclusively in the same plane as the translational direction. Balls may also be hit and imparted with sidespin or "slice" on the ball. Just as a baseball pitcher manipulates the amount and direction of spin and speed to create a repertoire of pitches, the tennis player can control these same parameters to position the point at which the ball will strike the court and the direction and angle in which the ball will move after the bounce.

Assume three tennis balls are struck with the same force but different spin -- one with underspin, one with no spin and the last with topspin. Due to the Magnus forces, a ball hit with underspin has a higher trajectory than a flat ball. The ball remains in the air longer and the trajectory is longer. Likewise, a ball hit with no or very little spin has a higher trajectory than a ball hit with topspin (Fig. 5.7).

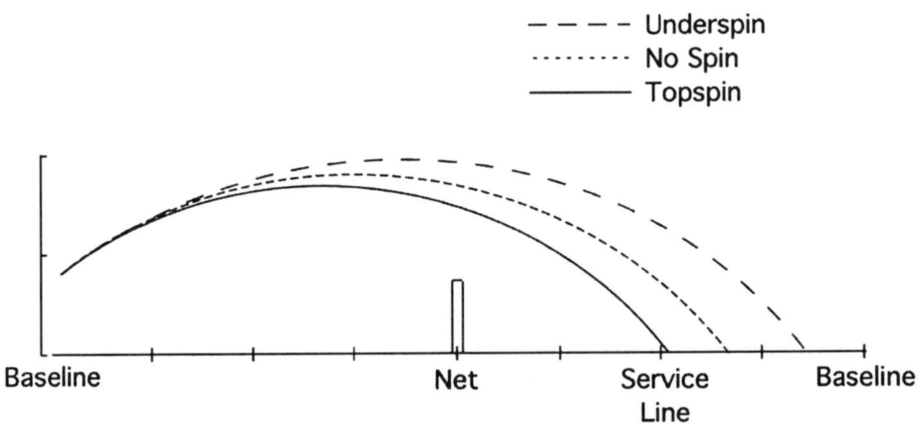


*Figure 5.7.* Effects of spin on tennis ball trajectories. Diagram courtesy Cislunar Aerospace, Inc.

In particular, tennis players utilize this knowledge during the serve (the first stroke used to initiate a point). Players have two chances to hit the ball into the service box to initiate the point. The ball must land into the opponent's service box, a rectangular area which extends 21 feet into the opponent's side of the court. Ideally, the player serving hits the ball with such speed and force that their opponent is unable to reach and return the ball and subsequently loses the point.

However, if the player serving is not successful on the first serve, the player will generally hit the ball with less force and with topspin or sidespin on the second serve to try and ensure that the ball lands inside the service box.

## 5.3.2 Court Surfaces

In certain competitions like the Davis Cup (a worldwide competition between national men's teams), countries even have the choice of the type of court material. In such tournaments, court selection can be a critical strategic decision, as a player may excel or tactically perform differently based on the surface material. Although outside the scope of this section, research has also documented a correlation between tennis player pain and injury and court surface materials.

### 5.3.2.1 Common Surfaces in Tennis

Tennis is played on a variety of surfaces both indoors and outdoors. Common surface materials include: natural turf, acrylic, Macadam and clay. Surface materials can be matched to playing style as well as climatic conditions of the region. For example, soft surfaces like clay create a slower game and the softer surface may reduce injury to the lower limbs (compared to hard courts). However, a clay court requires moisture as part of its maintenance procedure and is not suited to dry, hot climates, where water conservation is a major concern.

Court material diversity has supported the development of the sport, even in the poorest countries where crushed shells and dung have been used for tennis courts.

The tennis court surface itself absorbs a portion of the force converting it to other forms of energy. Subsequently, tennis court surfaces are rated not only for pace (speed) but also for the percentage of force reduction.

Laboratory tests have demonstrated force reductions compared to impact on concrete. Grass reduces the force by over 55 percent, clay and artificial turf about 25% and hard courts only about 5-10%. There is a correlation between pain, injury and surface hardness.

### 5.3.2.2 International Rules Affecting Tennis Ball Performance

The International Tennis Federation is the world governing body of the sport and establishes the "Rules of Tennis." These rules cover game logistics such as scoring and playing rules, as well as equipment specifications and regulations. In the case of tennis balls this includes performance tests, ball color and size specifications. While today's racquet regulations are very liberal with few restrictions, balls are closely regulated and must be certified annually by the ITF to be approved for play under the Rules of Tennis.

Racquets are not certified by the ITF. Approximately 170 different balls were certified for play under the Rules of Tennis in 2003.

Tennis balls may be pressurized or pressureless. A typical pressurized ball contains a hollow rubber-compound core, which holds a slightly pressurized gas and is covered with a felt fabric cover. The hourglass-shaped seam on the ball is a result of the adhesive drying during the curing process. Pressurized balls are stored in pressurized cans to retain pressure within the ball. Once the container is opened, the gases within the pressurized balls begin to leak very slowly through the core and fabric. The balls eventually lose their ability to bounce.

Upon examination, pressureless balls have thicker core material or have been filled with a micro-cellular material. Though pressureless balls wear from play, the balls do not lose bounce through gas leakage as pressurised balls do. Pressureless balls are often recommended for individuals who play infrequently.

Under the Rules of Tennis, tennis balls are classified as Type 1 (fast speed), Type 2 (medium speed), Type 3 (slow speed) and high altitude. Type 1 balls are intended for slow pace court surfaces such as clay. Type 2 balls, the traditional tennis ball, is meant for medium paced courts such as a hard court. Type 3 balls are intended for fast courts such as grass. High altitude balls are designed for play above 4000 feet (1219 meters). Accordingly, for each ball type, the Rules of Tennis stipulate specific size, ball deformation and bounce height.

Regardless of ball type, the tennis ball must have a uniform outer surface consisting of a fabric cover colored white or yellow. Any ball seams must be stitchless.

All tennis balls must weigh more than 1.975 ounces (56.0 grams) and less than 2.095 ounces (59.4 grams). The diameters of Type 1 and Type 2 balls must be between 2.575 inches (6.541 cm) and 2.700 inches (6.858 cm); Type 3 balls must be between 2.750 inches (6.985cm) and 2.875 inches (7.302 cm) in diameter (International Tennis Federation, 2003).

When dropped 100 inches (254 cm) upon a flat, rigid surface such as concrete, Type 1, 2 and 3 balls and pressureless high altitude balls must bounce more than 53 inches (134.62 cm) and less than 58 inches (147.32 cm). Pressurized high altitude balls must have a bounce of more than 48 inches (121.92 cm) and less than 53 inches (134.62 cm).

Type 1 balls must have a forward deformation of more than 0.195 inches (0.495 cm) and less than 0.235 inches (0.597 cm) and return deformation of more than 0.265 inches (0.673 cm) and less than 0.360 inches (0.914 cm) when a 18 lb. (8.165 kg) load is applied. Ball Types 2 and 3 must have a forward deformation of more than 0.220 inches (0.559 cm) and less than 0.290 inches (0.737 cm) and return deformation of more than 0.315 inches

(0.800 cm) and less than 0.425 inches (1.080 cm) under the same conditions. This test examines the ball's stiffness. When a ball is too stiff (and does not meet the minimum deformation) the ball will bounce too high; if the ball is too pliable or "soft" the ball will not bounce well.

Examining the rules, this series of tests do monitor these characteristics, but how does this relate to specific ball performance? Is it possible to pass these tests and yet have an undesirable tennis ball trajectory? This is the concern and more "dynamic tests" (tests which reflect how a tennis ball performs in practice) have been considered.

### 5.3.3 Factors Affecting Ball Trajectory

Once the tennis ball is struck by the racquet, its trajectory is dictated by physical mechanics and aerodynamics. Several theoretical and mathematical models have evolved to predict tennis ball motion. The ball's trajectories through the air, prior to and after court impact, are governed by aerodynamic forces which vary with spin (rotational velocity), translational velocity, the ball's physical characteristics (size, weight, fabric cover) and environmental conditions (air temperature, pressure and humidity) which affect the air density and viscosity.

While significant progress is being made to quantify the aerodynamic lift and drag coefficients to predict the aerodynamics portion of the trajectory, the ball's impact with the court (as well as the racquet) is still an area of fundamental research.

Brody's (1984) initial work modeled the impact of the ball as a spring. This model served as a classical model for many years, although it was clearly known that the ball deformed when in contact with the court. Furthermore, depending on the court surface, the ball was thought to stick or slip on the surface. Most recently, Cross' model (Brody et al., 2003) has defined ball "biting," deformation which causes the underside of the ball to grip the court.

Courts are characterized as "fast" or "slow" based on the materials and condition of the surface. As the ball hits the court it rolls or skids across the surface. In general, if the ball skids on the court, the court is fast. The skidding reduces energy absorption by the court. If the ball rolls when in contact with the surface, the court is considered slow.

The key to understanding ball/court interaction are the concepts of coefficient of friction and the coefficient of restitution.

#### 5.3.3.1 Coefficients of Friction and Restitution

When two objects slide across one another a frictional force is exerted against those objects, tangent to the surfaces. A tennis ball and its interaction with the court is an example of this. The frictional force acts in a direction opposite to that of the ball motion.

The science of physics gives us the following equation: f = mN for objects that slide against one another; where f, the frictional force is equal to N the normal (upward force that the surface exerts on the ball) multiplied by m, the coefficient of friction (COF). The COF, m, is not a constant and varies by ball and surface type. The greater the friction between the ball and the court, the slower the ball will move once it bounces. In contrast, balls which skid on a surface do not generate as much friction and subsequently do not slow down as much after the bounce. So, the COF tells us how fast (or slow) a ball will reach you in the horizontal direction. The higher the COF is for a court, the slower the ball will travel after the bounce.

For ball/court interaction, the coefficient of restitution (COR) is a ratio of the vertical velocity after the bounce to the vertical velocity before the bounce. The COR indicates how high the ball will bounce.

### 5.3.4 Ball/Surface Interactions

In general, once a ball strikes the court, regardless of the type or amount of spin before the bounce (no spin, topspin or underspin), a ball develops topspin after the bounce.

To understand this, picture a ball which strikes the court with no rebound angle after the bounce. The ball hits the court and just starts to roll towards you. What type of spin would the ball need to keep rolling towards you? Topspin. As the ball moves forward and strikes the ground, the friction between the ball and the court slows the bottom of the ball. This means that the top of the ball moves forward and rotates the ball.

In exceptional cases, balls with heavy underspin have been observed to have underspin after the bounce, but this is a rare exception.

#### 5.3.4.1 Trajectories after Ball/Court Impact

Aside from environmental factors and atmospheric conditions like wind, altitude, temperature, humidity and air pressure, the ball's trajectory after its bounce is affected by:

- the velocity before bounce;
- the angle the ball strikes the court (angle of incidence or angle in);
- court surface material;
- ball spin type and rate;
- the behavior of the ball's materials.

These components affect the angle and velocity off the court, the distance the ball travels after the bounce and the maximum rebound height. With a variety of court surfaces available, ball/court interaction affects the amount of time a player has to reach the ball and the height the racquet can approach and strike the ball.

Tennis players who compete on a variety of surfaces are aware that grass courts are considered "fast" (you typically have less time to get to the ball and the angle after the bounce is typically low) and clay courts are "slow." Hard court speeds are somewhere in between.

Brody (1987) has noted that friction between the ball and the court:

- causes the angle out (the rebound bounce) to change from the angle in;
- the smaller the friction the smaller the rebound angle;
- the smaller the friction the faster the court;
- the larger the friction the greater the rebound angle;
- the greater the friction the slower the court.

In a study conducted as part of a cooperative educational project between NASA Ames Research Center and Cislunar Aerospace, Inc. (1998) ball/court interaction effects were investigated.

High-speed cameras captured ball action at 250 frames per second. Balls were marked with a black line around the ball's circumference so spin rates could be examined. Different types of spin and rates of spin (flat, low topspin, medium topspin, high topspin, medium underspin and high underspin) on the four major tournament court surfaces (hard, red clay, green clay and grass) were used in this investigation. Wilson "US Open" balls were used on each court. In addition, court specific balls (Wilson clay court balls on the green clay, Roland Garros balls [the French Open ball] on the red clay and Slazenger [Wimbledon] balls for the grass court) were also tested. A ball machine able to control ball spin and speed was used to launch the balls.

Results for flat (no spin) balls demonstrate the trends expected (Fig. 5.8 and Table 5.1). Notice the difference between the rebound angle and the angle before the bounce. As expected grass courts were fastest and clay slowest. The grass surface has a tendency to let the ball skid.

*Table 5.1.* Average pre- and post- bounce angles on the major tennis court surfaces for flat (non-spinning) balls.

| Court Type | Ball Type | Angle In | Angle Out | Difference |
|---|---|---|---|---|
| Green Clay | Wilson US Open | 26.8 | 37.5 | 10.7 |
| Green Clay | Wilson Clay | 28.2 | 41.0 | 12.8 |
| Red Clay | Wilson US Open | 26.5 | 37.5 | 11.0 |
| Red Clay | Roland Garros | 24.3 | 34.7 | 10.4 |
| Hard | Wilson US Open | 23.9 | 32.9 | 9.1 |
| Grass | Wilson US Open | 24.9 | 29.4 | 4.5 |
| Grass | Wimbledon | 24.2 | 28.6 | 4.4 |

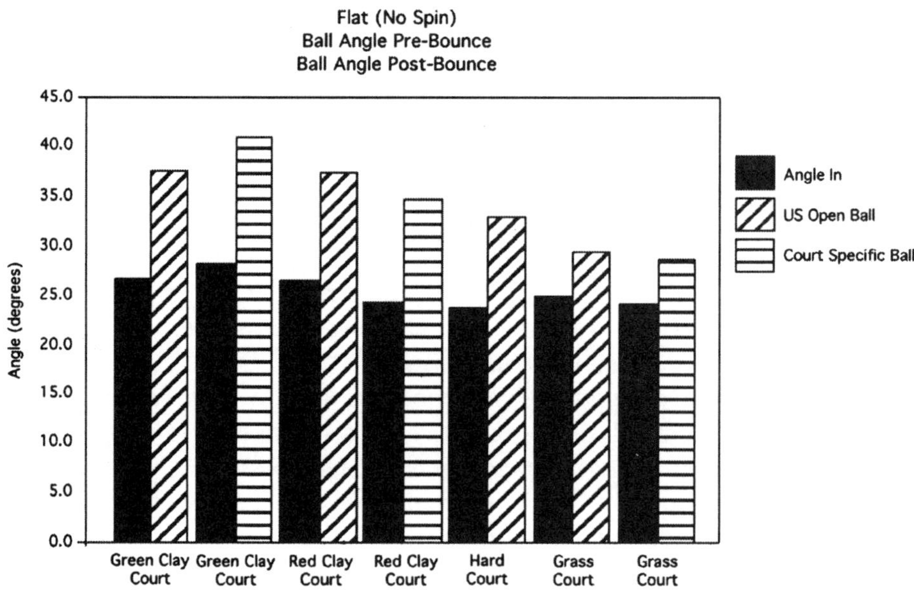


*Figure 5.8.* Comparison of pre- and post- bounce angles on the major tennis court surfaces for non-spinning balls. Diagram courtesy Cislunar Aerospace, Inc.

The trend expected and observed with the flat (no spin) continues with the topspin (Fig. 5.9a and Fig 5.9b). Based on the difference between the angle in and angle out, grass is fastest by far, then the hard court and the clay courts are slowest.

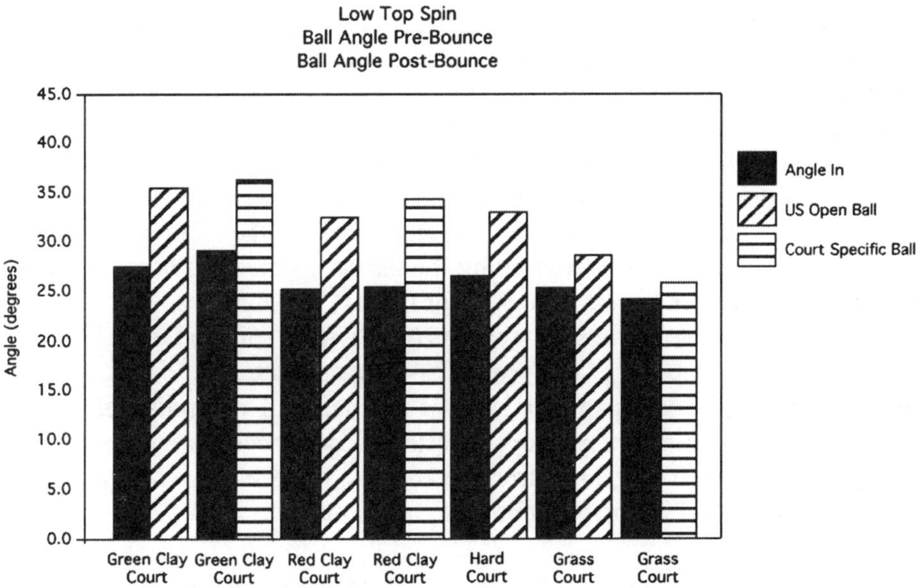


*Figure 5.9a.* Comparison of pre- and post-impact angles on the major tennis court surfaces for low topspin balls. Diagram courtesy Cislunar Aerospace, Inc.

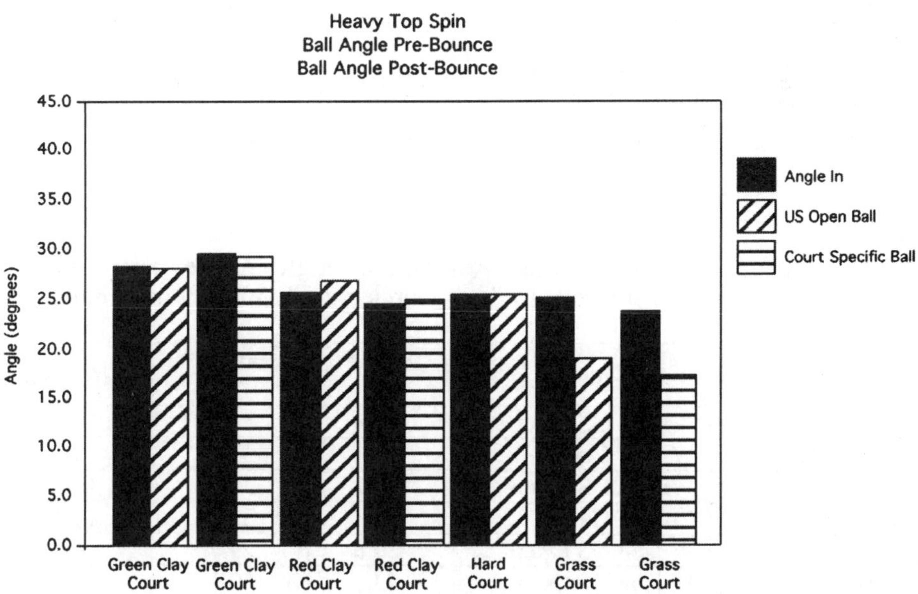


*Figure 5.9b.* Comparison of pre- and post-impact angles on the major tennis court surfaces for heavy topspin balls. Diagram courtesy Cislunar Aerospace, Inc.

However, there are some major differences. As the spin rate increases, the difference in the angles out is decreasing. While the angle out was always greater than the angle in for the flat (no spin) case, that trend disappears for the grass court as topspin increases. For heavy topspin, the angle out is significantly lower than the angle in.

Examining results for the other courts (red and green clay and the hard court) as the spin rate increases, the angle before the bounce is almost identical to the angle out. Determining the fastest court speed based on these numbers is no longer as easy as it was for the flat, no spin case.

In analyzing the underspin results (Table 5.2 and Table 5.3), it is important to remember that ball would have topspin after the bounce. While balls with no spin or topspin before the bounce will naturally follow the direction of flight and generate topspin after the bounce, balls with underspin before the bounce change spin direction after the bounce and have topspin.

There are several interesting observations to make:

*Table 5.2.* Average pre- and post- bounce angles on the major tennis court surfaces for medium underspin (1500 rev/min) balls.

| Court Type | Ball Type | Angle In | Angle Out | Difference |
|---|---|---|---|---|
| Green Clay | Wilson US Open | 25.1 | 39.9 | 14.8 |
| Green Clay | Wilson Clay | 25.9 | 41.7 | 15.8 |
| Red Clay | Wilson US Open | 23.7 | 37.9 | 14.2 |
| Red Clay | Roland Garros | 23.9 | 38.1 | 14.2 |
| Hard | Wilson US Open | 24.6 | 40.8 | 16.2 |
| Grass | Wilson US Open | 21.6 | 24.4 | 2.8 |
| Grass | Wimbledon | 22.9 | 26.1 | 3.2 |

*Table 5.3.* Average pre- and post- bounce angles on the major tennis court surfaces for heavy underspin (2500 rev/min) balls.

| Court Type | Ball Type | Angle In | Angle Out | Difference |
|---|---|---|---|---|
| Green Clay | Wilson US Open | 20.8 | 31.5 | 10.7 |
| Green Clay | Wilson Clay | 21.1 | 31.2 | 10.1 |
| Red Clay | Wilson US Open | 20.1 | 30.1 | 10.0 |
| Red Clay | Roland Garros | 22.1 | 33.2 | 11.1 |
| Hard | Wilson US Open | 20.6 | 29.7 | 9.1 |
| Grass | Wilson US Open | 23.1 | 29.1 | 6.0 |
| Grass | Wimbledon | 22.9 | 25.6 | 2.7 |

- The grass court balls did not behave like the balls on the slower courts.
- Medium underspin balls for the red and green clay and hard court had the steepest angles out and largest angle differences for any spin type and rate including heavy underspin.
- The heavy underspin balls had the lowest angles in.

Why would the grass court behave differently? Unlike the slower courts the grass surface allows the balls to skid. There is less friction and interaction between the ball and the grass court.

### 5.3.4.2 Ball Speed Variation With Spin

Table 5.4 summarizes the differences in velocity reduction after the bounce for a hard court for a wide range of direction and spin rates.

*Table 5.4.* Velocity Reduction for Different Spin Types on Hard Court.

| **Spin Type** | **Velocity Before Bounce (MPH)** | **Velocity After Bounce (MPH)** | **Percentage Reduction** |
|---|---|---|---|
| Flat (No Spin) | 32 | 20 | 38 |
| Topspin 900 rev/min | 35 | 24 | 31 |
| Topspin 1500 rev/min | 37 | 28 | 24 |
| Topspin 3000 rev/min | 35 | 31 | 11 |
| Underspin 1500 rev/min | 29 | 15 | 48 |
| Underspin 2500 rev/min | 28 | 16 | 43 |

Keep in mind that a low velocity reduction means that the velocity after the bounce remained closer to the value of the velocity before the bounce. Compare the flat average reduction to the topspin reductions. As more topspin is added, the balls' velocity after the bounce is not reduced as much. Observe the velocity reduction numbers drop from low topspin to heavy topspin.

A comparison of the medium underspin and heavy underspin results to those for the flat and topspin balls, demonstrated that medium underspin balls in this study reduced the ball's velocity the most.

Although the objective of this work was to summarize these parameters for educational purposes and to give players and coaches a feel for variations, these types of tests are also conducted to determine court pace (court speed).

These results were obtained using high-speed video cameras to capture the ball's behavior and computer software to calculate the different parameter data at a later time.

Today, a compact measuring device is used which is able to calculate ball size (radius), the angle and ball velocity before and after the bounce, contact time with the surface, and how far the ball slides when in contact with the surface. The Wassing sestée machine has been used by the International Tennis Federation, Lawn Tennis Association (the governing body for the game of tennis in Great Britain) and the French Tennis Federation.

## 5.4 CONCLUDING REMARKS

Engineers who apply their practice to sports have many of the same objectives as other sports professionals: performance enhancement, injury prevention and safety, increase enjoyment, ease of play and health benefits, extend longevity (to play the sport as long as you like) and maintain sport accessibility (throughout the human life cycle and any physical disability or challenge).

The practice of engineering in sports can manifest itself in several ways. Sometimes the bioengineer's work is quite obvious, as in the case of a new piece of ergonomically designed sports equipment or instrumentation of equipment to capture athletic performance data. Other times, the engineer's signature is not apparent, although many engineers develop industry standards, identify new materials for sports usage, perform equipment quality assurance tests, develop computer software for sports analysis tools, or conduct fundamental engineering research. Yet these activities are an integral part of sports engineering and have a profound impact on any sport.

New products drive the development of a sport and technology filters into a sport through equipment, facilities and performance enhancement. For many sports there is a balance between technology and maintaining the tradition and basic nature of the sport.

Fundamental research such as that presented in this chapter provides the knowledge to understand how technology affects a sport and is a proactive method of maintaining that balance. For example, wheelchair tennis players are allowed two bounces of the ball. However, many wheelchair players would like this rule changed to one bounce to push the limits of the game of wheelchair tennis. Is this wise? Is this feasible? Can all levels of wheelchair players realistically reach the ball after one bounce? Will changing this rule enhance the game by pushing the athletic limits of players or discourage players, result in a significant increase in injuries or risk their health? Analysis of ball/court interaction studies can assist in this assessment by determining ball speed after the first and second bounce.

Similarly, a manufacturer changes ball fabric or adds a coating on a tennis ball for durability. How will this affect the speed of the ball? Will the ball's aerodynamics, trajectory and speed be similar to other manufacturers' balls? Will the nature of the game be maintained?

Does the technology benefit or hurt the sport and those who play? This is a critical question which bioengineers working in sports must address continually.

## 5.5 REFERENCES


Achenbach, E., 1972, Experiments on the flow past spheres at very high Reynolds number, *Journal of Fluid Mechanics*. **54**: 565-575.

Achenbach, E., 1974, The effects of surface roughness and tunnel blockage on the flow past spheres, *Journal of Fluid Mechanics*. **65**: 113-125.

Brody, H., 1984, That's how the ball bounces, *The Physics Teacher*, **22**: 494-497.

Brody, H., 1987, *Tennis Science For Tennis Players*, University of Pennsylvania Press.

Brody, H., Cross, R., and Lindsey, C., 2003, The Physics and Technology of Tennis, Racquet Tech Publishing, Solana Beach, California, USA, pp. 359-367.

Chadwick, S. G., and Haake, S. J., 2000, Methods to determine the aerodynamic forces acting on tennis balls in flight, in: *Tennis Science and Technology, Proceedings of the 1st International Conference on Tennis Science and Technology*, S. J. Haake and A. Coe, eds., Blackwell Science, Oxford, U.K., pp.127-134.

Cislunar Aerospace, Inc., 1998, Ball/Court Interaction, http://wings.avkids.com/Tennis/Project/bounce-01.html, Cislunar Aerospace, Inc.

Dixon, S., 2003, Courts & Surfaces, in *The Science of Tennis CD-ROM*, A. Cooke ed., Lawn Tennis Association, London, U.K.

Guinness, 2000, *Guinness World Records, Millennium Edition*. Guinness World Records Ltd, Bantam, London, U.K.

Haake, S. J., Chadwick, S. G., Dignall, R. J., Goodwill, S., and Rose, P., 2000, Engineering tennis --- slowing the game down, *Sports Engineering*. **3**: 131 –143.

International Tennis Federation, 2003, Rules of Tennis, International Tennis Federation, London, U.K.

Mehta, R. D., 1985, Aerodynamics of Sports Balls. *Annual Review of Fluid Mechanics*. **17**: 151-189

Mehta R. D., and Pallis J. M., 2001, The Aerodynamics of a Tennis Ball. *Sports Engineering*. **4**: 177-189.

Newton, I., 1672, New theory of light and colours, *Philosophical Transactions of the Royal Society London*. **1**: 678-688.

Rayleigh, Lord, 1877, On the irregular flight of a tennis ball, *Messenger of Mathematics*. **7**: 14-16.

Stepanek, A., 1988, The aerodynamics of tennis balls --- the topspin lob, *American Journal of Physics*. **56**: 138-141.

# Chapter 6

# Shoe-Surface Interaction in Tennis

Sharon J. Dixon[1], Alison Cooke[2]

[1]*School of Sport and Health Sciences, University of Exeter, Exeter, EX1 2LU, PH: +44(0)1392 264712, FX: +44(0)1392 264726, EM: s.j.dixon@exeter.ac.uk*

[2]*Cooke Associates, Stonebridge Lane, Cambridge, PH: +44 (0)1223 882072, FX: +44 (0)1223 881442, EM: alison@cookeassociates.com*

## 6.1 INTRODUCTION

Tennis was originally played on grass courts, but developments in technology have allowed play to take place on a larger variety of surfaces than any other sport. In parks throughout the UK, players of ranging ability are playing on Macadam surfaces. In similar public park settings in the USA, the general public are found playing tennis on clay surfaces. At Wimbledon we see the famous, traditional turf surface being used. At the elite level across the World, players perform on natural turf, clay, acrylic hardcourts and carpet. Players at all levels have their favoured surface. Whilst hardcourt is the prevalent surface in professional tennis, we see players such as Tim Henman performing above par on the grass of Wimbledon, and Andre Agassi adapting effectively to clay courts.

In response to the existence of different playing surfaces, several sports shoe companies have developed surface-specific tennis shoes with characteristic design features, such as sole tread patterns and midsole cushioning devices. With the large range of surfaces and shoes, it is not clear which surface characteristics are 'best' for tennis and whether specific shoe characteristics are most appropriate for a defined surface. To address these questions, the playing requirements for a tennis court and shoe need to be considered. These include requirements in terms of allowing optimal performance and also minimising the chance of injury occurrence.

This chapter will focus on the design considerations of a tennis court surface and shoe that are influential to the player, both in terms of player performance and injury susceptibility. The playing requirements will be highlighted by identifying specific movements that are involved in a tennis match and considering how these may be influenced by a change in shoe or surface. Design characteristics likely to be influential regarding the likelihood of injury occurrence will be described, with summaries of the recent literature and suggestions for future research.

## 6.2 TENNIS MOVEMENTS

Tennis involves a wide range of movements and associated forces. There is the relatively slow application of force to the surface when performing movements such as the serve and when shuffling from side to side. In contrast, the sport also involves highly dynamic movements such as stopping when performing a running forehand foot plant and changing direction quickly to reach a shot. These movements result in different forces acting on the human body, and so also result in differing requirements from the playing shoe and surface.

The forces occurring during typical tennis movements have been measured using a force platform. This provides the resultant magnitude and direction of force applied to the surface throughout the movement. For example, a force graph obtained during a side ways shuffling movement can be compared with a force graph from the foot plant of a running forehand (Figs. 6.1a,b). The two graphs have different shapes, magnitudes of force and timing of force. The first peak on each graph has been termed the 'impact peak' and has been associated with the initial contact of the foot with the surface. For the shuffling movement illustrated, this impact peak has a magnitude in the region of 1300 N and occurs around 0.07 s following initial ground contact. For the same player performing a running forehand foot plant, the curve is characterised by an impact force that is at least twice as high as that observed for the shuffling movement, and this peak force occurs only 0.02s following initial contact with the ground. Characteristics of this impact force, particularly how quickly it is applied, have been associated with the occurrence of overuse injuries (Hrejlac, 2000). Therefore, one function of tennis shoes and the playing surface is to cushion this impact force, resulting in a reduction in magnitude or a slower application of force (lower loading rate). It is evident that this role is particularly important when movements such as the running forehand are performed, where relatively high forces occur over short time periods.

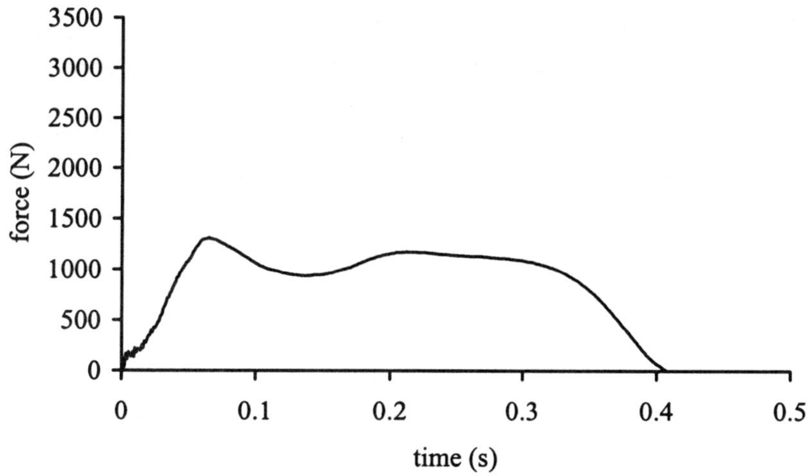


*Figure 6.1a.* Vertical ground reaction force for a sideways shuffling movement.

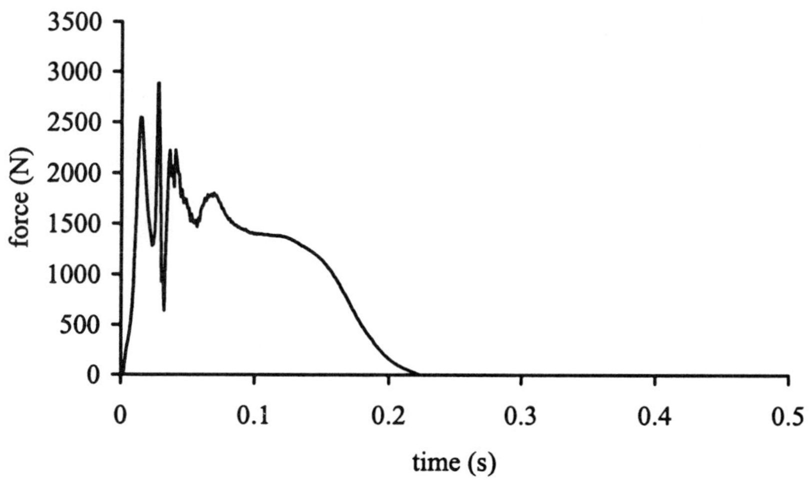


*Figure 6.1b.* Vertical ground reaction force for a running forehand foot plant.

Whilst the main component of force in running is in a vertical direction, forces are also applied horizontally, allowing braking and propulsion. In running, these are relatively low forces compared with the vertical force. When changing direction in shuffling (Fig. 6.2a) or when performing movements like the running forehand (Fig. 6.2b), relatively high horizontal forces are applied to the surface. These horizontal forces will influence the loads experienced by structures of the human body. The horizontal application of force is also likely to result in slippage occurring between the shoe and the surface; with different shoe-surface combinations allowing differing amounts of sliding.

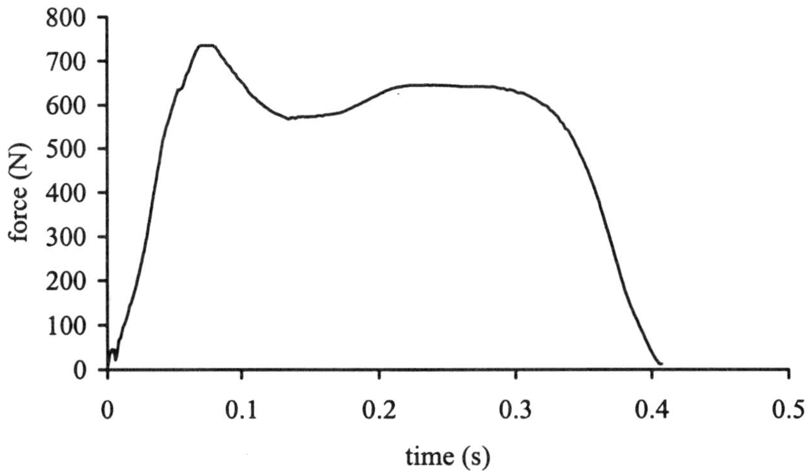


*Figure 6.2a.* Medial-lateral force component for a sideways shuffling movement.

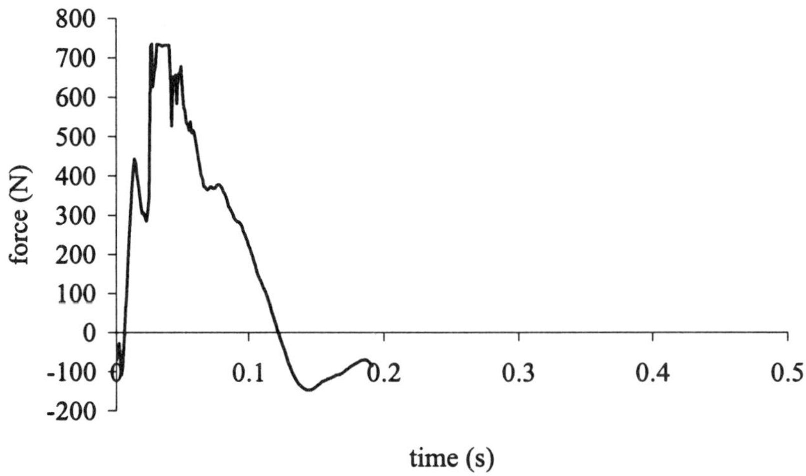


*Figure 6.2b.* Anterior-posterior force component for a running forehand foot plant.

# 6.3 DESIGN OF TENNIS SURFACES AND SHOES

## 6.3.1 Surfaces

As previously indicated, surfaces for tennis come in several distinct forms. These include natural grass, clay, hardcourts, carpets, porous Macadam and artificial turf. The pattern of layers for each of these surfaces tends to be similar, with a relatively thick aggregate layer, some form of cushioning layer and a top surface. For natural grass surfaces, the cushioning layer is in the form of soil (Fig. 6.3), whilst different types of shock pads can be introduced in artificial surfaces in an effort to provide some cushioning to the player. Extensive detail on the different constructions of tennis surfaces is documented by several authors (e.g. Seaward, 2000; Bull, 2000), and the associated advantages and disadvantages are described on the UK Lawn Tennis Association (L.T.A.) website (www.lta.co.uk). Some key features are summarised in Table 6.1.

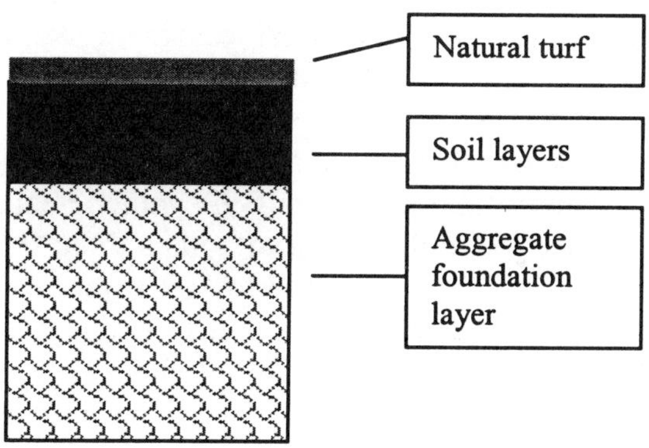


*Figure 6.3.* Layers of a natural turf tennis court.

Table 6.1. Key features of tennis playing surfaces

| Surface | Advantages | Disadvantages |
|---|---|---|
| **Natural Turf** | Relatively high cushioning<br>Aesthetically pleasing | Seasonal<br>High maintenance |
| **Clay** | High level of sliding<br>Comfortable to play on | Variable playing characteristics<br>Easily damaged |
| **Asphalt / Concrete** | Consistent characteristics<br>Low maintenance | Low cushioning ability<br>Reduced play when wet |

Table 6.1 (con't)

| | | |
|---|---|---|
| **Porous Macadam** | Relatively low cost<br>Low maintenance | Low cushioning ability<br>Temperature sensitive |
| **Artificial Turf** | Relatively high cushioning<br>Aesthetic appeal | Inconsistent characteristics<br>High maintenance |
| **Indoor Carpet** | | Limited to indoor use |

At least half of the ATP and WTA Tours are played on acrylic hardcourt surfaces, including the Australian and US Grand Slams. These surfaces have a foundation layer, an optional cushioning layer and different layers of impervious acrylic material (Fig. 6.4). According to the L.T.A. website, up to 80% of the tennis courts in the UK are porous Macadam. These typically comprise an aggregate foundation layer, two layers of modified bitumen and an acrylic or polyurethane colour coating. This top coating is applied for aesthetic purposes, rather than to give any additional cushioning to the combined surface. However, some coatings are claimed to increase traction between the foot (shoe) and the surface, to improve foot hold (and to influence ball bounce). Synthetic grass surfaces comprise a porous Macadam layer on an aggregate foundation layer. A tufted synthetic carpet is then laid in strips. Sand is applied to fill the carpet to within about three millimetres of the top of the pile. These surfaces tend to feel softer under foot than the acrylic hardcourt and Macadam surfaces and generally allow more sliding when changing direction. As found by Cross (1998), the sand in the surface influences both the sliding of the player and the ball interaction with the surface.

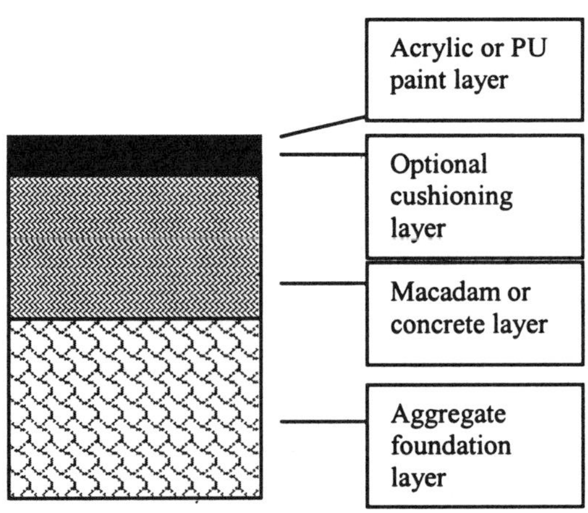


*Figure 6.4.* Layers of an acrylic hardcourt tennis court.

## 6.3.2 Shoes

Shoes for sport have been developed with the aims of providing protection from potentially injurious forces and to help optimise performance. Early research into the design of sports shoes focused primarily on shoes for running. Thus the main design criterion was the provision of adequate cushioning of impact forces. Medial stability and comfort have also been identified as important design considerations in running shoes (Reinschmidt and Nigg, 2000). When considering shoes for tennis, or other court sports, additional construction features have been introduced that aim to provide additional lateral support for stability of the foot, in an effort to reduce the chances of ankle inversion injuries (Reinschmidt and Nigg, 2000; Lafortune, 1997). In addition, the traction characteristics and torsional features of the shoe have been found to be important considerations in a court shoe (Stacoff, Steger, Stussi and Reinschmidt, 1996; Reinschmidt and Nigg, 2000).

All sports shoes have common key features. These include a midsole comprising cushioning elements, a relatively rigid heel counter, durable shoe outsole and a degree of medial and lateral heel flare (Figs. 6.5 and 6.6). For tennis shoes, the size of the toe box and reinforced toe bumpers have also been identified (Kulund, McCue, Rockwell and Gieck, 1979). Figure 6.6 illustrates a typical tennis shoe and running shoe. Key visible differences between these shoes include a lower heel height and less heel flare in the tennis shoe. These features have been included in the tennis shoe because of the potential influence of the moment arm of the subtalar joint on the stability of the rearfoot. The subtalar joint allows movement between the foot and the lower leg, primarily in the frontal plane. A greater height of the axis of this joint will increase instability. As illustrated in Fig. 6.7, a greater heel flare can also influence stability, by increasing the moment arm length of the subtalar joint, thus influencing the potential for ankle inversion.

Owing to the performance requirements of tennis shoes, a variety of outsole patterns and materials have been used in an effort to optimise grip with the surface. Outsole patterns are unlimited, with numerous different patterns being utilised (Fig. 6.8), whilst materials for outsoles include polyurethane and rubber (SATRA, 1990; van Gheluwe and Deporte, 1992).

Torsional stiffness of the shoe sole has also been suggested to be an important design consideration in a tennis shoe (Reinschmidt and Nigg, 2000). When running barefoot, relative movement of the rear and forefoot about the transverse talar and tarso-metatarsal joints occurs. When wearing shoes, the rear and forefoot are also connected by the shoe. The stiffness of the shoe sole can therefore influence the torisonal stiffness of the combined foot-shoe system (Stacoff, Kaelin, Stuessi and Segesser, 1989).

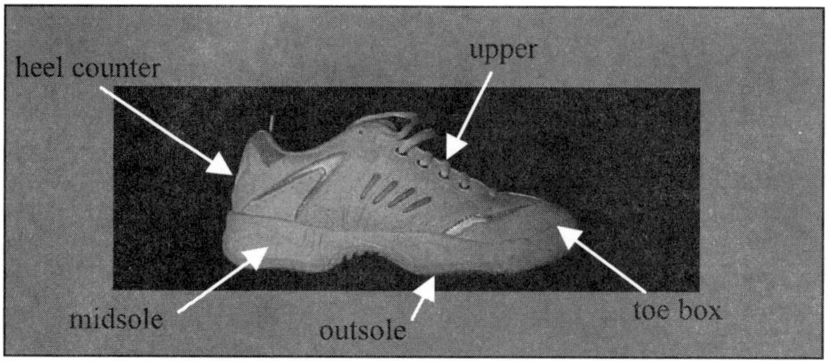


*Figure 6.5.* Components of a typical tennis shoe.

*Figure 6.6.* Rear view of a typical tennis shoe and running shoe (left shoes).

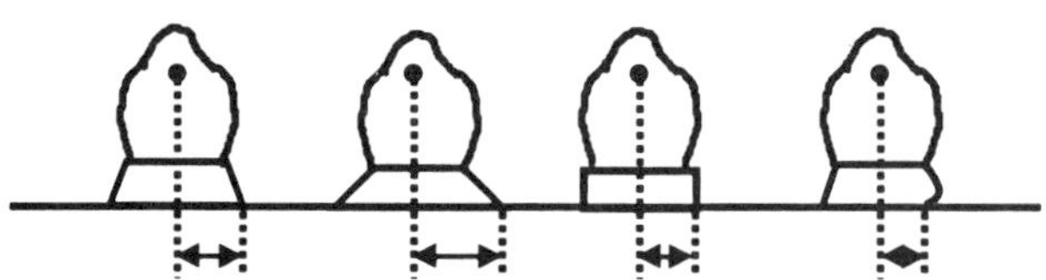

*Figure 6.7.* Influence of heel flare on the moment arm of the subtalar joint.

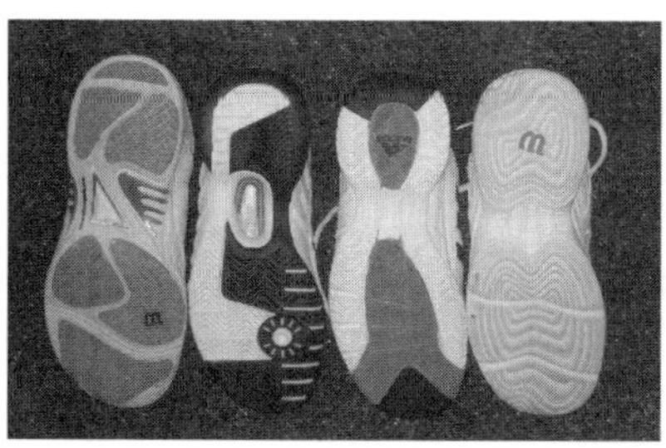

*Figure 6.8.* Sample soles of tennis shoes.

Although cushioning of impacts has not been highlighted as a key feature in tennis shoes to the same extent as in running shoes, different midsole cushioning systems have been utilised by different shoe companies, often claiming to provide effective cushioning of impact forces. Increasingly, cushioning systems that have been utilised in running shoe midsoles are being introduced in tennis shoes.

## 6.4 PLAYING SURFACES, SHOES AND INJURY

The properties of playing surfaces and shoes result in them providing differing amounts of cushioning, traction and stability, thought to effect both the influence on injury and the ability of a player to perform optimally.

The increase in tennis participation on artificial surfaces, such as acrylic hardcourts, Macadam and artificial turf, has been implicated as one factor contributing to an observed increase in overuse injuries (Nigg and Yeadon, 1990). It has been claimed that the introduction of artificial surfaces in sports has influenced both the type and the frequency of injury occurrence (Andreasson and Olofsson, 1983; James, Bates and Osternig, 1978). For example, the introduction of relatively hard court surfaces has been associated with an increased incidence of overuse injuries such as 'shin splints', Achilles tendinitis and patello-femoral pain (Bocchi, 1984; Chard and Lachmann, 1987). Bastholt (2000) reported injury data in relation to surface for three year's of play in the men's ATP tour, with four surfaces utilised: indoor carpet, clay, hardcourt and grass. It was found that the distribution of injury location was similar for the four surfaces, with up to 50% of injuries being to the lower extremity. Of these lower extremity injuries, the average number of treatments per match on grass was found to be more than double that on clay. The relative risk of lower extremity injury was also found to be significantly lower on hardcourt than on grass, but higher on hardcourt than on clay. Based on these data, it was concluded that playing on grass or hardcourt resulted in an increased number of lower extremity injuries than playing on clay. Bastholt (2000) discussed differences in style of play on different surfaces possibly contributing to these differences in injury treatments and also highlighted the relative hardness and roughness of hardcourt surfaces. However, these suggestions remain speculative, with no direct cause-effect yet being demonstrated. Despite there appearing to be a large range of cushioning abilities and traction characteristics for different tennis surfaces, the majority of information relating hard or rough playing surfaces with overuse injury has been anecdotal.

In support of the suggested relationship between tennis playing surface and injury occurrence, Nigg and Segesser (1988) reported a lower incidence of injury on clay surfaces than on artificial tennis court surfaces. These authors reported results of a questionnaire returned by 1003 players performing on a range of surfaces including clay, synthetic sand, a synthetic surface and asphalt. Differences in mechanical properties of the surfaces, including force reducing ability and coefficient of friction, were also reported (coefficient of friction is described in Section 6.5.2). The synthetic sand surface included a loose granular topping, whilst the synthetic surface had no granular topping. It was found that there were significantly more injuries reported by players performing on the synthetic surface than for those playing on clay or synthetic sand. Since the synthetic surface and synthetic sand surface had similar cushioning properties, the authors speculated that differences in cushioning of the surfaces were not the main cause of greater pain or injury on the synthetic surface. The main difference reported between the clay and synthetic sand surfaces and the synthetic and asphalt surfaces was in their frictional behaviour, with the clay and synthetic sand having a relatively low coefficient of sliding friction. Specifically, it was found that there was a lower incidence of injury to the lower limb when playing on surfaces that allow sliding. This was suggested to be the result of relatively high forces acting on the human body when a movement is stopped abruptly, such as when the front foot in a running forehand is fixed on the surface and no sliding occurs. If the surface allows a degree of sliding, lower forces are applied over longer time periods, which appears to reduce the chances of injury. The authors therefore speculated that tennis surfaces with relatively low frictional resistance were associated with lower injury incidence, and suggested an optimal coefficient of friction in the region of 0.5.

Lateral stability of a tennis shoe has been suggested to influence the likelihood of injury by controlling the degree of ankle inversion during stopping and turning movements. Excessive ankle inversion puts stress on the lateral structures of the ankle, resulting in ligament sprain, most typically to the anterior talotibular ligament (Hutson, 1982).

Unfortunately, the requirements of a shoe-surface combination for limiting the likelihood of injury and also allowing optimum playing performance are often conflicting. For example, a soft surface may have the potential to reduce the high loads acting on a player, but it may also make the surface more difficult to play on. It may end up being an 'energy sapping' surface or ball bounce may be compromised.

The amount of sliding allowed between a shoe and surface also appears to influence injury likelihood and performance in conflicting ways. When a player pushes off the ground to change direction, they ideally want their push-off foot to be fixed on the ground to allow maximum application of force and control of the movement. Any slipping will likely increase the time taken to perform the movement and thus reduce performance. Similarly, when stopping quickly, such as when planting the front foot in a running forehand, performance will likely be compromised if the player slides excessively and loses control of the movement.

## 6.5 MECHANICAL TESTING OF TENNIS SURFACES AND SHOES

The range of tennis surfaces available have differing properties in terms of the amount of cushioning of impact forces and the amount of sliding occurring between the foot (shoe) and the surface during typical tennis movements. Whilst it is possible to measure the forces acting between a player and a surface and also to measure the amount of sliding occurring during play, it is not practical or reliable to use tests with human subjects to quantify the properties of all playing shoe-surface combinations. As an alternative, mechanical tests have been developed to simulate the forces occurring between the player and the surface. These are used to test new surfaces and shoes to quantify their potential to reduce impact forces and to allow suitable sliding between the player and the surface.

### 6.5.1 Measurement of Force Reduction

A standard mechanical test using an 'Artificial Athlete' has been adopted to quantify the impact absorbing ability of tennis surfaces (Fig. 6.9, Artificial Athlete Berlin). The test was developed to simulate the impact of the human heel with the ground during running, to provide a mechanical test for detection of differences in cushioning ability between playing surfaces. The impact test involves the release of a 20 kg mass from a specified height onto a spring (Kolitzus, 1984). The spring is mounted on a metal foot and introduces compliance to provide a resulting impact force that is comparable in magnitude and timing to that measured for a running human. The peak force is measured and compared with the force when impacting a concrete surface. The potential ability of the surface to reduce impact force is then presented as a 'force reduction' relative to concrete. The International Tennis Federation (I.T.F.) categorises surfaces with a force reduction of 5-10% as providing low force reduction, 10-20% as medium and >20% as high (I.T.F., 1997). Example results for sample tennis surfaces tested using the Artificial Athlete Berlin are presented in Table 6.2 (from Dixon and Stiles, 2003). The range from 9.6% to 33.5% force reduction demonstrates the ability of these test procedures to detect differences between surfaces. Using the same test procedures, natural grass surfaces have been reported to provide a force reduction of around 50% (Lees and Nolan, 1998). Whilst these procedures provide an indication of the relative potential of tennis playing surfaces to reduce mechanical impacts, there is no evidence for stating a safe level of force reduction. For this reason, recommendations for preferred levels of force reduction are not stated by the I.T.F.

| Playing Surface | Force Reduction (%) |
|---|---|
| Concrete | 0 |
| Cushioned Acrylic Hardcourt with concrete base layer | 16.1 |
| Polyurethane hardcourt with concrete base layer, 4mm polyurethane | 9.6 |
| Polyurethane hardcourt with concrete base layer, 7 mm polyurethane | 26.2 |
| Sand-filled artificial turf surface with shockpad and concrete base layer | 33.5 |

*Table 6.2.* Force reduction for typical tennis surfaces (from Dixon and Stiles, 2003).

A standard mechanical drop test procedure is also available for testing the ability of shoes to reduce impact loads (ASTM, 1995). In this test, an 8.5 kg mass is released from a height of 5 cm to impact with the heel area inside the shoe. The peak deceleration during impact is measured and is presented as multiples of 'g' (the acceleration due to gravity). The lower the 'peak g' value, the greater cushioning potential of the shoe midsole. Typical peak g values for running shoes have been reported in the range 9 – 15 g (SATRA, 1990). In comparison, tennis shoes have been found to range from 9 to 21 g (SATRA, 1990; Dixon and Stiles, 2003). Whilst this test procedure is usually used to test isolated shoe samples, Dixon and Stiles (2003) also demonstrated that the placement of tennis playing surface materials underneath the shoe resulted in a reduction in peak deceleration value of approximately 1 g. This highlights that, in mechanical tests, tennis playing surfaces appear to provide additional cushioning to that provided by the shoe alone. Since tennis is played on a variety of surfaces with different cushioning potential, ideally the overall cushioning ability of typical shoe-surface combinations should be investigated.

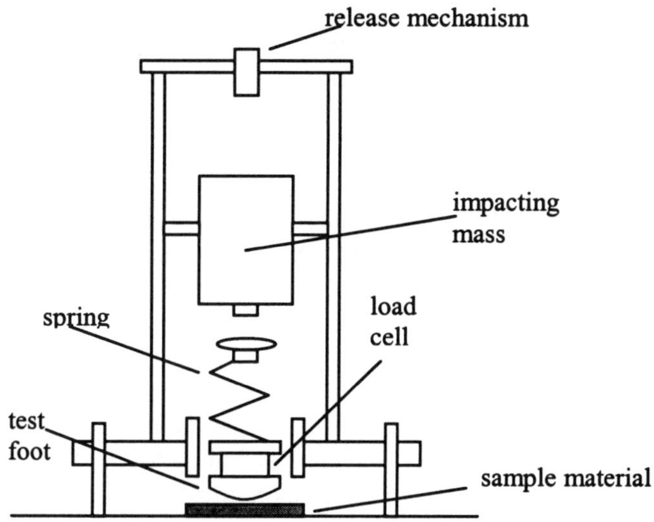


*Figure 6.9.* Artificial Athlete Berlin.

## 6.5.2 Measurement of Traction

A shoe-surface interface has both translational (sliding) and rotation friction characteristics. Translational friction is present when movement between the two contacting surfaces tends to occur in a straight line, for example when attempting to stop. The amount of twisting permitted between the shoe and surface is influenced by rotational friction. For example, if a player turns quickly and twists with their weight on the ball of their foot, rotational friction will oppose the ease of rotational sliding. If the rotational friction is high, this may stop the foot rotating easily and may result in high forces occurring at the ankle or knee.

Friction is the force acting between two contacting surfaces in a direction parallel to the area of contact. It is only present when there is a force acting in a direction tending to cause the movement of one body over the surface of another. If no relative movement between the two bodies occurs, the friction is known as static friction. An example would be a player pushing across the ground to change direction and no sliding occurring between the shoe and the surface. If there was no friction between the shoe and surface, sliding would occur and the player would not be able to change direction. Thus, a level of friction is needed for player movement on a surface.

The force opposing sliding movement between two bodies is influenced by the properties of both contacting surfaces. The frictional characteristics of the combined surfaces are represented by a constant value, known as the coefficient of friction. Thus, both the shoe and the court surface properties are influential in determining the amount of sliding that occurs. The higher the coefficient of friction between the shoe and the surface, the higher the force needed to be applied to cause sliding, and thus the less likely it is that sliding will occur.

Once sliding begins, friction is still present – it acts to slow down the rate of sliding. This is known as dynamic friction. The friction value that occurs just as sliding is about to commence is known as limiting friction. Once this limit has been exceeded and sliding occurs, the force required to maintain movement is lower than that needed to initiate motion. It is easier to keep a sliding object moving than it is to start sliding.

A mechanical test method used to measure translational friction is the British Transport and Road Research Laboratory portable skid resistance tester. A standard rubber foot is attached to the end of a pendulum. The pendulum is released from a horizontal position to slide over the court surface. A friction coefficient is determined by the maximum height attained by the foot following sliding over the surface (Fig. 6.10). By using a standard rubber foot rather than samples of typical shoe soles, this test does not provide specific friction values for the actual conditions occurring on the court surface during play. It does, however, provide a standard test procedure for comparison of different surfaces.

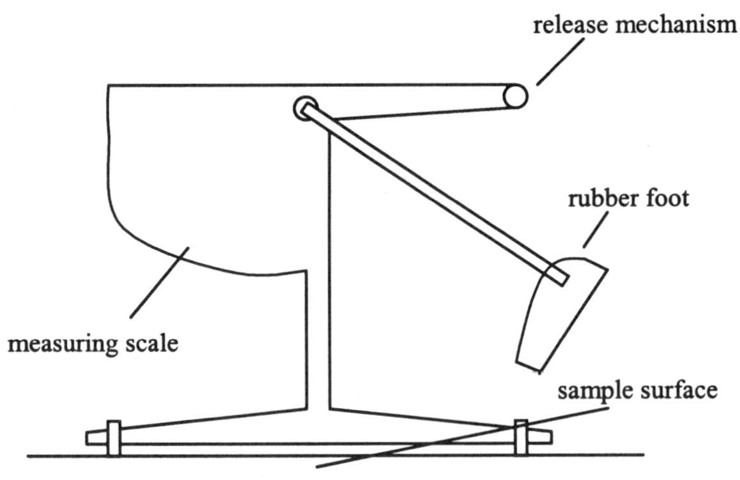


*Figure 6.10.* British Transport and Road Research skid resistance tester.

The rotational friction is influenced by the properties of the court surface and the shoe sole and also by the area of contact between the shoe and surface. Rotational friction can be quantified using a test device that measures the resistance to an applied constant torque (Stuttgart device), or the measurement of the torque required to rotate a weighted foot from a stationary position (BS7044). A typical test setup is illustrated in Fig. 6.11.

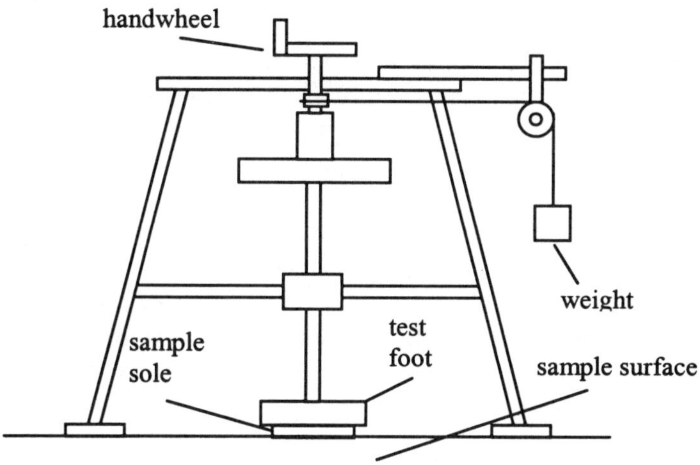


*Figure 6.11.* Artificial Athlete Stuttgart.

The mechanical test procedures described have been adopted by sports governing bodies, such as the I.T.F., because they provide repeatable tests that are relatively straightforward to perform. However, the relevance of the tests depends on whether the results correlate with human interaction with the tennis court surface.

# 6.6 TENNIS SURFACE AND SHOE RESEARCH

## 6.6.1 Tests Without Players

### 6.6.1.1 Mechanical Tests

One limitation of the majority of standard mechanical test procedures used for tennis surfaces is that they typically do not allow testing of the influence of different tennis shoes on the surface. However, efforts have been made to develop mechanical tests to include the shoe. For example, for frictional characteristics, a weighted shoe may be dragged across a playing surface. The vertical and horizontal force components can be measured using a force platform (Schlaepfer, Unold and Nigg, 1983) or a strain gauge system (Wojcieszak, Jiang and Frederick, 1997), allowing the calculation of friction values for different shoe-surface combinations.

Rotational friction has been quantified using similar procedures, with measurement of torque using a force plate (van Gheluwe, Deporte and Hebbelinck, 1983) or instrumented mechanical leg (Bonstingl, Moorhouse and Nichol, 1975). Using an instrumented leg to simulate human interaction with playing surfaces, Bonstingl and co-workers demonstrated that the torque required for foot rotation is influenced by the characteristics of the shoe sole and playing surface, the weight being supported and the stance adopted by the player. Such findings have led to the suggestion that the most appropriate approach for the assessment of the frictional characteristics of playing surfaces is the development of standardised test movements to be performed mechanically on the test surface using typical shoe sole materials (Dixon, Collop and Batt, 1999). Recent developments of robot systems that attempt to impact the shoe with the surface at typical velocities and angles occurring during sports movements may allow the development of an increased understanding of the factors influencing friction between a tennis shoe and court surface.

As well as frictional characteristics being influenced by both the shoe and the surface, so will the level of force reduction provided to the player. Whilst standard test procedures have been developed for the quantification of impact absorbing ability of shoes and of surfaces, these are rarely tested together. Using standard shoe testing procedures, Dixon and Stiles (2003) reported impact absorbing ability for combined shoe-surface combinations used in tennis. These tests highlighted that with a consistent shoe model it was not possible to differentiate between the impact reducing ability of the tested tennis surfaces. However, a change in shoe resulted in clear differences in impact absorbing ability, independent of the surface placed below the shoe. It was therefore concluded that, for the typical tennis shoes and surfaces studied, shoes had more potential to reduce mechanical impacts than tennis surfaces. It was suggested that combined shoe-surface combinations typical for the application should be tested to assess the resulting force reduction, in addition to testing the shoe or surface in isolation.

#### 6.6.1.2 Modelling Methods

As well as the testing of shoes and surfaces using mechanical tests, modelling methods have been utilised as an alternative approach to the investigation of the effect of defined changes in material properties. For example, McCullagh and Graham (1985) developed rheological models of materials used in sports shoes and surfaces. The authors claimed reasonable accuracy of the models, confirmed using the results of mechanical tests. Using this approach, it was demonstrated that the behaviour of typical shoe and surface materials could be modelled mathematically. The authors therefore suggested that such methods could potentially be used to develop shoes that are tuned for specific individuals for a defined surface. However, the modelling of materials used in playing surfaces is not as straightforward as McCullagh and Graham (1985) suggest. Shorten and Himmelsbach (2002) described how the non-linear behaviour of materials increases the complexities of models needed to accurately represent surface materials. These authors described how a power-law force displacement model could be used to provide an adequate representation of the impact reducing performance of surfaces with non-linearities.

Finite element models have also been used to investigate the influence of changes in surface construction on the surface behaviour. For example, Baroud, Nigg and Stephanyshyn (1999) described a three-dimensional model developed using parameters from existing sports surface samples. Finite element analysis was used to investigate the energy input, return and dissipation, using ground reaction force data from a running subject as input. The potential influence of controlled changes in surface properties could then be investigated. The development of models of increasing complexity allows the accurate modelling of surface materials and structures. This is likely to result in an increased use of such models to investigate the influence of defined changes on the combined behaviour of shoes and surfaces when used for sports.

### 6.6.2 Tests with Players

#### 6.6.2.1 Force Reduction

Although, as indicated by the results presented in Table 6.1, the results of the Artificial Athlete Berlin allow differentiation between tennis court surfaces, there is no evidence that a playing surface with ‘high’ force reduction will result in lower impact forces in running than a surface that is measured as having a ‘low’ force reduction value. In fact, several studies have demonstrated that similar impact forces occur during running on different surfaces, despite these clear differences in force reduction values in mechanical tests (Feehery, 1986; Nigg and Yeadon, 1987; Dixon, Collop and Batt, 2000). The tennis surfaces identified in Table 6.2 have also been tested during running (Dixon and Stiles, 2003). Eight players ran over the surfaces and impact forces were measured using a force platform under the surface sample (Fig. 6.12). Consistent with previous studies of playing surfaces, it was found that there was no difference in the impact force variables despite the change in surface. Specific variables measured included the magnitude and time of occurrence of peak impact force, and the rate of loading of impact force.

It has been suggested that the ground reaction force, measured using a force plate, is not sensitive to changes in running surface because this variable represents the acceleration of the whole body mass, rather than providing detail on the forces experienced by lower extremity structures. The development of models of the lower extremity has provided methods for estimating internal loads experienced by body structures (Morlock and Nigg, 1991). Using these methods, Cole, Nigg, Fick and Morlock (1995) demonstrated that the loads experienced at the ankle joint were not influenced by changes in the cushioning provided at the shoe-surface interface. In contrast, Krabbe, Farkas and Baumann (1992) reported that the estimation of internal loads highlighted differences in ankle joint loading on different surfaces, with frictional properties of the surfaces and running style of the subject being identified as influential factors. It therefore seems that our understanding of the force reducing ability of tennis court surfaces during human interaction needs to be improved if specific shoe-surface characteristics that are desirable are to be identified.

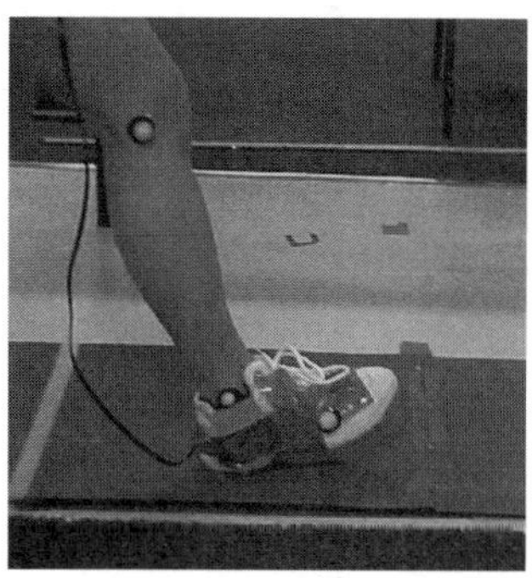

***Figure 6.12.*** Collection of biomechanical data for a tennis shoe-surface combination.

An interesting finding in the Dixon and Stiles (2003) study was that, although changes in the surface had no influence on impact force, a change in footwear was influential. Two different tennis shoes were worn on all the surfaces – one a basic model and the other a top of the range shoe from the same manufacturer. It was found that the impact force consistently exhibited a lower rate of loading for the basic shoe model. This result was the case for all the surfaces used. Therefore, for the typical tennis surface materials that were adopted in this study, the shoe had more potential to reduce impact force than the tennis court surface materials. Interestingly, the additional cushioning technology employed in the more expensive shoe was not found to have the desirable effect on impact forces during running.

Recent developments in technology allow the measurement of pressure distribution across the plantar surface of the foot using an insole system placed inside the shoe. The measurement of pressure distribution for different surfaces allows the detection of differences in contact area and distribution of resultant force, which cannot be obtained using a traditional force plate system. Using an insole system, Dixon and Stiles (2003) reported no difference in peak heel pressures for different tennis surfaces. However, more detailed analysis of pressure data in the future may reveal differences in distribution of plantar load for tennis movements performed on different surfaces that cannot be detected using a force plate system. In addition, it has been suggested that the measurement of pressure distribution may highlight specific differences between shoes that influence the comfort felt by the player (Llane, Brizuala, Dura and Garcia, 2002).

The majority of biomechanics research on playing surfaces and shoes has focused on running. However, as previously indicated, tennis is characterised by a range of movements. A number of studies have been described where attempts have been made to replicate typical tennis movements in a laboratory environment. For example, Luethi, Frederick, Hawes and Nigg (1986) compared the performance of a sideways shuffle in two different tennis shoe models. Both force plate and movement data were collected for 229 players. They reported that movement patterns during fast lateral movements are largely influenced by the model of shoe worn. However, no differences in ground reaction force data were reported for the different shoes. By using a single subject analysis, Hreljac (1998) revealed differences in ground reaction force values for three different tennis shoes during performance of a lateral movement. Preliminary work in the UK Exeter Biomechanics Laboratory has revealed that, despite no differences in peak impact force or rate of loading when typical tennis surfaces were compared during running (Dixon and Stiles, 2003), differences are revealed when a running forehand foot plant is performed on the same range of surfaces (Stiles, 2002). The future manipulation of playing surfaces for typical tennis movements such as shuffling, turning and stopping, may therefore reveal differences in force reduction that are not detected during running.

An important consideration when attempting to identify the most appropriate properties of a shoe and surface is the possibility of individual subject differences in behaviour. For example, in studies of running it has been found that a change in shoe or surface can influence ground reaction force differently for different athletes (Stergiou and Bates, 1997; Dixon et al., 1999). Different movement patterns in response to shoe or surface variation have also been identified (de Wit, 1995; Dixon et al., 1999). In addition, in a study of drop landings from 60 cm onto a force plate, it was found that the performance of basketball, running and volleyball footwear was subject-specific (Dufek and Bates, 1991). Thus, to further complicate the identification of optimal shoe-surface characteristics, it appears that single-subject analyses may be required to reveal appropriate combinations for an individual player.

#### 6.6.2.2 Traction

Subject tests have also been used to investigate the friction characteristics associated with typical movements, using sample shoe and surface conditions. For example, Stucke, Baudzus and Baumann (1984) used a force plate system to measure typical vertical and horizontal forces and movement patterns during starting, stopping and turning movements on three different surfaces. A video system was used to detect the start of sliding between the shoe and the surface, for differentiation of static and dynamic friction. It was found that subjects adapted to the change in surface by varying their movement patterns. For example, when performing a stopping movement on a cinder (clay) surface, the amount of knee flexion was found to remain reasonably constant throughout the movement, with sliding occurring between the shoe and surface. However, when performing the same task on an artificial surface, the high static friction resulted in negligible sliding occurring, and knee flexion was found to occur throughout the stopping movement. This adaptation to a change in surface by adjustment of movement may increase the energy cost of performing the task, it may reduce the effectiveness of the performance and it may also influence the chances of injury.

The ability of humans to adapt to different levels of friction provided between the shoe and surface has also been demonstrated in a reccnt study by Dura, Hoyes, Martinez and Lozano (1999). However, these authors suggest that performance is not influenced adversely. The performance of a 180-degree turn was compared on five different playing surfaces. It was found that, despite differences in coefficient of friction indicated by the mechanical test procedures illustrated in Fig. 6.11; the time taken to change direction on the different surfaces was comparable. For surfaces with relatively high coefficients of friction, the time taken in the braking phase of the turn was highest. This allowed time for more knee flexion than that observed for the surfaces with lower coefficients of friction. The authors suggested that this increase in knee flexion is a protective mechanism against potential high loads that may result from the limited sliding on these surfaces. For the surfaces with high coefficients of friction, the propulsion phase time was reduced, resulting in comparable total contact times for each of the five surfaces. It was also found that the maximum moment during turning was higher in surfaces with a higher coefficient of friction. The authors suggested that humans adapt to maintain forces and moment values below acceptable limits.

The relative contribution of the shoe and the surface to friction characteristics in tennis has been investigated using an open stance forehand (van Gheluwe and Deporte, 1992). Three typical tennis shoe models and four tennis surfaces with distinct traction characteristics were tested. It was found that, for the shoe and surface conditions tested, the surface characteristics were more influential than the shoe outsole on the level of friction. In particular, the surfaces described as having a 'fluid' top layer of rubber granules generally resulted in the lowest frictional forces and torques. Future research should identify the specific design characteristics of playing surfaces that influence the coefficient of friction during interaction with different shoes and for a range of tennis movements.

Using the present methodologies for characterisation of traction properties of different playing surfaces means that surface properties cannot be described in isolation. This makes it difficult to uniquely specify surface characteristics. The coefficient of friction is dependent on the properties of both the shoe and surface materials, and therefore cannot be used to present the properties for the surface. An alternative method to uniquely define surface characteristics is by the presentation of surface roughness parameters (Chang et al., 2001), since it has been demonstrated that surface roughness is highly correlated with slipperiness on a surface. Various stylus profilometers are available for the assessment of surface roughness. In addition, the recent development of laser scanning microscopes provides a non-contact method for quantification of surface roughness (Chang et al., 2001). The increased use of such methods to uniquely define surface characteristics should improve understanding of the specific surface properties that influence slip on playing surfaces and the features of shoes that are desirable on a particular surface.

#### 6.6.2.3 Stability and Torsion

In studies of running, the potential of variations in shoe geometry to influence rearfoot stability has been highlighted. It has been demonstrated that the shape of the lateral and medial heel flare can influence the initial rate of rearfoot movement and the peak value of rearfoot eversion, respectively (Nigg and Bahlsen, 1988). For tennis, Luethi et al. (1986) studied the influence of two different shoe constructions on rearfoot movement during a lateral sideways shuffle movement. One shoe was described as being soft and flexible, while the other test shoe was relatively hard and stiff. It was found that, immediately following ground impact, the rearfoot started to invert, with the total amount of inversion being markedly greater in the soft, flexible shoe compared with the hard, stiff shoe. Stacoff et al. (1996) investigated the influence of different tennis shoe sole properties on foot movement when performing a sideways cutting movement. Variations in sole hardness, thickness and torsional stiffness were found to influence the amount and velocity of inversion. Desirable shoe sole characteristics for maximal support were found to be a hollow inner core to the shoe sole and a high cut upper. These study results highlight how the design characteristics of a tennis shoe have potential to influence the rearfoot stability provided during lateral movements.

In a novel approach to the study of foot stability during sideways movements, Lafortune (1997) studied the effect of different court shoes on in-shoe pressures on the side of the shoe at the lateral heel area of the shoe counter. These pressure transducers therefore detected differences in the pressure applied to the lateral side of the shoe counter during the turning movement. It was found that changes in shoe resulted in different movement patterns and different in-shoe pressures, suggesting that the footwear tested exhibited large differences in stability. Similar studies with manipulation of shoes and surfaces may reveal differences in stability for different shoe-surface combinations.

#### 6.6.2.4 Performance Considerations

Performance considerations in surface design include the minimisation of energy cost when performing on the surface and the optimisation of movements such as turns and stops. In an early study of surface effects on performance, McMahon and Greene (1979) demonstrated that the energy cost of running and the time taken to cover a specified distance, could be influenced by track stiffness. However, this research involved an area elastic indoor running track surface, which was able to deform over a large area beyond the area of foot contact, and thus to experience relatively large deformations. The authors suggested that during deformation of the surface energy was stored that was then returned to the runner during the propulsion phase of the step. In contrast, typical tennis surfaces are point elastic in nature, deforming predominantly below the area of foot contact. These surfaces experience relatively small deformation during foot contact and thus do not have the same potential as area elastic surfaces to influence energy cost by storing and returning energy.

Whilst the storage and return of energy from tennis surfaces does not appear to be a realistic concept, it is possible that some surfaces have properties that reduce the energy cost of performing on them. To illustrate this possibility, Nigg and Anton (1995) demonstrated, using a mechanical model, that the work done during ground contact was higher for a more elastic shoe-surface combination than for a relatively viscous shoe-surface interface. This suggestion was subsequently supported by the demonstration of increased oxygen consumption when running in shoes with conventional elastic midsoles compared to those with increased viscoelastic properties (Stephanyshyn and Nigg, 1998). Nigg and Wakeling (2001) have suggested that this finding may be the result of a reduced energy cost of controlling tissue vibrations following ground impact when performing on a more viscous shoe-surface interface.

The design characteristics of a court surface and shoe most likely to influence performance in tennis are the traction characteristics. A high level of shoe-surface friction is required for optimum acceleration and changes in direction, but friction should be sufficiently low that loads acting on the body do not exceed thresholds for damage. Since the ideal coefficient of friction will vary for different applications, and tennis is characterised by a range of different movements, it is a difficult task to specify the most appropriate friction properties of a shoe-surface combination. However, Dura et al. (1999) conclude in their study of turning, that a coefficient of friction of around 0.4 is desirable. These authors suggest that this relatively low coefficient provides sufficient sliding to minimise the chances of injury, without sacrificing performance time in turning. This value of 0.4 is comparable to the value of 0.5 suggested by previous authors (Nigg and Segesser, 1986).

## 6.7 SUMMARY AND CONCLUSIONS

Based on research findings for tennis surfaces and shoes, it appears that the design characteristics of both can be influential on forces and movements during typical tennis movements. For surfaces used in tennis, the level of force reduction during human interaction has been found to be influenced primarily by the midsole properties of the footwear worn, rather than the surface properties (Dixon and Stiles, 2003). In contrast, the surface characteristics appear to be most influential when sliding or frictional properties are considered (van Gheluwe and Deporte, 1992). It has also been reported that the design of the shoe can influence the stability provided during side ways movements (Lafortune, 1997). Thus, if optimal levels of force reduction, coefficient of friction and shoe stability were known, shoes and surfaces could theoretically be designed accordingly.

Whilst the influence of shoe and surface characteristics on lower extremity biomechanics has received a reasonable level of attention, further investigation of relationships between biomechanical variables and injury are required before recommendations of optimal characteristics can be provided. In particular, it is suggested that the influence of shoe and surface design features on the internal loading of body structures should be investigated. Methods such as those described by Cole et al. (1995) could be used to investigate the influence of different shoe-surface combinations on the loads experienced by internal structures. Whilst the results presented by Bastholt (2000) are useful for identification of trends for different surfaces, the use of prospective studies of injury incidence are also required. This would require the identification of study subjects at a point in time and the subsequent monitoring of injury incidence, together with shoe and surface combinations used. It would then be possible to identify whether there are any differences between shoe-surface combinations used by injured verus uninjured players over a defined time period.

In order to increase understanding of the relationship between defined surface characteristics and biomechanical variables, it is suggested that standard methods be developed for the fundamental mechanical characterisation of surface material properties. The identification of fundamental mechanical characteristics of materials may provide additional insight into the human response to different surfaces. For example, it is possible that two materials with different viscosity provide the same peak force value in a standard drop test procedure, whilst being constructed from fundamentally different materials. Therefore, the human response to these surfaces is likely to differ, despite the same result in the drop test. The quantification of fundamental characteristics such as material hardness, dynamic stiffness and viscosity values should be used to uniquely define material properties, together with roughness parameters to represent friction. This will also provide data for the development of mathematical models of sports surfaces. It is suggested that such an approach is more likely than the present material test methods to reveal relationships between mechanical characteristics and human response.

The ability of some players to adapt to a change in surface, such as from hardcourt to clay, highlights the individual response of players to specific playing surfaces. It has also been found that subjects respond uniquely to a change in tennis shoe in terms of movement and force patterns (Hreljac, 1998), and that tennis shoe comfort is influenced by individual preferences (Llane et al., 2002). Whilst individual players are likely to benefit from prescription of footwear for a specific surface, desirable design criteria are not yet fully understood. Therefore, in addition to improving our understanding of shoe-surface interaction, a greater understanding of player-shoe-surface interaction is required. This is most likely to be achieved through collaborations between biomechanists and engineers.

## 6.8 ACKNOWLEDGEMENT

The authors would like to acknowledge the contributions of Victoria H. Stiles to the preparation of this paper, in particular for preparing the force plate data illustrated.

## 6.9 REFERENCES

Andreasson, G. and Oloffson, B., 1983, Surface and shoe deformation in sports activities and injuries, in: *Biomechanical Aspects of Sport Shoes and Playing Surfaces*, B. M. Nigg and B. A. Kerr, eds., University of Calgary, Calgary, Alberta, Canada, pp. 55-61.

American Society for Testing and Material Standards, 1995, *Test ASTM- F1614-95.*

Baroud, G., Nigg, B. M. and Stephanyshyn, D., 1999, Energy storage and return in sport surfaces, *Sports Engineering.* **2**: 173-180.

Bastholt, P., 2000, Professional tennis (ATP Tour) and number of medical treatments in relation to type of surface, *Med. and Sci. in Tennis.* **5**: 2-4.

Bocchi, L., Fontanesi, G., Orso, C. A. and Camurri, G. B., 1984, La patologia del piede nel tennis in rapporto al terreno di gioco, *Int. J. of Sport Trauma.* **6**: 325-332.

Bonstingl, R. W., Morehouse, C. A. and Nichol, B., 1975, Torques developed by different types of shoes on various playing surfaces, *Med. and Sci. in Sports.* **7**: 127-131.

Brown, R. P., 1987, Performance tests for artificial sports surfaces, *Polymer Testing.* **7**: 279-292.

British Standard 7044, 1990, Artificial Sports Surfaces, Part 2, Methods of Test, Section 2.2, Methods for determination of person/surface interaction, pp.1-7.

Bull, J. M., 2000, Ground conditions in tennis court construction, *Proc. 1st International Congress of Tennis Science and Technology*, Blackwell Science, Oxford, pp.209-218.

Chang, W. R., Kim, I. J., Manning, D. P. and Bunterngchit, Y., 2001, The role of surface roughness in the measurement of slipperiness, *Ergonomics.* **44**: 1200-1216.

Chard, M. D. and Lachmann, S. M., 1987, Racquet sports – patterns of injury presenting to a sports injury clinic, *Brit. J. of Sports Med.* **21**: 150-153.

Cole, G. K., Nigg, B. M., Fick, G. H. and Morlock, M. M., 1995, Internal loading of the foot and ankle during impact in running, *J. Applied Biomechanics.* **11**: 25-46.

Cross, R., 2001, The dual role of sand on a clay tennis court, *Physics Teacher.* **39:** 330-331.

De Wit, B., de Clercq, D. and Lenoir, M, 1995, The effect of varying midsole hardness on impact forces and foot motion during running, *J. of Applied Biomechanics.* **11**: 395-405.

Dixon, S. J., Batt, M. E. and Collop, A. C., 1999, Artificial playing surfaces research: a review of medical, engineering and biomechanical aspects, *Int. J. Sports Med.* **20**: 1-10.

Dixon, S. J., Collop, A. C. and Batt, M. E., 2000, Surface effects on ground reaction forces and lower extremity kinematics in running, *Med. and Sci. in Sports and Exercise.* **32**: 1919-1926.

Dixon, S. J. and Stiles V. H., 2003, Tennis shoe-surface interaction, *Sports Engineering,* accepted for publication.

Dufek, J. S and Bates, B. T., 1991, Dynamic performance assessment of selected sport shoes on impact forces, *Med. and Sci. in Sports and Exercise.* **9**: 1062-1067.

Dura, J. V., Hoyos, J. V., Martinez, A. and Lozano, L., 1999, The influence of friction on sports surfaces and turning movements, *Sports Engineering.* **2**: 97-102.

Feehery, R. V., 1986, The biomechanics of running on different surfaces, *Sports Medicine.* **3**: 649-659.

Van Gheluwe, B.and Deporte, E., 1992, Frictional measurement in tennis on the field and in the laboratory, *Int. J. of Sport Biomechanics.* **8**: 48-61.

Hreljac, A., Marshall, R. N. and Hume, P. A., 2000, Evaluation of lower extremity overuse injury potential in runners, *Med. and Sci. in Sports and Exercise.* **32**: 1635-1641.

Hreljac, A., 1998, Individual effects on biomechanical variables during landing in tennis shoes with varying midsole density, *J. Sports Sciences.* **16**: 531-537.

Hutson, M. A. and Jackson, J. P., 1982, Injuries to the lateral ligament of the ankle: Assessment and treatment, *Brit. J. Sports Medicine.* **4**: 245-249.

International Tennis Federation, 1997, *An Initial Study on Performance Standards for Tennis Court Surfaces,* I.T.F., London, UK.

James, S. L., Bates, B. T. and Osternig, L. R., 1978, Injuries to runners, *Am. J. Sports Med.* **6**: 40-50.

Kaelin, X., Denoth, J., Stacoff, A. and Stuessi, E., 1985, Cushioning during running – material tests verus subject tests, in: S. Perren, ed., *Biomechanics: Principles and Applications*, Martinus Nijhoff Publishers, London, pp. 651-656

Krabbe, B., Farkas, R. and Baumann, W., 1992, Stress on the upper ankle joint in tennis-specific forms of movement, *Sportverletz Sportschaden.* **6**: 50-57.

Kolitzus, H. J., 1984, Functional standards for playing surfaces, in: E.C. Frederick, ed., *Sport Shoes and Playing Surfaces: Biomechanical Properties,* Human Kinetics Publishers Inc., Champaign, IL, pp. 98-118.

Kulund, D. N., McCue, F. C., Rockwell, D. A. and Gieck, J. H., 1979, Tennis injuries: prevention and treatment. A review, *Am. J. Sports Medicine.* **7**: 249-253.

Lawn Tennis Association website. www.lta.co.uk/ retrieved October 2002.

Lees, A. and Nolan, L., 1998, The biomechanics of soccer: A review, *J. Sports Sciences.* **16**: 1-14.

Llane, S., Brizuela, G., Dura, J. V. and Garcia, A. C., 2002, A study of discomfort associated with tennis shoes. *J. Sports Sciences.* **20**: 671-679.

Luethi, S. M., Frederick, E. C., Hawes, M. R. and Nigg, B. M., 1986, Influence of shoe construction on lower extremity kinematics and load during lateral movements in tennis, *Int. J. Sports Biomechanics.* **2**: 166-174.

McCullagh, P. J. J. and Graham, I. D., 1985, A preliminary investigation into the nature of shock absorbency in synthetic sports materials, *J. Sports Sciences.* **3**: 103-114.

McMahon, T. A. and Greene, P. R., 1979, The influence of track compliance on running, *J. Biomechanics.* **12**: 893-904.

Morlock, M. M. and Nigg, B. M., 1991, Theoretical considerations and practical results on the influence of the representation of the foot for estimation of internal forces with models. *Clin. Biomechanics..* **6**: 3-13.

Nigg, B. M. and Segesser, B., 1988, The influence of playing surfaces on the load on the locomotor system and on football and tennis injuries, *Sports Medicine.* **5**: 375-385.

Nigg, B. M. and Yeadon, M. R., 1987, Biomechanical aspects of playing surfaces, *J. Sports Sciences.* **5**: 117-145.

Nigg, B. M. and Bahlsen, H. A., 1988, Influence of heel flare and midsole construction on pronation, supination, and impact forces for heel-toe running, *Int. J. Sport Biomechanics.* **4**: 205-219.

Nigg, B. M., 1990, The validity and relevance of tests for the assessment of sports surfaces, *Med.and Sci.in Sports and Exercise.* **22**: 131-139.

Nigg, B. M. and Anton, M., 1995, Energy aspects for elastic and viscous shoe soles and playing surfaces, *Med. and Sci. in Sports and Exercise.* **27**: 92-97.

Nigg, B. M. and Wakeling, J. M., 2001, Impact forces and muscle tuning: a new paradigm, *Exercise and Sports Sciences Reviews,* **29**: 37-41.

Reinschmidt, C. and Nigg, B. M., 2000, Current issues in the design of running and court shoes, *Sportverletz Sportschaden.* **14**: 71-81.

SATRA Footwear Technology Centre Bulletin, 1990, SATRA, Northants, UK.

Schlaepfer, F., Unold, E. and Nigg, B. M., 1983, The frictional characteristics of tennis shoes, in: B. M. Nigg and B. A. Kerr, eds., *Biomechanical Aspects of Sport Shoes and Playing Surfaces*, University of Calgary, Calgary, AB, Canada, pp. 153-160.

Seaward, E., 2000, Construction and maintenance of championship grass courts. *Proc. 1st International Congress of Tennis Science and Technology*, Blackwell Science, Oxford, pp. 201-208.

Shorten, M. R. and Himmelsbach, J. A., 2002, Shock attenuation of sports surfaces. *Proc. 4th International Conference of Engineering of Sport.* Blackwell Science, Oxford, pp.152-159.

Stacoff, A., Kaelin, X., Stuessi, E. and Segesser, B., 1989, The torsion of the foot in running, *Int. J. Sports Biomechanics.* **5**: 375-389.

Stacoff, A., Steger, J., Stussi, E. and Reinschmidt, C., 1996, Lateral stability in sideward cutting movements, *Med. and Sci. in Sports and Exercise.* **28**: 350-358.

Stephanyshyn, D. J. and Nigg, B. M., 1998, Influence of viscoelastic midsole components on the biomechanics of running. *Proc. 3rd World Congress of Biomechanics*, Sapporo, Japan, pg. 376b.

Stergiou, N. and Bates, B.T., 1997, The relationship between subtalar and knee joint function as a possible mechanism for running injuries, *Gait and Posture.* **6**: 177-185.

Stiles, V. H., 2002, Biomechanical response to different surfaces during a running forehand. *Proc. BASES student Congress*, London, p. 19.

Stucke, H., Baudzus, W. and Baumann, W., 1984, On friction characteristics of playing surfaces, in: E. C. Frederick, ed., *Sport Shoes and Playing Surfaces: Biomechanical Properties,* Human Kinetics Publishers Inc., Champaign, IL, pp. 87-97.

Wojcieszak, C., Jiang, P. and Frederick, E. C., 1997, A comparison of two friction measuring methods, *Proc. 3rd Symposium on Functional Footwear*, Tokyo, pp.32-33.

# Chapter 7

# Biomechanics of Tennis Strokes


Duane Knudson[1] and Bruce Elliott[2]

[1]*Department of Physical Education and Exercise Science, California State University-Chico, Chico, CA 95929-0330, PH: (530) 898-6069, FX: (530) 898-4932, EM: dknudson@csuchico.edu*

[2]*School of Human Movement and Exercise Science, The University of Western Australia, Perth, Australia 6009, PH; 61-8- 9380-2374, EM: Bruce.Elliott@uwa.edu.au*


## 7.1 INTRODUCTION

The primary goals of the sport biomechanist are to improve athletic performance and to reduce the risk of injury. Biomechanical research has typically involved descriptive studies of the kinematics and kinetics of sport skills, electromyographic studies of muscle activation, and modelling and simulation studies. A sport with a long history of biomechanics research is tennis. Engineering has played a prominent role in tennis research on rackets, balls, shoes, and playing surfaces.

While sport biomechanics plays a wide role in tennis development, this chapter will only discuss the biomechanical research on tennis groundstrokes. We will use the term "strokes" in the rest of this chapter to refer to the groundstrokes (forehand and backhand) of tennis. The biomechanical factors involved will be organized around the four major temporal phases of strokes: preparation, backswing, forward stroke, and follow-through. The research and theory behind stroke structure is reviewed and the application of biomechanical factors in an elite forehand drive will be presented. The chapter concludes with how these concepts can be used to optimise performance of a junior tennis player.

## 7.2 Fundamental Biomechanical Principles of Tennis Strokes

The tennis player must use a variety of strokes to win a match. While some biomechanical features of tennis strokes must remain flexible to adapt to the specifics of a particular shot, there are others that can be generalized for most strokes. These principles will be discussed in the four main temporal phases of a tennis stroke mentioned above.

### 7.2.1 Preparation

A tennis rally is an unpredictable struggle. The tennis player must react to the opponent's shot and select an appropriate stroke for the situation. The effectiveness of any stroke is dependent on the preparatory movements made by the player. The three biomechanical issues in preparing for a tennis stroke are readiness, stance, and grip.

#### 7.2.1.1 Readiness

A good ready position or "split step" is essential for a tennis player to move as quickly as possible to intercept an opponent's shot (Groppel, 1992). A split step involves squaring the player to the opponent and timing a small hop to coincide with the opponent stroking the ball (Fig. 7.1A). The biomechanical features of this movement can be illustrated with a similar movement, the vertical jump. If a player was to stand on a force platform and jump the kinematic and kinetic data would look like Fig. 7.1B. As the player flexes the knees there is a period of "unweighting" as the body is accelerated toward the court (Fig. 7.1B: area 2). The muscles of the anterior thigh (the quadriceps) and other lower extremity extensors are eccentrically activated (lengthened under tension) to reverse the direction of the vertical acceleration of the center of gravity of the body (Fig. 7.1B: area 3). The upward acceleration immediately continues with concentric action (muscle shortening) of the quadriceps to propel the body upward (Fig. 7.1B: area 4). This combination of an eccentric followed immediately by a concentric muscle action is a natural neuromuscular strategy that maximizes the initial muscle force in concentric action. This strategy is known as the Stretch-Shortening Cycle (SSC) (Komi, 1984). There are substantial benefits of a SSC with approximately 15-20% augmentation to final performance (ability to move quickly from the court or racket speed). This benefit is usually attributed to the storage and recovery of some of the elastic or strain energy in the muscle, potentiation of contraction, preloading, or reflex mechanisms (Gregor, 1997).

The benefits of a split step or SSC can be lost if there is too long a pause between the eccentric and concentric actions. The dissipation of the augmentation derived from the SSC, from various pauses, has been represented by a negative exponential equation with a 0.85 s half-life of decay for a weight lifting movement (Wilson et al., 1991). A pause of 0.35 seconds reduces prestretch augmentation by 25% while a pause of 0.9 seconds decreases augmentation by 52%. It is therefore important to reduce this pause as much as is possible.

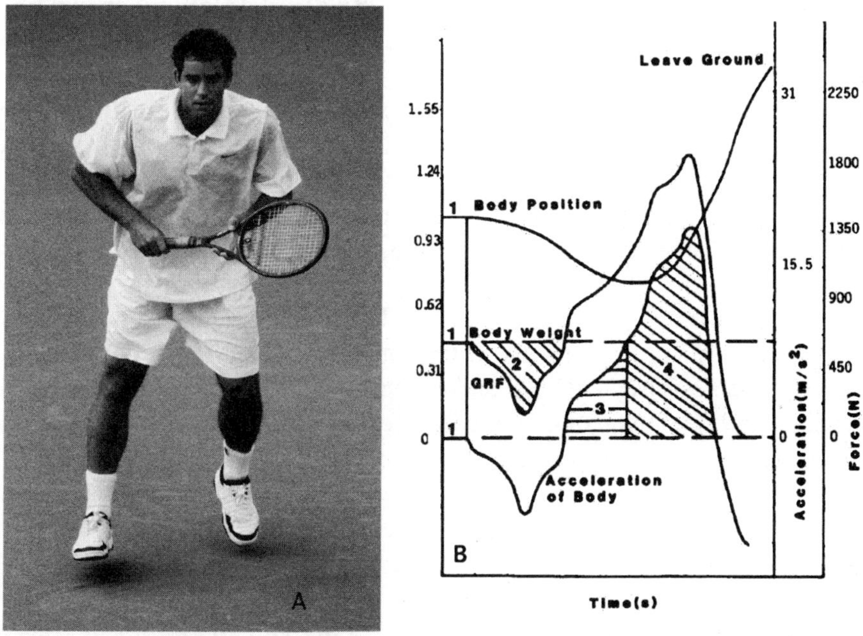


*Figure 7.1.* The skilled tennis player hops into a "split step" (A). This movement takes advantage of the stretch-shortening cycle of muscle action to move as quickly as possible. The vertical force, acceleration, and center of mass position for a vertical jump on a force platform (B). Photo by Lance Jeffrey.

The skilled tennis player "springs" out of the split step in the appropriate direction to intercept the opponent's shot. This involves synchronizing anticipatory and perceptual skills with the biomechanical skills of the split step and SSC. The steps or footwork used to position the body for a stroke have seen little systematic research (Bragg and Andriacchi, 2001). The common coaching opinion is that for short movements (< 2 m) the player should slide step, but turn and run for distances approximately farther than 2

m. This common knowledge is consistent with research on quick lateral movements in other sports (Cox, Noble, and Johnson, 1982).

#### 7.2.1.2 Stance

The stance of tennis strokes is important because much of the energy used to propel the arm and racket toward the ball is transferred from the ground up through the body (Groppel, 1992). By stance, we are referring to the position of the feet relative to the direction of the shot. There are three broad classifications of stances used in tennis strokes. The open stance has the feet roughly at right angles to the shot (see Fig. 7.13B near the end of the chapter). The square stance has the feet parallel and aligned with the shot, and the closed stance has the front foot beyond the parallel to the direction of the shot. The position of the feet have a major impact on the player's stability, ability to use body motion to swing the racket, and ability to recover for the next stroke.

The base of support for all stances should be shoulder-width or slightly wider apart to provide a good compromise between stability and mobility. Moderate amounts of hip and knee flexion allow for the quickest reaction and movement times (Yamamoto, 1996). The perplexing issue for coaches is that players are required to use different stances (and variations between them) to meet varying stroking and tactical demands. The use open stance forehands and backhands has increased recently possibly because of larger, lighter, and more powerful racket designs. This has fuelled concerns of sports medicine professionals that this increases the risk of overuse injuries by focusing stress on the trunk and upper extremity (Roetert et al. 1997). There has been a recent increase in the reports of stress fractures in the stroking arm of tennis players (Kibler and Safran, 2000; Wainger and Lombardo, 1995) however, the precise etiology of these injuries in unknown.

Traditional instruction has emphasized the square stances for tennis strokes. The logic has been that a square stance allows the player to shift their weight over a wide base of support in the direction of the shot. It was believed that an open stance stroke would limit the contribution of whole-body or linear momentum to creating racket speed. It was also hypothesized that the closed stance would limit the hip and trunk rotation or angular momentum that could be used to create racket speed. Unfortunately, biomechanical research on these issues is limited. It has been shown that racket paths near impact and racket speeds across stances are similar for skilled players, but intermediates are less consistent in creating desirable stroke paths using the open stance forehand (Knudson and Bahamonde, 1999). The advanced players in this study showed different patterns of trunk angular velocity between the square and the open stance forehand (Fig.7.2). In the square stance the trunk angular velocity increases up to impact, while in the open stance the trunk angular velocity peaks and decreases prior to impact. It is possible that stance affects how hip and trunk rotation contribute to racket speed. This transfer of energy across the body segments or the coordination of the kinematic chain (Steindler, 1955) of the tennis player will be reviewed in the forward stroke section.

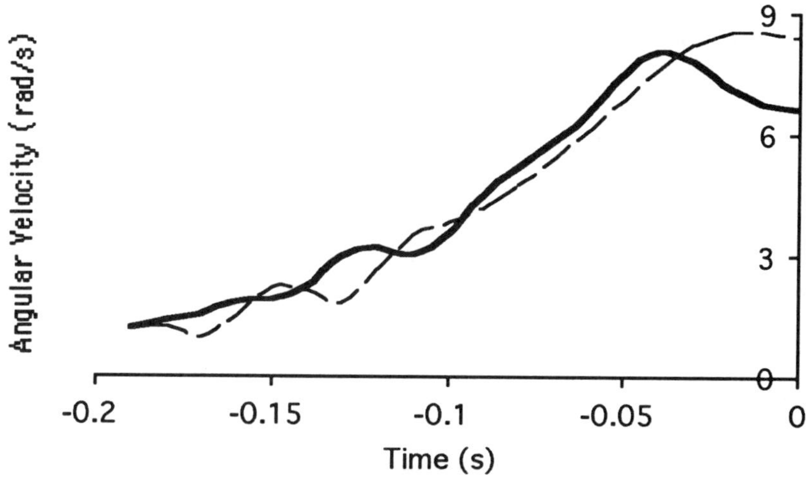


*Figure 7.2.* Difference in mean trunk axial angular velocity in the open (solid) and square (dashed) stance forehand of advanced players. Adapted from Knudson and Bahamonde (1999) with permission of Biology of Sport.

A few Master's theses have shown that using an open stance forehand does allow advanced players to recover laterally more quickly (Knudson, 1999b). Since there has been limited research on stance, it is advisable for coaches to primarily teach the square stance that uses both linear and angular sources of body movements to speed up the racket. As players progress towards advanced levels of play open stance strokes can be developed. More biomechanical research is needed in controlled conditions to document the differences in performance and injury risk.

### 7.2.1.3 Grip

Most tennis players change their grip on the racket as they move to prepare for a stroke. There are two biomechanical issues related to the player's grip on the racket. One issue involves the transmission of hand forces and impulse to the ball, and the other is related to wrist and hand motion that can contribute to the speed of the stroke. This section will discuss the latter issue, while the research on the former will be reviewed in the section on the forward swing.

In tennis strokes, the speed of the racket depends on the rotational velocities of the body segments and the corresponding instantaneous position of the racket. The eastern, and western style grips (Fig. 7.6) used in strokes favors wrist motion and consequently racket motion, over the larger grip forces that can be created using a power grip (Knudson, 1991b). Elliott et al. (1997) showed that significant variations were evident in the contribution of the hand segment to racket speed for different tennis grips. The need to maintain a racket-face perpendicular to the court and develop a large vertical velocity vector through impact in topspin strokes certainly has been shown to favor the western style grips (Takahashi et al., 1996). A more eastern style grip racket permits the player to hit "through the ball" with increased consistency for a hip height impact because the palm of the hand is directly behind the handle (Knudson and White, 1989).

## 7.2.2 Backswing

Effective strokes must have a well-timed backswing to position the body and racket for the forward stroke. The backswing serves to increase the distance over which the speed of the racket can be increased, to contribute to SSC muscle actions to generate racket speed, and to position the racket for the forward stroke. These issues will be discussed in three areas: the pattern of the backswing, the storage of elastic energy in the SSC, and the range of motion used in the backswing.

### 7.2.2.1 Pattern

Tennis players draw the racket backward in a loop to ready the racket for the forward stroke (Fig. 7.3). The biomechanical advantages of this kind of backswing are the smooth generation of racket speed and the use of gravity to assist in racket acceleration. A large loop backswing also provides distance and time over which the racket can be accelerated. The size of the loop and the use of gravity varies, depending on the amount of upward movement of the racket prior to dropping the racket downward and into the forward stroke. Large loop backswings effectively generate racket head speed but have the disadvantage of being vulnerable to variations in bounce or timing. A smaller loop backswing may create less racket head speed, but it can be executed quicker and may be more easily adjusted to unusual bounces. Unfortunately, these hypotheses have not been examined by biomechanical studies.

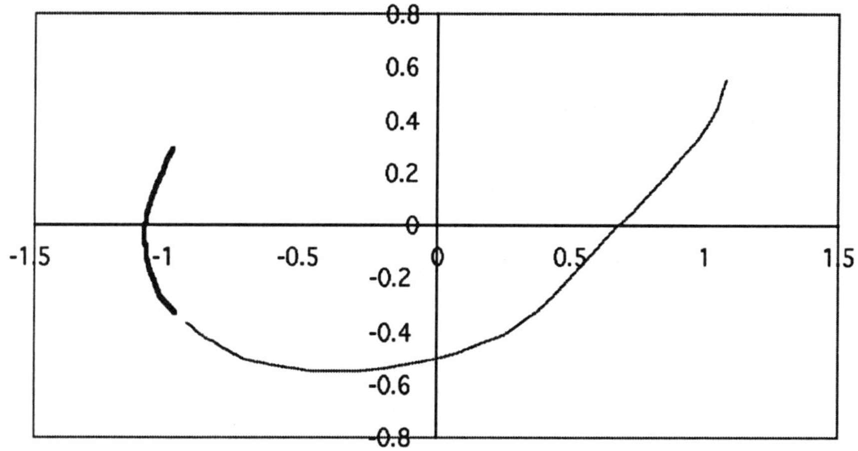


*Figure 7.3.* Side view of the path of the tip of the racket in a forehand showing the end of the backswing (thick) and forward swing (thin). Two-dimensional projection of 3D position data in meters. Adapted from Knudson (1988).

### 7.2.2.2 Storage of elastic energy (SSC)

Another benefit of a loop backswing may be to contribute to the SSC action of the major muscles accelerating the arm and racket. If the backward and downward motion of the racket is timed to coincide with the initial forward motion of the proximal segments of the body, this increases the eccentric stretch of the muscles that may be used to accelerate the arm and racket. A smooth and well-timed backswing can store more elastic energy in these muscles that contribute to a more forceful concentric action in the SSC (Fig. 7.4). This hypothesis is consistent with the increase in pelvic axial rotation torque observed near the initiation of forward racket motion (Fujisawa et al., 1997; Iino and Kojima, 2001). Players who prepare early and position the racket ready for the forward swing, will often rapidly move the racket-arm unit backwards prior to moving it forward to the ball in an endeavor to use the SSC to benefit performance while also being prepared for the desired return. This hypothesized benefit of the backswing requires further confirmation with biomechanical research since few tennis studies have focused on preparatory movements.

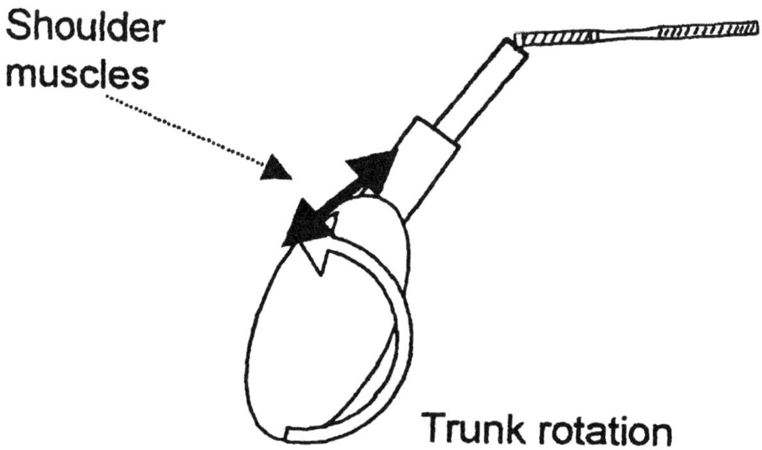


*Figure 7.4.* Top view schematic of the storage of elastic energy (SSC) in the chest/shoulder muscles (e.g. pectoralis major, anterior deltoid) from the combined effect of the final stages of the backswing and initiation of forward trunk rotation. Adapted from Bahamonde (1999) with permission of R. Bahamonde.

### 7.2.2.3 Range of Motion

Whatever the pattern of backswing and its potential contribution to the storage of elastic energy in the muscles, the joint range of motion commonly used in strokes has been documented by biomechanics research. Trunk rotation at the completion of the backswing in forehands and backhands are usually between 110 and 120 degrees from parallel to the baseline (Elliott et al., 1989a, 1989b; Reid and Elliott, 2002; Takahashi et al., 1996) for the single-handed stroke. This position typically includes about 30 degrees of trunk twist (separation angle - shoulders rotated beyond the hips) that stretches the trunk rotators (Reid and Elliott, 2002; Takahashi et al., 1996). The level of trunk twist is less for the two-handed backhand stroke and the trunk rotation closer to 20 degrees. In forehands and one-handed backhands the elbow is flexed by 30 to 80 degrees and the wrist is usually hyperextended by 20 to 60 degrees (beyond a natural alignment of the hand and forearm). Contrary to popular coaching cues to "point the racket at the back fence," rackets are rotated beyond this point by 45 to 90 degrees (see Fig.7.12 near the end of the chapter). That is, the racket is rotated by 225-

270 degrees from a position where it was pointed at the opponent at the completion of the backswing in all styles of single-handed strokes (Elliott et al., 1989a, 1989b; Reid and Elliott, 2002; Takahashi et al., 1996). Rotation in the double-handed backhand is marginally less than reported above.

## 7.2.3 Forward Stroke

Immediately following the backswing is the forward stroke phase of the tennis stroke. The forward stroke must generate both racket speed and accurate racket movement to intercept the ball. The generation of racket speed is the major determining factor in the speed that can be imparted to the ball (Cross, 2002; Elliott et al., 1997; Plagenhoef, 1970). The major biomechanical issues in the forward stroke are coordination to accurately generate racket speed, recovery of elastic energy in the SSC, and creating the desirable racket path to impact. This section concludes with a short review of technical difficulties in making accurate kinematic measurements near tennis impacts that affects the biomechanical measurements of the forward stroke.

### 7.2.3.1 Coordination

Coordination is the precisely timed segment movements that combine to create the forward stroke. Early high-speed film studies of professional tennis players showed a variety of techniques (Plagenhoef, 1970) for each tennis stroke. More recent research has classified forward stroke coordination as either a single unit or multisegment style (Ariel and Braden, 1979; Elliott et al., 1989a). It can be concluded that tennis strokes tend to fall somewhere on a continuum between simultaneous and sequential coordination strategies (Hudson, 1995). More simultaneous coordination of the trunk, arm, and racket will promote stroke accuracy over speed, while more sequential coordination of segments favors racket speed over accuracy. The sequential coordination of the multisegment style uses more degrees of freedom (dof) of the joints of the upper extremity to create racket speed.

Traditional tennis strokes in the 20$^{th}$ century tended to use simultaneous coordination. Players were instructed to stroke from the shoulder and use a firm wrist to minimize wrist motion. This coordination was appropriate because of the smaller racket heads required more stroke accuracy, and the gradual acceleration of the racket suited the more massive wooden rackets. In the last 20 years more players have adopted more sequential coordination in strokes. Larger headed and lighter rackets has made this a good coordination strategy. The rest of this section will focus on the sequential coordination used in the modern or multisegment stroke technique.

The sequential coordination in a modern tennis stroke can be easily observed in the linear or angular kinematics of the major joints of the trunk and upper extremity. Figure 7.5 shows the 3D resultant velocity of the racket center of mass, wrist, and elbow in a forehand drive for the forward swing to impact. The sequential peaking of resultant velocities is most prominent in shots like the serve (Bahamonde, 2000; Elliott, 1988).

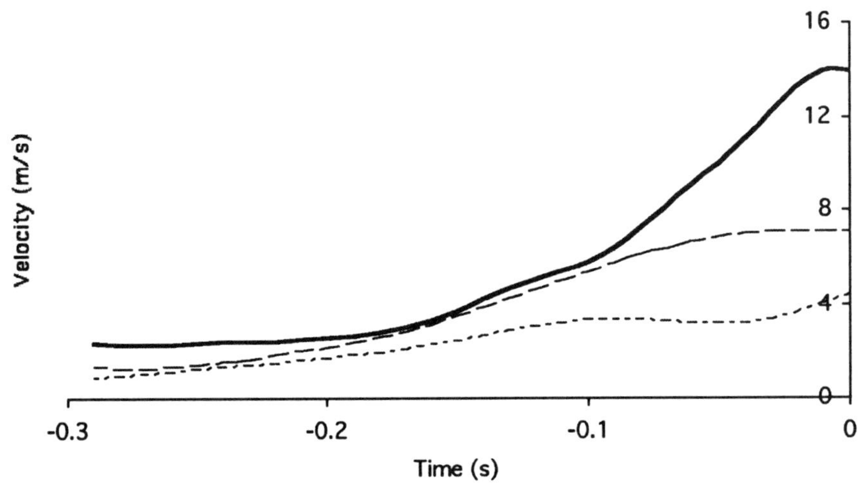


*Figure 7.5.* Three-dimensional resultant velocities of the upper extremity peak sequentially from proximal to distal in the forehand drive for racket center of mass (line), wrist (dash), and elbow (dotted). Adapted from Knudson (1988).

Two biomechanical issues related to optimising the sequential coordination of tennis strokes are the player's grip and the mechanism of the transfer of mechanical energy through the body segments that maximizes racket speed. Research on the former is clear, while the mechanism of optimal coordination is still unknown. Knowledge of these issues is necessary for the most effective coaching and physical conditioning of tennis players.

Greater detail on the sequential coordination of the multisegment stroke can be illustrated by the angular velocities of the rotations of the racket-arm in the forehand stroke (Fig. 7.6). It is evident that internal rotation of the upper arm is important contributor to the development of racket speed, because this large rotation occurs late in the forward swing phase of the stroke (Takahashi et al., 1996). In Fig.7.5, the long axis rotations such as shoulder internal rotation and forearm pronation are included in the increases in resultant velocity of the joint centers.

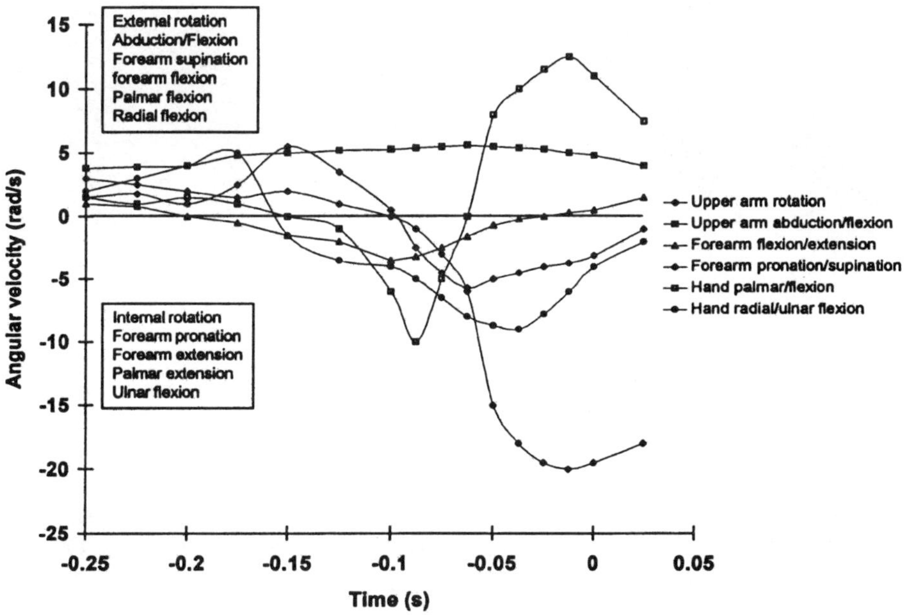


*Figure* 7.6. Racket-arm angular velocities in the tennis forehand. Reprinted fromTakahashi et al. (1996) with permission of Australian Journal of Science, Medicine, and Sport.

The grip used by a tennis player affects the kinds and amounts of wrist motion that can be used to accelerate the racket (Elliott and Christmass, 1995; Elliott et al., 1997). The eastern (A) and western (B) methods of holding the racket for a forehand stroke are shown in Fig. 7.7. Players with a western style grip use palmar flexion (direction of palm of hand) to help generate the vertical velocity vector of the racket head (needed in a topspin stroke), while those with an eastern grip use radial flexion (toward thumb) to this end. The contribution of the hand segment (with its 2 degrees of freedom) to the forward velocity vector of the racket in the forehand is created primarily by ulnar flexion (toward little finger) for those with a western grip, while those with an eastern grip use more palmar flexion to generate this vector.

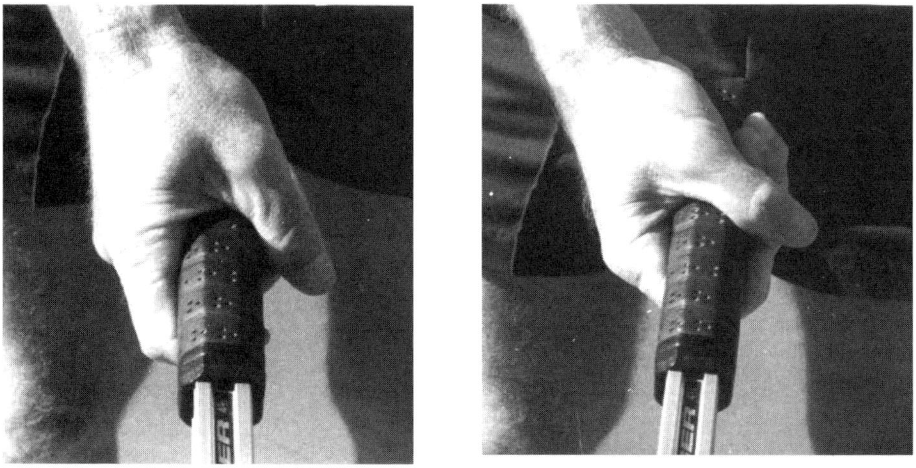

A B

*Figure 7.7.* The eastern (A: palm behind racket) and western (B: palm more under grip) forehand methods of holding the racket.

Grip most strongly affects the vertical component of racket velocity at impact in forehands, with western grips creating higher contributions (20%) to vertical velocity than eastern (5%) grips (Elliott et al., 1997). The horizontal adduction of the upper arm accounts for 20 to 30% of racket speed at impact, with shoulder internal rotation making the largest (30 to 40%) contribution to racket velocity (Elliott et al., 1997). The elbow makes little contribution to racket speed in one-handed backhands (Elliott et al., 1989b; Wang et al. 1998) and forehands (Takahashi et al. 1996), and one style of the two-handed backhand (Reid and Elliott, 2002). Trunk axial rotation accounts for about 10% of racket speed at impact, but his may underestimate the importance of the hips and trunk that transfer energy from the lower extremity to the upper extremity. This will now be discussed based on the kinetic studies of strokes.

Documenting which anatomical motions occur in a tennis stroke represents the kinematic contributions of various movements. These are important studies for understanding what joint movements are used in strokes and the extent of their variability (Knudson, 1990). A weakness of this approach is that the percentage contributions to racket speed vary depending on when in the forward stroke these calculations are made (Roberts, 1991). This is one reason that some biomechanists prefer to examine kinetic variables (forces, torques, EMG) to study coordination.

The causes of the movement kinematics represent the kinetic level of biomechanical analysis. Some studies have applied inverse dynamics to document the net joint forces and torques acting in tennis strokes. Energy from the lower extremities is transferred through the trunk by torques in the hips and trunk (Fujisawa et al., 1997; Iino and Kojima, 2001). Peak axial hip and trunk torques occur near or 30 to 40 ms after the initiation of the forward stroke (Fujisawa et al., 1997; Iino and Kojima, 2001), these torques (20 –50 Nm) are strong contributors to racket acceleration. This is consistent with the report of strong trunk muscle activation in the forward stroke of the forehands (Knudson and Blackwell, 2000). Exactly how energy is transferred through the trunk to the upper extremity is not clear, but there have been two coordination styles for trunk axial rotation noted in forehands (Knudson and Bahamonde, 1999). However, the role of hip and trunk rotation in transferring energy to the upper extremity and racket in sequential coordination warrants further research.

The net joint moments in the upper extremity of tennis strokes have not been studied as extensively as the serve (Bahamonde, 2000; Buckley and Kerwin, 1988). Bahamonde and Knudson (1998) reported that the joints likely to make the largest contributions to racket speed in the forehand (based on peak net joint moments) are shoulder horizontal adduction (35 – 90 Nm), shoulder internal rotation (20 - 40 Nm) and wrist flexion (10 - 25 Nm). Slightly smaller peak torques were reported by Fujisawa et al. (1997). These kinetic data are consistent with observations of a proximal to distal sequence of muscle activation (Blackwell and Knudson, 2002; Van Gheluwe and Hebbelinck, 1986) and the large angular velocities of these movements prior to impact in Fig. 7.6.

Kinetic biomechanical analyses also suffer from several weaknesses. First, these analyses are more difficult and highly influenced by errors in experimental data. Second, it is difficult to use kinetic data to account for the segment accelerations they create (Sprigings et al., 1994; Zajac and Gordon, 1989), especially in three-dimensional movements like tennis strokes. Third, kinetics are more variable than kinematics and a variety of kinetic patterns can create essentially identical kinematic patterns of movement (Winter, 1984). Biomechanists disagree on whether forces acting across joints that create segmental interaction torques originate proximally (Roberts, 1991) or distally (Putnam, 1991). Others have noted that joint movements that do not appear to make substantial contributions to racket speed, may in fact be quite important in positioning the limb so that other joints or torques may contribute to the movement of interest (Bahamonde, 2000; Springings et al., 1994). Defining optimal coordination in tennis strokes or conditioning programs will be difficult until these theoretical issues have been resolved.

#### 7.2.3.2 Recovery of elastic energy (SSC)

Whatever the primary kinetic origin of sequential coordination is, it is likely that SSC muscle actions also contribute to creating high racket speeds. The sequential movement of several body segments allows each segment to recover some elastic energy in the muscles that were eccentrically stretched across these segments. Fujisawa and associates (1997) showed the importance of a vigorous trunk rotation, which created a lag between the trunk and the upper arm, thus enhancing the SSC muscle action in the hitting- arm movement in the early forwardswing. Later in the stroke, the lag in wrist motion as the arm accelerates facilitates SSC activity in the forearm muscles (Kleinoder and Mester, 1998; Knudson 1990).

#### 7.2.3.3 Paths

The trajectory or path of the racket at the moment of impact with the ball affects the velocity and spin imparted to the ball. The forces of gravity and air resistance, as well as the geometry of the court make certain racket paths well suited for specific tennis techniques. Biomechanics research has documented these typical racket paths for flat, topspin, and underspin strokes.

A common feature of the forward stroke is a change in racket trajectory as it approaches the ball. Most players begin the forward stroke at a moderate angle to the horizontal that is then adjusted to the final path through the impact point with the ball (Elliott et al., 1989a, 1989b). In Figure 7.3 there is this shift in racket path near the 0.4 m point in the racket's horizontal motion. This forehand in this figure is a flat to moderate topspin stroke.

Research has shown that flat forehands and backhands typically have racket paths of 20 to 30 degrees above the horizontal near impact (Elliott et al. 1989a; Knudson and Bahamonde, 1999; Takahashi et al., 1996). Strokes with topspin have racket paths between 35 and 45 degrees (Elliott et al., 1989a,b, Reid and Elliott, 2002), while topspin lobs have paths between 50 and 70 degrees (Takahashi et al., 1996). Since ball spin does not need to be reversed and more horizontal trajectories are needed in backspin strokes, the typical racket paths in backspin strokes to impact are between –15 to –25 (below the horizontal) degrees (Elliott and Christmass, 1995; Elliott and Marsh, 1989; Groppel, 1975). These paths are general guidelines for tennis coaches because the height of impact, speed of the ball, position on the court, and tactical objective of the stroke all affect the optimal stroke path for a particular shot.

In the double-handed backhand the two hands may be used, as a force couple, to rapidly alter the orientation of the racket. Coaches believe that this enhances the ability of the two-handed backhand player to delay the commitment to stroke direction, which adds to stroke disguise. This belief has been confirmed by the observation of a later peak horizontal acceleration of the racket for the two-handed compared with single-handed topspin lob observed by Reid and Elliott (2002) (Fig. 7.8). Both methods of holding the racket produced similar racket speed at impact. However, no differences were reported in the horizontal acceleration of down-the-line and across-court strokes.

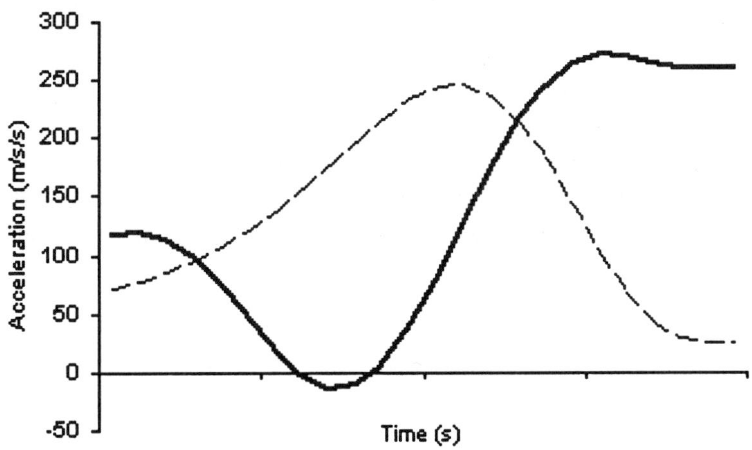


*Figure 7.8.* Mean horizontal racket-tip acceleration for the topspin down-the-line stroke for single-handed (dashed) and double-handed (solid) players. Adapted from Reid and Elliott (2002) with permission of Sports Biomechanics.

### 7.2.3.4 Impact Kinematics and Data Smoothing

A problem in documenting the path of the racket at impact is the large acceleration created by impact. Biokinematic data are routinely smoothed to eliminate very small digitising errors that grow considerably larger when the data are differentiated to obtain velocity and acceleration (Wood, 1982). Common smoothing methods are low-pass digital filters, Fourier series, and splines. Unfortunately, the sharp deviations in ball and racket motion created by impact are similar to the digitizing errors. Most biomechanical studies of tennis have smoothed data through impact that creates systematic errors in the data before and after impact (Knudson and Bahamonde, 2001). The error may be small for position measurements (< 6%) but grows considerably for velocity (> 50%) and acceleration time derivatives (Knudson and Bahamonde, 2001). These errors have a significant impact on the accuracy of angular kinematic and inverse dynamics calculations. Figure 7.11 illustrates the high frequency behaviour of the wrist flexion/extension angle in the forehand that is lost if data are smoothed through impact.

Figure 7.9 illustrates the false peak resultant velocity of the racket before impact created by smoothing position data through impact. Using a linear or fifth order polynomial estimation of the impact point and extrapolation reduces errors in impact position to less than one percent (Knudson and Bahamonde, 2001). These authors noted that any smoothing method (digital filter, cubic or quintic spline) were effective using the extrapolation techniques. Linear extrapolations are effective for accurate velocity calculations, while the quintic extrapolations are more appropriate for calculating acceleration near impact in tennis. Future research should not smooth kinematic data through impact, but use either extrapolation (Knudson and Bahamonde, 2001) or data separation techniques (Knudson, 1990) to make accurate measurements of the kinematics at impact in tennis.

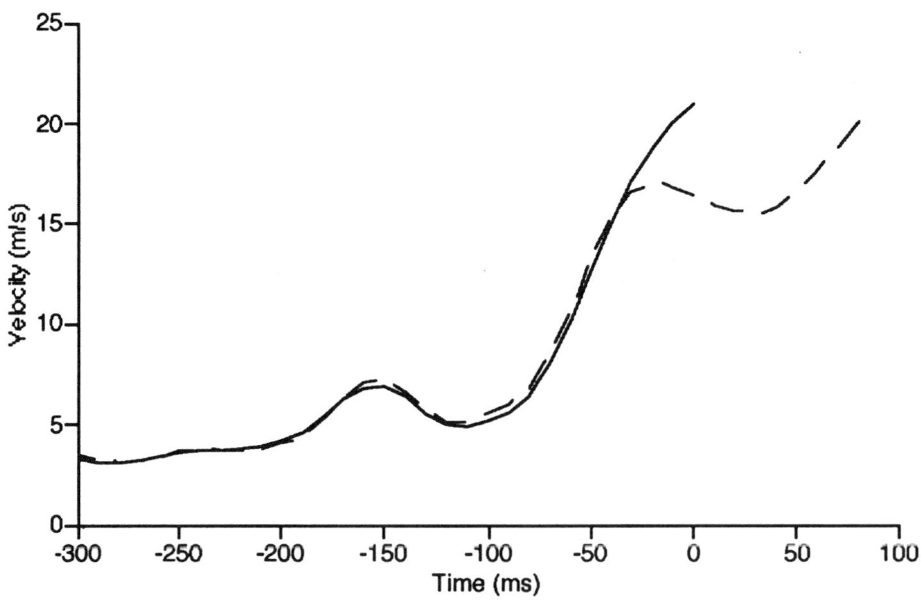


*Figure 7.9.* The resultant velocity of the racket in a forehand drive calculated by smoothing through impact with a quintic spline (dashed) and using a 5 point linear extrapolation and quintic spline smoothing (solid). Reprinted from Knudson and Bahamonde (2001) with permission of Journal of Sports Sciences.

The kinematics of the racket at the moment of impact are critical in tennis strokes. The position and orientation of the racket face determines the initial direction of the shot (Brody, 1987a; Blievernicht, 1968; Elliott et al., 1989b) while the velocity of the racket is the primary determining factor is the speed of ball rebound (Cross, 2002; Elliott et al., 1997; Plagenhoef, 1970). The racket face is nearly vertical (angle to vertical < 10 degrees) at impact for most strokes, although topspin may have the racket face slightly "closed" (tilted forward) while backspin may be slightly "open" (Elliott and Christmas, 1995; Elliott and Marsh, 1989; Elliott et al., 1989a; Takahashi et al. 1996). The wrist is hyperextended at impact for shots hit down-the-line (Fig. 7.10A)(Elliott et al., 1989a, 1989b; Knudson, 1990). This wrist angle would be reduced for shots hit across-court (Fig. 7.10B). The position at impact for the one-handed backhand is illustrated in Fig. 7.10C.

A B C

*Figure 7.10.* Forehand impact positions for down-the-line (A) and across-court (B) forehands and a down-the-line backhand (C).

At impact the upper trunk alignment has rotated passed the hip alignment such that it is almost parallel with the baseline in the forehand stroke and approximately 70 degrees to the baseline in the single-handed backhand stroke (Fig. 7.10). These different angles are caused by the position of the racket-shoulder to the impact situation. The angle of trunk twist at impact is approximately -10 degrees (shoulders now in front of hips) for all strokes. While multisegment and single unit forehands begin the forward swing with significantly different elbow angles (Elliott et al. 1989a), the elbow angle at impact in tennis strokes is similar. Elbow angles at impact are within 15 to

40 degrees of full extension (Elliott and Christmas, 1995; Elliott et al. 1989a, 1989b; Knudson 1990) (Figs. 7.10 & 7.12).

#### 7.2.3.5 Grip Firmness

Tennis coaches have traditionally believed that grip firmness was an important technique point for increasing the speed of ball rebound off the racket. Plagenhoef (1970) even hypothesized and reported data that grip forces could increase the "striking mass" of the racket. The very short duration (4-8 ms) of a tennis impact, however, provides little opportunity for hand forces to increase the impulse applied to the ball during impact. The biomechanical significance of grip forces at and during impact in tennis is still controversial and unclear. Initial studies of static impacts showed no difference in ball rebound speed when a racket was freely standing or rigidly clamped, but these studies involved impact centrally located on the racket face that are not common in tennis play (Knudson, 1991c). Elliott (1982) was the first to show that restraining forces at the racket grip do affect ball velocity in more realistic impact models (dynamic and with varying impact locations on the racket face).

More recently, experimental studies of hand forces and biomechanical modelling have cast doubt on the mechanical significance of hand forces in tennis strokes. First, measurements of forces on the hand have shown that gripping forces in tennis strokes are small and only increase in the 50 ms prior to impact up to a maximum of 62% of maximum gripping force (Knudson, 1991; Knudson and White, 1989). The forces measured on the hand just before impact are not correlated with the peak acceleration of the racket (Knudson, 1988) because of the other impact factors and the large variability in the kinetics of impact (Knudson and White, 1989). It is possible that hand forces at the moment of impact modify the mechanical response of the tennis racket providing a small increase in ball rebound speed, but the consensus of biomechanics research suggests tennis rackets mechanically behave more like free objects than rigidly clamped objects (Brody, 1987b; Casolo and Ruggieri, 1991; Hatze, 1993; Knudson, 1997). Most tennis players will naturally learn to use enough grip force to control the racket during the stroke and only increase hand force just prior to impact. Grip forces at impact are important in strokes where impact occurs off-center on the racket face (Elliott, 1982; Knudson, 1991c, 1997). Consequently, coaches should teach player to maintain a relaxed during strokes but rapidly increase grip pressure at impact. It is possible that the effective mass of the racket/hand system can be increased to provide small improvements in racket stability and ball rebound. However, these hand forces do not significantly increase the impulse applied to the racket during impact or make the racket behave as a rigidly clamped object.

## 7.2.4 Follow-Through

The final temporal phase of tennis strokes is the follow-through. The major biomechanical issues in the follow-thorough are the range of motion used, safely dissipating the kinetic energy of the stroke, and recovering for the next stroke.

### 7.2.4.1 Range of Motion

The kinematics of the follow-through of tennis strokes have not been explicitly studied. Imaging studies of strokes (e.g. Elliott et al., 1989a, 1989b, Elliott and Christmass, 1995; Takashashi et al., 1996) have noted that the racket arm continues horizontal adduction/internal rotation (forehand) or horizontal abduction/external rotation (one-handed backhand) to a position where the upper arm is parallel to the court or higher. These authors also note that the trunk and lower extremities also continue to rotate into the follow-through possibly decreasing the loading on the upper extremity muscles. It is important that players allow these movements to gradually slow so that the arm and racket finishes the stroke on the opposite side of the body (Figs. 7.12 and 7.13).

### 7.2.4.2 Energy dissipation

Tennis instruction traditionally has advocated elongating the follow-through of strokes to reduce the risk of injury and increase stroke accuracy. The evidence for dissipating the forces of impact and the kinetic energy of the racket and arm over a long time and distance to reduce the peak loads on the tissues are well established. Unfortunately, there is a lack of research on relationship of the follow-through to stroke accuracy.

The energy of impact that is not transferred to the ball or damped in the racket, must be dissipated by the body. The shock wave created by impact creates peak forces and vibrations on the hand for about 40 ms after impact (Plagenhoef, 1979; Knudson, 1991b; Knudson and White, 1989). These peak forces and shock vary widely (CV between 26 and 140%) and depend on many factors including impact location, preimpact grip force, and racket vibration (Knudson, 1991c; Nab et al., 1998). These forces momentarily reverse wrist motion (Knudson, 1990; Knudson & Blackwell, 1997) creating an eccentric stretch of the wrist flexors (forehands) or extensors (one-handed backhands). Figure 7.11 illustrates this rapid wrist motion in a tennis forehand. A recent modelling study (Riek et al., 1999) of the forearm musculature supports the hypothesis that the less extended wrist of novices (Blackwell and Cole, 1993) and repetitive eccentric forearm muscle actions (Knudson, 1991b) may be primary causes of tennis elbow. Once this immediate impulsive loading is dissipated, the body still must absorb the remaining kinetic energy of the racket and upper extremity.

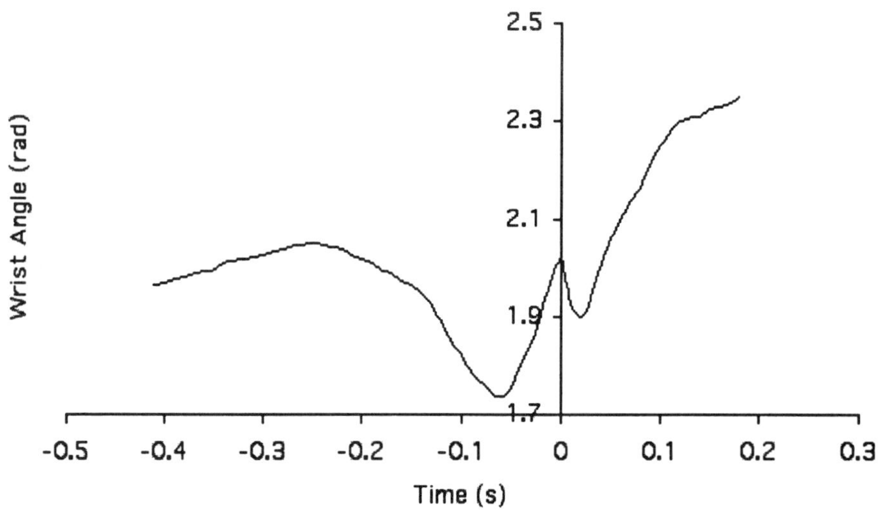


*Figure 7.11.* Electrogoniometer measurements of wrist angle showing reversal of wrist motion by the forces of impact in a forehand drive. Adapted from Knudson and Bahamonde (2001) with permission of Journal of Sports Sciences.

A long follow-through is important to slowly dissipate this remaining kinetic energy. After impact, tennis strokes can have 50 to 88% of the resultant velocity and consequently kinetic energy as they had before impact (Elliott and Christmass, 1995; Elliott et al., 1989a, 1989b). Gradually slowing down the motion of the whole body assists in minimizing the eccentric muscle forces required to dissipate the kinetic energy. The large joint motions briefly described in the previous section allow this energy absorption to be completed with the lowest risk of injury to the musculoskeletal system. For example, players using an open stance forehand cannot dissipate racket energy with forward body motion, so the follow-through is typically exaggerated to wrap the arm and racket around the body (Fig. 7.13).

### 7.2.4.3 Recovery

An important part of the follow-through is to immediately transition into the next stroke. This recovery usually involves moving on the court to bisect the angle of the opponent's potential shots and preparing the racket and body for the split-step. Coaching literature suggests that the footwork used in recovery be similar to the sliding and running used in moving to intercept a stroke. Use of an open stance forehand stroke may result in significantly faster recovery to the opposite side of the court, but not to the center of the court (Hopkins, 1981). There has been very little biomechanical research on this aspect of tennis play.

## 7.3 APPLICATION TO THE FOREHAND DRIVE

The application of the biomechanical factors reviewed in Section 7.2 can be easily observed in elite tennis players. Figure 7.12 illustrates the open stance forehand technique of an elite tennis player in competition. We will review the biomechanical factors that make this an effective stroke. Careful observation of Figure 7.12 suggests that this player effectively utilizes the following biomechanical concepts: A large loop, effective SSC muscle actions, a western grip and racket path to create topspin, and a long follow-through that transitions to recovery for the next shot.

The photo sequence does not allow us to comment on much of the preparation phase of the stroke. Note that the player used a large loop backswing. The player also uses an effective SSC of the trunk axial rotation (photos 1 to 3) while racket is still looping backward and downward. This vigorous trunk rotation is an important contributor to racket speed. This open stance stroke does not allow for forward linear motion of the body to contribute to racket speed, but the upward leg drive does contribute to the vertical velocity of the racket that is used to create topspin on the ball.

The ballistic nature of the stroke makes it difficult for coordination issues to be inferred from photos 3 and 4. It is clear that upward path of the racket created a topspin stroke, and the player has an effective follow-through that dissipates the kinetic energy of the racket and prepares for the next stroke (photo 5). Note how the racket rotates completely around the player's body. As his feet land the player is in a position for another SSC to begin to recovery movements for the next stroke (photo 6).

*Figure 7.12.* The open stance forehand technique of Andy Roddick. Photos by Lance Jeffrey.

## 7.4 OPTIMIZING PLAYER DEVELOPMENT

Traditionally, tennis professionals have visually identified errors in stroke technique and provided corrective feedback to help players improve. This approach is useful but is not as effective as what is known as qualitative analysis (Knudson and Morrison, 2002). To truly optimise the development of motor skills, like tennis strokes, the professional must be skilled at four key tasks of qualitative analysis. The four tasks of the Knudson and Morrison (2002) model of qualitative analysis are preparation, observation, evaluation/diagnosis, and intervention. In the preparation task the tennis pro gathers the relevant knowledge about the stroke and characteristics of the player and situation. In the observation task, all the senses (not just vision) gather relevant information about performance. For the third task, the analyst evaluates the strengths and weaknesses of the performance and prioritises their importance (diagnosis). The fourth task is the actual intervention to help the player improve. The qualitative analysis process is usually repeated to monitor progress in practice.

We will provide an example of how stroke biomechanics are used in the qualitative analysis of stroke performance for a 15-year old competitive junior tennis player. There are several publications that provide other examples of qualitative analysis of tennis strokes (Knudson, 1991a, 1999a; Knudson and Shriver, 2001). The mechanical factors emphasised in the section on fundamental biomechanical principles can be used as a checklist to evaluate stroke technique. This section will focus on those biomechanical principles, although the ideal approach would be to integrate all the relevant information on the performer and task in the analysis. Much of the value of the qualitative analysis of the tennis professional is this integrated and interdisciplinary evaluation and diagnosis of factors that affect stroke performance. Novice analysts do not see all the factors that affect performance or how they interact.

As preparatory movements are not included in the figure, minimal comment will be made on their efficiency, although this certainly must occur in the qualitative analysis of strokes. This young player, who holds the racket with a western style grip, clearly uses an open stance and is generally well balanced in preparation for the stroke.

In the backswing (Figs. 7.13A & B):

- A looped swing has been used to increase the racket speed.
- The range of motion is such that the racket is almost parallel with the baseline. This may prove a problem in the service return and where quick forward movement is required.
- The shoulders are rotated beyond the hips so that an appropriate separation angle is created, placing the trunk muscles on a stretch for a SSC.

*Figure 7.13*. Forehand drive of a high performance junior tennis player (A & B Backswing position, C impact, and D follow through).

In the forward swing (Figs. 7.13B to C):

- The forward swing is not started from a leg drive against the ground. This young player needs to practice "driving from the right leg" to develop a more effective body rotation that can transfer energy to the racket arm for a faster and safer stroke. It is believed that using the large muscles of the lower extremity to assist with racket acceleration will decrease the loading of the upper extremity.
- A smooth coordinated movement sequence was observed on the video from which these images were taken. The position of the upper arm in Fig. 7.13D shows that this segment has not internally rotated as much as elite players, limiting a major source of stroke power (racket-face would be about parallel to court if upper arm had internally rotated vigorously).
- Racket speed has been generated primarily through the development and transfer of angular momentum rather than a combination of angular and some linear momentum. This action is more the structure of a fully developed open stance stroke, where the majority of momentum generation is from angular momentum transfer.

Follow through:

- The position shown in Fig. 7.13D, shows that a full follow-through has been adopted where the segments would have been permitted to slow gradually. Preparation for movement to the next stroke is also evident.

To optimise the forehand development of this player, the tennis pro must diagnosis performance. Both weaknesses identified, lack of right leg drive to initiate the stroke and limited shoulder internal rotation, are important contributors to racket speed. A tennis professional would likely priortize the leg drive over the shoulder motion. Greater leg action may diminish the influence of poor shoulder internal rotation, and often improvements in earlier phases of a movement affect technique in the phases that follow (Knudson and Morrison, 2002). Biomechanical research is critical to understanding how technique factors interact and the diagnosis of tennis performance. Unfortunately, there have been few biomechanical studies of the effect of various modifications of tennis technique on performance that would inform stroke diagnosis. Once right leg drive has been selected the professional must select the intervention strategy that is likely to be most effective for this player.

Competitive junior tennis players tend to be motivated to improve and eager to emulate top players. The tennis pro would likely select a two-phase intervention strategy. First, they might reinforce player effort or some strength or technique point the player had been working on. Then the pro gives a cue to focus the player's attention on the leg drive. The pro might say "Great effort and racket preparation. On the next few strokes concentrate on starting the stroke with a vigorous push of your right leg. Think about driving the ball with your legs." Note how the feedback focuses the player's attention on the new technique and summarizes the intervention in a short cue that can be used in subsequent practice. The tennis professional then returns to observation as the practice or play continues. Good intervention in qualitative analysis utilizes many strategies (feedback, modified practice, manual guidance, computer-enhanced video replay etc.) beyond traditional corrections (Knudson and Morrison, 2002). Showing this young player a split-screen images of his leg drive versus an elite model (Fig. 7.12) and focusing his attention on the right leg would likely motivate him to work hard on this technique change.

## 7.5 SUMMARY

Sport biomechanics is concerned with improving sport performance and decreasing the risk of injury. A great deal of biomechanical research has focused on the sport of tennis, with engineering contributing to improvements in courts, rackets, balls, and shoes. This chapter focused on the biomechanical principles of tennis groundstrokes with reference to improving performance. In preparing for strokes, players have biomechanical options related to the stance and grip to use. The backswing of strokes tend to have a "loop" that increases the range of motion used to accelerate the racket and load the trunk muscles. This eccentric stretching of major propulsive muscles prior to their concentric shortening is a common neuromuscular strategy called the stretch-shortening cycle. The forward swing can use a variety of body segment movements and coordination, but effective strokes require high racket speeds and a nearly vertical racket face at impact. There is an opportunity for future research for bioengineers to improve our understanding of how muscles create and transfer energy through the body to the racket and how grip forces modify the mechanical response of the racket at impact. There is also a need for the development of models and algorithms that simulate the complex three-dimensional movements in tennis to help answer theoretical issues in racket speed development. The follow-through phase of strokes serves to decrease the risk of injury by dissipating the kinetic energy of the racket over a large range of motion. Good tennis players also use the follow-through to transition or prepare for the next stroke. The chapter concluded with two case studies of how these biomechanical principles can be used in the qualitative analysis of the tennis forehand of an elite and a junior player.

## 7.6 REFERENCES

Ariel, G., and Braden, V., 1979, Biomechanical analysis of ballistic vs tracking movements in tennis skills, in: *Proceedings of a national symposium on the racquet sports*, Groppel, J. ed., University of Illinois Press, Champaign, IL, pp. 105-124.

Bahamonde, R., 1999, Producing an 'explosive' forehand and backhand, in: *Applied proceedings of the XVII international symposium on biomechanics in sports: tennis.* Elliott, B., Gibson, B., and Knudson, D. eds., Edith Cowan University, Perth, WA, pp. 17-25.

Bahamonde, R., 2000, Changes in angular momentum during the tennis serve, *J. Sports Sci.* **18**: 579-592.

Bahamonde, R., and Knudson, D., 1998, Kinetic analysis of the open stance and square stance tennis forehand, Paper presented to the *4th International Conference on Sports Medicine and Science in Tennis*, Miami, FL.

Blackwell, J. R., and Cole, K. J., 1993, Wrist kinematics differ in expert and novice tennis players performing the backhand stroke: implications for tennis elbow, *J. Biomech.* **27**: 509-516.

Blackwell, J. R., and Knudson, D. 2002, Effect of the type 3 (oversize) tennis ball on serve performance and upper extremity muscle activity, *Sports Biomech.* **1**: 187-192.

Blievernicht, J. G., 1968, Accuracy in the tennis forehand drive: cinematographic analyis, *Res. Quart.* **39**: 776-779.

Bragg, R. W., and Andriacchi, T. P., 2001, The lateral reaction step in tennis footwork. in: Proceedings of poster sessions: *XIX International Symposium on Biomechanics in Sports*, J. R. Blackwell, ed., University of San Francisco, San Francisco, CA, pp. 34-27.

Brody, H., 1987a, *Tennis Science for Tennis Players*, University of Pennsylvania Press, Philadelphia.

Brody, H., 1987b, Models of tennis racket impacts, *Int. J. Sport Biomech.* **3**: 293-296.

Buckley, J. P., and Kerwin, D. G., 1988, The role of biceps and triceps brachii during tennis serving, *Ergonomics*, **31**: 1621-1629.

Casolo, F., and Ruggieri, G., 1991, Dynamic analysis of the ball-racket impact in the game of tennis, *Meccanica*, **26**: 67-73.

Cox, R. H., Noble, L., and Johnson, R. E., 1982, Effectiveness of the slide and cross-over steps in volleyball blocking—a temporal analysis, *Res. Q. Exerc. Sport*. **53**: 101-107.

Cross, R., 2002, Customizing a tennis racquet by adding weights, *Sports Eng.* **4**: 1-14.

Elliott, B., 1982, Tennis: the influence of grip tightness on reaction impulse and rebound velocity, *Med. Sci. Sport Exerc.* **14**: 348-352.

Elliott, B. C., 1988, Biomechanics of the serve in tennis—a biomedical perspective, *Sports Med.* **6**: 285-294.

Elliott, B., and Christmass, M., 1995, A comparison of the high and low backspin backhand drives in tennis using different grips, *J. Sports Sci.* **13**: 141-151.

Elliott, B., and Marsh, T., 1989, A biomechanical comparison of the topspin and backspin forehand approach shots in tennis, *J. Sports Sci.* **7**: 215-227.

Elliott, B., Marsh, T., and Overheu, P. 1989a, A biomechanical comparison of the mulitsegment and single unit topspin forehand drives in tennis, *Int. J. Sport Biomech.* **5**: 350-364.

Elliott, B. C., Marsh, A. P., and Overheu, P. R., 1989b, The topspin backhand drive in tennis: a biomechanical analysis, *J. Hum. Mov. Stud.* **16**: 1-16.

Elliott, B., Takahashi, K., and Noffal, G., 1997, The influence of grip position on upper limb contributions to racket head velocity in a tennis forehand, *J. Appl. Biomech.* **13**: 182-196.

Fujisawa, T., Fuchimoto, T., and Kaneko, M., 1997, Joint moments during tennis forehand drive: an analysis of rotational movements on a horizontal plan, in: *Abstracts of XVI ISB Tokyo Congress, International Society of Biomechanics*, Tokyo, pp. 354.

Groppel, J. 1975, *A Kinematic Analysis of Topspin and Backspin Techniques in the Tennis Forehand Drive*, Unpublished master's thesis, University of Illinois.

Groppel, J., 1992, *High Tech Tennis*, 2nd ed., Leisure Press, Champaign, IL.

Gregor, R. J., ed. 1997, Mechanics and energetics of the stretch-shortening cycle. [Special issue], *J. Appl Biomech.* **13:** 387-496.

Hatze, H., 1993, The relationship between the coefficient of restitution and energy losses in tennis rackets, *J. Appl. Biomech.* **9**: 124-142.

Hudson, J. L., 1995, Core concepts in kinesiology. *J Phys. Ed. Rec.Dance* **66**: 54-55, 59-60.

Hopkins, P. W., 1981, *A Comparison of Movement Times Between the Open and Closed Stance for the Tennis Forehand Groundstroke,* Unpublished Master's thesis, Penn State University.

Iino, Y., and Kojima, T., 2001, Torque acting on the pelvis about its superior-inferior axis through the hip joints during a tennis forehand stroke, *J. Hum. Mov. Stud.* **40**: 269-290.

Kibler, W. B., and Safran, M. R., 2000, Musculoskeletal injuries in the young tennis player, *Clin. Sports Med.* **19**: 781-792.

Kleinoder, H. , and Mester, J., 1998, Stress for the hitting arm by wrist and forearm actions in tennis, [abstract] *Int. J. Sports Med.* **19**: S58.

Knudson, D., 1988, *An Analysis of Grip Forces and Three-Dimensional Accelerations in the Tennis Forehandd Drive,* Doctoral dissertation, University of Wisconsin-Madison.

Knudson, D., 1990, Intrasubject variability of upper extremity angular kinematics in the tennis forehand drive, *Int. J. Sport Biomech.* **6**: 415-421.

Knudson, D., 1991a, The tennis topspin forehand drive: Technique changes and critical elements, *Strategies* **5**: 19-22.

Knudson, D., 1991b, Forces on the hand in the tennis one-handed backhand, *Int. J. Sport Biomech.* **7**: 282-292.

Knudson, D., 1991c, Factors affecting force loading in the tennis forehand, *J. Sports Med. Phys. Fit.* **31**: 527-531.

Knudson, D., 1997, Effect of grip models on rebound accuracy of off-center tennis impacts, in: *Biomechanics in Sports XV*, J. Wilkerson, K Ludwig, and W. Zimmerman, eds., Texas Woman's University, Denton, TX, pp. 483-487.

Knudson, D., 1999a, Using sport science to observe and correct tennis strokes, in: Applied *Proceedings of the XVII International Symposium on Biomechanics in Sports, TENNIS*, B. Elliott, B. Gibson, and D. Knudson, eds., Edith Cowan University, Perth, Western Australia, pp. 7-16.

Knudson, D., 1999b, An open and square stance case, *TennisPro,* **8**: 10-11.

Knudson, D., and Bahamonde, R., 1999, Trunk and racket kinematics at impact in the open and square stance tennis forehand, *Biol. Sport*, **16**:3-10.

Knudson, D., and Bahamonde, R., 2001, Effect of endpoint conditions on position and velocity near impact in tennis, *J. Sports Sci.* **19**: 839-844.

Knudson, D., and Blackwell, J., 1997, Upper extremity angular kinematics of the one-handed backhand drive in tennis players with and without tennis elbow, *Int. J. Sports Med.* **18**: 79-82.

Knudson, D., and Blackwell, J., 2000, Trunk muscle activation in open stance and square stance tennis forehands, *Int. J. Sports Med.* **21**: 321-324.

Knudson, D., and Morrison, C. S., 2002, Qualitative analysis of human movement, *2nd ed. Human Kinetics*, Champaign, IL.

Knudson, D., and Shriver, P., 2001, Self-analysis, in: World-class tennis technique P. Roetert and J. Groppel, eds., *Human Kinetics*, Champaign, IL, pp. 247-262.

Knudson, D., and White, S. C., 1989, Forces on the hand in the tennis forehand drive: application of force sensing resistors, *Int. J. Sport Biomech.* **5**: 324-331.

Komi, P.V. 1984, Physiological and biomechanical correlates of muscle function: effects of muscle structure and stretch-shortening cycle on force and speed, *Exerc. Sports Sci. Rev.* **12**:81-121.

Nab, D., Hennig, E. M., and Schnabel, G. 1998, Ball impact location on a tennis racket head and its influence on ball speed, arm shock, and vibration. in: *Proceedings II of the XVI ISBS Symposium,* H.H. Riehle and M.M. Vieten eds., Universitatsverlag Konstanz, Konstanz, pp. 229-232.

Plagenhoef, S. 1970, *Fundamentals of Tennis*. Prentice-Hall, Englewood Cliffs, NJ.

Plagenhoef, S. 1979, Tennis racket testing related to tennis elbow. in: *A National Symposium on the Racket Sports,* J. Groppel ed., University of Illinois Press, Champaign, IL, pp. 291-310.

Putnam, C., 1991, A segment interaction analysis of proximal-to-distal sequential segment motion patterns, *Med. Sci. Sports Exerc.* **23**: 130-144.

Reid, M., and Elliott, B., 2002, The one- and two-handed backhands in tennis, *Sports Biomech.* **1**: 47-68.

Riek, S., Chapman, A. E., and Milner, T., 1999, A simulation of muscle force and internal kinematics of extensor carpi radialis brevis during backhand tennis stroke: implications for injury, *Clin. Biomech.* **14**: 477-483.

Roberts, E. M., 1991, Tracking velocity in motion, in: *Biomechanics in Sports IX,* C.L. Tant, P. E. Patterson, and S. L. York, eds., Iowa State University, Ames, IA, pp. 3-25.

Roetert, E. P., Ellenbecker, T. S., Chu, D. A., and Bugg, B. S., 1997, Tennis-specific shoulder and trunk strength training, *Strength Cond.* **19**: 31-39.

Sprigings, E., Marshall, R., Elliott, B., and Jennings, L., 1994, A three-dimensional kinematic method for determining the effectiveness of arm segment rotations in producing racquet-head speed, *J. Biomech.* **27**: 245-254.

Steindler, A., 1955, *Kinesiology of the Human Body Under Normal and Pathological Conditions,* Charles C. Thomas, Springfield, IL.

Takahashi, K., Elliott, B., and Noffal, G., 1996, The role of upper limb segment rotations in the development of spin in the tennis forehand, *Aus. J. Sci. Med. Sport.* **28**: 106-113.

Van Gheluwe, B., and Hebbelinck, M., 1986, Muscle actions and ground reaction forces in tennis, *Int. J. Sport Biomech.* **2**: 88-99.

Wang, L., Wu, H., Su, F., and Lo, K., 1998, Kinematics of upper limb and trunk in tennis players using single handed backhand stroke, in: *Proceedings of XVI International Symposium on Bbiomechanics in Sports*. Riehle, H. and Vieten, M. eds., University of Konstanz, Konstanz, pp. 273-276.

Waninger, K. N. and Lombardo, J. A., 1995, Stress fractures of index metacarpal in an adolescent tennis player, *Clin. Sports Med.* **15**:63-66.

Wilson, G., Elliott, B., and Wood, D, 1991, The effect on performance of imposing a delay during a stretch-shorten cycle movement, *Med. Sci. Sport Exerc.* **23**: 364-370.

Winter, D. A., 1984, Kinematic and kinetic patterns of human gait: variability and compensating effects, *Hum. Mov. Sci.* **3**: 51-76.

Wood, G. A., 1982, Data smoothing and differentiation procedures in biomechanics, *Exerc. Sports Sci. Rev.* **10**: 308-362.

Yamamoto, Y., 1996, The relation between preparatory stance and trunk rotation movements, *Hum. Mov. Sci.* **15**: 899-908.

Zajac, F. E, and Gordon, M. E., 1989, Determining muscle's force and action in multi-articular movement, *Exerc. Sports Sci. Rev.* **17**: 187-230.

# Chapter 8

# Optimizing Ball and Racket Interaction

Howard Brody[1] and Paul Roetert[2]

[1] *Dept. of Physics, University of Pennsylvania, Philadelphia, PA 19104, PH: (215) 898-8430, FX: (215) 898-2010, EM: brody@dept.physics.upenn.edu*

[2] *USA Tennis High Performance Program, 7310 Crandon Blvd, Key Biscayne, FL 33149, PH: (305) 365-8730, FX: (305) 365-8786, EM: roetert@usta.com*

## 8.1 INTRODUCTION

There are many ways to optimize the ball-racket interaction, from finding the best location on the head to impact the ball, to determining the optimum racket head speed and orientation, to choosing the best racket and strings for your game and the court conditions. What may be an optimum for one player may not be best for another player. However, once these choices are made, the player must still be able to perform the required stroke with the required precision and be in the correct court position. This is because, unlike baseball where the pitcher is required to deliver the ball to a specific location, in tennis, the opponent is often trying hard to deliver the ball as far away from you as possible.

In this article, general guidelines are presented for selecting racket weight, balance, head size, and string tension, based on scientific principles. In addition, the choice of shots (how hard to hit the ball, how much spin to put on the ball, going cross-court or down-the line) is also analyzed with the aim of optimizing the percentage of points won.

## 8.2 SWEET SPOTS

The optimal location on the racket to strike the ball is determined by the racket parameters, the type or style of stroke, and the speed of the incoming ball. It is generally assumed that the player tries to hit the ball at the "sweet spot" of the racket, but exactly where that point is located, is usually found by trial and error. Which raises an interesting question. Why can't the racket manufacturer paint a bright spot on the strings of the racket to indicate where the sweet spot is located? The player then could concentrate on having the racket meet the ball at that spot.

This does not work because there are actually three separate sweet spots on a tennis racket (Brody, 1987), each one referring to separate physical phenomena. When you ask a player where the sweet spot is (or what defines the sweet spot), the answer you usually get is "It's the place where it feels good when you hit the ball." An analysis of the technology of tennis shows that the three sweet spots are:

1. The node of the lowest vibrational mode of the racket (the ball impact location that produces minimum vibration also known as the NODE).

2. The ball impact location where the shock or jar due to ball impact is a minimum (the center of percussion or COP).

3. The ball impact location on the racket that gives you the maximum rebounding ball speed (the maximum of the coefficient of restitution or COR).

Since these three spots may be located at different places on the racket head, it is not possible for the manufacturer to show the player a specific spot to hit the ball.

Although some research has been done on player preference for a particular sweet spot over another one (the node seemed to be preferred over the COP – Hatze, 1994a), a thorough examination of this problem has yet to be done.

### 8.2.1 The Node

A racket is not a perfectly rigid body, but has some flexibility to it. When a ball hits the racket, the frame deforms and then oscillates. There is one particular location on the head of a racket where, when the ball impacts, the result is a minimum of vibrations. This location is called the node, and since vibrations are a minimum at this location, it clearly feels good when the ball hits there. For a uniform rod, this point is about one fifth of the rod length from the end.

Since a tennis racket may be considered to be somewhat similar to a uniform rod in its mass and flexibility distribution, the node of a typical 27 inch long tennis racket will be about 5 to 6 inches from the tip (Brody et al., 2002b). This point will vary from racket to racket depending on the mass distribution and the flexibility distribution. (There is a second node in a tennis frame that is located in the handle about 5 to 6 inches from the butt end, but since players usually don't hit the ball there, it can be neglected.).

The farther the impact location is from the node, and the harder the ball is hit, the larger is the amplitude of vibration. Stiffer rackets have smaller amplitude of vibration when the ball misses the node because they deform less due to the ball impact. Racket manufacturers have attempted various strategies to damp out the vibrations of the frame quickly, but it turns out that the human hand damps out the oscillations of a typical frame in just a few cycles of vibration, particularly if the hand is holding the grip firmly. A typical, modern, graphite reinforced composite racket will have a principal mode of oscillation in the 125 to 225 cycles per second range. The stiffer and lighter the frame is, the higher its frequency of oscillation (Brody et al., 2002c).

### 8.2.2 The Center of Percussion

When a ball strikes a tennis racket at its balance point (also known as the center of mass or center of gravity), the racket recoils to conserve linear momentum. However, since the balance point is usually located quite near the racket's throat, very few players strike the ball there. When the ball hits the racket in the middle of the head (which is usually well above the balance point) the racket still recoils to conserve linear momentum, but it now twists or rotates about its center of mass to conserve angular momentum. If the ball is approaching from the left and moving to the right, the racket center of mass will recoil to the right (linear momentum) and the head will rotate to the right (angular momentum), which causes the handle to rotate to the left. With the entire racket moving to the right, but the handle rotating to the left, it is possible for the two motions to exactly cancel each other at one point. If your hand were located at this point, then because of the cancellation of the two motions, your hand would feel no shock or jar from the ball impact.

There is a unique point on the head of the racket where, when the ball impacts, the two motions cancel at your hand, and this is called the center of percussion (COP). It is actually a conjugate pair of points – the grip location and the ball impact location. If the racket is held in a slightly different location on the handle, the location of the center of percussion point in the head moves correspondingly. If the ball impacts above the COP, the racket handle will tend to jump out of your hand. If the ball impacts below the COP, the racket handle will tend to bang into your hand. A ball impact at or close to the COP will feel good, in that there is neither a tendency for the racket to pull out of your hand or push against your hand.

The center of percussion ball impact location is basically determined by the mass distribution of the racket and where on the handle you grip the racket. You can find the COP distance from your hand by turning the racket into a pendulum and pivoting it at the location of your grip. The distance from the pivot to the COP impact location is 9.73 inches (or 24.8 cm.) times the period of the pendulum in seconds, squared. If it takes 1.3 seconds for the pendulum to complete a full cycle, the distance from the pivot point to the COP will be 9.73 * 1.3 *1.3 inches (= 16.4 inches) (Brody, 1979).

### 8.2.3 Maximum of the Coefficient of Restitution

This is the location on the head of the racket where you get the most "power" or rebounding ball speed. However, unlike the node and the COP, it is not a point determined only by the racket parameters. The maximum power point is determined by the type of stroke you hit, the relative ball-racket head speed, and a number of the racket parameters, such as racket mass, mass distribution, flexibility, string pattern, and string tension. Unlike the other two sweet spots that clearly feel good when you hit the ball at one of them, this location makes you feel good since you are getting the most rebounding ball speed with the least effort and the ball seems to fly off of the strings. In addition, some players feel that this is the most important of the three sweet spots, in that it may give you more control over the ball's flight.

When rackets are tested for power in the laboratory, balls are fired at a free racket at rest and the ratio of ball rebound speed to incident ball speed is determined for various points on the head. This ratio is called the apparent coefficient of restitution (ACOR) since the recoil speed of the racket is neglected (Hatze 1994b). The value of the ACOR tends to maximize near the balance point (center of mass) and falls off as the ball impact location moves away from balance point. The reason for this comes from the basic physics of the interaction. When a ball impacts at the center of mass, no energy goes into racket rotation, since the racket just recoils and does not rotate. The further the impact point is from the center of mass the greater is the impulsive torque tending to rotate the racket about its center. As more energy goes into racket rotation less goes into the ball's rebound. For a free racket at rest, this leads to lower values of ACOR as the impact point moves away from the balance point and toward the tip of the frame.

If, when you hit the ball, the racket were translated (straight line motion only) the balance point would be the maximum power location since all points in the racket are moving with the same speed. But a racket is swung, not translated, so the tip is moving faster than the throat, which in turn moves the effective power point up higher on the head. There is a simple formula for rebounding ball speed involving racket head speed, ACOR, and incoming ball speed (Brody, 1995). It is:

$$V(\text{hit ball}) = ACOR * V(\text{incident ball}) + (1 + ACOR) * V(\text{racket}) \quad (8.1)$$

Where V(racket) is the speed of the racket at the impact point (Brody, 1995).

The variation with respect to location of V(racket) depends on the nature of the swing (wristy, etc). The ACOR also varies with location and depends somewhat on the racket construction. As the formula shows, V(hit ball) also depends on the ratio of incoming ball speed to racket head speed. A swing which uses a great deal of wrist action will have a much larger V(racket) near the tip, as compared to the throat, and will move the maximum power location up higher in the head. The incoming ball speed is multiplied by ACOR, while the factor multiplying the racket head speed is (1 + ACOR). This moves the maximum power point down toward the throat as the incoming ball speed increases or as the racket head speed decreases. As the incoming ball speed decreases or the racket head speed increases, the maximum power point moves upward toward the tip. The limit of all of these factors is the serve with a swing having a great deal of wrist action, no incoming ball speed and a large racket head speed. For a typical serve, the maximum power point is up toward the tip of the racket, well above the center of the head. The extra height above the ground this gives for the ball impact location also increases the window of acceptance, as will be shown later.

Therefore, based on the above analysis, you should try to hit groundstrokes closer to your hand on fast courts (grass, etc) and hit your shots further out on the racket when you are playing on slow (clay) courts. All of this is fine in theory, hitting the ball at one of the sweet spots, or closer to your hand, if you can do it. In baseball, the player is fixed in position, and the pitcher must deliver the ball within a confined region. In tennis, your opponent often tries to hit shots as far away from you as possible. The tennis player must be able to perform the required stroke with the required precision, often on the dead run. Proper positioning and movement (excellent footwork) on the court are critical to good ball-racquet interaction.

## 8.3 CONTROL AND UNIFORMITY OF RESPONSE

Regardless of how good a player you are, you will on occasion not hit the ball at the exact point on your racket head that you are aiming for. If you are not a particularly good player, then you are likely to strike the ball over a sizeable portion of the head. How the racket responds to hits that miss the intended impact point may determine whether your ball lands in the court or out of bounds. Let us assume that your intended impact point is exactly at the center of the head of the racket, but you occasionally miss that location.

### 8.3.1 Miss-hits Along the Principal or Main Axis

If the maximum of the COR is near the middle of the head, which means that you can strike a ball above or below it, an unexpected benefit and decrease of unforced errors is obtained. As an example of this benefit, let us assume that the COR maximum is near the throat and not the center of the head. It is unlikely that you will aim to hit the ball at this location because it is close to the frame. Let us assume you again try to hit your shots at the center of the head. Then there is a strong correlation between where on the head the ball strikes and the outgoing ball speed. If you aim to hit the ball at the center of the head, and attempt to have your shots land fairly deep, then balls hitting below the center (closer to the maximum COR point) will often sail long since they come off the strings with higher speed. You can compensate for this by hitting your normal (center of the head impact) shots somewhat shallower. This will reduce you error rate, but also reduce your effectiveness because your opponent will be able to pounce on your shorter shots.

If the maximum of the COR point is near the center of the head and that is where you attempt to hit balls, aiming your shots deep does not result in errors if the ball is struck above or below the aimed point on the strings. Ball impacts along the axis but above or below the intended impact point (maximum COR) will now come off of the strings with slightly less speed and therefore will not sail long. Your error rate due to shots landing long will be reduced and, in effect, you have a racket that corrects for some of your miss-hits.

### 8.3.2 Miss-hits Perpendicular to the Principal or Main Axis

When the ball impact point is not along the main axis of the racket, there is a degradation of power (the rebound ball speed is reduced) and the racket will twist, changing the angle at which the ball exits the strings. Both of these effects can easily lead to balls that end up in the net, or at least not where they were intended to go. What can be done to reduce the effect of the racket twisting on off-axis impacts?

It is often stated that you must grip the racket handle very tightly to keep the frame from twisting in your hand on off-center ball impacts. When you calculate the magnitude of the impulsive torque (quick twisting) that a ball hitting an inch or two off-axis produces, you discover that your hand is not strong enough to prevent the racket from twisting (Brody et al., 2002a). Yet you rarely have the racket twist out of your hand. Watch young children playing tennis. They have very weak grips and they miss the axis more than they strike it, yet the racket does not twist out of their hands. The previous two arguments seem to contradict each other – you cannot prevent the racket from twisting, yet it rarely twists out of your hand.

What is actually happening is that the racket does twist, but the hand and arm-twist with it. High-speed videos show that on off-axis impacts, both the racket and arm twist, but then they recover so quickly that an observer or player often does not realize that it has happened. This twisting occurs whether the hand is gripping the handle loosely or very tightly. On occasion, when the racket grip material has gotten moist and slippery and the hand muscles have become tired, the racket will rotate out of the hand, but it is not the usual case.

What can be done to reduce the twisting of the racket (besides not hitting the ball off- axis)? There is a property of the racket called the polar or roll moment of inertia, and it is the magnitude of this property that determines how much the racket will twist when subjected to a torque. If this polar moment is increased, the racket will twist less on off-axis impact and will be more stable against them. As a general rule, a racket with a wider (as opposed to thicker) head will have a larger polar moment, and this stability factor goes roughly as the square of the head width (Brody, 1985). If you switch from a mid-size racket to a frame that is 20% wider, the polar moment (hence the stability against twisting) will increase by about 40%. This was one of the advantages of the original Prince over-size racket. It was 25% wider, so it was 50% more resistant to twisting. The other advantage of the oversize Prince was that by enlarging the head size, the sweet spots came out closer to the center of the head, rather than down near the throat, making them easier to play with.

## 8.4 STRINGS AND STRING PLANE DEFORMATION

Most players think about strings in terms of tension alone. However, it is string plane deformation that determines how a racket plays. String plane deformation is a measure of how much the string plane deforms when subjected to a given force. The deformation depends not only on the string tension, but the string density (strings/inch), the size of the head, and the properties of the string (gauge, gut vs synthetics, etc). The larger or greater the string plane deformation, the softer the racket plays. If all other parameters are held constant, lower string tension leads to more string plane deformation. A more open string pattern or a larger head will also lead to a greater string plane deformation, if all else is held constant.

It is string plane deformation rather than the tension that you want to keep the same if you change rackets and want to keep identical playing characteristics. As a general rule, if you switch to a racket with a different head size (but the same string spacing), you should adjust the tension in proportion to the average length of strings in the new frame (Brody et al., 2002d).

Many players judge their string tension by listening to the pitch or frequency of the sound their strings make when plucked. This will work fine as long as they are using the same frame and same strings. If they switch to a thicker or thinner string, the pitch will change even if the tension remains the same. For a given tension, a thicker (lower gauge) string will vibrate at a lower frequency and a thinner string (higher gauge) will vibrate at a higher frequency. The frequency or pitch also goes inversely with the string length. A longer string vibrates more slowly for a given tension.

String plane deformation determines the dwell time of the ball on the strings, which is the "feel" you get when you hit the ball. A smaller string plane deformation reduces the dwell time of the ball on the strings. A larger string plane deformation tends to give slightly more power for the following argument.

### 8.4.1 Strings and Power

The principal purpose of the strings is to take the kinetic energy of the ball-racket system, store it as energy of deformation, and the give it back to the ball as efficiently as possible. When a ball deforms and stores energy, it only returns about half of the energy it absorbed. This is written into the rules of tennis (ITF 2002). There are no rules about the strings. Strings have been designed to return about 95% of the energy that they absorb

(Brody, 1989). When a ball impinges on the strings, where do you want the energy stored – in the ball, which dissipates half of the incoming energy or in the strings, which give back 95% of the energy they absorb? The answer is obvious. In order to get the strings to absorb a larger fraction of the incident energy, they must deform more and the ball deform less. Increasing the string plane deformation (lowering the string tension) accomplishes this, as long as the tension is not reduced too much, which allows the strings to move within the string plane and dissipate energy through the friction of one string rubbing on another one.

Since, when playing on slow courts (clay, for example), there is an appreciable loss of ball speed due to the bounce, players must swing harder to attempt to maintain the same pace that they would get on faster courts. A second method to make up for the energy the ball loses at the bounce is to string the rackets at slightly lower string tension. Conversely, on fast courts (such as grass) where there is less loss of ball speed on the bounce, it may make sense to string the racket a bit tighter, since by doing so, control may be improved.

### 8.4.2 Strings and Control

Where the correlation between string tension and power is fairly clear, both experimentally and theoretically, the relationship between string tension and control is not as obvious. There is anecdotal evidence that tighter strings lead to better control, but there is not much scientific work to back it up. There is one argument and some experimental data that tends to back up the anecdotal opinion of the players.

As previously noted, if a ball impacts a racket away from the principal axis of the frame, the racket will tend to twist and the ball will come off of the strings at a somewhat errant angle. The more the racket twists before the ball leaves the strings, the larger is this effect. Since lower string tension (greater string plane deformation) leads to greater dwell times of the ball on the strings, the racket has more time to twist before the ball exits, and as a result the ball will leave the strings at a greater errant angle.

Where it is relatively easy to measure the ball speed leaving a racket (hence the power of a racket), it is not easy to measure the control a player has with respect to the string tension. The authors of this chapter know of only one experiment where players were asked to hit a target using identical rackets but strung at different tensions (private communication, Victor Gut Company). The result of this experiment, conducted about 30 or 40 years ago, showed that better players seemed to have better control at higher string tension and players of lower ability had better control at lower string tension.

## 8.5 ACCEPTANCE WINDOWS

Once you have optimized where on the head to hit the ball and you have selected the correct frame and strings, you still have to decide where to aim your shot, how hard to hit it, and how much spin to apply. These decisions can be made easier by using the concept of windows of vertical acceptance and computer generated ball trajectories. In general, in making these decisions, there is a trade-off between hitting aggressive shots and reducing the probability of making unforced errors, as will be explained below. However, since most points in tennis end in an unforced error rather than a winning shot, the choice of shots is often unambiguous.

### 8.5.1 Trajectories

The trajectory of a tennis ball through the air can be calculated using Newton's Laws of Motion, once its initial conditions are specified. These initial parameters include the ball impact height above the ground, ball vector velocity as it leaves the strings, ball spin, location on the court where the ball was struck, and the atmospheric conditions as well as the drag and lift coefficients of the ball. If the parameters of the ball-court interaction (coefficient of restitution, coefficient of friction) are also known, then the subsequent ball trajectory after the bounce can be determined. It is a rather straightforward task to write a computer program which, given the initial ball conditions as input, will calculate the position, velocity and spin of the ball at any later time (programmed in QBASIC). With the information that this type of program can generate, various shot making tactics can be investigated and a form of risk-benefit analyses can be done in order to optimize the chances of winning a point.

It is also possible to work backwards and determine the racket head velocity and orientation needed to produce the desired initial ball parameters previously calculated.

### 8.5.2 Groundstrokes

Fig. 8.1 shows the trajectory of a typical groundstroke hit at 65 miles/hour from the baseline. To simplify the calculation and the display, this computer-generated shot was "hit" down the middle, but it could just as well have been hit cross-court by inputting an initial transverse angle and a different hitting location. This particular shot was launched at a height of 3 feet above the ground from the baseline and at an initial angle of 7 degrees above the horizontal and with no initial spin. As you can see from the figure, the ball easily clears the net and bounces well within the opposite baseline. It is possible to instruct the computer to find the minimum initial angle for a ball that will just clear the net or the maximum initial angle for a ball so that it lands on the baseline (keeping all the other initial conditions constant). These two resulting ball trajectories for a 65 mph initial ball velocity are shown in Fig. 8.2. The difference in these two initial angles is a window of vertical angular acceptance that a player must hit within, if the 65 mph shot is to be good. The program also can calculate the corresponding physical window above the net that the ball must pass through in order to land in the court or the corresponding region of the court surface that is accessible to the ball, given the minimum and maximum angles previously calculated. These minimum and maximum angles can be calculated for ball speeds other than 65 miles/hour and the results are shown in Fig. 8.3.

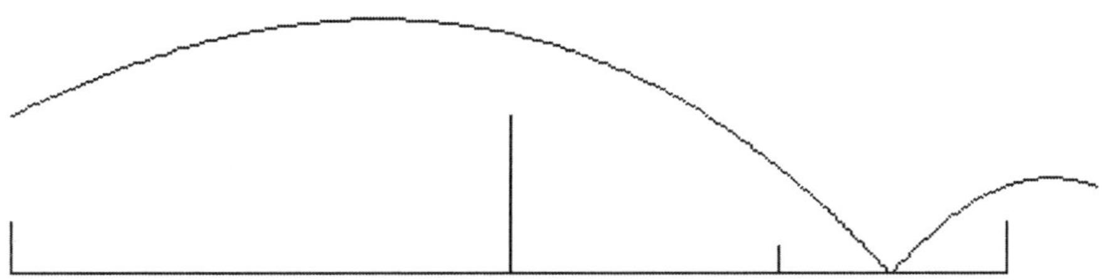

*Figure 8.1.* Trajectory of a 65 mph groundstroke hit at 7 degrees above horizontal.

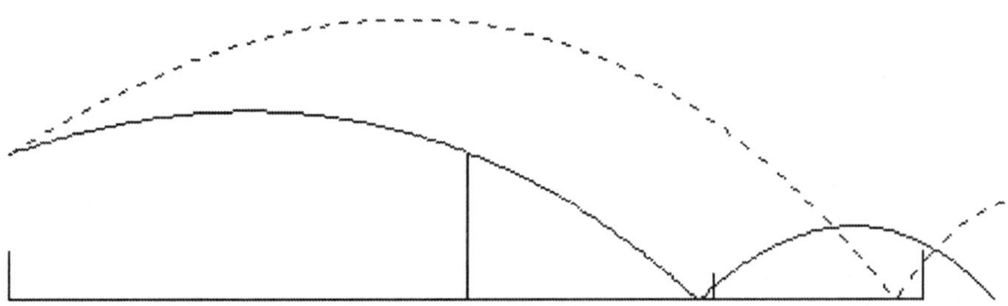

*Figure 8.2.* Trajectory of two 65 mph groundstrokes, one that just clears the net and one that lands just in front of the baseline.

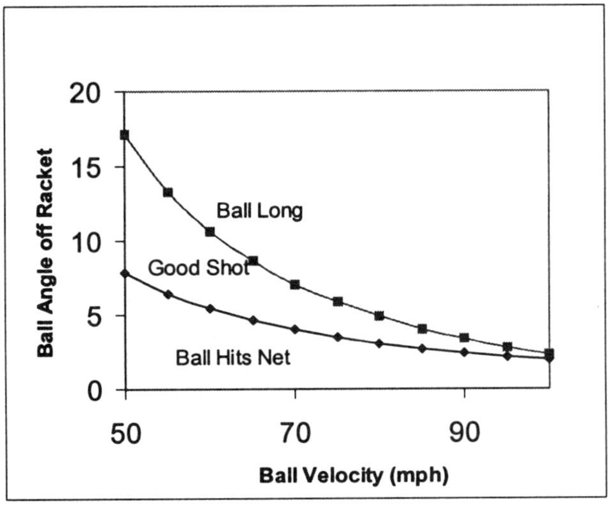


*Figure 8.3.* The minimum angle for a ball to just clear the net and the maximum angle so the ball just lands in the court, both for various values of initial ball speed.

The size of the vertical angular acceptance window (the difference between the upper and lower curve in Fig. 8.3) can be calculated for various initial conditions that you can select when you hit a shot. These input values that you can select are the impact height, ball speed off of the racket, initial horizontal angle and spin imparted to the ball as well as the location on the court where the ball is struck. It is your choice when you are playing, whether to hit the ball harder, hit with more or less spin, go cross-court or down-the-line, etc. The size of the window that corresponds to one of these choices is a measure of how likely it is for a given shot to be good and not result in an error. You are selecting the vertical angular acceptance window size by your choice of how you hit your shot. It is clear that with a larger window, you are less likely to make an error, so if that were all there was, you would strive to hit shots that had the maximum acceptance window.

When you hit a shot, you aim it at a particular location on the court, at a particular initial angle, or for a particular clearance of the net, but your aim is not perfect. You will often end up hitting the ball a bit lower or higher than you desired. If the variance in the vertical angle from shot to shot is large compared to the widow through which you are trying to hit that shot, you will make many errors. If, on the other hand, the variation in vertical angle of your shot is small compared to the window through which you are trying to pass, you will make few errors and you will be known as a steady player. As a general rule, a more aggressive shot will have a smaller window of acceptance and lead to more errors. You can reduce the stroke-to-stroke variation in your shots by practice (lots of practice), improving your concentration, taking lessons, and by following accepted biomechanical stroking rules when you swing, but that is not what this article is about.

If your window for a particular shot is many times larger than your shot-to-shot variation, then increasing your window size will not appreciably reduce your error rate. If your variation in vertical angle from stroke to stroke for a given shot is larger or comparable in size to the calculated size of the window for that shot, then an increase in window size will make a big difference in your ability to reduce your error rate. However, on occasion, what you must do to increase the window size for a particular shot (such as reducing your ball speed or aiming shorter) may actually be advantageous to your opponent, in which case it is not a good tactic to employ.

## 8.5.3 Vertical Window of Acceptance Versus Ball Speed

Let us examine the size of the vertical acceptance window for a typical groundstroke hit from the center of the baseline, waist high, with no spin, as the ball speed off of the racket is varied. This is shown in Figs. 8.4a and b and as you can see, the acceptance falls quickly as the ball is hit harder. Figure 8.4a shows the vertical acceptance angle or vertical angular window versus ball speed. (Figure 8.4a was obtained by calculating the difference between the upper curve and lower curve in Fig. 8.3). The angular

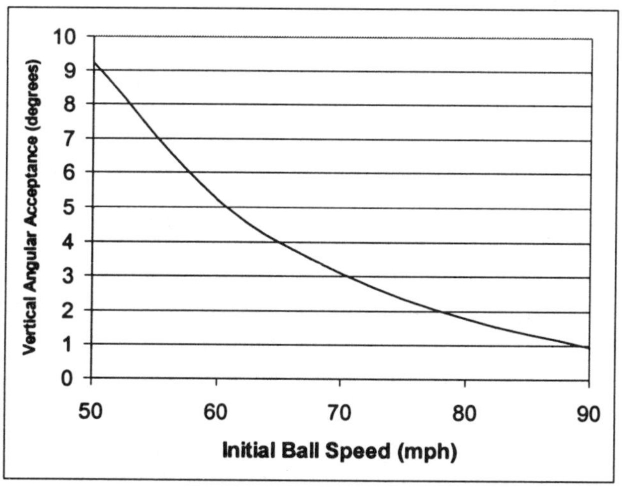


*Figure 8.4a.* The difference in angle between the ball that just clears the net and the ball that lands on the baseline as a function of initial ball speed. This is the vertical angular acceptance window for a groundstroke.

acceptance is an important parameter since the crucial variation in stroking from shot-to-shot is in the vertical angle that the ball comes off of the racket. However, tennis players do not have an inherent feeling for angular acceptance, where they can more readily visualize a physical window above the net. Therefore, Fig. 8.4b displays the window (in inches) above the net that the ball must be hit through, in order for the shot to be good. If the ball leaves the strings at 60 miles an hour, this window above the net is almost 43 inches high, which is larger then the net height itself. Most players feel that they could easily hit the ball through such a large window and not make an error. At 90 miles per hour, the window above the net is about 7.5 inches, or about 3 tennis ball diameters in size. To get a ball to go through such a small window is a very difficult task even for a very skilled player, so errors are likely to occur.

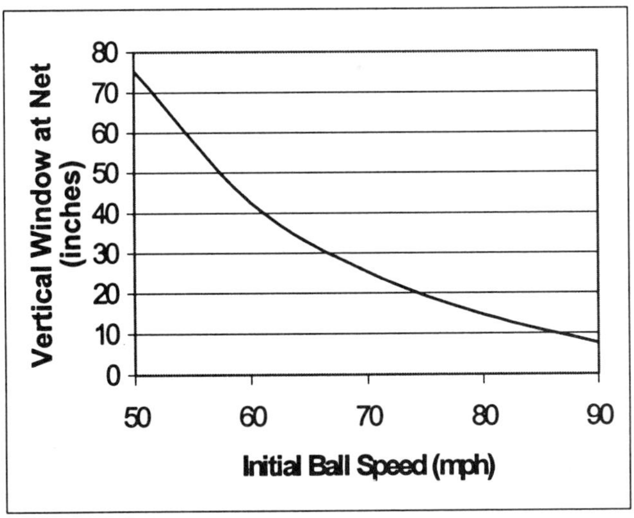


*Figure 8.4b.* The window above the net through which a ball must pass if it is to be good, as a function of ball speed.

There is a rather simple explanation for this reduction of the acceptance window with ball speed. The principal reason that a tennis ball in its trajectory can clear the net and still bounce within the baseline is GRAVITY. If there were no gravity acting on the ball its trajectory would be a straight line. The geometry of the court is such that under those conditions (no gravity) no ball hit waist high from the baseline could land inside the playing area if it had to go over the net. Gravity pulls the ball down into the court and slower balls get pulled down more than faster balls for two reasons. High speed balls have greater momentum than low speed balls, and the force of gravity will not change the direction a high momentum ball as much as a low momentum ball. In addition, a slowly moving ball spends more time in its flight, giving the force of gravity more time to pull the ball down into the court.

As Fig. 8.4a illustrates, the vertical angular acceptance window decreases as the speed of the ball leaving the racket increases. Just hitting the ball a little harder and increasing the ball speed from 60 to 70 miles per hour will reduce the acceptance window by 40%. This means it is more difficult to get a hard-hit shot to go in, just because of Newton's Laws and the geometry of the court. In addition, the control of the racket head may deteriorate somewhat when a faster swing is attempted, leading to a second reason why harder shots seem to result in errors more often. Yet players persist in hitting the ball hard for a number of reasons even though they know that they are less likely to get it to go in. A player will often hit hard in hopes of hitting a winner or a shot that sets up an easy winner, rather than give their opponent a chance to hit a winner (which will often end up as an error by the opponent). A passing shot is a good example where the baseline player tries to pass the net player by hitting the ball hard and low over the net. In addition, there seems to be a certain sense of satisfaction that many players get from hitting the ball hard and having it go in, and that pleasure is greater than the pain they suffer when the shot does not go in.

You as a player decide on the size of the acceptance window you are playing into by how hard you hit the ball. It is clear that one winning strategy is to hit the ball sufficiently hard so the size of the window of acceptance is not smaller than some measure of the variance of angles of your shots, yet does not give your opponent easy points to play.

As was illustrated, increasing the acceptance by not hitting the ball as hard has its advantages and disadvantages. Are there other ways to increase the acceptance window for your shots that do not have disadvantages? One can determine this by holding the ball speed constant and varying another parameter, such as hitting height, location on the court, cross-court versus straight, or by applying spin and examining the resultant window size.

### 8.5.4 Vertical Window of Acceptance Versus Hitting Height

There is a strong correlation between the height at which a ball is hit and the probability of that shot being good. Figure 8.5 shows how the vertical angular acceptance increases as you raise the impact location on a groundstroke, keeping all else the same. A hit at head height has twice the window that an impact somewhat below the knees has. However, not too many people can physically hit a groundstroke when the ball is at head height, so they will usually back up and allow the ball to fall into their comfort zone. This figure shows you that if you have the ability to hit high shots you should not wait for the ball to fall in its trajectory, but you should, whenever possible, hit the ball when it is at the peak of its rebound height.

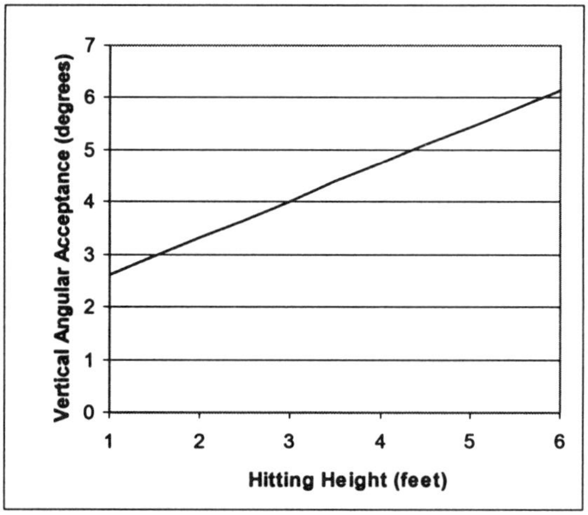


*Figure 8.5.* The vertical angular acceptance for a 65 mph groundstroke hit from the baseline as a function of the ball impact height.

This is also a strong argument when returning a lob (after it bounces) with an overhead stroke, rather than wait for the ball to come down to waist height and hit a traditional groundstroke. In fact, when you hit the ball overhead, you can then hit it much harder and still have a bigger window than if you returned it somewhat slower, but with a groundstroke. This is a win-win situation. By hitting it overhead and harder, your opponent has more difficulty in returning it and at the same time, your window of acceptance is larger.

This figure also shows one of the reasons why the half volley is such a difficult shot to master. Since the impact point is usually very close to the ground, the acceptance window for a half volley is greatly reduced compared to a normal groundstroke, hit at waist or chest height.

### 8.5.5 Vertical Window of Acceptance Versus Location on the Court

You can move in and hit your shots inside the baseline or you can back up and hit from well behind the baseline. How does this affect the vertical acceptance window? Figure 8.6 shows the window for shots hit from various depths. As you can see, there is some change in the vertical acceptance, but the effect is not as pronounced as the change due to impact height or ball speed. The most interesting (and counterintuitive) thing shown by this figure is that the acceptance window seems to increase as you back away from the baseline, as opposed to moving inside the baseline. This is because, the further behind the baseline you are when you hit a groundstroke, the more the ball's trajectory is falling when it crosses the net and the easier it is then for the ball to land in the court. This is not the case for a typical serve. This is also one of the reasons why hitting "on the rise" is risky, compared to backing up and waiting for the ball to come down.

You know quite well that shots hit from inside the baseline tend to be more forcing and stronger, since they give the opponent less time to get to the ball. However, from a defensive standpoint, retreating behind the baseline to return a shot gives you more time and also increases your window to hit through, as long as you do not let the ball fall too low before you hit it.

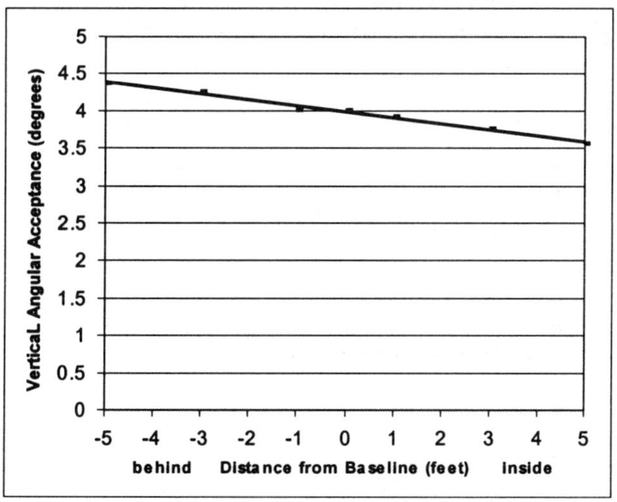


*Figure 8.6.* Vertical angular acceptance versus distance behind or inside baseline for a 65 mph groundstroke.

## 8.5.6 Vertical Window of Acceptance Versus Return Angle

You are in the corner of the court, on the baseline. You can hit a shot down-the-line or cross-court. Which selection has a larger vertical window of acceptance? Listen to most TV commentators and they will tell you to hit "cross-court" because the net is higher down-the-line compared to being lower in the middle. While this is true that a cross-court shot is safer, they are not giving the real reason. It is obvious that the net is lower at the center, but it is also further away from the corner than the net is when you hit down-the-line. This means when you go cross-court you must lift the ball a little more to go the extra distance and clear the net at the center compared to down-the-line. This results in making the minimum angle to just clear the net the same cross-court as it is down-the-line, so the TV commentators have the wrong reason. The vertical acceptance window is greater cross-court compared to down-the-line because the court is much deeper cross-court, due to the Pythagorean theorem. Exactly how much advantage cross-court gives you can be calculated, and it depends strongly on how hard you hit the ball.

Figure 8.7 concerns a shot hit from the corner of the court at 45 miles/hour. It shows the minimum vertical angles need to clear the net and the maximum vertical angles not to go over the baseline, both as a function of horizontal hitting angle. (Zero degrees is down-the-line while 19 degrees and greater is cross-court.) Subtracting the minimum vertical angle (lower curve) from the maximum vertical angle (upper curve) at the same horizontal angle, gives the vertical acceptance window at that horizontal

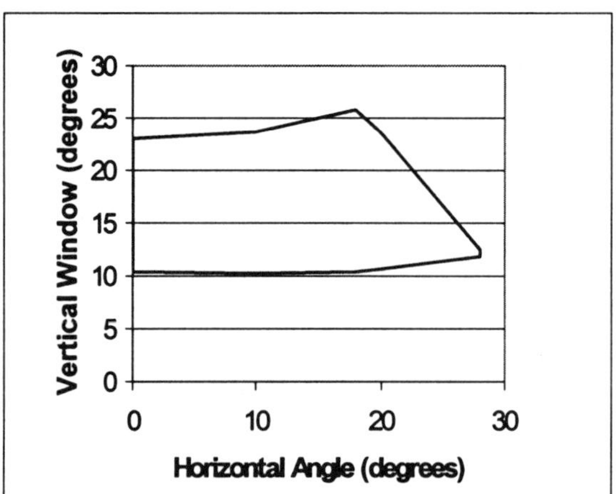


*Figure 8.7.* The angular limits on a "good" 45 mph groundstroke, hit from the corner of the court. Zero degrees is down-the-line and 19 degrees is cross-court.

angle. As Fig. 8.7 shows, the minimum angles needed to just clear the net between zero and 19 degrees are essentially constant, while the maximum vertical angle line increases as the shot is hit cross-court.

Figure 8.8 subtracts the clearing the net angle from the just hitting the baseline angle, giving the vertical angular acceptance as a function of horizontal angle for three different initial ball speeds, 45, 65, and 90 miles per hour. Several things are clear from the computer-generated data shown in this figure. It is obvious that the vertical acceptance window closes down as the ball speed increases, independent of angle. It is also obvious that you have a better chance to get a shot to go in if you hit is cross-court (19 degrees) rather than down-the line (zero degrees). Figure 8.9 displays the ratio of cross-court vertical acceptance angle to the down-the-line acceptance angle as a function of ball speed. From 45 to 70 miles/hour, the cross-court shot has 20 to 50% larger vertical acceptance, than the down-the-line shot, but if you hit the ball even harder, this ratio increases very quickly. If you hit an 80 mph groundstroke, you are more than twice as likely to get it to go in if you hit cross-court. If, in the unlikely event that you attempt to hit a 90 mph groundstroke, Figs. 8.8 and 8.9 show that you have almost no chance of getting it to land in the court if you go down-the-line. The harder you hit the ball, the more important it is to go cross-court if you want to avoid making an error.

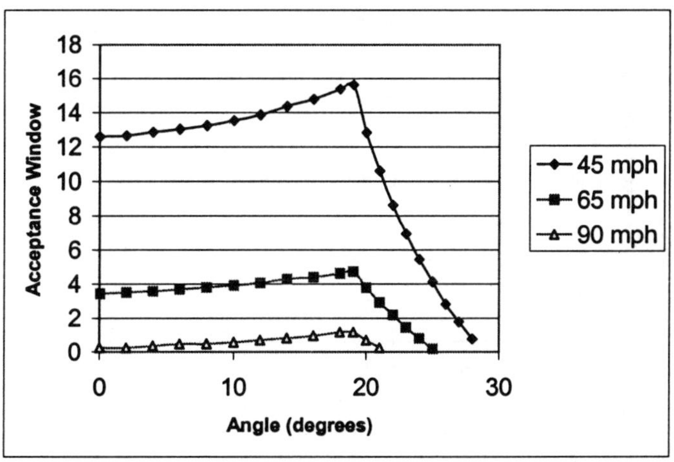


*Figure 8.8.* The vertical angular acceptance for groundstrokes of 45, 65, and 90 mph as a function of angle for shots hit from the corner.

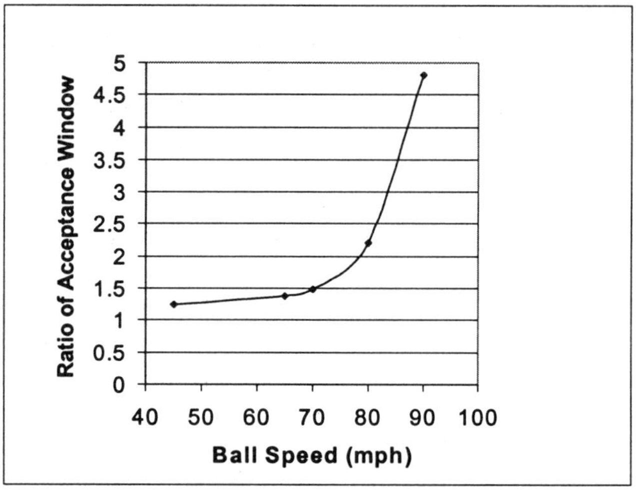


*Figure 8.9.* The ratio of the acceptance windows cross-court to down-the-line as a function of ball speed.

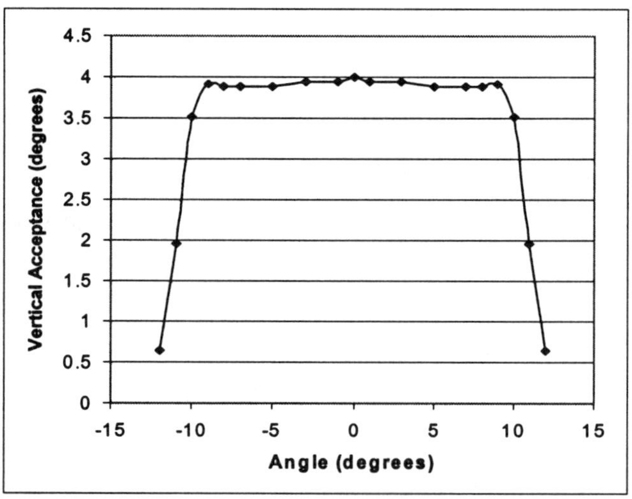


*Figure 8.10.* The vertical angular acceptance as a function of angle for a groundstroke hit from the center of the baseline

If, instead of returning a shot from the corner of the court, you find yourself right in the middle of the baseline where should you hit the ball to maximize the vertical acceptance window, down-the-middle or into the corner? Down-the-middle the net is lower, but the court is shorter. Hitting into the corner, the net is somewhat higher, but the court is a little deeper. Figure 8.10 displays the results of the computer calculation, and it shows that to first order it makes no difference in the size of the vertical window. The chances of making an error in depth are the same regardless of whether you aim for the middle or the corner. Of course, the chances of making an error in width are clearly greater when you go for the corner, but that is not what is being considered here.

## 8.5.7 Vertical Window of Acceptance Versus Spin

Spin is applied to a ball by having the racket moving upward at the impact time (topspin) or downward while swinging forward at the contact time (backspin). The spin on a ball markedly changes the ball's trajectory, so it also greatly affects the vertical angular acceptance as is shown in Fig. 8.11. Topspin produces a downward force on the ball (due to the Magnus effect, Brody et al., 2002e) that acts in addition to gravity, opening up the acceptance window. It tends to make the ball dive. Backspin produces an upward force on the ball, opposing gravity, and closes down the acceptance window. It tends to make the ball float longer than a flat (no spin) shot

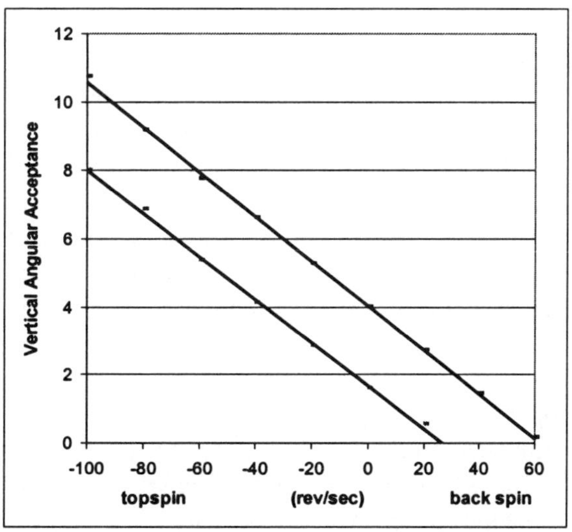


*Figure 8.11.* The vertical angular acceptance for a 65 mph and an 80 mph groundstroke hit from the baseline as a function of spin on the ball.

would. The obvious question then is: If topspin substantially opens up the acceptance window, then why isn't everyone taught from day one to hit with appreciable topspin? Why do some people hit with backspin (chop or slice) when the vertical acceptance window for such a shot is greatly reduced?

There are a number of answers to the previous questions, some dealing with the biomechanics of stroke production and some with the physics of the ball bounce and ball-racket interaction. If, when you attempt to apply spin to a ball, your ability to control the direction of the ball off your racket changes more than the increase in your acceptance window size due to the spin you have applied, then the problem is not as simple as has been indicated. There is no scientific information in the literature (and no anecdotal information in print either) concerning vertical angular control (the variation from shot to shot) and type of stroke being hit.

A tennis novice or recreational player may have absolutely no control over the direction of the ball when attempting to hit with topspin, yet have reasonable control when chopping the ball or hitting a flat (no spin) shot. A good player, who has practiced hitting topspin shots for years, may have control over that shot that is comparable to a flat shot. Why is this the case? Are not topspin and back spin symmetric in their stroking, with one having the racket moving up and the other moving down? No! This may not be the case with respect to the biomechanics of the swing, and it is absolutely different with respect to the ball-racket interaction.

The reason that there is an asymmetry between topspin groundstrokes and backspin groundstrokes is that the ball has bounced before being struck. When a ball bounces it acquires a great deal of topspin (the exact amount of spin acquired depends on the ball speed and trajectory, the court surface and the initial ball spin) and that spin interacts with the strings on the racket. To put backspin on your shot, if the ball is coming at you with topspin, you must only turn the ball's direction around, since it is already spinning the way you want it to. (A ball coming toward you with topspin is spinning in the same sense as a ball with backspin moving away from you.) To put topspin on your return, if the ball is coming at you with its own topspin, requires that you must turn around the ball's direction and in addition reverse the ball's spin direction. This requires you to swing the racket almost twice as hard as you would if you were attempting to put backspin on your shot.

When you swing a racket appreciably harder, you may not have the control you might otherwise have. In addition, to get the racket up to the required speed to hit an effective topspin shot requires much more preparation than is required for a slice or chop. It is no wonder that beginners will often start off slicing at the ball and rarely begin with topspin strokes. They can wait until the last instant and then just push at the ball with a slight downward motion of the racket, rather than get their body into the correct position, wind up, shift their weight, and do all the other things needed to hit a good topspin groundstroke.

## 8.6 CONCLUSIONS

The ultimate objective of the game is to win the last point of a match. Success in tennis often depends on performing the required skills better than your opponent. Unfortunately, we cannot control how well our opponent executes his or her shots. However, there are some things we can control.

The initial section of this chapter analyzed some of the characteristics of the equipment used in tennis. The different types of sweet spots were discussed, specifically how they relate to the technology of racket design. Two of the sweet spots (node and COP) are related to the feel a player has when the ball impacts. Their location is determined by the physical parameters of the racket, such as length, weight distribution, stiffness, etc. and for most traditional racket designs, they are located close to each other. The third sweet spot (maximum power or COR) is determined by many of the same physical parameters, and in addition, by the details of the stroke being executed and the relative ball-racket velocity. To get all three sweet spots to be located in the same place would require a racket design that is specifically engineered for a given player. Although racket technology has improved tremendously over the past 30 years, the concept of designing rackets to match individual players is presently not cost effective. The next frontier in tennis technology will be to match the specific player and game style characteristics to the most appropriate equipment.

In the second section of the chapter we analyzed the types of shots a player can select. Several variables come into play when considering shot selection. Player position on the court, goal of the shot, racket head speed, impact height, racket face angle at contact, swing pattern (spin) all have an impact on the final outcome of the shot. From the results obtained here, it is clear that by reducing struck ball speed, a player can increase the window of acceptance for those shots. This will reduce the number of unforced errors made. On the other hand, reducing the ball speed allows the opponent to hit more winners and also to make fewer errors. Each individual player must select how hard to hit the ball to optimize the chances of winning the point. While computer predictions can guide the player, the final decisions must be made by trial and error.

Future research should focus on vertical angular control (the variation from shot to shot) and the type of stroke being hit. In addition, based on the information presented in this chapter, it is quite clear that technique and tactics go hand in hand. Neither should be taught or coached in isolation. Coordinating positioning on the court, swing patterns and use of spin with proper placement, height over the net and depth of shot, shows us the delicate balance between technique and tactics in the game of tennis.

## 8.7 REFERENCES

Brody, H., 1979, Physics of the Tennis Racket, *American Journal of Physics*, **47**: 482-487.

Brody, H., 1985, The Moment of Inertia of a Tennis Racket, *The Physics Teacher*, **23**: 213-216.

Brody, H., 1987, *Tennis Science for Tennis Players*, University of Pennsylvania Press, Philadelphia, PA, pp. 23-25.

Brody, H., 1989, More Bounce to the Ounce, *TennisPro,* **2**: 34.

Brody, H., 1995, How Would a Physicist Design a Tennis Racket? *Physics Today*. **48**: 26-31.

Brody, H., 2002a, Myths and Misconceptions in Tennis, *TennisPro* **11**: 10-11.

Brody, H, Cross, R, Lindsey, C, 2002b, *The Physics and Technology of Tennis,* Racquet Tech Publishing, an imprint of the USRSA, Solana Beach, CA, pg. 78.

Ibid 2002c, pg 88.

Ibid 2002d, pg 244.

Ibid 2002e, pg 384.

Hatze, H., 1994a, Impact probability distribution, sweet spots, and the concept of an effective power region in tennis rackets, *J. Applied Biomech*. **10**: 43-50.

Hatze, H., 1994b, The relationship between coefficient of restitution and energy loss in tennis rackets, *J. Applied Biomech*. **10**: 124.

ITF 2002 Rules of Tennis, *International Tennis Federation,* Roehampton, London, U.K.

# III

# BASEBALL

# Chapter 9

# Biomechanics of Pitching

Nigel Zheng[1], Glenn S. Fleisig[2], Steve Barrentine[3], and James R. Andrews[4]

[1]*American Sports Medicine Institute, 1313 13th Street South Birmingham, AL 35205, PH: (205) 918-2118, FX: (205) 918-0800, EM: nigelz@asmi.org*

[2]*American Sports Medicine Institute, 1313 13th Street South Birmingham, AL 35205, PH: (205) 918-2139, FX: (205) 918-0800, EM: glennf@asmi.org*

[3]*American Sports Medicine Institute, 1313 13th Street South Birmingham, AL 35205, PH: (205) 918-2119, FX: (205) 918-0800*

[4]American Sports Medicine Institute, 1313 13th Street South Birmingham, AL 35205, PH: (205) 918-0000, FX: (205) 918-0800

## 9.1 INTRODUCTION

### 9.1.1 Baseball Pitching

Baseball pitching is one of the most demanding activities in sports on the human body. In an activity where maximum speeds have been measured at 100 mph, the demand on the throwing arm is great, especially at the shoulder and elbow. While maximizing the speed of the ball is not the ultimate goal of pitching, it often improves the chances for getting the hitter out. A good fastball is typically the first pitch a young pitcher will learn to throw, followed by change-up, curveball and slider (Andrews and Fleisig, 1996).

Pitching plays an important role in the success of a baseball team. It has been estimated that anywhere from 60 to 75% of the outcome of any given baseball game is determined by the pitching (Kindall, 1993). A pitcher with good mechanics has the same body movement for the same type of pitch. Pitching requires such demands on the body that the osseous adaptation occurs to allow greater external rotation and less internal rotation of the shoulder (Crockett HC, 2002). Such adaptation results from playing many years of baseball, from youth baseball, to high school, college and professional.

### 9.1.2 Sports Medicine

Injuries in baseball have been one of the driving forces for many advancements in sports medicine. Injuries to the throwing arm of a baseball pitcher are common, including a tear of the rotator cuff of the shoulder. (Altchek and Dines, 1995; Fleisig et al., 1995a; Miniaci et al., 2002; Tibone et al., 1986). A tear of the ulnar collateral ligament (UCL) is also one of the more common injuries of baseball pitchers. The surgical reconstruction of torn UCLs began in the 1970s (Andrews, 1985; Indelicato et al., 1979; Jobe et al., 1986; Jobe and Nuber, 1986). Related to this, there was a need for a better understanding of pitching biomechanics. However, information on pitching biomechanics was limited due to the rapid motions involved and the lack of adequate methods for quantification. Recent advances in high-speed video and computerized motion analysis has enabled biomechanical studies of pitching to be conducted (Zheng and Barrentine, 2000). Over the past several years, the understanding of injuries in baseball pitching has been significantly improved (Dillman et al., 1993; Fleisig et al., 1995a; Fleisig et al., 1996c; Fleisig et al., 1999).

### 9.1.3 Overview of Chapter

This chapter is intended for bioengineers, orthopaedic surgeons, physical therapists, athletic trainers, and advanced clinicians working in this area. First the equipment for data collection will be briefly discussed, followed by data collection for loads, body movement and major muscle activities. Data reduction and data analysis will include reviews of the engineering principles involved, and equations used for kinematic and kinetic analysis. Over 700 throwing athletes, which include over 500 baseball pitchers (range in age from 10 to 36 years) have been analyzed at the American Sports Medicine Institute (ASMI). Results from these pitchers will be used in a discussion of the biomechanical differences among four basic pitches and among four different skill levels (from youth to professional). In addition, these results will be used to discuss the relationship between pitching kinematics, ball velocity and potential mechanisms of injury. Through this chapter, readers will gain an understanding of the biomechanics of baseball pitching, including the methods used in the analysis of baseball pitching, the interpretation of biomechanical data based on baseball pitching, and the understanding of common injury mechanisms in baseball pitching.

## 9.2 EQUIPMENT AND DATA COLLECTION

Three types of data can be collected for baseball pitching: motion data (how the body moves during pitching), electromyography ("EMG", how the major muscles work) and external forces (interactions between the body and the pitching mound or with the ball during pitching). We are going to discuss equipment and data collection for each of these.

### 9.2.1 Motion Data Collection

#### 9.2.1.1 Equipment for Motion Data Collection

Dynamic features like fast shoulder internal rotation velocity make photogrammetric reconstruction the favorite method of collecting motion data for baseball pitching. Photogrammetric reconstruction is a process to recreate a three-dimensional object from two or more two-dimensional projections of the object, which can be images created by cameras.

In order to uniquely track the location and orientation of a rigid body in space, at least three non-collinear markers on the body are identified on images from two or more cameras at the same time. Although the human body is not rigid, the human musculoskeletal system can be decomposed into a series of jointed segments, which can be approximated as rigid bodies. More markers on each segment require more cameras in order to see each marker by at least two cameras at any time. By knowing the locations and orientations of each rigid body, the six motions (three translations and three rotations) can be determined for the two adjacent jointed segments. Fortunately we are only interested in the major joint motions, which allow us to reduce the number of markers.

Manual digitization is a tedious process of extracting coordinates from the images recorded by the cameras. For some cases, such as during competition, manual digitization is the only choice available since no active or passive makers are attached to the subject. For example, biomechanists videotaped baseball pitching during the 1996 Atlanta Olympic Games (Escamilla et al., 2001). Two synchronized high-speed video cameras (Peak Performance Technologies, Inc., Englewood, CO) were used to collect data at a rate of 120 Hz. The limited number of cameras and digitizing through clothing made digitizing a more difficult task.

Markers attached to pertinent landmarks make digitizing much easier, although these markers are not located on the joint center. They also make automatic digitizing and real-time digitizing possible since the whole body is represented by a series of small markers. Two types of markers are available: active and passive. Active markers require an external energy supply. Light-Emitting Diodes (LED) that emit infrared light are usually used as active markers. Active markers are turned on in sequential manner, allowing the cameras to identify the momentary spatial location of the one marker that is on at any instant. Active marker systems require that wires be attached to each marker, which makes measurements cumbersome and difficult, especially for baseball pitching. Long and loose wires are needed to allow stretching of the pitching arm but they fly and slap against the subject. Passive retro-reflective markers reflect light in the direction from which it comes. This is the same principle demonstrated by traffic signs, which appear to be lit up when illuminated by the headlights of approaching vehicles. The cameras simultaneously pick information from the passive markers that appear on the screen of a video monitor as bright spots. At ASMI, six electronically synchronized 240 Hz charged coupled device cameras are used to transmit pixel images of the reflective markers attached onto the baseball pitcher directly into a video processor without being recorded onto video (Motion Analysis Corporation, Santa Rosa, CA).

After the projections of the markers on the cameras have been identified, the photogrammetric approach can be applied to calculate the 3-D spatial coordinates of each marker. The most common approach used to calculate the 3-D coordinates of each marker is based on direct linear transformation (DLT) (Abdel-Aziz and Karara, 1971; Shapiro, 1978; Walton, 1981). The determination of the parameters necessary for the implementation of the DLT algorithm requires a calibration procedure, which can be performed prior to or after data has been collected. A cubic frame is often used with numerous reflective markers attached. Since the global coordinates of these markers are known, data collected from the cubic frame is used to calculate the camera positions. Because the equations are linear, a minimum of 6 markers is required to determine the camera parameters. More markers are often used to provide a degree of redundancy to the process of parameter determination. A secondary step in the calibration process is the wand calibration. A 1 m stick or wand with 2 markers attached at the ends and one attached off the center is used for the procedure. The objective of wand calibration is to fill the entire capture volume with images of the wand in order to expand the calibrated volume defined by the cube. For baseball pitching, the capture volume should cover the areas where the pitcher's hands and feet would reach during pitching. With wand calibration, the initial camera positions calculated during cube calibration are refined. The position of each camera in space (x, y, z) and its orientation with respect to the global X, Y, Z axes are determined. Additional parameters are also determined including the position of the camera image sensor with respect to the camera lens, the effective focal length of the lens and a measure of the geometric lens distortion of the lens. These camera internal parameters are difficult to measure but necessary for photogrammetric reconstruction. The cubic frame and wand are shown in Fig. 9.1.

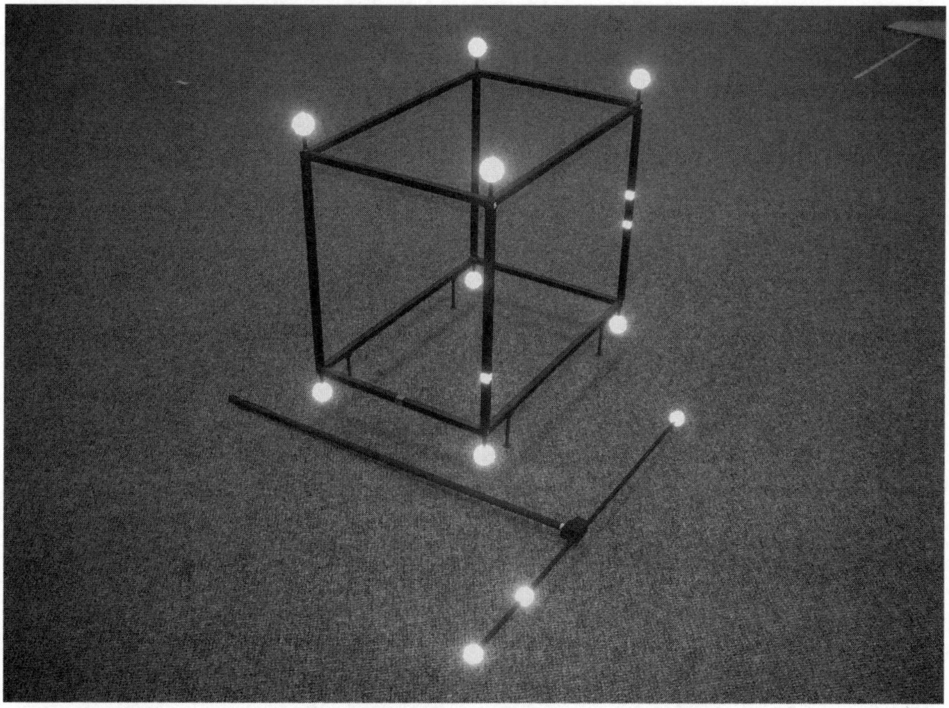

*Figure 9.1.* A cubic frame and wand used for calibration.

### 9.2.1.2 Marker Set-up for Baseball Pitching

A rigid body has six degrees-of-freedom: three translations (X, Y and Z) and three rotations. When one point of the body is fixed, it has only three degrees-of-freedom (three rotations). The upper arm can be considered as such an object when the humeral head is fixed relative to the trunk. A rigid body with two points of the body fixed has only one degree-of-freedom (the rotation about the axis of these two points). The forearm can be considered as this type of rigid body when the axis of the elbow is fixed relative to the upper arm. If we put one marker on each shoulder at the tip of the acromion and one on each hip at the greater trochanter, we are able to determine the position of the trunk in three-dimensional space. After knowing the positions of the shoulders and hips, two rotations at the shoulder or the hip can be determined by knowing the position of the elbow or knee. A third point is needed to determine the rotation about the long axis of the upper arm or thigh. Since both the elbow and knee are considered as hinge joints, the rotation about the upper arm axis or thigh axis can be determined by knowing the position of the wrist or ankle if three markers on these joints (shoulder, elbow and wrist or hip, knee and ankle) are not on the same line.

*Figure 9.2.* A marker set-up for baseball pitching.

A marker set consisting of fourteen reflective markers has been used in certain throwing analysis (Fleisig et al., 1996d; Fleisig et al., 1999). These markers were attached bilaterally to the distal end of the mid-toe, lateral malleolus, lateral femoral epicondyle, greater trochanter, lateral tip of the acromion, lateral humeral epicondyle and wrist. In earlier ASMI studies, a reflective marker was placed on the ulnar styloid process of non-pitching wrist and a reflective band approximately one cm wide was placed around the pitching wrist. More recently, with improved camera image resolution, two reflective markers have been used to replace the reflective band around the pitching wrist (Fig. 9.2).

### 9.2.2 Equipment and EMG Data Collection

There are several devices available for EMG data collection, but not many are suitable for dynamic motions like baseball pitching. The range of motion and rapid acceleration of upper extremity joints make the task of keeping surface or wire electrodes attached to the throwing arm troublesome. As a result, there are few publications on muscle activities during baseball pitching. Furthermore, most of the studies on muscle activity during pitching have focused on the temporal aspects of activity, such as onset and offset of activity, and not the magnitude of muscle activity (DiGiovine, 1992; Jobe et al., 1983; Jobe et al., 1984; Watkins et al., 1989; Werner et al., 1993; Yamanouchi, 1998).

The procedure for collecting EMG data during baseball pitching is similar to that used for other activities. The various uses of surface EMG in the field of biomechanics were reviewed by De Luca (De Luca, 1997). As previously mentioned, a challenge with collecting EMG data is the placement of the electrodes, either surface or fine wire. The dynamic movement of the pitching arm creates difficulty in keeping the electrode on the belly of the muscle throughout the motion and therefore contributes to the signal artifact. The EMG signal is unstable if a wrapping tape is put around the electrode due to varying pressure from the tape during the pitching motion. Collecting EMG data is relatively easier for trunk and leg muscles.

### 9.2.3 Equipment and Force Data Collection

Two types of external forces are applied in pitching: ground reaction forces from the pitching mound and resistive forces from the ball. MacWilliams and his colleagues studied ground reaction forces during baseball pitching using force platforms (MacWilliams et al., 1998). No study has been conducted directly measuring the resistance forces from the ball.

## 9.3 BIOMEDICAL ENGINEERING PRINCIPLES

Before we analyze collected data, let us review the biomedical engineering principles we are going to use. First we want to know how the body moves during pitching. Second we would like to find out how much force is applied to a joint. Third we need to understand how strong these joints are and why injuries occur. With this understanding, we would like to know how to fix injured joints or tissues and get the pitcher back to competition. The goal is to prevent injuries during pitching. Kinematic and kinetic analyzes are the two most common biomechanical approaches used to analyze pitching.

### 9.3.1 Kinematics

Kinematics is the study of body movement with an emphasis on the analysis and description of 'how' the body moves rather than what causes the movement. Here we would like to know how each body segment moves during baseball pitching. We also would like to know their relative motions, which define the motion of the joint connecting two adjacent segments.

In determining kinematic parameters, the initial step is to define a spatial reference system. In the ASMI studies the global reference system is defined by a vertical axis, Z, a horizontal axis in the direction of pitching (to home plate), X, and a horizontal axis perpendicular to the X direction, Y (parallel to the line connecting first and third base). Fig. 9.3 depicts this convention. The origin of the coordinate system is based on the calibration cube previously discussed. Let's call this coordinate system the global coordinate system since it is still relative to the lab or baseball field.

Second, a local reference or coordinate system is defined for each body segment. For the trunk, the X axis is defined from the leading shoulder to the throwing shoulder, the Z axis is defined as pointing to the superior, and the Y axis is defined as pointing to the anterior for the right-handed pitcher and to the posterior for the left-handed pitcher (Fig. 9.4).

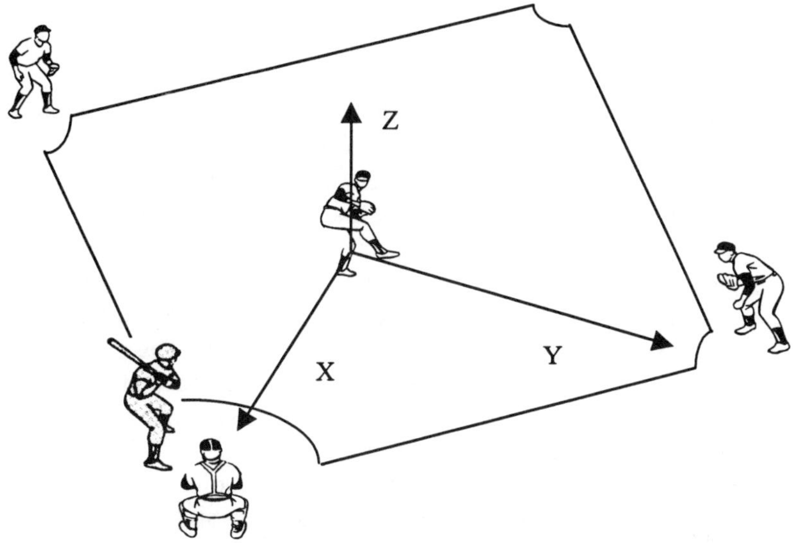


*Figure 9.3.* A global reference system for baseball pitching.

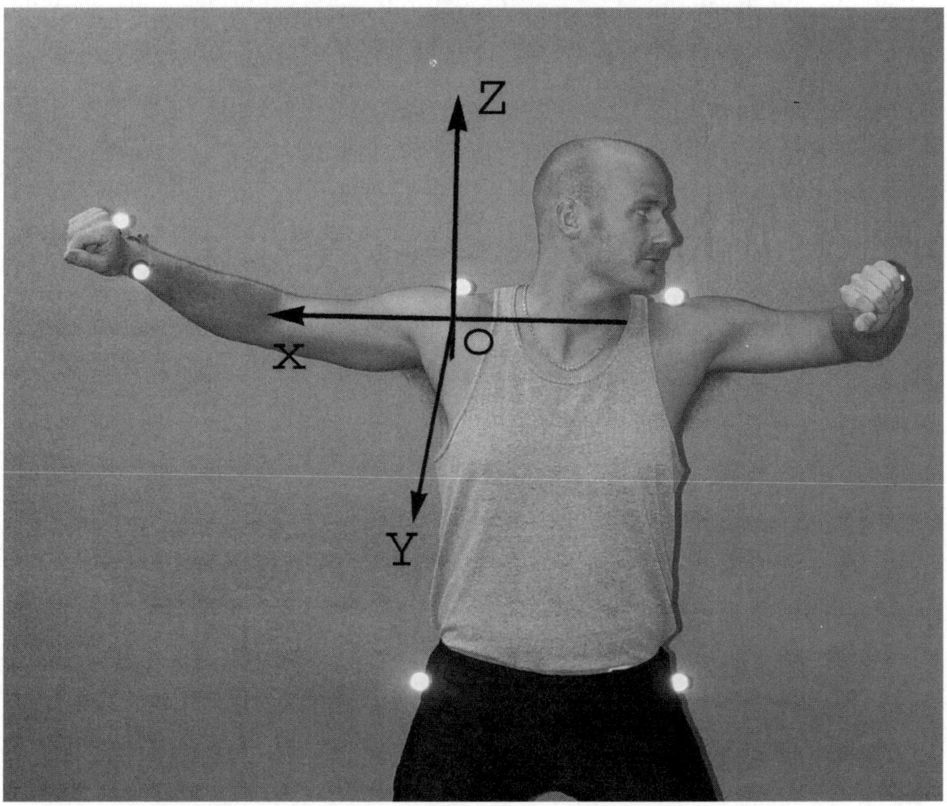


*Figure 9.4.* A local reference system at the throwing shoulder.

#### 9.3.1.1 The Shoulder

The shoulder has four separate articulations – the glenohumeral joint, the sternoclavicular joint, the acromiocalvicular joint, and the coracoclavicular joint. The glenohumeral joint is the articulation between the humerus head and the glenoid fossa of the scapula, and is typically considered as the major shoulder joint. It is the ball and socket joint, which has three rotations and no translations. Clinically, shoulder motions are defined as flexion / extension, abduction / adduction, horizontal abduction /adduction, and external rotation / internal rotation. The first three rotations are not independent, as only two of them are needed to determine the position of the humerus. External/internal rotation is needed to determine the rotation of the humerus about its long axis.

To avoid any confusion caused by the order of rotations, shoulder motion is defined as follows (Fig. 9.5). Shoulder abduction is defined as the angle between the humerus and the inferior direction of the trunk (represented by the line connecting the middle point of the two shoulder markers and the middle point of the two hip markers) in the trunk's frontal plane. Shoulder horizontal abduction is defined as the angle between the humerus and a line connecting the two shoulder markers in the trunk's transverse plane. Third, shoulder external rotation is defined as the rotation of the upper arm about its own long axis.

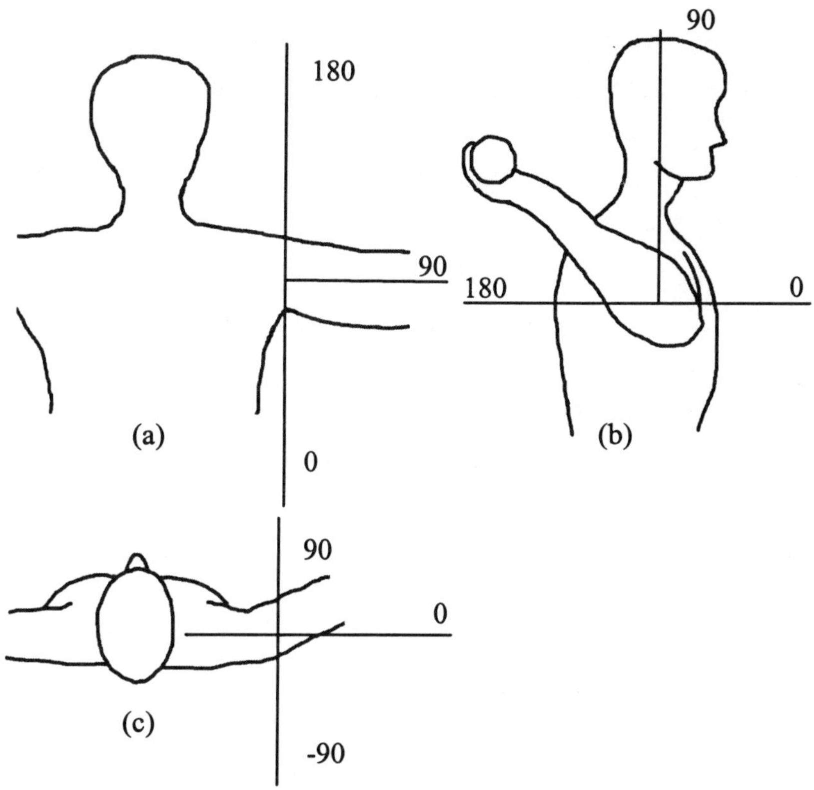


*Figure 9.5.* Motion definitions at the throwing shoulder (in degrees): (a) shoulder abduction, (b) shoulder external rotation and (c) shoulder horizontal adduction.

Thus, the local coordinate system at the shoulder is defined as follows:

X axis: $\vec{I}_{sx} = (\vec{V}_{sh-t} - \vec{V}_{sh-l}) / \left|\vec{V}_{sh-t} - \vec{V}_{sh-l}\right|$ (9.1)

Y axis: $\vec{I}_{sy} = \vec{I}_{trunk} \times \vec{I}_x / \left|\vec{I}_{trunk} \times \vec{I}_x\right|$ (9.2)

Z axis: $\vec{I}_{sz} = \vec{I}_{sx} \times \vec{I}_{sy}$ (9.3)

where $\vec{V}_{sh-t}$ is the vector of the throwing shoulder in the global system, $\vec{V}_{sh-l}$ is the vector of the leading shoulder in the global system,

$$\vec{I}_{trunk} = (\vec{V}_{sh-t} + \vec{V}_{sh-l} - \vec{V}_{hip-t} - \vec{V}_{hip-l}) / \left|\vec{V}_{sh-t} + \vec{V}_{sh-l} - \vec{V}_{hip-t} - \vec{V}_{hip-l}\right|$$

, $\vec{V}_{hip-t}$ is the vector of the hip on the throwing side, and $\vec{V}_{hip-l}$ is the vector of the hip on the leading side.

If $\vec{V}_{el-t}$ represents the vector of the throwing elbow and $\vec{V}_{w-t}$ represents the vector of the throwing wrist, $\vec{V}_{ua-t} = \vec{V}_{el-t} - \vec{V}_{sh-t}$ represents the vector of the upper arm, and $\vec{V}_{fa-t} = \vec{V}_{w-t} - \vec{V}_{el-t}$ represents the vector of the forearm. Then the shoulder angles can be determined as follows:

Shoulder abduction

$$\alpha = 180 - \cos^{-1}(\vec{V}_{ua-t} \bullet \vec{I}_{sz} / |\vec{V}_{ua-t}|) \tag{9.4}$$

Shoulder horizontal abduction

$$\beta = \begin{cases} \tan^{-1}(\vec{V}_{ua-t} \bullet \vec{I}_{sy} / \vec{V}_{ua-t} \bullet \vec{I}_{sx}) & right-handed \\ -\tan^{-1}(\vec{V}_{ua-t} \bullet \vec{I}_{sy} / \vec{V}_{ua-t} \bullet \vec{I}_{sx}) & left-handed \end{cases} \tag{9.5}$$

Shoulder external rotation

$$\gamma = \begin{cases} \tan^{-1}(\vec{V}_{fa-t} \bullet \vec{I}_{uaz} / \vec{V}_{fa-t} \bullet \vec{I}_{uay}) & right-handed \\ 180 - \tan^{-1}(\vec{V}_{fa-t} \bullet \vec{I}_{uaz} / \vec{V}_{fa-t} \bullet \vec{I}_{uay}) & left-handed \end{cases} \tag{9.6}$$

where $\vec{I}_{uay} = (\vec{I}_{trunk} \times \vec{V}_{ua-t}) / |\vec{V}_{ua-t}|$,

$\vec{I}_{uaz} = (\vec{V}_{ua-t} \times \vec{I}_{uay}) / |\vec{V}_{ua-t}|$.

### 9.3.1.2 The Elbow

The elbow includes three articulations: the humeroulnar, humeroradial, and proximal radioulnar joints. These three joints are enclosed in the same joint capsule and connected by the anterior and posterior radial collateral, and the ulnar collateral ligaments. The humeroulnar joint is considered to be the elbow joint, which is a classic hinge joint. The elbow joint permits flexion and extension (Fig. 6).

Elbow flexion angle can be determined from the upper arm and forearm vectors.

$$\theta = \cos^{-1}(\vec{V}_{ua-t} \bullet \vec{V}_{fa-t} / |\vec{V}_{ua-t}| \bullet |\vec{V}_{fa-t}|) \tag{9.7}$$

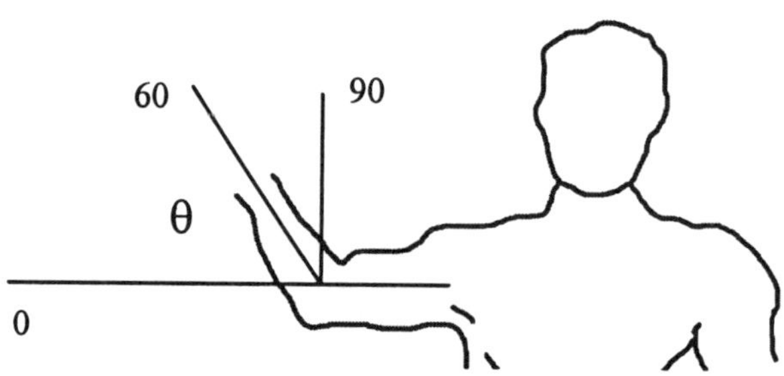


*Figure 9.6.* Motion definition at the elbow (in degrees).

#### 9.3.1.3 The Wrist and Forearm

The wrist is composed of radiocarpal and intercarpal articulations. The major wrist joint refers to the articulation of the radius with three of the proximal carpal bones, which is the condyloid joint. The movements of the wrist include flexion/extension and radial/ulnar deviation (Fig. 9.7 (a) and (b)). Pronation and supination of the forearm involve rotation of the radius around the ulnar (Fig. 9.7 (c)). Although they are not wrist movements, they have direct effect on the wrist position in the global system. In order to determine the wrist motion, extra markers are needed. Markers on the head of the third metacarpal on the dorsal aspect of the throwing wrist and on the radial and ulnar styloid processes are the minimal requirements.

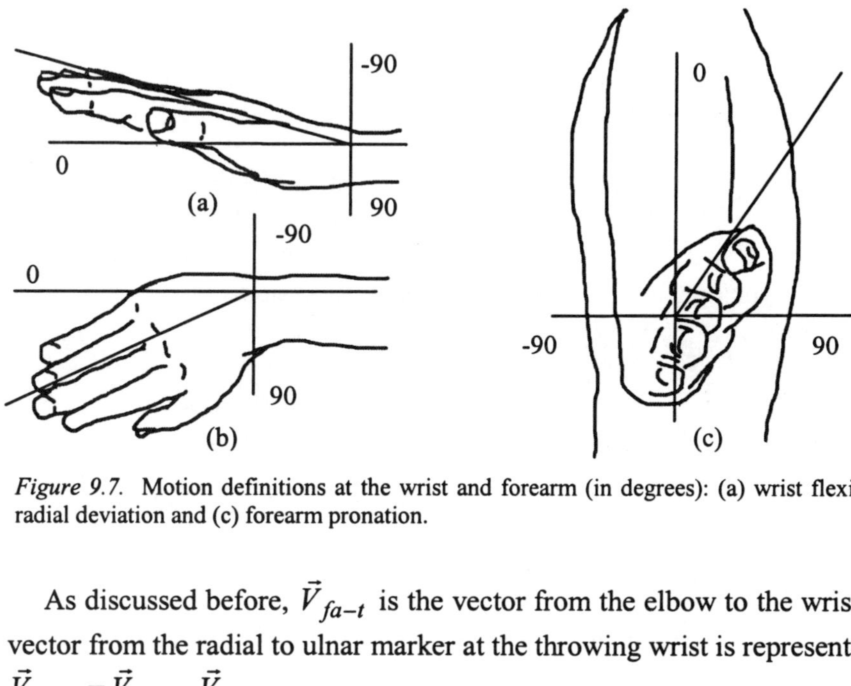


*Figure 9.7.* Motion definitions at the wrist and forearm (in degrees): (a) wrist flexion, (b) radial deviation and (c) forearm pronation.

As discussed before, $\vec{V}_{fa-t}$ is the vector from the elbow to the wrist. The vector from the radial to ulnar marker at the throwing wrist is represented as $\vec{V}_{ru-t} = \vec{V}_{u-t} - \vec{V}_{r-t}$.

The vector from the wrist to the marker on the hand is represented as

$$\vec{V}_{wh-t} = \vec{V}_{h-t} - 0.5(\vec{V}_{r-t} + \vec{V}_{u-t})$$

If the X axis ($\vec{I}_{wx}$) of the wrist coordinate system is defined as the unit vector of $\vec{V}_{ru-t}$, the Y axis ($\vec{I}_{wy}$) as the unit vector of $\vec{V}_{fa-t}$, and the Z axis ($\vec{I}_{wz}$) is the unit vector of $\vec{V}_{ru-t} \times \vec{V}_{fa-t}$. Then the wrist flexion ($\phi$) and ulnar deviation ($\varphi$) can be determined as:

$$\phi = \begin{cases} \tan^{-1}(\vec{V}_{wh-t} \bullet \vec{I}_{wz} / \vec{V}_{wh-t} \bullet \vec{I}_{wy}) & right-handed \\ -\tan^{-1}(\vec{V}_{wh-t} \bullet \vec{I}_{wz} / \vec{V}_{wh-t} \bullet \vec{I}_{wy}) & left-handed \end{cases} \tag{9.8}$$

$$\varphi = \begin{cases} \tan^{-1}(\vec{V}_{wh-t} \bullet \vec{I}_{wx} / \vec{V}_{wh-t} \bullet \vec{I}_{wy}) & right-handed \\ -\tan^{-1}(\vec{V}_{wh-t} \bullet \vec{I}_{wx} / \vec{V}_{wh-t} \bullet \vec{I}_{wy}) & left-handed \end{cases} \tag{9.9}$$

The pronation and supination of the forearm can be determined from vector of the upper arm, vector of the forearm and the vector from the radial to the ulnar markers at the wrist.

$$\psi = \begin{cases} 90 - \cos^{-1}((\vec{V}_{ua-t} \times \vec{V}_{fa-t}) \bullet \vec{I}_{wx} / |\vec{V}_{ua-t} \times \vec{V}_{fa-t}|) & right-handed \\ \cos^{-1}((\vec{V}_{ua-t} \times \vec{V}_{fa-t}) \bullet \vec{I}_{wx} / |\vec{V}_{ua-t} \times \vec{V}_{fa-t}|) - 90 & left-handed \end{cases} \tag{9.10}$$

### 9.3.1.4 The Spine and Trunk

Coaches frequently use the position of the trunk at certain instances of the pitching motion for instructional purposes. The trunk vector discussed above can be used to determine the trunk forward ($\xi$) and side tilt ($\zeta$). They describe the trunk position in the global coordinate system.

$$\xi = \tan^{-1}(\vec{I}_{trunk} \bullet \vec{I}_{gx} / \vec{I}_{trunk} \bullet \vec{I}_{gz}) \tag{9.11}$$

$$\zeta = \begin{cases} \tan^{-1}(\vec{I}_{trunk} \bullet \vec{I}_{gy} / \vec{I}_{trunk} \bullet \vec{I}_{gz}) & right-handed \\ -\tan^{-1}(\vec{I}_{trunk} \bullet \vec{I}_{gx} / \vec{I}_{trunk} \bullet \vec{I}_{gz}) & left-handed \end{cases} \tag{9.12}$$

Typically, the motion of the spine during baseball pitching is not analyzed relative to a specific level of the vertebral column (thoracic, lumbar, etc.). Instead, overall spine motions based on the shoulder markers and hip markers are determined. A local reference system is developed at the pelvis with the X - axis pointing from leading hip to the throwing hip.

$$\vec{I}_{px} = (\vec{V}_{hip-t} - \vec{V}_{hip-l}) / |(\vec{V}_{hip-t} - \vec{V}_{hip-l})|$$

$$\vec{I}_{py} = \vec{I}_{trunk} \times \vec{I}_{px}$$

$$\vec{I}_{pz} = \vec{I}_{px} \times \vec{I}_{py}$$

The orientation of the X - axis of the local reference system at the shoulder discussed before can be used to determine the spine lateral bending ($\delta$) and axial rotation ($\omega$).

$$\delta = \tan^{-1}(\vec{I}_{sx} \bullet \vec{I}_{pz} / \vec{I}_{sx} \bullet \vec{I}_{px}) \tag{9.13}$$

$$\omega = \begin{cases} \tan^{-1}(\vec{I}_{sx} \bullet \vec{I}_{py} / \vec{I}_{sx} \bullet \vec{I}_{px}) & right-handed \\ -\tan^{-1}(\vec{I}_{sx} \bullet \vec{I}_{py} / \vec{I}_{sx} \bullet \vec{I}_{px}) & left-handed \end{cases} \tag{9.14}$$

#### 9.3.1.5 Positions, Velocities and Accelerations

We have discussed the equations to calculate the major joint angles during baseball pitching. These angles are relative motion of one body segment to another. They describe the position of joints in motion or stationary, such as full extension or full flexion. The human body can be considered as a kinetic chain. The angle of one joint will change the position of adjacent segments and other joints in the chain. For example, more abduction of the shoulder will lead to higher position of the elbow. More horizontal abduction of the shoulder will lead to the elbow position more toward to the anterior and less to the side. Coaches will often instruct the pitcher to raise the elbow higher and bring the elbow to the front, rather than telling the pitcher to have more abduction or horizontal abduction of the shoulder. Other instructional cues may be the pitcher's eyes, head and shoulders as references, such as 'high as your eyes,' 'as wide as your shoulders,' etc.

After determining the positions of body segments and angles of joints, we need to calculate linear and angular velocities and accelerations. Velocity is change in position with respect to time. With a camera sampling frequency of 240 Hz, the time between sequential images is 0.004 seconds. The velocity between two sequential data points will be the distance moved divided by 0.004 seconds. Any error in position data will lead to exponential errors in velocity and acceleration data derived from the position data, even with filtered and smoothed position data. Thus, a five-point derivative is used to calculate the velocity and acceleration to avoid this problem and reduce the amount of error. The velocity of location $x$ at time frame $i$ is determined based on two data points before ($i-1, i-2$) and two data points after ($i+1, i+2$) the position data $\vec{p}_x$:

$$\vec{v}_x[i] = \frac{1}{12}\left(-\vec{p}_x[i+2] + 8 * \vec{p}_x[i+1] - 8 * \vec{p}_x[i-1] + \vec{p}_x[i-2]\right) * f \tag{9.15}$$

where $f$ stands for the sampling frequency. For the same reason the acceleration of variable $x$ can be determined from its velocity data $\vec{v}_x$.

$$\vec{a}_x[i] = \frac{1}{12}\left(-\vec{v}_x[i+2] + 8 * \vec{v}_x[i+1] - 8 * \vec{v}_x[i-1] + \vec{v}_x[i-2]\right) * f \tag{9.16}$$

## 9.3.2 Kinetics

Kinetics is the study of forces and moments of force applied to a body. Here we are interested in the forces and moments of force applied to the elbow, shoulder and other joints during baseball pitching. If we have a full description of body movements (positions, velocities and accelerations), accurate anthropometric measurements, including body segment's mass, moment inertia, location of the mass center and external forces applied to the body, the forces and moments of forces applied to the joint can be calculated. During baseball pitching, external forces include gravitational force, ground reaction force and the ball's resistive force. An inverse dynamic model is used to perform such a calculation.

Let's assume that the hand and ball are one body before ball release. Fig. 9.8 shows the free-body diagram. Resultant force ($\vec{R}_{w-h}$) and moment of force ($\vec{M}_{w-h}$) acting at the wrist to the hand can be determined using the Newton's second law.

$$\sum \vec{R}_i + m\vec{g} = m\ \vec{a} \tag{9.17}$$

$$\sum (\vec{M}_i + \vec{R}_i \times \vec{d}_i) = I\ \vec{\alpha} \tag{9.18}$$

where $m$ is the mass of a body segment, $\vec{g}$ is the acceleration due to the gravity, $\vec{a}$ is the acceleration vector of the segment's mass center, $I$ is the moment inertia, $\vec{d}_i$ is the vector of the moment arm of force $\vec{R}_i$ about the mass center, and $\vec{\alpha}$ is the angular acceleration vector.

For the hand-ball free-body diagram, from Eqs. 9.17 and 9.18, we have:

$$\vec{R}_{w-h} = m_{h+b}(\vec{a}_{h+b} - \vec{g}) \tag{9.19}$$

$$\vec{M}_{w-h} = I_{h+b}\vec{\alpha}_{h+b} - \vec{R}_{w-h} \times \vec{d}_{Rw-h} \tag{9.20}$$

where $m_{h+b}$ is the mass of the hand and ball before ball release and the mass of the hand after ball release.

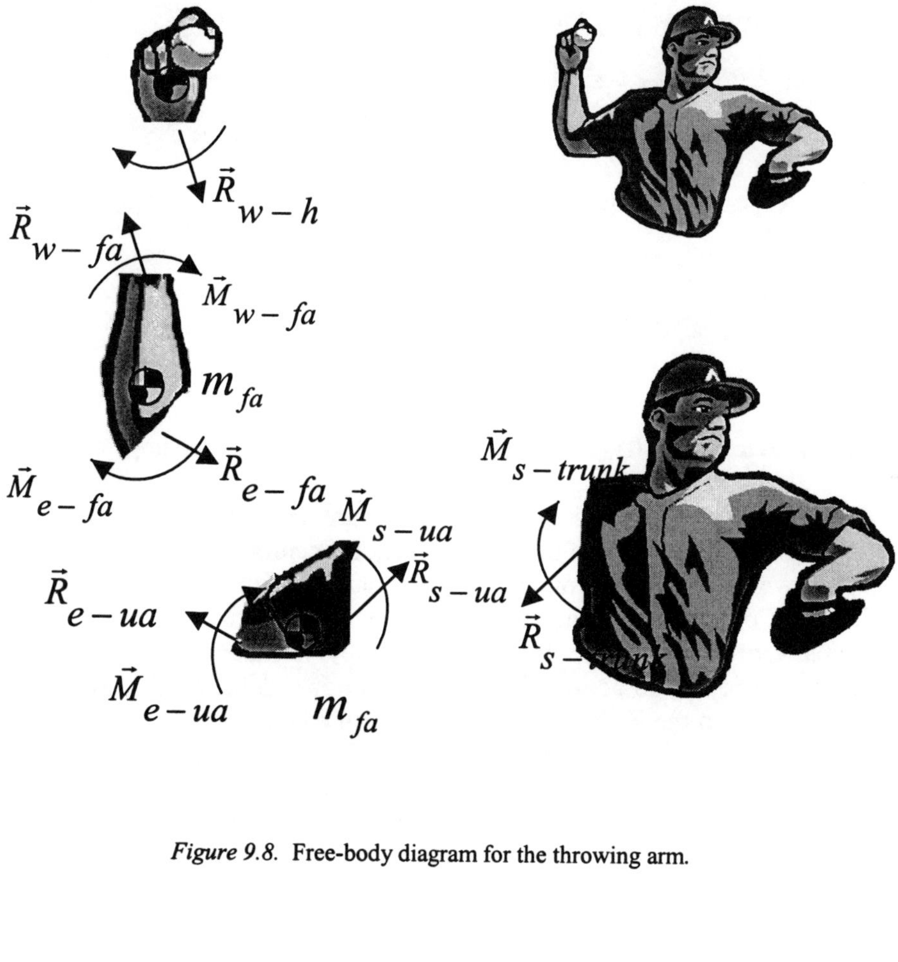


*Figure 9.8.* Free-body diagram for the throwing arm.

In accordance with Newton's third law, there are equal and opposite forces and moments of forces acting at the wrist to the forearm ($\vec{R}_{w-fa} = -\vec{R}_{w-h}$, $\vec{M}_{w-fa} = -\vec{M}_{w-h}$). Again using Newton's second law, the resultant force ($\vec{R}_{e-fa}$) and moment of force ($\vec{M}_{e-fa}$) acting at the elbow to the forearm can be determined with known $\vec{R}_{w-fa}$ and $\vec{M}_{w-fa}$.

$$\vec{R}_{e-fa} = m_{fa}(\vec{a}_{fa} - \vec{g}) - \vec{R}_{w-fa} \tag{9.21}$$

$$\vec{M}_{e-fa} = I_{fa}\vec{\alpha}_{fa} - \vec{R}_{w-fa} \times \vec{d}_{Rw-fa} - \vec{R}_{e-fa} \times \vec{d}_{\text{Re}-fa} - \vec{M}_{w-fa} \tag{9.22}$$

For the same reason we can determine the resultant force ($\vec{R}_{s-ua}$) and moment of force ($\vec{M}_{s-ua}$) at the shoulder to the upper arm.

$$\vec{R}_{s-ua} = m_{ua}(\vec{a}_{ua} - g) - \vec{R}_{e-ua} \tag{9.23}$$

$$\vec{M}_{s-ua} = I_{ua}\vec{\alpha}_{ua} - \vec{R}_{e-ua} \times \vec{d}_{\mathrm{Re}-ua} - \vec{R}_{s-ua} \times \vec{d}_{Rs-ua} - \vec{M}_{e-ua} \tag{9.24}$$

The force and torque at the shoulder applied to the trunk will be determined:

$$\vec{R}_{s-trunk} = -\vec{R}_{s-ua} \tag{9.25}$$

$$\vec{M}_{s-trunk} = -\vec{M}_{s-ua} \tag{9.26}$$

## 9.3.3 Data Filtering, Smoothing and Modification

During the data collection process, there are several potential sources of artifact noise. The position of a marker or joint center is reconstructed using several two-dimensional images. Noise could be from optoelectric devices, calibration processes, or human error in manual digitizing. Therefore it is essential that raw data be filtered and smoothed. Generally human motion data are considered to be low frequency. Positional data are digitally filtered independently in the X, Y and Z direction with a Butterworth low-pass filter. Qualitative evaluation of displacement, velocity and acceleration data during baseball pitching indicated that a ratio of 12 for the sampling frequency over low-pass cut-off frequency is effective at rejecting noise and passing data (Barrentine et al., 1998; Fleisig et al., 1996d; Fleisig et al., 1999). For a sampling frequency of 240 Hz, this is equivalent to a second-order, low-pass cut-off frequency of 20 Hz. The data are then passed through the filter a second time in reverse order to eliminate phase distortion. This eventually created a fourth-order, zero-phase shift, double-pass filter with a new cut-off frequency of 16 Hz. Sixteen hertz is greater than the cut-off frequency commonly used for other human motions such as gait analysis.

For some joints and segments, the locations of markers are used to directly calculate the angles and positions. For the throwing shoulder, elbow and wrist, modifications from digitized locations of their markers are needed to estimate their joint centers. The modifications can be done with additional markers aligned to the joint center from sagittal, frontal and transverse views. The subject is filmed with these extra markers in a static position, and the extra markers are then removed for the baseball pitching trials. Data from the extra markers are used to determine the relationship of the other markers and the joint center.

Instead of establishing the relationship between a marker and joint center for each subject, Dillman et al. (1993) developed equations to calculate the position of joint centers from the markers based on subject's humerus ($l_{humerus}$) and radius lengths ($l_{radius}$) (Dillman et al., 1993). As described by Fleisig et al. (Fleisig et al., 1996d), the procedure is based on manually digitized data determined from three subjects using a Peak Performance motion measurement system. The data was collected with a two-camera system recorded at 500 Hz (Kodak Ektapro 1000 system, Eastman Kodak Company, San Diego, CA). Reflective markers with a radius of 0.019 meters were placed on the shoulder at the lateral tip of the acromion, and on the elbow at the lateral humeral epicondyle. The distance between the shoulder joint center and the shoulder marker ($\left|\vec{S}_{m-jc}\right|$) was ($0.019 + l_{humerus}/6.05$). The direction of each component for this vector was 0.413 in the anterior direction, 0.903 in the inferior direction, and 0.121 in the lateral direction. According to Eqs. 9.1 to 9.3, $\vec{I}_{sx}$ is the lateral component of the unit vector of the throwing shoulder, $\vec{I}_{sy}$ is the anterior component of the unit vector of a right throwing shoulder and the posterior component of a left throwing shoulder, and $\vec{I}_{sz}$ is the superior component of a unit vector of the throwing shoulder. The modification vector (from the marker to the joint center) can be expressed in the global reference system as:

$$\vec{S}_{m-jc} = \begin{cases} \left|\vec{S}_{m-jc}\right| \begin{bmatrix} \vec{I}_{sx} & \vec{I}_{sy} & \vec{I}_{sz} \end{bmatrix} \begin{bmatrix} 0.121 \\ 0.413 \\ -0.903 \end{bmatrix} & right-handed \\ \left|\vec{S}_{m-jc}\right| \begin{bmatrix} \vec{I}_{sx} & \vec{I}_{sy} & \vec{I}_{sz} \end{bmatrix} \begin{bmatrix} 0.121 \\ -0.413 \\ -0.903 \end{bmatrix} & left-handed \end{cases} \tag{9.27}$$

The distance between the elbow joint center and the marker ($\left|\vec{E}_{m-jc}\right|$) was ($0.019 + l_{radius}/8.70$). The direction of this vector was 0.800 in the medial direction, 0.521 in the anterior direction, and 0.296 in the distal direction. The medial, distal and anterior unit vectors at the elbow can be defined as:

$$\vec{I}_{e-m} = \vec{V}_{fa-t} \times \vec{V}_{ua-t} \Big/ \left|\vec{V}_{fa-t} \times \vec{V}_{ua-t}\right| \tag{9.28}$$

$$\vec{I}_{e-d} = \vec{V}_{fa-t} \Big/ \left|\vec{V}_{fa-t}\right| \tag{9.29}$$

$$\vec{I}_{e-a} = \vec{I}_{e-d} \times \vec{I}_{e-m} \tag{9.30}$$

The modification vector (from the elbow marker to the joint center) can be expressed in the global reference system as:

$$\vec{E}_{m-jc} = \begin{cases} \left|\vec{E}_{m-jc}\right| \begin{bmatrix} \vec{I}_{e-m} & \vec{I}_{e-a} & \vec{I}_{e-d} \end{bmatrix} \begin{bmatrix} 0.800 \\ 0.521 \\ 0.296 \end{bmatrix} & right-handed \\ \left|\vec{E}_{m-jc}\right| \begin{bmatrix} \vec{I}_{e-m} & \vec{I}_{e-a} & \vec{I}_{e-d} \end{bmatrix} \begin{bmatrix} -0.800 \\ 0.521 \\ 0.296 \end{bmatrix} & left-handed \end{cases} \tag{9.31}$$

## 9.4 MOTIONS OF JOINTS DURING BASEBALL PITCHING

### 9.4.1 Six Phases of Baseball Pitching

Overhead or three quarter baseball pitching motion is often divided into six phases: wind-up, stride, arm cocking, arm acceleration, arm deceleration and follow-through (Fig. 9.9) (Dillman et al., 1993; Fleisig et al., 1995a; Werner et al., 1993). Wind-up starts when the athlete initiates the first motion and ends with maximal knee lift of leading (stride) leg. There is no stressful movement for the arms in this phase. The stride phase begins at the end of the wind-up and ends when the leading foot first contacts the ground. In this phase the leading leg moves toward home plate and the two arms separate from each other. The arm cocking phase begins at the time of lead foot contact and ends at maximal shoulder external rotation. The upper body is rotated to face home plate while the throwing arm (humerus) is externally rotated in preparation for fast internal rotation. The arm acceleration phase begins at maximal shoulder external rotation and ends at ball release. In this very rapid and short phase, the elbow quickly extends and the shoulder rapidly internal rotates. From the time of ball release to the time of maximum internal rotation of the shoulder is the arm deceleration phase. During this phase the throwing arm moves forward, down and across the body. This natural occurrence in baseball pitching is believed to minimize injury potential at the elbow and shoulder after ball release. The follow-through begins at the time of maximum shoulder internal rotation and ends when the pitcher reaches a balanced position.

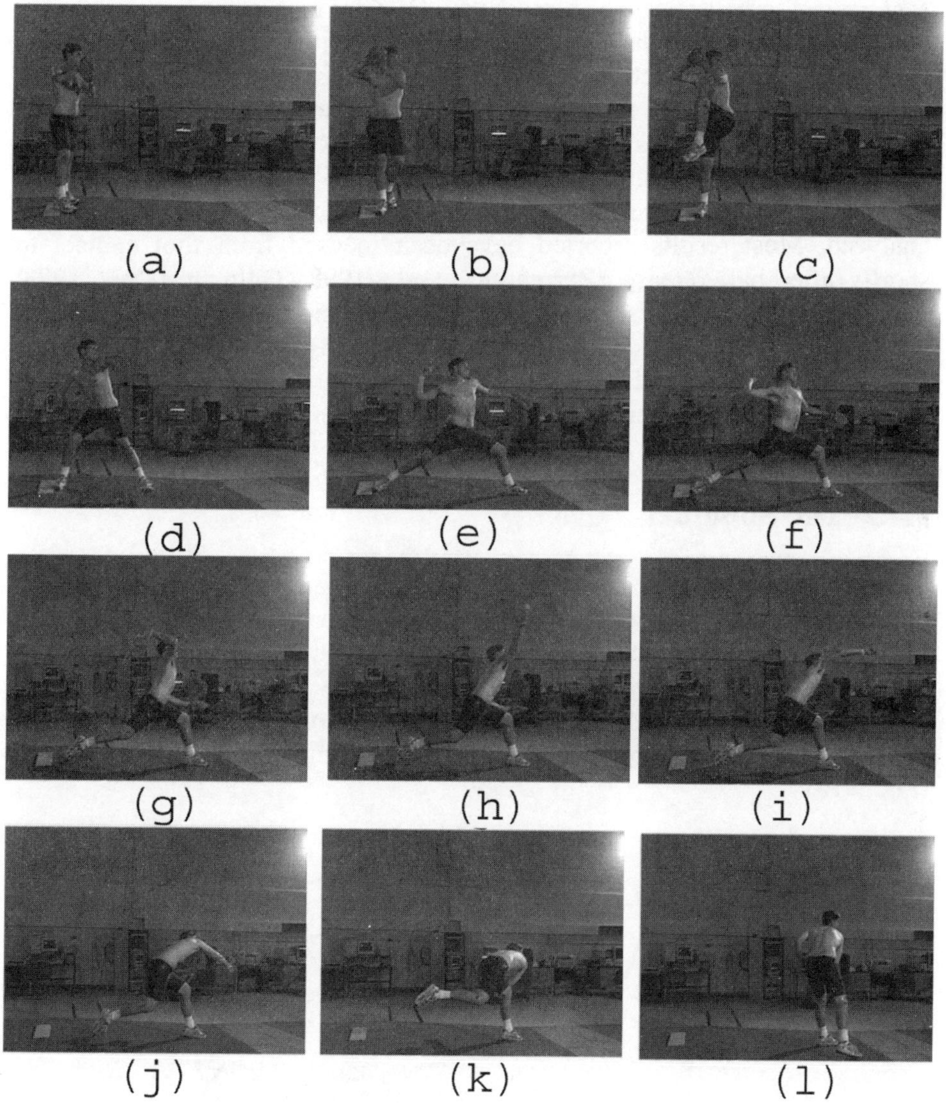


*Figure 9.9.* Six phases of baseball pitching: (a) to (c) winding-up, (c) to (e) stride, (e) to (g) arm cocking, (g) to (h) arm acceleration, (h) to (j) arm deceleration and (j) to (l) follow-through. Some key events are (e) foot contact, (g) maximum shoulder external rotation and (h) ball release.

In order to compare variables among different pitches, kinematic and kinetic variables are time-normalized from leading foot contact (as 0%) and ball release (as 100%). The time from leading foot contact to ball release is approximately 0.14 seconds, while the entire motion is approximately 3 seconds in length. The majority of the time is during the wind-up phase and follow-through. These phases have the greatest amount of variability from pitcher to pitcher and pitch to pitch. While the time interval from the leading foot contact to ball release is very short, it is more consistent from pitcher to pitcher and pitch to pitch. Zheng et al. (1998) reported the time interval from the leading foot contact to ball release for 26 highly skilled adult male pitchers averaged 0.139 seconds with a standard deviation of 0.002 seconds (Zheng et al., 1998). It is a better choice to use time from foot contact to ball release than the entire pitching motion for normalizing variables. Data from the stride phase to the follow-through phase have been digitized and analyzed. Most results reported have been focused from foot contact to shortly after ball release (Barrentine et al., 1998; Dillman et al., 1993; Escamilla et al., 1994; Escamilla et al., 2001; Fleisig et al., 1995a; Fleisig et al., 1996d; Fleisig et al., 1999; Zheng et al., 1998). Here the joint motions during pitching, except the wrist and the forearm, are presented from 26 high skilled adult male pitchers collected at ASMI. Their ball speed ranged from 37.5 m/s (84 mph) to 40.7 m/s (91 mph).

## 9.4.2 The Shoulder

Figs. 9.10, 11 and 12 show the shoulder motion during baseball pitching, normalized from foot contact to ball release. The radius of a circle represents the time in milliseconds. Curves show the shoulder motion versus time during a pitch for a right-handed pitcher. The throwing shoulder was abducted to about 90 degrees during the wind-up phase and remained relatively constant at approximately 100 degrees until the ball release phase (Fig. 9.10).

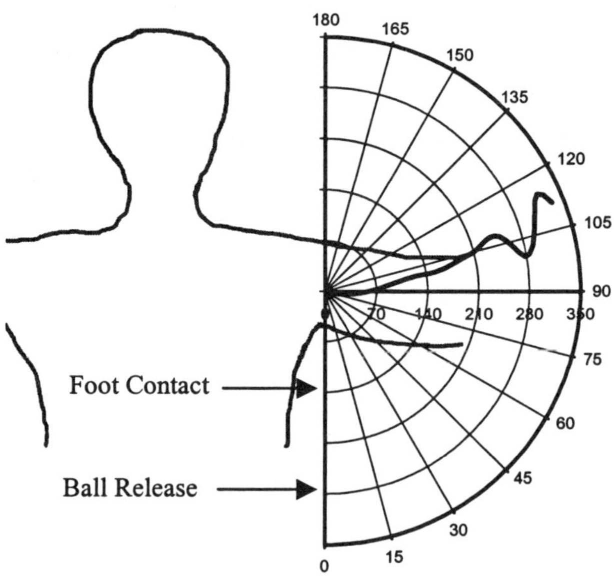


*Figure 9.10.* Shoulder abduction in degrees, time in milliseconds.

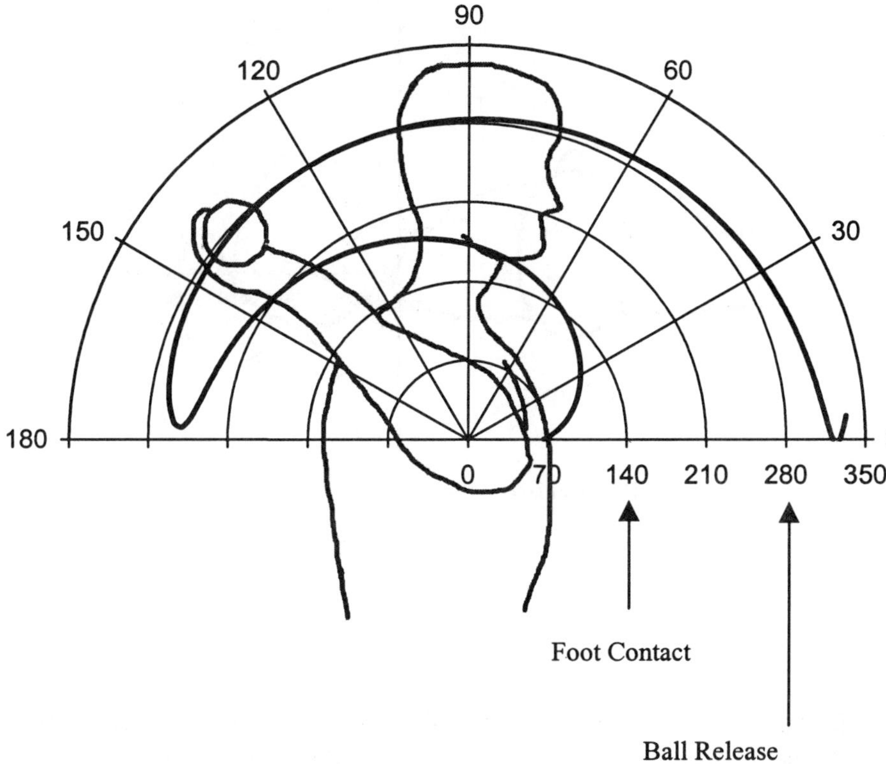


*Figure 9.11.* Shoulder external/internal rotation.

The shoulder was externally rotated about 50 degrees at foot contact, and continued to rotate to approximately 180 degrees of external rotation during arm cocking phase (Fig. 9.11). Because of how it is calculated, this maximum external rotation is actually a combination of glenohumeral rotation, sternoclavicular motion, and extension of the spine. From maximum external rotation the throwing shoulder started internal rotation. It internally rotated about 60 degrees in less than 10 milliseconds during the arm acceleration phase, and continued its internal rotation during the arm deceleration phase, eventually reaching zero degrees shortly after ball release. The maximum internal rotation velocity is typically 7000 degrees/second, making it one of the fastest human motions in sports.

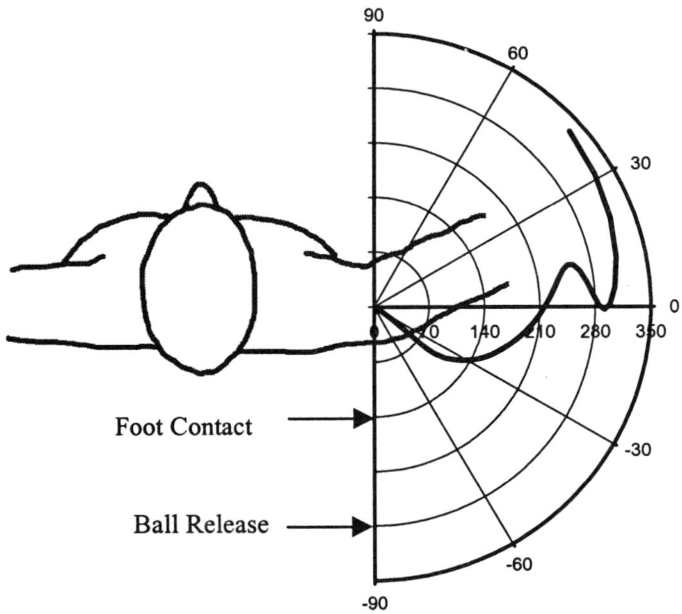


*Figure 9.12.* Shoulder horizontal adduction.

The throwing shoulder is horizontally abducted during the stride phase and reaches approximately 30 degrees of horizontal abduction at foot contact (Fig. 9.12). It horizontally adducted during arm cocking and abducted during arm acceleration phase, reaching approximately zero degrees horizontal adduction at ball release. After ball release, the arm continues to horizontally adduct reaching 40 degrees of horizontal adduction during the arm deceleration phase.

## 9.4.3 The Elbow

The elbow was flexed about 90 degrees during the arm cocking phase and extended to about 20 degrees flexion during the arm acceleration phase (Fig. 9.13). A typical elbow extension velocity is approximately 2400 degrees/second.

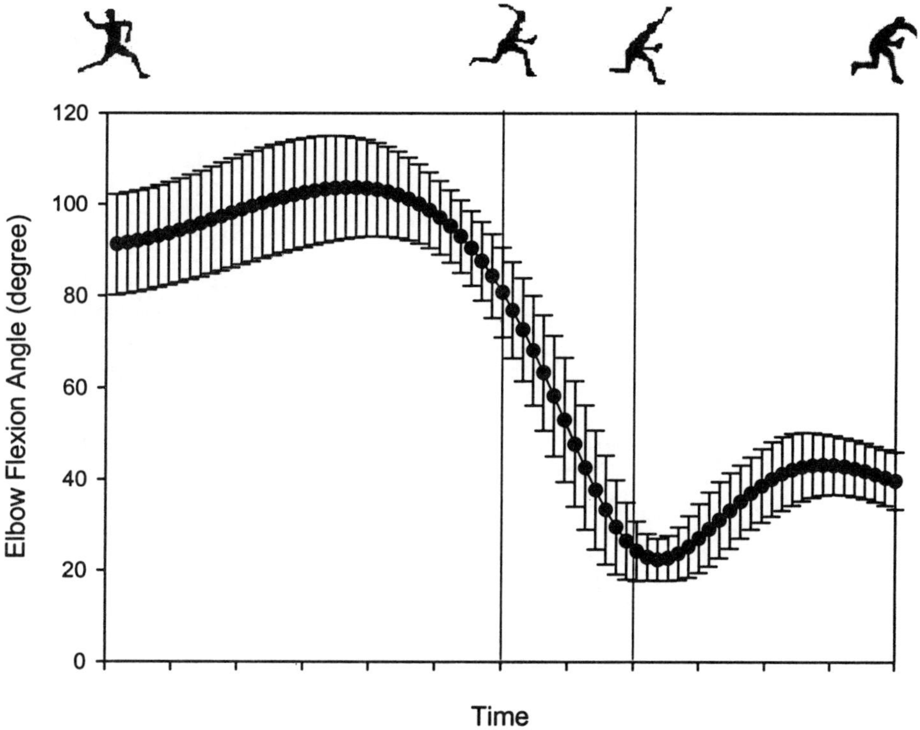


*Figure 9.13.* Elbow flexion angle during baseball pitching.

## 9.4.4 The Wrist and Forearm

Barrentine et al. (1998) compared wrist and forearm motions during baseball pitching (Barrentine et al., 1998). Data were collected from eight healthy collegiate baseball pitchers. The average ball speed was $34 \pm 2$ meters/second for the fastball pitch. The wrist was extended about 35 to 40 degrees during the arm cocking phase, and reached 40 degrees at the time of maximum shoulder external rotation (Fig. 9.14). The wrist flexed during the arm acceleration phase and reached approximately zero degrees at the time of ball release.

The wrist was in radial deviation approximately 5 degrees during the arm cocking phase (Fig. 9.15). It deviated to the ulnar side during the arm acceleration phase and reached approximately 20 degrees of ulnar deviation.

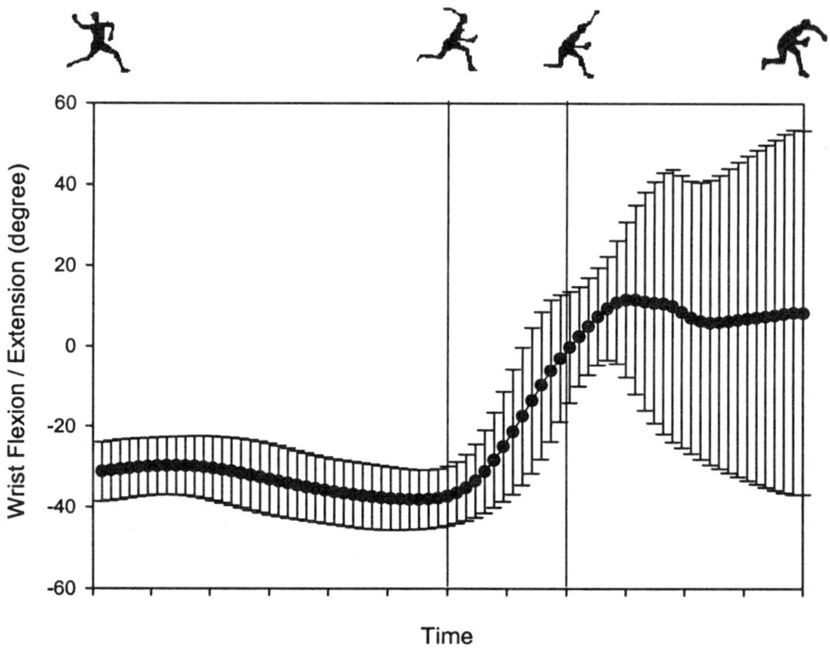


*Figure 9.14.* Wrist flexion during baseball pitching.

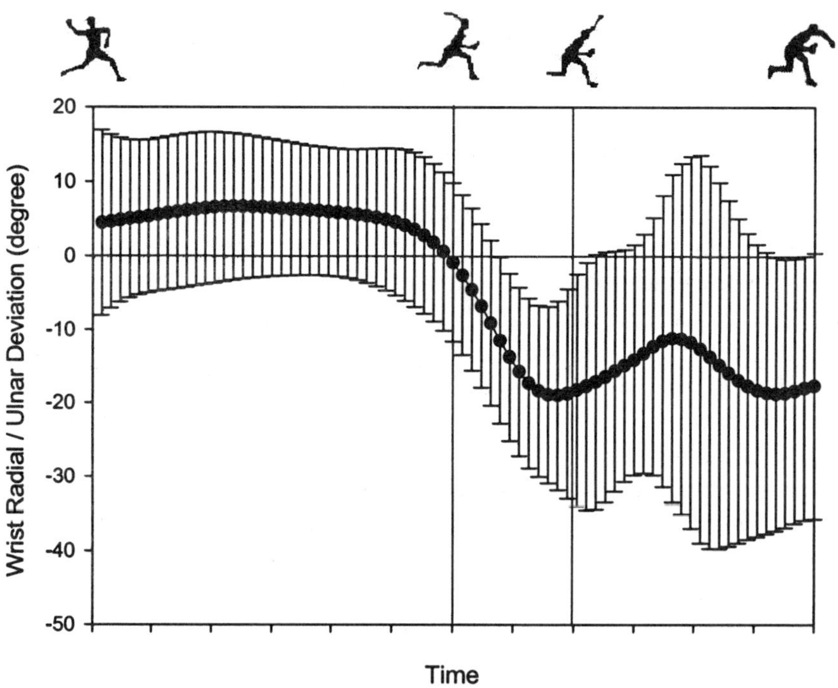


*Figure 9.15.* Wrist radial and ulnar deviation during baseball pitching.

The forearm had minimal pronation during the arm cocking phase and reached approximately 24 degrees of pronation at the time of ball release (Fig. 9.16). It reached almost 60 degrees at the moment of shoulder maximum internal rotation and the end of the arm deceleration phase.

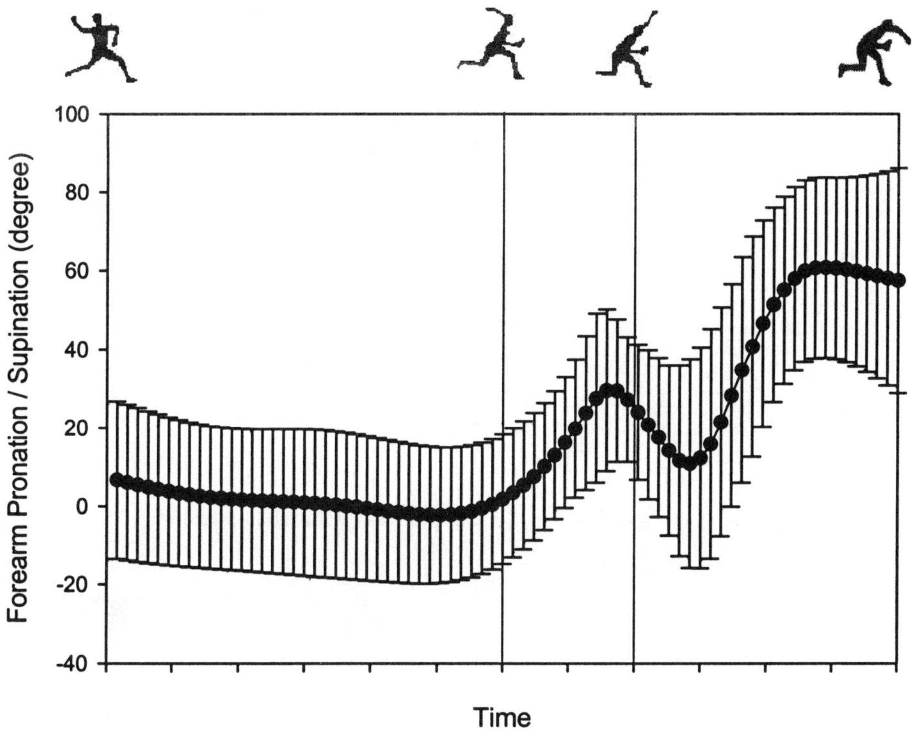


*Figure 9.16.* Forearm pronation and supination during baseball pitching.

## 9.4.5 The Spine

The axial rotation of the spine decreased during the stride phase, reaching its minimum at the contact ($-51.4 \pm 9.6^{o}$) (Fig. 9.17). It increased after foot contact, reaching the neutral position at ball release. The lateral bending of the spine increased during the arm cocking phase, reaching the neutral position at foot contact (Fig. 9.18). It kept bending to the same direction (to the leading side to raise the throwing arm) after foot contact, reaching ($-28.2 \pm 6.2^{o}$) at ball release.

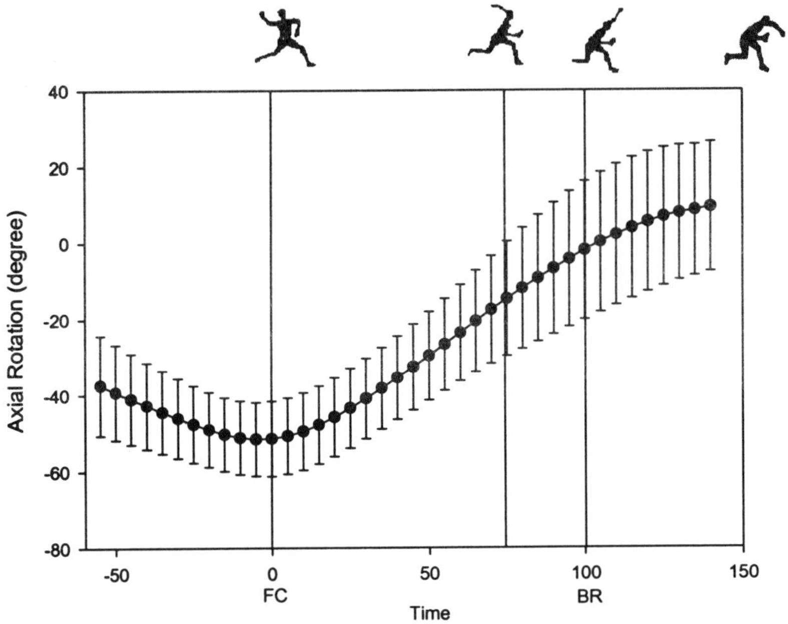


*Figure 9.17.* Axial rotation of the spine during baseball pitching.

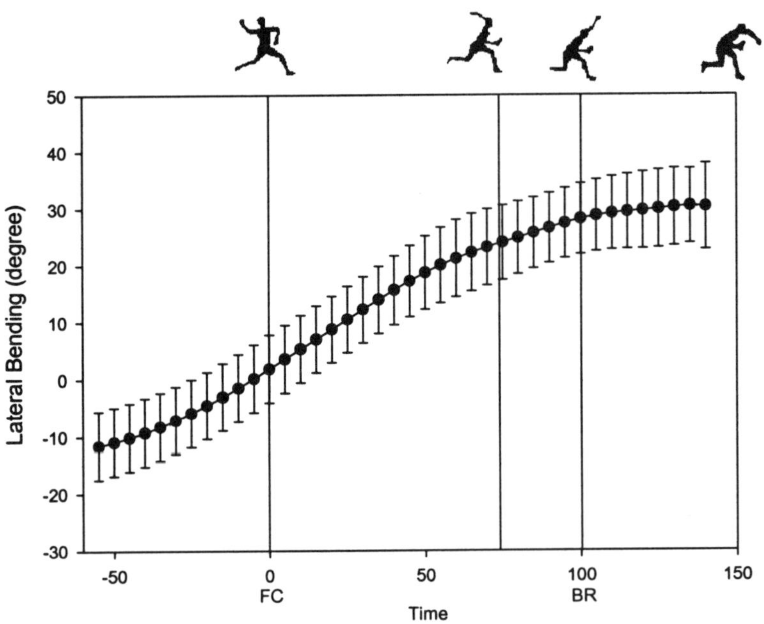


*Figure 9.18.* Lateral bending of the spine during baseball pitching.

## 9.4.6 Other Joint Motions

Other joint motions include the leading knee flexion angle and trunk tilt angles. The leading knee was flexed 38 ± 12° at foot contact and 46 ±10° at ball release. The trunk forward tilt angle was 41 ± 7° at ball release and its side tilt was 40 ± 9.6° at ball release.

# 9.5 FORCES AND TORQUES DURING BASEBALL PITCHING

Since no wrist motion was determined for most subjects collected at ASMI, the estimated force and torque at the elbow and shoulder presented here are based on assumption that no motion occurred at the wrist. The motion of the hand and ball was assumed to be the same as the wrist of the pitching arm. Kinetics equations discussed in Section 9.3.2 were used to calculate the forces and torques applied to the forearm at the wrist by the hand and ball before ball release and by the hand after ball release, to the forearm at the elbow and to the upper arm at the shoulder. The mass of a baseball is about 0.14175 kg. Inertial property parameters were obtained from previous cadaveric studies by Dempster (1955) and Clauser et al. (1969) (Dempster, 1955; Clauser et al, 1969). Table 9.1 lists these parameters used in calculating forces and torques. Moment of inertia values were scaled by the pitcher's height (h, in meters) and mass (m, in kg).

$$k_T = m * h^2 /(74.2 * 1.755^2)$$

$$k_L = m^2 * h /(74.2^2 * 1.755)$$

Table 9.1 Anthropometric Data Used in Calculating Forces and Torques

| Segment | Mass/Mass$_b$ | $CM_P/L_S$ | $I_T$ $(kg \cdot m^2)$ | $I_L$ $(kg \cdot m^2)$ |
|---|---|---|---|---|
| Hand | 0.006 | 0.506 | | |
| Forearm | 0.016 | 0.430 | 0.0076*$k_T$ | 0.0011*$k_L$ |
| Upper arm | 0.028 | 0.436 | 0.0213*$k_T$ | 0.0024*$k_L$ |

*Notes: Mass$_b$ is the total mass of the body, $CM_p$ is the distance of the center of the mass to the proximal end of the segment, $L_S$ is the length of the segment, $I_T$ is the moment of inertia about the transverse axis and $I_L$ is the moment of inertia about the longitudinal axis, $k_T$ is scale factor for $I_T$ and $k_L$ is the scale factor for $I_L$.*

### 9.5.1 Forces and Torques at the Elbow

The force applied to the forearm at the elbow can be decomposed into three parts: medial force, anterior force and proximal force. Fig. 9.19 shows these three forces during baseball pitching.

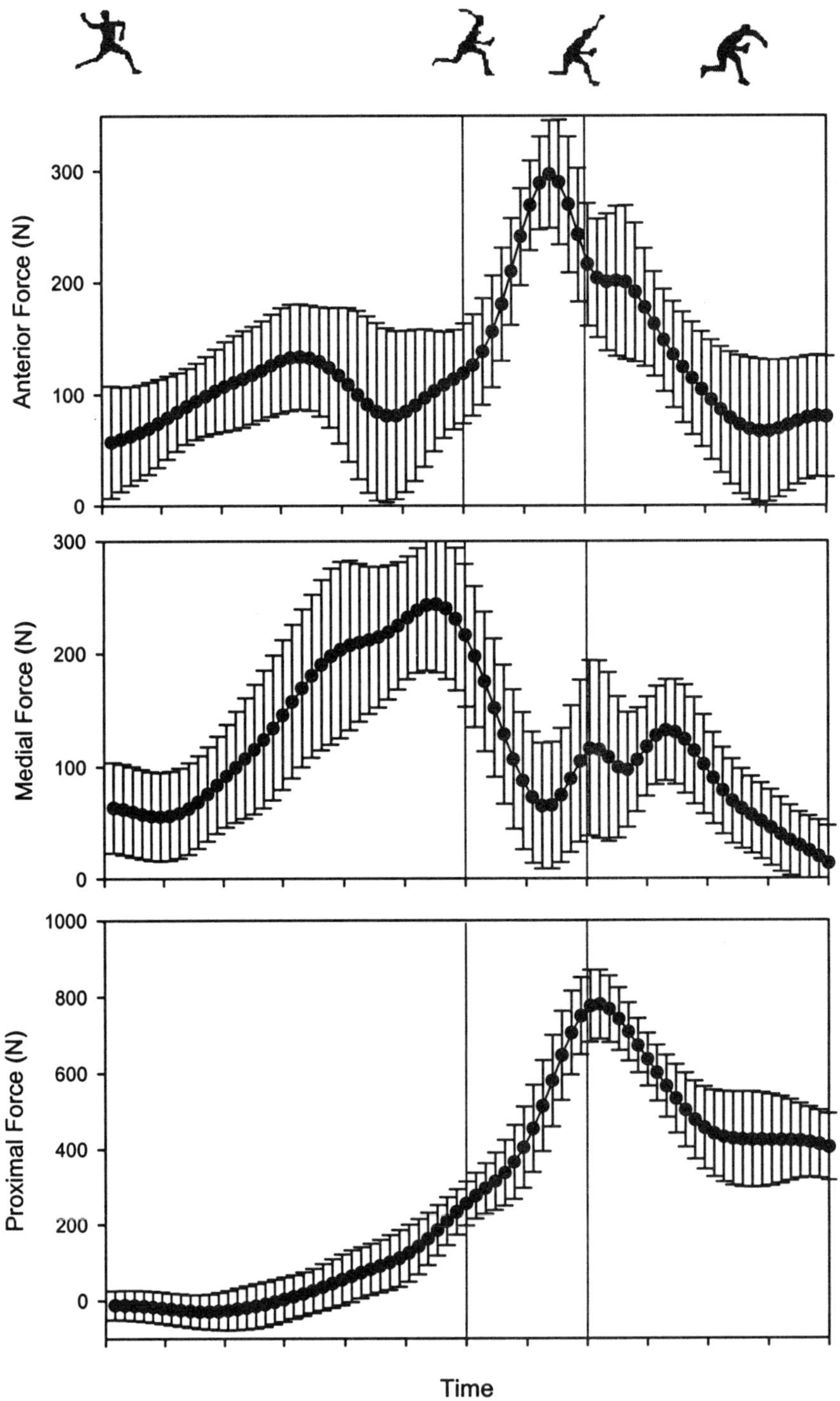


*Figure 9.19.* Forces applied to the forearm at the elbow during baseball pitching.

A maximum medial force of 250 N occurred during the arm cocking phase. A peak anterior force of 300 N occurred in the middle of the arm acceleration phase. A maximum proximal force of 800 N occurred right after ball release, during the arm deceleration phase.

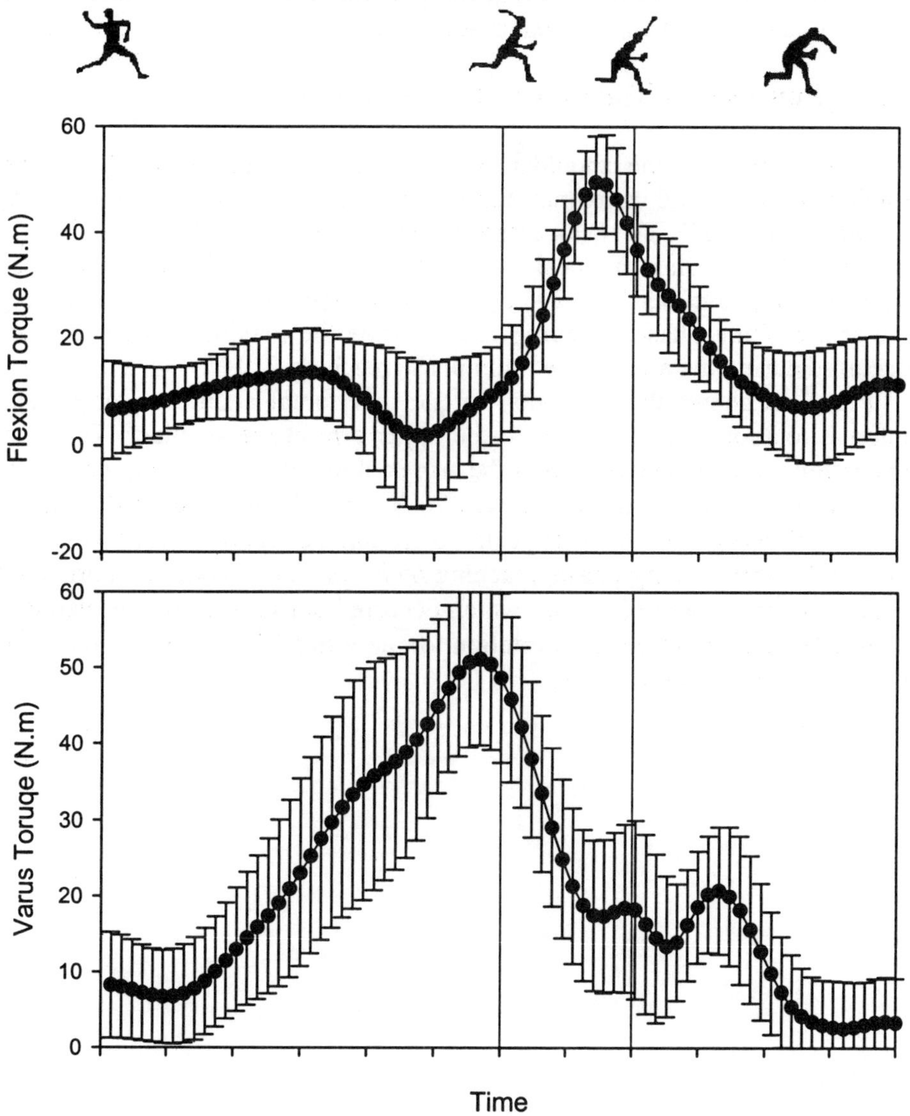


*Figure 9.20.* Torques applied to the forearm at the elbow during baseball pitching.

Two torques applied to the forearm at the elbow were determined: the flexion torque and varus torque (Fig. 9.20). No torque in the axial rotation was determined. The maximum flexion torque, 50 Newton meters ($N \cdot m$), occurred in the middle of the arm acceleration phase. The maximum varus torque of 52 $N \cdot m$ occurred during the arm cocking phase, just prior to the instant of shoulder maximum external rotation.

## 9.5.2 Forces and Torques at the Shoulder

Three forces at the shoulder were determined: the anterior/posterior, superior/inferior and proximal forces applied to the upper arm at the shoulder (Fig. 9.21). The peak anterior shear force was 280 N during the arm cocking phases, and the peak posterior force occurred during the arm deceleration phase, reaching 295 N. The peak superior force occurred during the arm cocking phase, reaching 250 N while the maximum inferior force occurred during the arm deceleration phase reaching 230 N.

Fig. 9.22 shows the abduction torque, horizontal adduction torque and internal rotation torque applied to the upper arm at the shoulder. The peak abduction torque occurred during the arm cocking phase, reaching 45 $N \cdot m$. The maximum adduction torque was 70 $N \cdot m$, which occurred during the arm deceleration phase. The peak horizontal adduction torque occurred during the arm cocking phase, reaching 65 $N \cdot m$. The maximum horizontal abduction torque was 63 $N \cdot m$, which occurred during the arm deceleration phase. The maximum internal rotation torque was 52 $N \cdot m$, which occurred during the arm cocking phase.

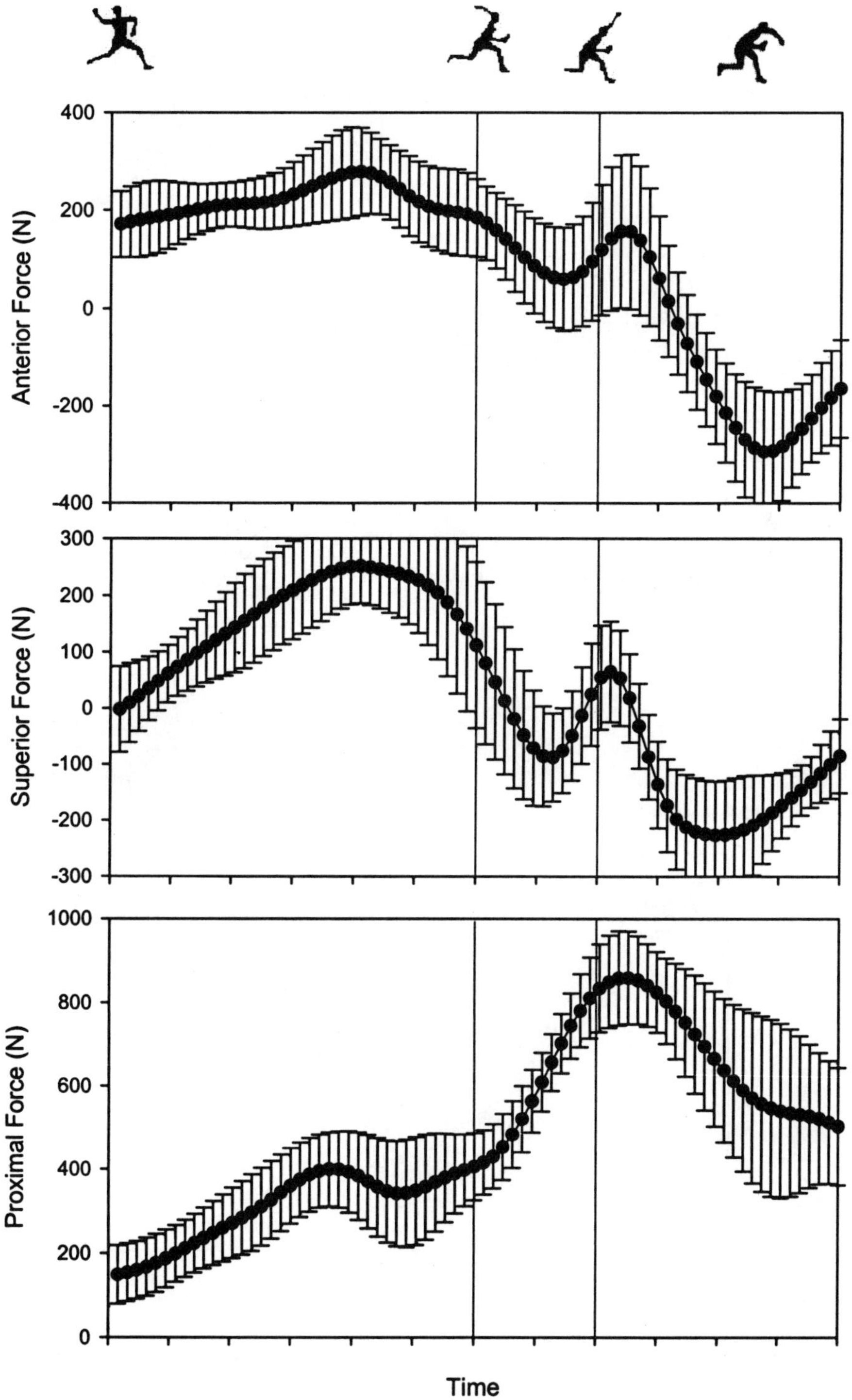


*Figure 9.21.* Forces applied to the upper arm at the shoulder during baseball pitching.

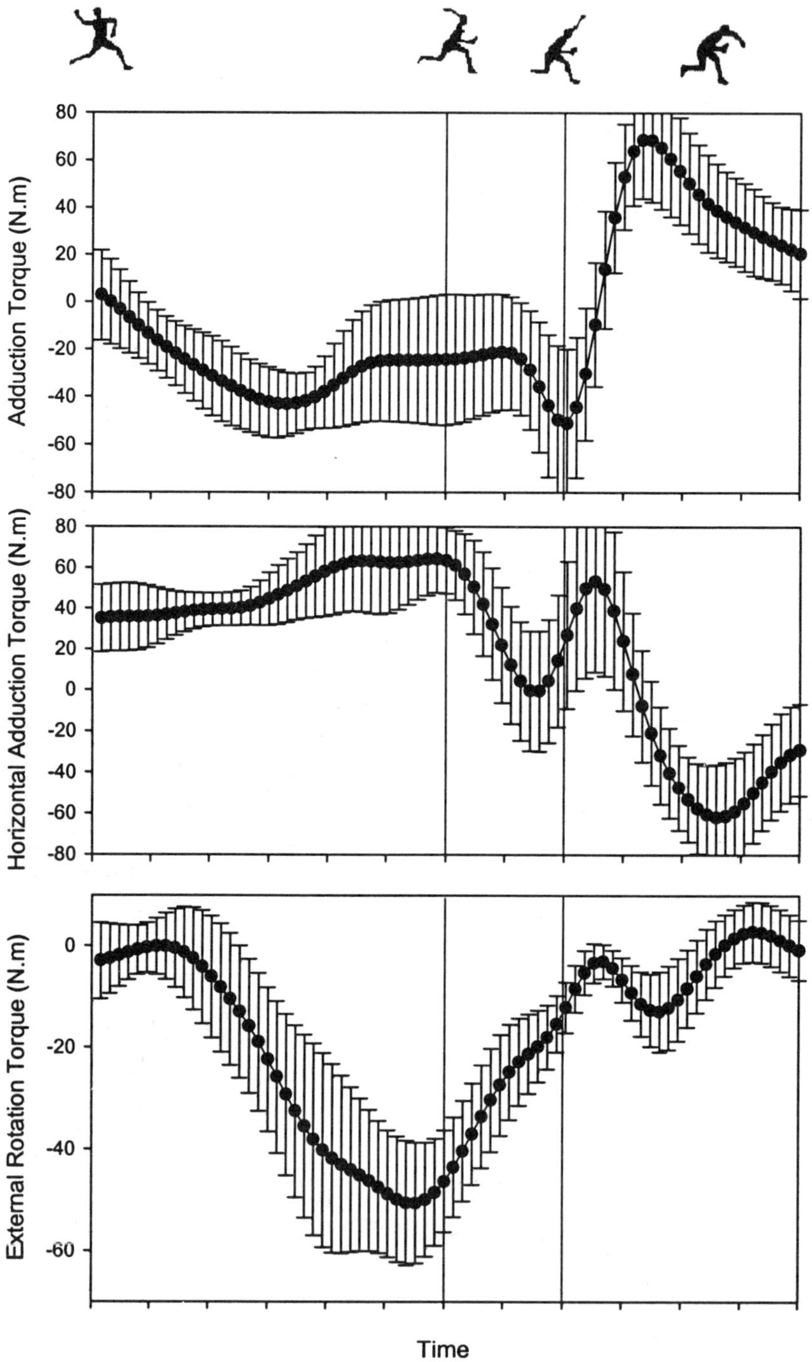


*Figure 9.22.* Torques applied to the upper arm at the shoulder during baseball pitching.

## 9.6 DIFFERENT PITCHES, PITCHING LEVELS AND BALL SPEED

After we described the normal pitching kinematics and kinetics, comparative questions were raised by physicians, coaches and others related to baseball pitching. What is the difference between throwing a fastball and throwing a change-up? At what age is it safe for someone to start throwing off speed pitches such as a curveball? Does a curveball require different mechanics? Does a ten year old have a similar pitching motion as a professional? If the height of a pitching mound increased from 10 inches to 13 inches, does a pitcher need to adjust his pitching mechanics? Are there any differences between pitching from a mound and throwing from flat ground? Some efforts have been made at ASMI to find answers for some of these questions.

### 9.6.1 Pitching Kinematics of Different Pitches

The most effective and used pitch is the fastball pitch. The data presented above are based on fastball pitches. Now we are going to discuss the differences between throwing a fastball and throwing other pitches like the curveball, slider or change-up.

#### 9.6.1.1 The Curveball or Slider

As we know the purpose of a good pitch is to not allow the hitter to get a hit or at least minimize the hitter's ability to get a hit. A good fastball pitch is designed to pass the hitter before the swing of the bat. The good curveball or slider breaks down, left, or right before the batter swings. There are obvious differences in how the ball is gripped for each pitch. Barrentine et al. (1998) studied the differences of the wrist and forearm movements between the fastball pitch and the curveball pitch (Barrentine et al., 1998). They found the wrist was less extended and the forearm was more supinated during the arm cocking phase for the curveball pitch compared to the fastball pitch. Escamilla et al. (1998) also found differences between the fastball, curveball and slider pitches (Escamilla et al., 1994). The authors reported the pitchers generated lower pelvis and upper torso angular velocity, higher shoulder horizontal adduction angle, and more knee flexion at ball release while throwing the curveball pitch compared to the fastball. No significant differences were found between the fastball and slider pitches except for the ball speed. The pitchers generated greater shoulder internal rotation velocity, less shoulder abduction, less lateral trunk tilt at ball release and higher ball velocity for the slider pitch compared to the curveball pitch.

#### 9.6.1.2 The Change-Up

An effective change-up can deceive a batter and cause a weak, off-balance swing. A pitcher is normally taught to throw a change-up with the exact same motion of the fastball pitch. Ideally, the resulting ball speed will be approximately 75 percent as fast as the fastball. Barrentine et al. (1998) reported that no significant difference was found between the fastball and change-up pitch relative to the kinematic variables at the wrist and forearm (Barrentine et al., 1998). The change-up ball speed was significantly lower at approximately 85% of the fastball. Escamilla et al. (1998) found that several kinematics variables were significantly different between the change-up and fastball (Escamilla et al., 1994). For the change-up, a pitcher had less lead leg knee flexion at foot contact, lower pelvis and upper torso angular velocities, higher shoulder horizontal adduction, lower elbow extension angular velocity, lower shoulder internal rotation velocity, and higher shoulder abduction. Additional differences included greater lead leg knee flexion at ball release, less lateral trunk tilt, more elbow flexion and more shoulder horizontal adduction at ball release.

### 9.6.2 Kinematics and Kinetic Differences among Pitching Levels

Fleisig et al. (2000) reported the differences of pitching mechanics between four skill levels. Two hundred thirty one youth, high school, college and professional level pitchers participated in the study. The ball speed significantly increased from the youth (28 m/s) and high school (33 m/s) groups to the college (35 m/s) and professional (37 m/s) groups. The study found that there were no significant differences among the four levels for 10 of 11 kinematic position parameters and all six temporal parameters. The study supported a commonly used philosophy by coaches that a child should be taught the same pitching mechanics as an adult. All eight kinetic parameters were found to be significantly different among all four levels. During the arm cocking phase, elbow varus torque, shoulder internal rotation torque and shoulder anterior force increased as the skill level progressed. Similarly, elbow flexion torque during the arm acceleration phase increased from level to level. During the arm deceleration phase, elbow proximal force, shoulder proximal force, shoulder posterior force, and shoulder horizontal abduction torque increased from level to level. The study suggested that a pitcher should learn proper mechanics as early as possible, and build strength as the body matures.

### 9.6.3 Relationship between Pitching Kinematics and Ball Speed

It is believed that optimal pitching mechanics allow the pitcher to maximize ball velocity. Matsuo et al. (2001) used data collected at ASMI from 127 healthy college and professional pitchers to investigate the relationship between certain pitching mechanics and ball velocity (Matsuo et al., 2001). Mean ball speed for the group was 36.1 m/s with the standard deviation (SD) of 1.9 m/s. Subjects were grouped relative to ball speed, with 29 subjects in the high velocity group (more than one SD above mean ball speed) and 23 subjects in the low velocity group (with ball speed more than one SD below the mean ball speed). The average ball velocity for the high velocity group (38.4 m/s) was 5.2 m/s higher than for the low velocity group. The high velocity group had significantly longer arm measurements (humerus length and radius length) and was significantly taller than the low velocity group. Among 12 kinematic parameters, four parameters had significant differences between the high and low velocity groups. The high velocity group had lower maximum leading leg knee flexion angular velocity, higher maximum shoulder external rotation angle, higher leading leg knee extension angular velocity and more trunk forward tilt at the instant of ball release. Among 9 temporal parameters only two were significantly different between the high and low velocity groups. The maximum elbow extension angular velocity and maximum shoulder internal rotation angular velocity occurred later in the low velocity group.

Stodden et al. (2001) investigated the relationship between pelvis and upper torso kinematics to ball velocity using data collected at ASMI (Stodden et al., 2001). From a database of pitchers, 19 elite baseball pitchers that had more than 1.8 m/s ball speed variation during 10 fastball pitch trials were included in the study. Among 12 pelvis and upper torso kinematic variables analyzed, five variables demonstrated a strong relationship with the variations in ball velocity within individual pitchers: pelvis orientation at maximum shoulder external rotation and at ball release, upper torso orientation at maximum shoulder external rotation, average pelvis velocity during arm cocking and average upper torso velocity during arm acceleration.

### 9.6.4 Comparison of Pitching from a Mound and Throwing from Flat Ground

Most overuse throwing injuries in baseball involve the pitcher. Throwing from a flat ground is commonly used for training and rehabilitation protocols. Fleisig et al. (1996) investigated kinematic and kinetic differences between throwing from a mound and throwing from a flat ground (Fleisig et al., 1996a). Twenty-seven healthy college baseball pitchers were included in the study comparing pitchers while throwing from a mound, a 180 feet crow-hop throw, 120 feet crow-hop throw and 60 feet crow-hop throw. Contrary to the belief that pitching and flat ground throwing mechanics are the same, several significant differences were found. Throwing from flat ground corresponded with a shorter stride and less shoulder external rotation at foot contact. The drop of the mound appears to give the pitch more time to stride forward a greater distance, and more time to externally rotate the shoulder. Kinematic and kinetic values for the throwing arm and body during arm cocking and acceleration were similar between mound and flat ground throwing. At the instant of ball release, a pitcher's trunk was in a more vertical position when throwing from flat ground. However, relative to the throwing surface, the trunk angle was the same for 60 feet throwing from flat ground and pitching from a mound. The trunk was 3 degrees more upright for the longer distance throws, allowing the athlete to throw the ball with a slightly more upward trajectory in order to get more distance. During the arm deceleration phase, proximal force at both elbow and shoulder were less in long distance flat throws than in 60 feet mound or flat throws. This may be related to the low incidence of throwing injuries in non-pitchers.

### 9.6.5 Comparison of Full-effort and Partial-effort Pitching

Baseball pitchers often use partial-effort pitching during training, warm-up, and rehabilitation. The intent of these throws is to progressively increase the loads on the athlete's body to the high levels produced in full-effort pitching while reinforcing proper timing, coordination, and movement patterns. Fleisig et al. (1996) reported kinematic and kinetic differences between full-effort and partial-effort baseball pitching (Fleisig et al., 1996b). Twenty-seven healthy college baseball pitchers were used in this study. Subjects were instructed to throw with 100% effort, 75% effort and 50% effort from a mound. Although 75% and 50% effort are vague instructional cues, they are often used by coaches, therapists, trainers and pitchers. At 75% effort, pitchers produced approximately 90% of the ball velocity, 90% of the arm and trunk speed, and 85% of the forces and torques at the shoulder and elbow when compared to those during 100% effort. At 50% effort, pitchers produced approximately 85% of the ball speed, 85% of the arm and trunk speed, and 75% of the forces and torques at the shoulder and elbow when compared to those during 100% effort. Reduced effort pitching also corresponded with reduced arm rotation during arm cocking, increased shoulder horizontal adduction, and a more upright trunk with less knee flexion and forward trunk tilting at ball release. The ball speed consistency for an individual pitcher at partial-effort demonstrated that the instruction is not vague to these pitchers. The ball speed dropped from 35.3 m/s at full effort to 32.6 m/s at 75% effort and 30.4 m/s at 50% effort. Pitching with partial-effort appears to be useful for a pitcher in training or rehabilitation when he does not want to put high loads on his arm as normally occurs during full-effort throwing.

## 9.6.6 Comparison of Baseball Pitching and Football Passing

Many talented athletes are both the quarterback on their school's football team and a pitcher on their baseball team. In order to understand whether participation in both activities is beneficial or detrimental to the athlete's performance and safety, Fleisig et al. (1996) conducted a comparison of kinematic and kinetic variables for baseball pitching and football passing (Fleisig et al., 1996d). Twenty-six quarterbacks (13 collegiate, 13 high school) and 26 baseball pitchers (13 collegiate and 13 high school) participated in the study. There were several kinematic differences between baseball pitching and football passing, although they displayed a similar throwing motion. As expected, the ball speed during pitching was significantly higher (35 m/s) than during football passing (21 m/s). At the instant of foot contact, the pitcher had a longer stride, less shoulder horizontal adduction, less shoulder external rotation, more elbow flexion and leading knee flexion. During the arm cocking phase, pitchers demonstrated higher pelvis and upper torso angular velocity, less shoulder horizontal adduction and elbow flexion. Pitchers had higher maximum shoulder external rotation, and higher elbow extension velocity during the arm acceleration phase. At ball release, pitchers had less shoulder horizontal adduction, less elbow flexion, less trunk forward tilt, more trunk side tilt and leading knee flexion. Pitchers also had higher maximum shoulder internal rotation velocity. Fewer kinetic variables were found to be significantly different between baseball pitching and football passing. This may be a balance between greater ball weight and less ball speed for the football. However, pitchers were found to have higher proximal force applied to the upper arm by the trunk at the shoulder, higher proximal force applied to the forearm at the elbow, and higher shoulder adduction torque during the arm deceleration phase. Football passing did not produce greater forces or torques at the shoulder and elbow even though the ball was heavier. To decelerate the arm, pitchers put higher forces and torques at the shoulder and elbow. The results from this study may explain differences in performance and injury rates between the two sports.

## 9.7 PITCHING MECHANICS AND INJURY MECHANISMS

It is believed that the extreme range of joint movement and forces and torques applied during pitching are the causes of certain injuries to pitchers (Dillman et al., 1993; Fleisig et al., 1995a; Zheng et al., 1998; Zheng et al., 1999; Zheng and Barrentine, 2000). Proper pitching mechanics may enable a pitcher to achieve good performance with minimum chance of injury. Overuse injuries to the throwing arm of pitchers is the most common type of injury in baseball (Andrews and Fleisig, 1998). Overuse injury is a result of cumulative microtrauma from the repetitive dynamic overhead throwing motion used to pitch a baseball. Here we are going to discuss the pathomechanics of elbow and shoulder injuries, linking the motion, force and torque to a specific injury.

### 9.7.1 Pathomechanics of Elbow Injuries

The elbow is one of the most commonly injured joints in pitchers. Injuries could occur at any area of the joint: anterior, lateral, posterior and medial compartment. A repetitious overhead throwing motion can produce incomplete elbow extension because of anterior capsule strains, flexor-pronator strains, bicipital tendonitis, and the presence of intra-articular loose bodies. The anterior capsular sprain is often caused by traumatic hyperextension of the elbow, resulting in anterior elbow pain. As we have discussed before, a pitcher extends his elbow at speeds of over 2400 m/s. A pitcher with good mechanics does not reach full extension of the elbow. However, if hyperextension stress is added to generalized joint laxity, that is common in youth, insufficient strength and improper mechanics combine to produce marked symptoms of anterior compartment pain and tenderness.

At the end of the arm cocking phase, a large elbow valgus torque is applied. High compressive force between the radial head and humeral capitellum with high tension on the ulnar collateral ligament produce a varus torque to counterbalance such large valgus torque at the elbow. This compression in the lateral compartment of the elbow can lead to avascular necrosis, osteochondritis dissecans, or osteochondral chip fractures. In the immature skeleton, this compression force in conjunction with valgus medial stretching can result in lateral elbow problems. Osteochondrosis and osteochondritis dissecans of the capitellum and head of the radius frequently are considered a part of Little League elbow syndrome, which commonly occurs in 10-15 year-old baseball pitchers. Extension traction in the arm deceleration and follow-through phases of pitching can cause injuries like avulsion of the lateral apophysis, a fragment of the condyle, which is the origin of the extensor forearm muscles or the radial collateral ligament.

Incomplete elbow flexion frequently indicates posterior elbow problems in the athlete (Andrews and Whiteside, 1993). The posterior compartment is susceptible to injury during the arm acceleration phase, throughout which significant varus torque is produced. The olecranon wedges up against the medial wall of the olecranon fossa of the humerus in repetitive throwing motion, with the formation of osteophytes that may later develop into loose bodies. Loose bodies often migrate to the anterior, causing incomplete elbow extension. Valgus Extension Overload (VEO) is a throwing-related syndrome that is a common cause of pain in the posterior elbow. This condition is commonly seen in baseball pitchers as pitching causes elbow tension and bony compression. This compression leads to bony changes that are painful and restrict the range of motion. Because of pain with throwing, pitchers gradually become ineffective. This abutting leads to bony protuberance at the posterior olecranon tip causing softening of the articular cartilage and loose body formation. The bony changes in the posterior elbow will prevent complete straightening of the elbow. Therefore during the follow-through phase of pitching, when the arm is straightening out, further impingement and symptoms will occur. Forceful contraction of the triceps during the arm acceleration phase can yield contraction apophysitis and non-union of the secondary ossification center of the olecranon. Avulsion of the triceps at its origin may occur in the stronger pitcher.

Common origins of medial elbow pain are medial epicondylitis, ulnar collateral ligament sprain, flexor-pronator mass strain, and ulnar nerve neuritis. As we have mentioned before, near the end of the arm cocking phase, the shoulder is in extreme external rotation, the elbow is flexed approximately 90 degrees, and the forearm is in a valgus position. To resist valgus stress, high tension is developed in the ulnar collateral ligament at the end of the arm cocking phase. This high tension is near the ultimate tensile strength of the ligament. The UCL is the primary stabilizer of the elbow. Rupture of this ligament used to be the end of any thrower's career until the development of the "Tommy John" surgery. This revolutionary surgery was first performed on a professional baseball pitcher named Tommy John in 1974 by Dr. Frank Jobe. The palmaris longus tendon from the left (throwing) forearm was used to reconstruct the UCL in the left elbow and allowed John to return to the Major Leagues. Repetitive tension that the UCL and joint capsule exert on the ulnar can cause spur formation, which can compress the ulnar nerve.

The term Little League elbow has been used to describe multiple medical problems in the adolescent baseball player's elbow. It is currently being used to describe injury to the growth plate on the medial epicondyle. It results in inner elbow pain when throwing a baseball. Young athletes sustain a different set of sports injuries than mature athletes because of their growing areas of the skeleton. At the ends of long bones in children, such as the humerus in the arm, there are growth plates. These areas "close" at an average age of 14.6 years for girls and 16.5 years for boys. These areas are the weakest part of the muscle-ligament-bone complex. It is this weak area that is prone to injury due to repetitive stress during sports in active adolescent's pitchers. The muscles that flex the wrist attach to the inner elbow. During pitching these muscles pull at the growth plate causing separation of the open growth plate as well as small tears in the muscles causing inflammation. The tension that the flexor muscle group exerts on the elbow is at its maximum in the early acceleration phase of the throwing motion. Since Little League elbow is a syndrome of a child with immature growth plates, it is most common in the 9 to 12 year old group. Pitchers are affected more than any other position because of the number of balls that they throw. This condition is an overuse injury caused by a variety of factors. Examples include excessive throwing, usually without proper preseason conditioning, poor pitching technique and inadequate warm-up. Physical factors such as anatomic abnormalities, lack of flexibility, and muscle imbalance may contribute as well.

## 9.7.2 Pathomechanics of Shoulder Injuries

The shoulder is the other commonly injured joint in pitchers. Rotator cuff muscles resist distraction, horizontal adduction and internal rotation at the shoulder during arm deceleration. The posterior shoulder muscles are very active in producing proximal force, posterior force, and horizontal abduction torque during this phase (DiGiovine, 1992; Glousman et al., 1988; Jobe et al., 1983; Jobe et al., 1984). A force over 800 N is generated during the arm deceleration phase to pull the arm proximally, posteriorly and inferiorly. Also at this phase the posterior shoulder muscles (i.e. teres minor, infraspinatus, and posterior deltoid) produce a torque up to $100 N \cdot m$ in adduction and horizontal abduction. The shoulder capsule is at risk of tensile failure under such large forces and torques since the capsule is supported posteriorly by the tendons of the infraspinatus and teres minor. High forces generated by these muscles and the shoulder motions during arm deceleration cause rotator cuff tear. Andrews and Angelo (1988) have reported that most rotator cuff tears in throwers are located in the area from the mid-supraspinatus posterior to the mid-infraspinatus (Andrews and Angelo, 1988). Translation, subluxation, and compression of the humeral hcad can induce forceful entrapment of the labrum between the humeral head and the glenoid rim, resulting in labral tearing. The rapid rotations and large forces and torques produced at the shoulder during throwing make humeral translation possible. Capsular laxity, muscle weakness and fatigue affect the ability to maintain joint stability and increase the potential damage caused by humeral translation. Degeneration of the labrum may result from repetitive translation and subluxation of the humeral head during arm acceleration and deceleration.

Extreme external rotation during throwing makes the shoulder prone to injury. As the abducted arm externally rotates at the shoulder, the humeral head translates anteriorly and the posterior rotator cuff becomes impinged between the glenoid labrum and the humeral head. Such impingement may cause degeneration of both the superior labrum and the rotator cuff, which has been called "over-rotation" injury (Zheng et al., 1999).

During the arm deceleration phase, the shoulder is abducted, horizontally adducted, and internally rotated. The inability to generate sufficient muscle force and torque can lead to translation of the humeral head and generate high tensile force on the capsule. Superior translation of the humerus can cause impingement of the greater tuberosity, rotator cuff muscles, or biceps against the inferior surface of the acromion or coracoacromial ligament, which may lead to supraspinatus, infraspinatus, and bicipital tendonitis, or even to abrasion wear.

Another type of shoulder injury is a tear of the superior labrum in the anterior and posterior direction, often called SLAP lesion. Andrews et al. (1985) observed such injury among 73 baseball pitchers and other throwing athletes (Andrews et al., 1985). They believed that the injury resulted from repetitive overuse throwing and not from a traumatic episode. A specific injury mechanism was proposed in which forces imparted by the long head of the biceps brachii, particularly during arm deceleration, tear the labrum away from the glenoid (Fleisig et al., 1995b). Biceps force is particularly strong during the arm deceleration phase as the biceps contracts to both decelerate elbow extension and resist glenohumeral distraction (Fleisig et al., 1996c). Laxity in the shoulder joint may result in increased shoulder proximal force needed, further increasing the demand on the biceps tendon labrum complex.

## 9.7.3 Implications for Conditioning and Rehabilitation

Because of the involvement of the full body in the throwing motion, it is important to condition the musculature of the upper and lower body as well as the trunk to reduce the risk of injury. The kinetic chain principle implies that weakness in any segment may result in a deficiency in performance and higher risk of injury. When athletes have an injury or a physical limitation in one area, they will try to compensate by increasing the demands on other segments. Before allowing a pitcher to return to competition from an injury it is important to evaluate his mechanics after rehabilitation to insure proper mechanics and avoid new injury.

Kinetic analysis of baseball pitching reveals the high stresses on some tissues. A pitcher must be carefully examined before he returns to competition. If a pitcher returns to competition too soon, the strength of the reconstructed (e.g. UCL) or repaired structure (e.g. rotator cuff) may not be able to accommodate the dynamic activity, during which high stresses are applied to these structures. A good rehabilitation program must be followed after surgery to expedite recovery.

## 9.8 FATIGUE AND FUTURE STUDIES

Fatigue is considered one of the major causes of injuries among baseball pitchers. One of the main responsibilities of a baseball pitching coach is to determine when a pitcher has become fatigued, as a fatigued pitcher is believed to be less effective and at higher risk of injury. Sports medicine clinicians share this concern about fatigue in pitching, as the vast majority of pitching injuries are attributed to overuse (Fleisig et al., 1995b). There is not enough attention given to fatigue and the affects of fatigue. It may be due to the difficulty of defining fatigue during pitching and the lack of a quantitative description. ASMI conducted a study in 1996 to measure kinematic and electromyographic changes in baseball pitching during a simulated game (Barrentine et al., 1997; Takada et al., 1997). Ten subjects participated in a study investigating the effects of fatigue on pitching. The study included a simulated game with each pitcher paired against another pitcher for 18 half innings or until unable to continue. Two kinematic changes often observed by coaches were found with fatigue: a lower elbow (less shoulder abduction) and a less flexed leading knee at ball release. Some electromyographic changes were also noticed but no significant differences were found. Further research on muscle activities and joint kinetics may provide more insight about the fatigue and warning sign for overuse injury. Other future studies include computer modelling of pitching motion and biomechanical model analysis of injury mechanisms.

## 9.9 SUMMARY

As presented in this chapter, the understanding of the motion in pitching a baseball has benefited tremendously from biomechanical analysis. Innovations in scientific instrumentation has enabled researchers to quantify the extreme angular velocities that occur during baseball pitching, including 7000 °/s for internal rotation at the shoulder and 2400 °/s for clbow extension. Along with the dynamic angular movements that occur during pitching are the stresses on the pitcher's body and the potential for overuse injury. Biomechanical analysis techniques have also provided the basis for estimating the forces and torques that are experienced at the shoulder and elbow joints during pitching. Using inverse dynamic models, the forces and torques applied to the elbow and shoulder have been estimated. The proximal forces exerted to resist distraction of the arm and forearm at the shoulder and elbow during pitching has been estimated to be near 100 % of the pitcher's body weight for each pitch. From the estimation of stresses on throwing, two critical instants of the pitching motion have been identified. These instances occur near the end of the arm cocking phase and near the time of ball release. The motions, forces, and torques that occur at these instants are being linked to the pathology of common throwing injuries.

The knowledge of the kinematic and kinetic factors gained from biomechanical analyses has been combined with anatomical considerations to improve the understanding of injury mechanisms. Injuries to the UCL have been attributed to the high forces and torques that are experienced at the elbow during arm cocking. The stress due to these forces and torques is exacerbated by the extreme external rotation at the shoulder and extension at the elbow creating a valgus extension overload that occurs at this time. Impingement injuries at the shoulder and elbow have been attributed to the forces and torques that occur during arm cocking. Rotator cuff tears, and SLAP lesions at the shoulder have also been attributed to these factors. The time near ball release and initial arm deceleration is also considered a critical instant related to potential injury mechanisms. The forces and torques exerted to resist distraction at the shoulder and elbow have been attributed to causes of shoulder pathology including SLAP lesions. The dual role of the biceps tendon complex in controlling elbow extension and distraction of the arm at the shoulder during arm deceleration, have been associated with SLAP lesions.

Biomechanical analysis of the throwing motion has improved our understanding of injury mechanisms, and has provided a basis for improved prevention and treatment of injury. Trainers, therapists and coaches have gained valuable knowledge in determining proper rehabilitation and prevention protocols for the throwing athlete. Surgical treatment has also benefited from an understanding of the stresses that are experience at the various joint structures during pitching. Applications in all of these areas will continue to improve as scientific methods of analyzing human movement continue to be refined. This includes the application of computer modelling techniques to analyze specific joint structures and related joint actions that occur during human movement. These advances will benefit pitchers at all levels of competition by improving their performance and preventing injuries.

## 9.10 REFERENCES

Abdel-Aziz, Y. I., and Karara, H. M., 1971, Direct linear transformation: From comparator coordinates into objects coordinated in close-range photogrammetry, *American Society of Photogrammetry Symposium on Close-Range Photogrammetry*, Falls Church, VA, pp. 1-19.

Altchek D. W., and Dines, D. M., 1995, Shoulder Injuries in the Throwing Athlete, *J. Am. Acad. Orthop. Surg.* **3**: 159-165.

Andrews, J. R., 1985, Bony injuries about the elbow in the throwing athlete, *Instr. Course Lect..* **34**: 323-331.

Andrews, J. R., and Angelo, R. L., 1988, Shoulder arthroscopy for the throwing athlete, *Techniques in Orthopaedics.* **3**: 75-81.

Andrews, J. R., Carson, W. G., and McCleod, W. D., 1985, Glenoid labrum tears related to the long head of the biceps, *The American Journal of Sports Medicine.* **13**: 337-341.

Andrews, J. R., and Fleisig, G. S., 1998, Prevention of throwing injuries, *Journal of Orthopaedic and Sports Physical Therapy.* **27**: 187-188.

Andrews, J. R., and Whiteside, J. A., 1993, Common elbow problems in the athlete, *The Journal of Orthopaedic and Sports Physicial Therapy.* **17**: 289-295.

Andrews, J., and Fleisig, G., 1996, How many pitches should I allow my child to throw? *USA Baseball News, April 5.*

Barrentine, S. W., Takada, Y., Fleisig, G., Zheng, N., and Andrews, J. A., 1997, Kinematic and EMG changes in baseball pitching during a simulated game, *American Society of Biomechanics Annual Meeting*, Clemson, SC, pp. 9-27.

Barrentine, S., Matsuo, T., Escamilla, R., Fleisig, G., and Andrews, J., 1998, Kinematic analysis of the wrist and forearm during baseball pitching, *Journal of Applied Biomechanics.* **14**: 24-39

Clauser, C. E., McConville, J.T., and Young, J. W., 1969, *Weight, Volume, and Center of Mass of Segments of the Human Body*, AMRL-TR-69-70, Wright-Patterson Air Force Base.

Crockett, H. C., Gross, L. B., Wilk, K. E., et al., 2002, Osseous adaptation and range of motion at the glenohumeral joint in professional baseball pitchers, *The American Journal of Sports Medicine.* **30**: 20-26.

De Luca, C. J., 1997, The use of surface electromyography in biomechanics, *Journal of Applied Biomechanics.* **13**: 135-163.

Dempster, W. T., 1955, *Space requirements of the seated operator*, WADC-TR-55-159, Wright-Patterson Air Force Base,

DiGiovine, N. M., 1992, An electromyographic analysis of the upper extremity in pitching, *Journal of Shoulder and Elbow Surgery.* **1**: 15-25.

Dillman, C. J., Fleisig, G. S., and Andrews, J. R., 1993, Biomechanics of pitching with emphasis upon shoulder kinematics, *J Orthop. Sports Phys. Ther.* **18**: 402-408.

Escamilla, R. F, Fleisig, G. S., Alexander, E., and Andrews, J. A., 1994, A kinematic and kinetic comparison while throwing different types of baseball pitches, *Medicine and Science in Sports and Exercise.* **26**: S175.

Escamilla, R. F., Fleisig, G. S,, Zheng, N., Barrentine, S. W., and Andrews, J. R., 2001, Kinematic comparisons of 1996 Olympic baseball pitchers, *J Sports Sci.* **19**: 665-676.

Fleisig, G., Escamilla, R. F., Barrentine, S. W., Zheng, N., and Andrews, J. A., 1996, Kinematic and kinetic comparison of baseball pitching from a mound and throwing from flat ground, *Twentieth Annual Meeting of the American Society of Biomechanics*, Georgia Tech., Atlanta, Georgia, pp. 153-154.

Fleisig, G., Zheng, N., Barrentine, S. W., Escamilla, R. F., and Andrews, J. A., 1996, Kinematic and kinetic comparison of full-effort and partial-effort baseball pitching, *Twentieth Annual Meeting of the American Society of Biomechanics*, Georgia Tech., Atlanta, Georgia, pp. 151-152.

Fleisig, G. S., Andrews, J. R., Dillman, C. J., and Escamilla, R. F., 1995, Kinetics of baseball pitching with implications about injury mechanisms, *Am. J. Sports Med.* **23**: 233-239.

Fleisig, G. S., Barrentine, S. W., Escamilla, R. F., and Andrews, J. R., 1996, Biomechanics of overhand throwing with implications for injuries, *Sports Med.* **21**: 421-437.

Fleisig, G. S., Barrentine, S. W., Zheng, N., Escamilla, R. F., and Andrews, J. R., 1999, Kinematic and kinetic comparison of baseball pitching among various levels of development, *J. Biomech.* **32**: 1371-1375.

Fleisig, G. S., Dillman, C. J., Escamilla, R. F., and Andrews, J. R., 1995, Kinetics of baseball pitching with implications about injury mechanisms, *American Journal of Sports Medicine.* **23**: 233-239.

Fleisig, G. S., Escamilla, R. F., Andrews, J. R., Matsuo, T., Satterwhite, Y., and Barrentine, S. W., 1996, Kinematic and kinetic comparison between baseball pitching and football passing, *Journal of Applied Biomechanics.* **12**: 207-224.

Glousman, R., Jobe, F., Tibone, J., Moynes, D., Antonelli, D., and Perry, J., 1988, Dynamic electromyographic analysis of the throwing shoulder with glenohumeral instability, *J. Bone Joint Surg. Am.* **70**: 220-226.

Indelicato, P. A., Jobe, F. W., Kerlan, R. K., Carter, V. S., Shields, C. L., and Lombardo, S. J., 1979, Correctable elbow lesions in professional baseball players: a review of 25 cases, *Am. J. Sports Med.* **7**: 72-75.

Jobe, F. W., Moynes, D. R., Tibone, J. E., and Perry, J. 1984, An EMG analysis of the shoulder in pitching. A second report, *Am. J. Sports Med.* **12**: 218-220.

Jobe, F. W., and Nuber, G., 1986, Throwing injuries of the elbow, *Clinics in Sports Medicine.* **5**:621-636.

Jobe, F. W., Stark, H., and Lombardo, S. J., 1986, Reconstruction of the ulnar collateral ligament in athletes, *Journal of Bone and Joint Surgery.* **68**: 1158-1163.

Jobe, F. W., Tibone, J. E., Perry, J., and Moynes, D., 1983, An EMG analysis of the shoulder in throwing and pitching. A preliminary report, *Am. J. Sports Med.* **11**: 3-5.

Kindall, J., 1993, *Sports Illustrated Baseball: Play the Winning Way*, Sport Illustrated Books, Lantham, MD.

MacWilliams, B. A., Choi, T., Perezous, M. K., Chao, E. Y., and McFarland, E. G., 1998, Characteristic ground-reaction forces in baseball pitching, *Am. J. Sports Med.* **26**:66-71

Matsuo, T., Fleisig, G. S., Escamilla, R. F., Barrentine, S. W., and Andrews, J. R., 2001, Comparison of kinematic and temporal parameters between different pitch velocity groups, *Journal of Applied Biomechanics.* **17**:1-13

Miniaci, A., Mascia, A. T., Salonen, D. C., Becker, E. J., 2002, Magnetic resonance imaging of the shoulder in asymptomatic professional baseball pitchers, *Am. J. Sports Med..* **30**: 66-73.

Shapiro, R., 1978, Direct linear transformation method for three-dimensional cinematography, *Research Quarterly.* **49**:197-205

Stodden, D., Fleisig, G. S., McLean, S. P., Lyman, S. L., and Andrews, J. A., 2001, Relationship of pelvis and upper torso kinematics to pitched baseball velocity, *J. Applied Biomechanics.* **17**: 164-172.

Takada, Y., Barrentine, S. W., Fleisig, G., Zheng, N., and Andrews, J. A.,1997, Kinematic and electromyographic changes with fatigue, *XVIth International Society of Biomechanics.*

Tibone, J. E., Elrod, B., Jobe, F. W., Kerlan, R. K., Carter, V. S., Shields, C. L., Jr., Lombardo, S. J., and Yocum, L., 1986, Surgical treatment of tears of the rotator cuff in athletes, *J. Bone Joint Surg. Am.* **68**: 887-891.

Walton, J. S., 1981, *Close Range Cine Photogrammetry: A Generalized Technique for Quantifying Gross Human Motion.* Penn State University.

Watkins, R. G., Dennis, S., Dillin, W. H., Schnebel, B., Schneiderman, G., Jobe, F., Farfan, H., Perry, J., and Pink, M., 1989, Dynamic EMG analysis of torque transfer in professional baseball pitchers, *Spine.* **14**: 404-408.

Werner, S. L., Fleisig, G. S., Dillman, C. J., Andrews, J. R., 1993, Biomechanics of the elbow during baseball pitching, *J. Orthop. Sports Phys. Ther.* **17**: 274-278.

Yamanouchi, T., 1998, EMG analysis of the lower extremities during pitching in high-school baseball, *Kurume Med. J.* **45**: 21-25.

Zheng, N., and Barrentine, S. W., 2000, Biomechanics and motion analysis applied to sports, *Phys. Med. Rehabil. Clin. N. Am.* **11**: 309-322.

Zheng, N., Fleisig, G., and Andrews, J. A., 1999, Biomechanics and injuries of the shoulder during throwing, *Athletic Therapy Today.* **4**:6-10.

Zheng, N., Fleisig, G., Young, J. L., and Andrews, J. A., 1998, Spine kinematics during baseball pitching with implications about injury mechanisms, *The Third North American Congress on Biomechanics*, Waterloo, Ontario, Canada, pp. 479-480.

# Chapter 10

# The Rising Fastball and Other Perceptual Illusions of Batters

Terry Bahill[1] and David G. Baldwin[2]

*[1]Systems and Industrial Engineering, University of Arizona, Tucson, AZ 85721- 0020; PH: (520) 621-6561; FX: (520) 621-6555; EM: terry@sie.arizona.edu*

*[2]HylaPark, Inc., Carlsbad, CA 92009; PH: (760) 931-8078; EM: david_baldwin@adelphia.net*

## 10.1 INTRODUCTION

Sometimes athletes develop effective strategies although their underlying assumptions are fallacious. Baseball provides an example in which misunderstood physical and psychological phenomena are the basis of strategies of both the pitcher and the batter.

With limited knowledge of the physics of a pitch's trajectory, baseball players argue that a high, hard fastball will sometimes accelerate and rise suddenly as it approaches the plate. Batters claim the ball can jump a foot or more just before the pitch reaches the strike zone. The batter is unable to adjust to this apparent rapid change in trajectory, as the ball seems to hop over the bat.

This perceived rise occurs on fastballs that have exceptional speed, are belt high or higher, and, most often, are on the inside half of the strike zone. Batters say that the pitcher "pumped it up," or "put a little extra on that one." In describing Dwight Gooden, Tony Gwynn states, "He rears back and throws you that high-rising fastball whenever he needs a big pitch" (Gwynn and Rosenthal, 1992). The rising fastball is often referred to as "smoke," "cheese," or "express."

Pitchers have devised various means for producing and enhancing the rise effect. The pitch is thrown with an overhand or three-quarter delivery so that a rapid backspin is imparted to the ball (Seaver and Lowenfish, 1984). The ball is usually gripped with the index and middle fingers extending across four seams (House, 1994). Each seam is a ridge that provides high impedance to the airflow around the ball when the seam is perpendicular to the direction of flight. Backspin with four seams perpendicular to the trajectory is thought to give the ball maximum "lift." In general, pitchers believe that this lift accounts for the rise experienced by batters (e.g., see Seaver and Lowenfish, 1984, p. 99 and Flatow, 1988, p. 93). For a literature review of the rising fastball, see McBeath (1990).

A phenomenon that batters experience far more often than the rising fastball, however, is a pitch that appears to break sharply downward as it nears the plate. Several kinds of pitches behave in this way – the overhand or three-quarter curveball, the sinking fastball (thrown three-quarter or sidearm, with strong pressure applied by the index finger), the split-finger fastball or forkball (thrown directly overhand), and several kinds of change ups. We will refer to these pitches collectively as "breaking pitches." A breaking pitch will be defined as any pitch that seems to exhibit an abrupt drop near the plate.

Batters describe these pitches with phrases such as, "the bottom dropped out of it" or "it fell off the table." Gwynn uses the metaphor, "a dead fish." He writes that hitting a sinker is "like hitting a rock – it won't go anywhere" (Gwynn and Rosenthal, 1992).

Breaking pitches should be thrown very low in the strike zone or even a little below the knees. If the batter makes contact with such a pitch, he is most likely to hit the ball on the ground. The objective of an experienced pitcher is to "take the sting out of the ball," and forcing the batter to hit a groundball is the surest way to do that (Curt Simmons, personal communication).

Nearly all major league pitchers have at least one breaking pitch in their repertoire. Pitching coaches consider a breaking pitch to be an essential ingredient for an effective strategy. Unlike the rising fastball, some kind of breaking pitch can be learned by any pitcher.

Thus, pitchers rely on sudden, rapid vertical movement of their pitches and batters perceive that movement and must contend with it to be successful. . Even great batters (e.g., Aaron and Cohen, 1974) and great pitchers (e.g., Palmer and Cohen, 1975) are convinced that a high fastball can rise and a breaking pitch can drop sharply. But is it just an illusion?

## 10.2 THE RISING FASTBALL

According to principles of physics, a rising fastball is impossible. Yet, most batters claim it exists and can describe their frustration in attempting to hit this pitch. This widespread belief has persisted in spite of the recognition by a few coaches that such a pitch cannot occur (House, 1994; Thrift and Shapiro, 1990; Mike Scioscia quoted by Will, 1990, p. 91). Our simulation and model explain this contradiction. Although the numbers given in this paper are for professional baseball players, the simulation and model apply to all players, right down to Little Leaguers. They can also be extended to cricket (Regan, 1992). This section on the rising fastball is based on Karnavas, Bahill and Regan (1990) and Bahill and Karnavas (1993).

The rising fastball could be defined variously as a pitch where the ball (1) jumps up, right in front of the plate, (2) crosses the plate above the pitcher's release point, (3) is going upward when it crosses the plate, or (4) falls less than would be expected due to gravity. By definitions (2) and (3), a rising fastball could be thrown by a sidearm baseball pitcher or a softball pitcher, but not with an overhand or three-quarter pitching delivery. For example, an overhand pitch is released about six feet above the ground; if the ball crossed the plate higher than this it would not be a strike. And in order for the ball to be going upward when it crosses the plate, it would have to fall initially, and then near the end of its flight, experience an upward force that is greater than that of gravity. A force that opposes gravity is produced by the backspin on an overhand fastball. However, the maximum spin rate ever measured for a human pitch, 2300 revolutions per minute (rpm), would only create a force two-thirds that of gravity (Watts and Bahill, 2000). So, although a fastball's lift due to backspin may not overcome gravity, it does make it fall less than would be expected due to gravity, which is definition (4). All fastballs fit this definition, making it trivially simple and uninteresting. Therefore, for the rising fastball we will only consider overhand pitches that conform to definition (1).

### 10.2.1 The Simulation of the Rising Fastball

Figure 10.1a and Table 10.1 show our simulations of 95 and 90 mile per hour (mph) fastballs. Our simulations include the effects of gravity and the aerodynamic forces of lift and drag. In these simulations, both pitches were launched one degree upward with 1500 rpm of backspin. As shown in Fig. 10.1d, the distance between the front of the pitcher's rubber and the tip of the plate is 60.5 feet. But the pitcher releases the ball about 5 feet in front of the rubber. Therefore, for these simulations, the pitcher's release point was 55.5 feet away from the tip of the plate. The bat hits the ball about 1.5 feet in

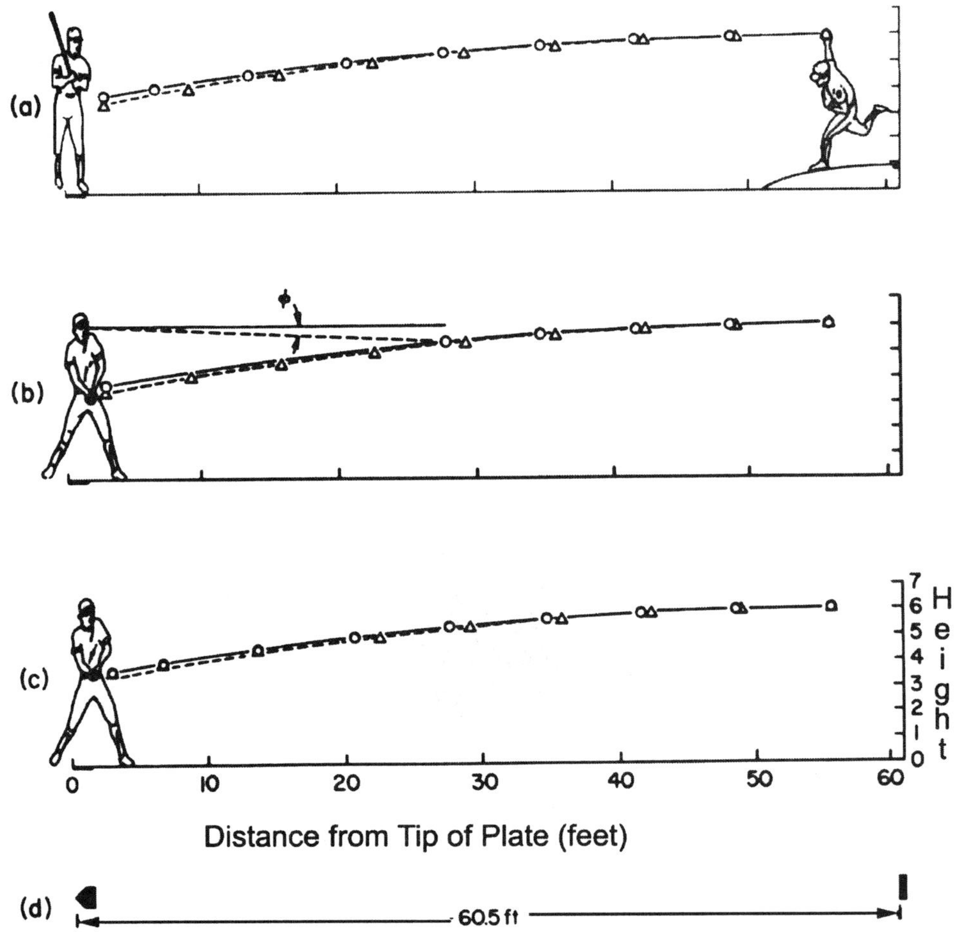


*Figure 10.1* (a) Computer simulation of the trajectory of a 95-mph fastball (solid line and circles) and a 90-mph fastball (dashed line and triangles). The slower pitch takes longer to get to the plate and therefore drops more. (b) Computer simulation of the trajectory of a 95-mph fastball (solid line and circles), and the batter's mental model of this trajectory (dashed line and triangles) when the batter underestimated the speed of the pitch by 5 mph. (c) The same simulation as Fig. 10.1b, except that when the ball was 20 feet in front of the plate, the "batter" realized his mental model was wrong and corrected it, thus putting his mental model triangles on the 95 mph trajectory. (d) Physical dimensions for adult baseball. Adapted from Karnavas et al. (1990), pg. 924, with permission of © IEEE).

| Table 10.1. Trajectories of Fastballs | | | | | | |
|---|---|---|---|---|---|---|
| Time Since Release (msec) | 95 mph fastball | | | 90 mph fastball | | |
| | Distance (ft) | Height (ft) | Speed (mph) | Distance (ft) | Height (ft) | Speed (mph) |
| 0 | 55.5 | 6.00 | 95.0 | 55.5 | 6.00 | 90.0 |
| 50 | 48.6 | 5.86 | 93.3 | 49.0 | 5.86 | 88.5 |
| 100 | 41.8 | 5.67 | 91.7 | 42.5 | 5.68 | 87.0 |
| 150 | 35.2 | 5.43 | 90.2 | 36.2 | 5.44 | 85.6 |
| 200 | 28.6 | 5.15 | 88.6 | 30.0 | 5.16 | 84.3 |
| 250 | 22.2 | 4.83 | 87.2 | 23.8 | 4.84 | 82.9 |
| 300 | 15.8 | 4.46 | 85.7 | 17.8 | 4.47 | 81.6 |
| 350 | 9.6 | 4.05 | 84.4 | 11.9 | 4.05 | 80.4 |
| 400 | 3.4 | 3.59 | 83.1 | 6.0 | 3.59 | 79.2 |
| 404 | 3.0 | **3.56** | 83.0 | | | |
| 426 | | | | 3.0 | **3.33** | 78.6 |

front of the batter's head, which was assumed to be aligned with the front of the plate. The plate measures 17 inches from the back tip to the front edge. So, in these simulations the point of bat-ball collision was 3 feet in front of the tip of the plate, which is represented in the bottom rows of Table 10.1. The pitcher's release point was assumed six feet high. Later, in Table 10.3, we will show how sensitive our conclusions are to these numbers.

## 10.2.2 The Perceptual Model

The illusion of the rising fastball could be the result of the batter underestimating the speed of the pitch. For example, suppose the pitcher threw a 95-mph fastball, but the batter underestimated its speed and thought it was only going 90 mph. The batter would expect to hit the ball 3.33 feet above the ground (from the sixth column, 426-msec row of Table 10.1). But if the batter were to take his eye off the ball (indicated by the absence of some circles in the actual 95 mph pitch of Fig. 10.1b) and look at his bat in order to see the expected bat-ball collision, then the next time the batter saw the ball it would be 3.56 feet above the ground (third column, 404 msec row of Table 10.1), three inches above his bat. The batter might explain this by saying, "The ball jumped three inches right in front of the plate." Should the batter underestimate the speed at 80 mph, the batter would say the ball jumped one foot. The manager in the dugout, the catcher and the center field TV camera, having different perspectives, would not see this jump.

### 10.2.2.1 Four Types of Eye Movements

The purpose of the eye-movement systems is to keep the fovea, the region of the retina with the greatest visual acuity, on the object of interest. To accomplish this task, the following four types of eye movements work in harmony: *saccadic* eye movements, which are used in reading text or scanning a roomful of people; *smooth-pursuit* eye movements, used when tracking a moving object; *vergence* eye movements, used when looking between near and far objects; and *vestibulo-ocular* eye movements, used to maintain fixation during head movements. These four types of eye movements have four independent control systems, involving different areas of the brain. Their dynamic properties, such as latency, speed and bandwidth are different, and they are affected differently by fatigue, alcohol, drugs and disease. For simplicity, none of the other neural systems associated with vision or eye movement will be discussed in this paper.

The specific actions of these four systems can be illustrated by the example of a duck hunter sitting in a rowboat on a lake. He scans the sky using saccadic eye movements, jerking his eyes quickly from one fixation point to the next. When he sees a duck, he tracks it using smooth-pursuit eye movements. If the duck lands in the water right next to his boat, he moves his eyes toward each other with vergence eye movements. Throughout all this, he uses vestibulo-ocular eye movements to compensate for the movement of his head caused by the rocking of the boat. Thus, all four systems are continually used to move the eyes.

### 10.2.2.2 Eye Tracking Strategies

Batters use one of two strategies in tracking the pitch (Bahill and LaRitz, 1984). *The optimal learning strategy,* which allows the batter to see the ball hit the bat, is: track the ball over, roughly, the first two-thirds of its trajectory with smooth pursuit eye movements, make a saccadic eye movement to a predicted point of bat-ball collision, continue to follow the ball with peripheral vision letting the ball catch up to the eye, and finally, at the end of the ball's flight, resume smooth pursuit tracking with the images of the ball and bat on the fovea. It is called the optimal learning strategy, because the batter observes the ball, makes a prediction of where it will contact the bat, sees the actual position of the ball when it contacts the bat, and uses this feedback to learn to predict better the next time the pitcher throws that pitch. *The optimal hitting strategy*, which does not allow the batter to see the ball hit the bat, is: track the ball with smooth pursuit eye movements and fall behind in the last five feet. It is called the optimal hitting strategy, because the batter keeps his eye on the ball longer, which should allow him to make a more accurate prediction of where the ball will cross the plate. We have no evidence that batters voluntarily switch between these two strategies.

It is not surprising that batters do not verbalize their strategies. As you are reading this paper, you typically make three saccades to the right and one to the left for each line of type. But you are probably not capable of explaining your scanning patterns to someone else. Likewise, the batter is probably not capable of stating whether he is using the optimal learning strategy or the optimal hitting strategy.

With the optimal learning strategy, a batter would perceive more rising fastballs because his eyes would not be on the ball (which is why we removed some circles from Fig. 10.1b) when the batter started his swing. This occurs about 150 msec before bat-ball contact, when the ball is about 20 feet in front of the plate. Therefore, during the pitch the batter could not discover inaccuracies in his estimation of pitch speed and make adjustments. Whereas, with the optimal hitting strategy, a batter would perceive fewer rising fastballs because his eye would be on the ball when it was 15 to 25 feet from the plate. Therefore, the batter could sense inaccuracies in his speed estimation and, as shown in the last 150 msec of Fig. 10.1c, make an appropriate correction, such as checking his swing.

The batter performs a different operation in each third of the pitch's flight: observe, compute and swing. For the rising fastball, as the batter observes the ball for the first third of its flight, he underestimates the pitch-speed and, therefore, miscomputes the height of the bat-ball collision. Then, at the start of his swing, the batter takes his eye off the ball to look at the estimated bat-ball collision point. When the ball comes back onto his fovea it is higher than the batter thought it would be.

In the next section, we develop a neurophysiological model to investigate the perceptual illusion of the rising fastball. This model of the human visual processing is based on (Todd, 1981: Regan, 1986: Karnavas, Bahill and Regan, 1990: Regan, 1992: Bahill and Karnavas, 1993) and uses only experimentally verified primary visual processes.

### 10.2.3 The Neurophysiological Model

The batter must precisely predict the location of the ball in four-dimensional space: $t$, the time coordinate and the $x$, $y$ and $z$ spatial coordinates shown in Fig. 10.2. To hit the ball the batter must predict *when* and *where* it will cross the plate. Bunting the ball is analogous to catching the ball with a bat. In this act, the batter does not need to predict **both** when and where; because when is not important. The batter has only one task:

predicting *where* the ball will be. And this task is much simpler; thus, most bunt attempts make contact with the ball.

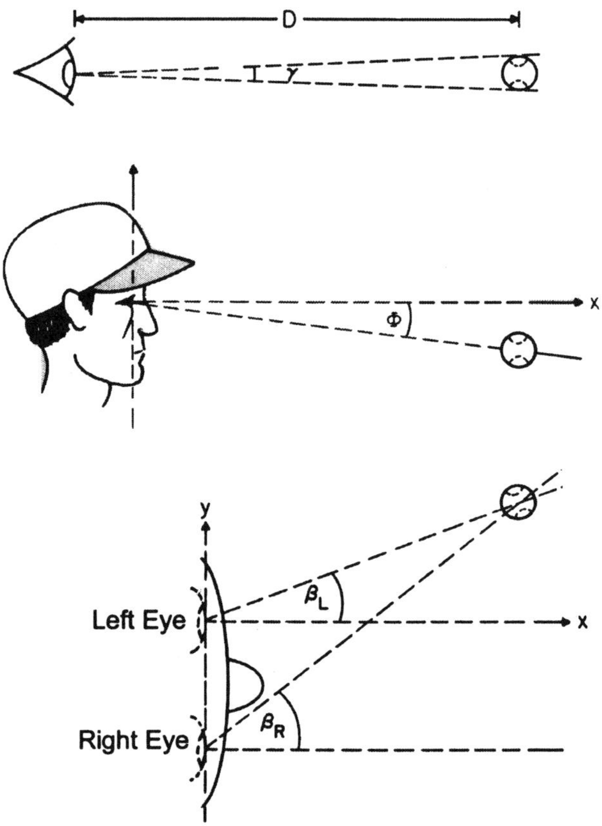


*Figure 10.2.* Visual system parameters used by the batter: (top) angular size of the baseball, $\gamma$, the batter has no direct means to sense the distance to the ball (*D*); (middle) distance of the ball's image off the batter's fovea, $\Phi$; and (bottom) horizontal angle of the right and left eyes, $\beta_R$ and $\beta_L$. Adapted from Karnavas et al. (1990), pg. 925, with permission of © IEEE).

### 10.2.3.1 Predicting When

Let us see how the batter who is swinging at a pitch can judge *when* the ball will be crossing the plate. In his novel The Black Cloud, Sir Fred Hoyle (1957) showed that the time until contact with an object moving along the line-of-sight can be approximated by the equation

$$\text{time until contact} \approx \frac{\gamma}{d\gamma / dt} \tag{10.1}$$

where $\gamma$ and $d\gamma / dt$ are respectively the object's angular size and rate of change of angular size. Lee (1976) subsequently gave the label tau ($\tau$) to the right side of the equation and suggested that estimations of time until contact that are used to control motor actions are based on this variable. For example, it has been shown that birds use $\tau$ when diving into the water to catch prey, and athletes use $\tau$ when jumping to hit a dropped ball, adjusting strides when running hurdles and timing their swings in table tennis; for these tasks time until contact is judged with an accuracy around 2 to 10 msec. Cricket players time their swings with an accuracy of ±5 msec (Regan, 1992). Top sports players estimate the time until contact of the bat and ball with an accuracy of better than three msec (Regan, 1997; Gray and Regan, 1998). The batter's calculation of time until contact has three sources of error. First, Eq. 10.1 is only an approximation because it uses the approximation $Tan\Psi = \Psi$. Second, the ball is not headed directly at the batter's eye. In our simulations, these two sources produced errors of about one msec at the start of the swing. The third source of error, which results from the batter hitting the ball 1.5 feet in front of his eyes, produces a constant 11 msec of error in underestimating the time of arrival at impact.

The human visual system can implement Eq. 10.1. First, there is psychophysical evidence that the human brain contains units tuned to size ($\gamma$), and size-tuned neurons have been found in monkey visual cortex. Second, psychological studies have shown that the visual system has specialized "looming detectors" that compute $d\gamma / dt$ independent of the object's trajectory (Regan and Beverley, 1978, 1980). Furthermore, specific brain neurons are sensitive to changing-size, $d\gamma / dt$ (Regan and Cynader, 1979; Saito et al., 1986). Using these two pools of neurons, the brain could compute $\tau$.

Is the movement of the baseball within physiological thresholds? For objects subtending less than 1.5 deg, $d\gamma / dt$ as low as 0.02 deg/sec can be detected (Regan and Beverley, 1978). Lee (1976) suggested a threshold of 0.08 deg/sec. When the pitcher releases a 95 mph fastball, $\gamma$ is ¼ deg and $d\gamma / dt$ is 0.66 deg/sec. Therefore $d\gamma / dt$ is well above visual threshold from the moment the ball leaves the pitcher's hand. The value of $\gamma$ remains below 1.5 deg until the ball is ten feet from the tip of the plate. Figure 10.3 shows how $\gamma$ and $d\gamma / dt$ change during the flight of the ball. The arrows indicate the points where they cross their thresholds.

In this section, we proposed a model for estimating time until contact. We showed that the human brain has neuronal circuitry that could do the needed computations. And we showed that the parameters of a pitch are within physiological thresholds for these neuronal circuits. Next, we assert that people can accurately estimate time until contact using Eq. 10.1 alone. In experiments reported by Gray and Regan (1998), observers using only Eq. 10.1 were able to estimate time until contact with errors of 2 to 12%. As we have stated before, the pitch can be divided into thirds. During the first third the batter gathers sensory information; during the middle third a computation is made of where and when the ball will make contact with the bat; and the final third is the swing of the bat. With this scenario, the batter would have to use information gathered in the first third of the pitch to estimate time until contact. For a 95 mph fastball with a duration of 404 msec, this would be 268 msec before contact. A 2% estimation error corresponds to a temporal error of 5 msec, which is within the required temporal error of 9 msec computed by Watts and Bahill (2000). The 2% error is at the low end of the 2 to 12 % range given by Gray and Regan. It has been shown before (Bahill and LaRitz, 1984) that professional athletes usually perform at the optimal end of the range of values found in laboratory experiments. Using other estimators for time until contact might improve accuracy, as discussed below.

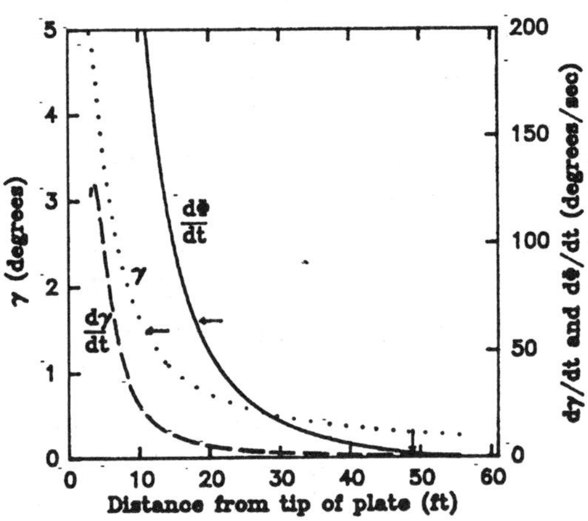


*Figure 10.3*. Physiological clues to ball position, $\gamma$ , $d\gamma\ /\ dt$ and $d\Phi\ /\ dt$, as functions of distance to the tip of the plate, for a 95 mph fastball. Adapted from Bahill and Karnavas (1993) with permission of Journal of Experimental Psychology, Human Perception and Performance.

Time until contact can be estimated using Eq. 10.1. It can also be estimated using binocular disparity ($\delta$ ) according to the equation

$$\text{time until contact} \approx \frac{I}{Dd\delta / dt} \tag{10.2}$$

where I is the interpupillary separation and D is the distance to the ball (Gray and Regan, 1998). When a 90 mph fastball is 40 feet from the plate, $d\delta / dt$ is about 0.8 deg/sec, which is above discrimination threshold. Gray & Regan (1998) found that estimates of time until contact for a simulated approaching ball are more accurate when based on a combination of Eqs. 10.1 and 10.2 than for either cue alone suggesting that a hitter might estimate time until contact of an approaching ball more accurately if both these sources of information were used. It is not known whether or how they are combined during baseball batting, however.

We conclude that from the instant the ball leaves the pitcher's hand, the batter's retinal image contains accurate cues for time until contact, and that the human visual system is capable of utilizing these cues. Evidentially most batters estimate when the ball crosses the plate with an accuracy of ± 9 msec, otherwise the ball would be hit foul. Next, we want to consider the more difficult issue of judging *where* the ball will be at the time of bat-ball contact

### 10.2.3.2 Predicting Where

To hit the ball the batter must predict both *when* and *where* it will cross the plate. Let us now see how the batter can judge *where*.

It is important to note that the batter has no direct visual access to the *x*, *y* and *z* spatial coordinates: the batter's judgments must be based entirely on retinal image data. The relevant retinal parameters are the angular size of the ball, $\gamma$ , and the angular distance of the ball's image off the fovea, $\Phi$, shown in Fig. 10.2, and their time derivatives $d\gamma / dt$ and $d\Phi / dt$. We assume the batter's eyes are in the x-y plane as shown in Fig. 10.2, although the Pittsburgh Pirate studied by Bahill and LaRitz (1984) actually rotated his head 23° in pitch and 12° in roll.

The batter can estimate the ball's height at contact from the ball's vertical speed and the time until contact. The batter can estimate the ball's vertical speed from the retinal velocity and the distance to the ball. The smallest value of retinal velocity, $d\Phi / dt$, that can be detected is about 0.02 deg/sec. Over a range of moderate speeds, between about 2 and 64 deg/sec, the discrimination of differences in $d\Phi / dt$ is within 5% (McKee, 1981; Orban et al. 1984). Figure 10.3 shows that for a 95 mph pitch the value of $d\Phi / dt$ reaches 0.02 deg/sec before the ball has traveled one foot, reaches 2 deg/sec 49 feet from the tip of the plate, and reaches 64 deg/sec 18 feet from the tip of the plate. Therefore, the batter can make discriminations of $d\Phi / dt$ that diminishes only after the swing has begun. (The swing starts when the ball is about 19 feet away from the bat-ball contact point.)

Now comes the crucial element in our explanation of the illusion of the rising fastball. Although retinal image information provides an accurate cue for the ball's time until contact, it provides poor cues to the absolute distance to the ball (*D* in Fig. 10.2a) and for its line-of-sight speed. Classical stereoscopic depth perception is of little help in this regard: although the stereoscopic depth system provides a precise indication of relative depth (i.e. the difference between the x-axis distances of two objects imaged near the fovea), it provides little indication of absolute distance. In tracking the pitched ball, the batter has one object, the ball, imaged on his fovea. Therefore, the batter cannot measure the distance to the ball or the pitch speed; the batter can only estimate them.

Our psychophysical explanation for the rising fastball is as follows. The batter can estimate pitch speed and the time since the ball left the pitcher's hand. The batter can use these data in conjunction with his experience to estimate the distance to the ball ($\hat{D}$). The "hat" on top of a symbol indicates that it is an estimate. The batter can then use this estimate for distance and the ball's retinal image velocity $d\Phi / dt$ to estimate the vertical velocity $d\hat{z} / dt$. From the vertical velocity and the time until contact ($\tau$) the batter can estimate how far the ball will fall in the last part of its flight, and therefore estimate the height of the ball when it crosses the front edge of the plate. For example, suppose the pitcher enhances the batter's expectations with a series of 90 mph pitches, and then throws an unusually fast 95 mph pitch. Assume the batter uses a 90 mph mental model to interpret the retinal image information about the 95 mph pitch. Suppose the batter tried to estimate the ball's vertical speed 250 msec after the ball left the pitcher's hand. If the actual pitch were a 95 mph fastball, it would be 20.7 feet from his eye, subtracting 1.5 feet (the distance between the tip of the plate and the batter's eye) from the x distance of 22.2 in Table 10.1. Its vertical velocity of -8 ft/sec (from velocity = *gt*) would, at this distance, produce a retinal velocity of -21 deg/sec. However, if the batter thought the pitch was a 90-mph fastball, this would translate to 22.3 feet away at 250 msec after release. At this distance, a retinal image velocity of -21 deg/sec would indicate that the vertical velocity was about -8.6 ft/sec. So the batter would think the ball was falling faster than it really was and would probably swing under the ball. Therefore, if the batter made a saccadic eye movement to a predicted point of bat-ball contact, this point would be below the ball when the ball caught up with the eye, and the ball would seem to have jumped upwards, in this example by three inches. To be sure, this error of visual judgment could be avoided if the batter had an accurate visual cue to the ball's absolute distance

($D$) or its speed, but, as we have seen, the batter is essentially "blind" to these two important parameters. Blind means he as no direct optical sense for these two parameters: he must estimate them using other retinal image data.

This model can be summarized with Fig. 10.4 and Eq. 10.3,

$$\hat{z}_{\text{fall}} = (\hat{D}_0 - \hat{t}_{\text{sinceRelease}}\hat{s})\left(\frac{d\Phi}{dt}\frac{\gamma}{d\gamma/dt}\right) \tag{10.3}$$

where $\hat{z}_{\text{fall}}$ is the estimate of how much further the ball will fall before contact, $\hat{D}_0$ is the estimated distance between the ball and the batter's eye at the time of release, $\hat{t}_{\text{sinceRelease}}$ is the estimated time since release, $\hat{s}$ is the estimated pitch speed, $d\Phi/dt$ the retinal image velocity, $\gamma$ is the retinal image size, and $d\gamma/dt$ is the rate of change of retinal image size. The model of Fig. 10.4 shows how the batter's brain can predict how much further the ball will fall before its contact with the bat. Summation ($\Sigma$ symbol in a circle), division ($\div$ sign in a pentagon) and multiplication ($\times$ sign in a pentagon) can be accomplished by brain circuitry. Symbols with hats (^) are estimates.

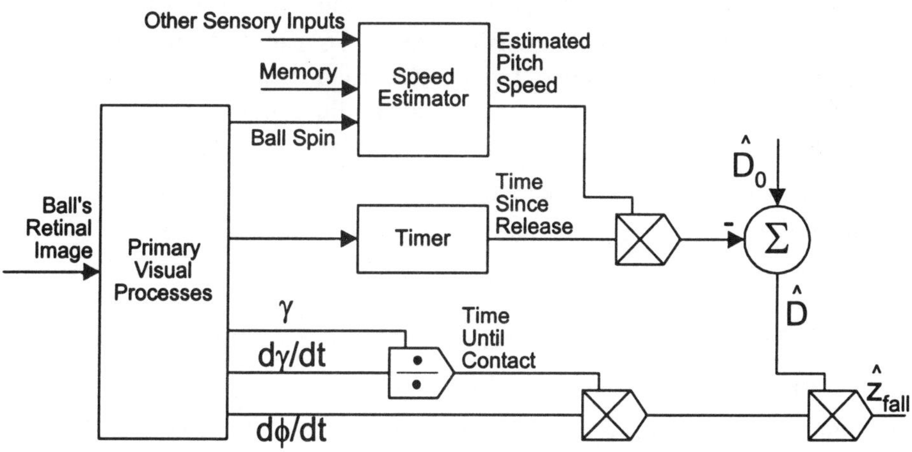


*Figure 10.4.* A model for a batter estimating the height of the ball at the time of its collision with the bat. The input to the system is the ball's retinal image and its output is the estimate of how much further the ball will fall before contact with the bat. Adapted from Bahill and Karnavas (1993) with permission of Journal of Experimental Psychology, Human Perception and Performance.

The first term of Eq. 10.3 is the estimated distance to the ball at the moment of prediction, $\hat{D}$, and the last term is $\tau$ from Eq. 10.1, so Eq. 10.3 can be rewritten as

$$\hat{z}_{\text{fall}} = \hat{D}\frac{d\Phi}{dt}\tau \tag{10.4}$$

Overestimating any one of these three terms could produce the illusion of the rising fastball. But, as we have already shown, batters can accurately perceive $d\Phi / dt$ and estimate $\tau$ ; the difficulty is in estimating the distance to the ball. And in this model, they use estimated pitch speed to estimate the distance to the ball. That makes the speed estimator in Fig. 10.4 the crucial element of this model. In order to estimate the height of the ball at the time of contact, the batter must be able to estimate the pitch speed. But pitch speed cannot come from the primary visual processes. The speed estimator receives inputs from the primary visual processes, such as ball spin. The speed estimator probably uses memory and other sensory inputs such as the motion of the pitcher's arms and body, or a sign from a base runner on second who has stolen the catcher's signals.

The speed estimator might be most accurate just after the pitcher releases the ball, for at this point, the batter should be able to estimate its distance $\hat{D}_0$ quite well. And the distance $\hat{D}_0$ combined with the time until contact would provide the speed. But at this time, the estimation of the ball's height at the time-of-contact may still be in error, because retinal image velocity $d\phi / dt$ is inaccurate until the ball is 49 feet from the tip of the plate.

We considered many other signals that might be used to help estimate the speed of the ball and its distance, but we discarded them because they were inappropriate. For example, the retinal image could contain a cue to line-of-sight speed $dD / dt$ because $dD / dt = (D / \gamma)(d\gamma / dt)$. And, as we have already shown, the batter has visual clues for $\gamma$ and $d\gamma / dt$, but unfortunately the batter has no reliable visual cue to absolute distance $D$, so the batter is unable to take advantage of the geometrical fact expressed in this equation. We considered many techniques for estimating the distance to the ball. (1) The vergence eye movement system (the difference between $\beta_L$ and $\beta_R$ in Fig. 10.2c) cannot help, because vergence changes do not contribute to motion-in-depth sensation (Regan et al., 1986). (2) The differences in the batter's lines-of-sight for the two pitches of Fig. 10.1 do not differ by one degree until the ball is 6 feet from the batter, and extra-retinal sensation of eye position is only good to about one degree (Matin and Kibler, 1966; Steinbach, 1970). (3) Parallax is often used to help judge distance. However, the batter only starts his translational head movements in the last third of the ball's flight (Bahill and LaRitz, 1984; Watts and Bahill, 2000) and this is too late to help him. (4) We have already commented that stereoscopic depth perception only gives clues about the relative depth of two objects and not about absolute distance to a single object, so it cannot help and is, in fact, not necessary for good hitting. Indeed, according to Kara (1990), an ophthalmologist who examined him, Babe Ruth was amblyopic and never had more than 20/200 vision in his left eye. Therefore, we conclude once again that the batter has no information to determine accurately the ball's absolute distance (*D*) from his eye.

We have said before that the batter must predict both *when* and *where* the bat-ball collision will occur. To hit a line drive in fair territory, the batter must estimate *when* to within ± 9 msec and *where* to within ± ½ inch (Watts and Bahill, 2000). Comparing these numbers to the 22-msec and 2.8 inch differences between the 90 and 95-mph fastballs, we see that estimating *where* requires greater percent accuracy. But batters seem to be more accurate at estimating *when*. Indeed few line drives are hit into foul territory (accurate position but poor timing), whereas many foul tips and pop ups are hit into foul territory (poor accuracy and poor timing). Fig. 10.4 suggests an explanation: *when* ($\tau$) can be computed from primary visual processes, whereas the pitch speed, and therefore *where*, can only be estimated. So *where* is the crucial parameter: it requires greater precision, yet it cannot be calculated from primary visual processes, it can only be estimated.

Although we developed this model to explain the illusion of the rising fastball, it might also be applied to other pitches. For example, the change up is only effective if it fools the batter and makes him overestimate the pitch speed. According to our model, this would make the batter swing over the ball. Our model predicts that effective change ups to right-handed batters should result in ground balls to the left side of the infield.

The existence of the speed estimator in Fig. 10.4 enables predictions that (1) large variations in speed from pitch to pitch should lead to large spatial errors in the swing, (2) unexpectedly fast pitches (e.g., when a fast pitch follows a sequence of slow pitches or a fast pitch is thrown on an unexpected count) should cause the batter to swing under the ball, (3) unexpectedly slow pitches (e.g. a change up) should cause the batter to swing over the ball and (4) the sequence of recent events should influence performance. All of these predictions from the model were supported by Gray (2002). They also are manifest in the philosophies of great pitchers – Warren Spahn is quoted by Will (1990) as saying, "Hitting is timing. Pitching is upsetting timing."

## 10.3 EXPERIMENTAL ASSESSMENT OF THE MODEL

To help assess the performance of the model Bahill and Karnavas (1993) ran experiments using a mechanical pitching machine. They threw 450 pitches to seven subjects: three adults and four boys aged 9, 11, 11 and 13. Nominally, the speed of the machine was set for 50 mph, but occasionally it threw a fast pitch at 55 mph. The number of normal pitches between these fast pitches was randomly chosen from amongst 3, 4, 5, and 6. An observer (who did not know the pattern of normal and fast pitches) recorded the relationship of the bat and ball when the ball crossed the plate. They averaged the results of the fast pitches and of the two pitches before and after, as shown in Fig. 10.5. These data show that on the fast pitches the batters swung below the ball, which is just what would happen if they underestimated the speed of the pitch. It was statistically significant that when the unusually fast pitch was thrown, the batters underestimated the speed and swung under the ball. This is precisely the illusion of the rising fastball.

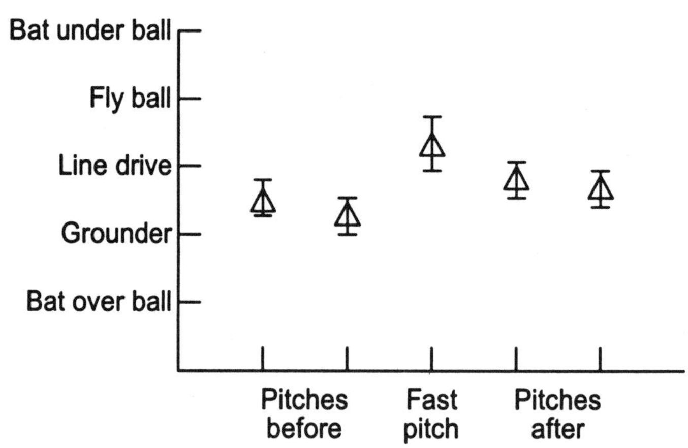


*Figure 10.5.* Averaged data from seven batters showing that when an unusually fast pitch was thrown most batters swung under the ball. The triangles are the mean values and the vertical bars are the 95% confidence intervals.

## 10.4 THE BREAKING CURVEBALL

There is no longer a controversy about whether or not a curveball curves; it does, because it obeys the laws of physics. These laws say that the spin of the ball causes the curve (Watts and Bahill, 2000; Fig. 10.6). Should this spin be horizontal (as on a toy top) the ball curves horizontally (of course it also drops due to gravity). If it is top spin, the ball drops more than it would due to gravity alone. Else, if it is somewhere in between, the ball both curves and drops. In baseball, most curveballs curve horizontally and drop vertically. The advantage of the drop is that the sweet area of the bat is about four inches long but only an inch high. Thus, a vertical drop would be more effective at taking the ball away from the bat's sweet area than a horizontal curve. We now want to present the principles of physics that explain why the curveball curves.

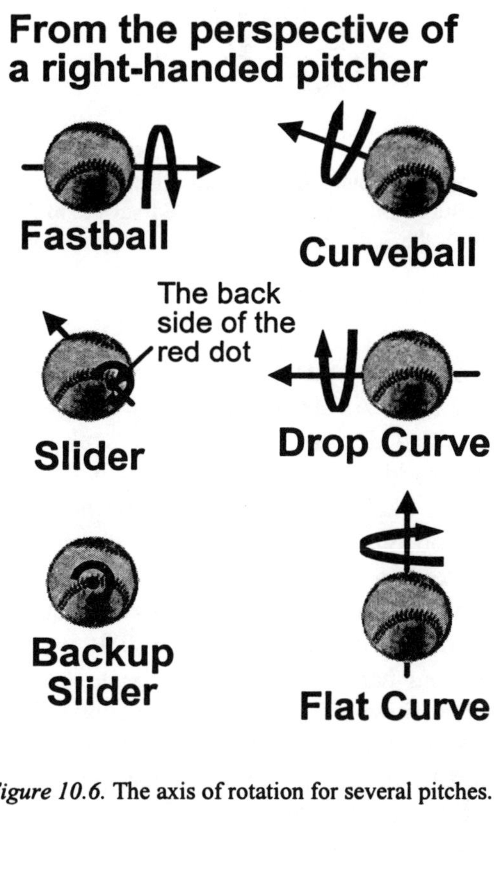


*Figure 10.6.* The axis of rotation for several pitches.

The first part of our explanation invokes Bernoulli's Principle. When a spinning ball is placed in moving air, as shown in Fig. 10.7, the movement of the ball and its seams slows down the air flowing over the top of the ball and speeds up the air flowing underneath the ball. Now, according to Bernoulli's Equation, the point with lower speed (the top) has higher pressure, and the point with higher speed (the bottom) has lower pressure. This difference in pressure pushes the ball downward.

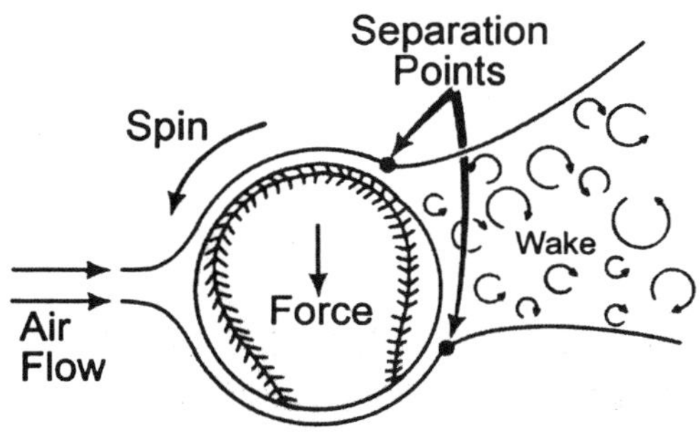


*Figure 10.7.* Air flows smoothly around the ball until it gets to the separation points. There the airflow changes into a turbulent wake that is deflected upward in this figure. If this were a top view, it would explain the curve of a ball. If it were a side view, it would explain the drop of a ball.

Our second (and probably better) explanation involves the principle of Conservation of Momentum. As shown in Fig. 10.7 the wake of turbulent air behind the ball is deflected upward. You can see such wakes behind boats. In a boat, swinging the rudder to the right deflects water to the right, and to conserve momentum the back of the boat must be pushed to the left. You can feel this force if you put your hand out the window of a moving car. (Make sure the driver knows you are doing this!) Tilt your hand so that the wind hits the palm of your hand at an angle. This deflects the air downward, which causes your hand to be pushed upward. Tilt your hand so that the wind hits the back of your hand. This deflects the air upward, which causes your hand to be pushed downward. Now let us relate this to the spinning baseball in Fig. 10.7. Before the ball interacts with the air, all the momentum is horizontal. Afterwards the air in the wake has upward momentum. The principle of conservation of momentum therefore requires that the ball have downward momentum. Therefore, the ball will curve downward.

There are several ways to shift the wake of turbulent air behind a baseball. The wake is shifted by the spin on a curveball. The friction that slows down the flow of air over the top of the ball causes the air to separate from the ball sooner on the top than on the bottom, as shown in Fig. 10.7. This shifts the wake upward, thus pushing the ball downward. For nonspinning pitches, such as the knuckle ball and the scuff ball, when the seams or the scuff are near the bottom separation point they create turbulence, which delays the separation, as shown in the bottom of Fig. 10.7, again shifting the wake upward, and pushing the ball downward.

So, when the pitcher puts spin on the ball, the wake of turbulent air behind the ball is moved to one side, causing the ball to curve, thereby confounding the poor batter who is trying to hit it.

Now let us return to the breaking curveball. Baseballs (except for knuckleballs etc.) follow smooth parabolic trajectories. The 90-mph fastball of Table 10.1 and Fig. 10.1 falls more than two and one-half feet in its flight to the plate. A plot of this vertical distance as a function of time would be parabolic. The ball falls 3.8 inches in the first 100 msec, 6.2 inches in the second 100 msec, 8.3 inches in the third 100 msec, and 10.3 inches in the fourth 100 msec. The ball drops more in each time period, but it follows a smooth parabolic trajectory. Now let us see how the ball drops with the addition of a vertical force due to top spin on the ball. Table 10.2 shows the results of simulations of 80 and 75-mph drop curves. We are going to use the term *drop curve* for a pitch with pure top spin. Both were launched at an angle of 2.5 degrees up with 1900 revolutions per minute (rpm) of pure top spin. We used the following formula from Watts and Bahill (2000) for the downward force due to spin: $F = \frac{1}{2}\rho\pi R^3\omega v$, where $\rho$ is the air density, $R$ is the radius of the baseball, $\omega$ is its spin rate and $v$ is its velocity.

Let us look at the 75-mph pitch. The ball falls 2.4 inches between 100 and 200 msec, 7.7 inches in the next 100 msec, 12.7 inches in the next 100 msec, and 17.5 inches in the next 100 msec. Once again, the ball drops more in each time period, but it still follows a smooth parabolic trajectory. Therefore, this is a gradual curve rather than a sharp break. This conclusion is shared by Adair (p. 31, 1990) who stated, "Like its cousin, the curve ball, the hopping fastball follows a smooth arc on its way to the plate, but half of the hop deviation occurs in the last 15 feet of that flight." A breaking pitch would have to drop the same amount as this ball in the early 100-msec epoch, but more than this in the last 100 msec. It would be impossible to hit the ball if it really did jump or break just before reaching the plate. For example, a foul tip changes the ball's trajectory abruptly in the last few feet before the catcher and the umpire. As a result, these masked men cannot predict the ball's trajectory and they must wear protective clothing.

| Table 10.2. Drop curve Trajectories | | | | |
|---|---|---|---|---|
| | 80 mph | | 75 mph | |
| Time Since Release (msec) | Distance (ft) | Height (ft) | Distance (ft) | Height (ft) |
| 0 | 55.5 | 6.00 | 55.5 | 6.00 |
| 50 | 49.7 | 6.20 | 50.0 | 6.18 |
| 100 | 44.0 | 6.28 | 44.7 | 6.25 |
| 150 | 38.3 | 6.24 | 39.4 | 6.20 |
| 200 | 32.7 | 6.09 | 34.1 | 6.05 |
| 250 | 27.2 | 5.83 | 29.0 | 5.78 |
| 300 | 21.9 | 5.47 | 23.9 | 5.41 |
| 350 | 16.5 | 4.99 | 18.8 | 4.93 |
| 400 | 11.3 | 4.41 | 13.9 | 4.35 |
| 450 | 6.11 | 3.73 | 9.0 | 3.67 |
| 480 | 3.0 | 3.27 | | |
| 500 | | | 4.2 | 2.89 |
| 513 | | | 3.0 | 2.68 |

We are now ready to explain the apparent abrupt break of some pitches. Suppose the pitcher threw the 75-mph drop curve of Table 10.2, it would drop 25 inches in the last 150 msec before contact. However, if the batter overestimated the pitch speed and thought it was going 80 mph, then the batter would expect it to fall 21 inches in the last 150 msec before contact. Thus, if the batter took his eye off the ball 150 msec before the projected time of contact, and saw it again when it arrived at his bat, he would think that it broke downward 4 inches. Therefore, we suggest that the apparent, abrupt-break of some pitches might be a result of the batter's **over**estimation of the speed of the pitch in his mental model. This is the opposite of the explanation for the rising fastball.

Of course, the illusion would be greatly enhanced if the batter made a mistake in estimating both the speed and the type of the pitch. If the pitch were a 95 mph fastball with 1600 rpm of backspin and the batter thought it was a 75 mph drop curve with 1900 rpm of top spin, then the illusionary jump would be one foot. Similarly, if such a drop curve were mistaken for such a fastball, then the illusionary break would be one foot.

In summary, we think every pitch (except a nonspinning pitch such as the knuckle ball) follows a smooth trajectory. The apparent, abrupt jumps and sharp breaks right before the plate are perceptual illusions, caused by the batter using the optimal learning strategy of tracking and making mistakes in his mental model of the pitch. We think that in the first third of the ball's flight, batters gather data and form mental models of the pitch. In the middle third, batters compute when and where the ball will cross the plate. Then they start their swings. In the last third of the ball's flight, the batters either observe errors in their mental models so that they can track the next pitch better (the optimal learning strategy), or they update their mental models and perhaps check their swings (the optimal hitting strategy).

## 10.5 THE EFFECTS OF SPIN AND SPEED

Batters say that the ball hops, drops, curves, breaks, rises or tails away. They say that the pitcher throws a fastball, curveball, drop curve, flat curve, slider, backup slider, change up, split fingered fastball, splitter, forkball, sinker, two seam fastball or four seam fastball. This sounds like a lot of variation. However, no matter how the pitcher grips or delivers the ball, once it is in the air (excepting the knuckleball), its motion depends only on its (1) speed, (2) rate of rotation and (3) direction of the axis of rotation. As shown in Fig. 10.6 and previously discussed, the curveball has top spin and side spin, so from the perspective of a right-handed pitcher, it drops and curves right to left. The fastball has predominately backspin, which gives it lift and decreases the fall due to gravity. But if it were not thrown with an overhand delivery, then it would also have some lateral deflection. The slider is thrown like a football, which causes it to spin like a bullet. If its axis of rotation is coincident with the direction of motion, then the spin causes no deviation. Therefore, the slider is usually thrown so that the axis of rotation is pointed up and to the left. This causes the ball to drop and curve right to left. Rotation about this axis allows the batter to see a red dot on the top-right-side of the ball. For the backup slider, the axis of rotation is coincident with the direction of motion: the spin causes no horizontal curve and the batter sees a red dot right in the middle of the ball. The drop and curve of all these pitches is infinitely variable depending on the axis of rotation.

The black arrows in Fig. 10.6 indicate the vector moment of the spin. You can determine the direction of the spin-induced deflection using the right-hand rule. Point the thumb of your right hand in the direction of the black arrow and point your index finger in the direction of motion of the pitch. Bend your middle finger so that it is perpendicular to your index finger. Your middle finger will be pointing in the direction of the *spin-induced* deflection (of course the ball also drops due to gravity).

Besides spin, speed also characterizes the movement of the ball. The purpose of changing pitch speeds is not to fool the *when* computer, but rather to fool the speed estimator and consequently the *where* computer.

A 90 mph fastball is in the air 426 msec, so it drops 2.8 feet due to gravity (½ $at^2$) but the backspin lifts it 1 foot, producing a total drop of 1.8 feet. Whereas a 75 mph drop curve is in the air 511 msec, so it drops 4 feet due to gravity and the top spin drops it an additional 1.6 feet, producing a total drop of 5.6 feet. (These numbers are different from those in Tables 10.1 and 10.2 because here both of these pitches were launched horizontally, that is with a launch angle of zero.)

So changing from backspin to top spin drops the ball 2.6 feet. Whereas changing the speed from 90 mph to 75 mph drops the ball 1.2 feet. These are very large distances compared with the vertical margin of error on the sweet area of the bat.

For these simulations of Section 10.5, the pitch was launched horizontally to avoid confounding effects of an upward launch. However, in order to cross the plate in the strike zone almost all pitches much be launched with an upward angle. Fastballs by Randy Johnson are possible exceptions. He is tall enough and fast enough to pitch some fastballs with a downward launch angle and still have them go through the strike zone. Such fastballs must seem strange to a batter.

## 10.6 SENSITIVITY ANALYSES

To help validate our simulation and model we will now investigate how well our conclusions hold up under parametric sensitivity analyses. First, we will show the results of a sensitivity analysis of our simulation (Fig. 10.1 and Tables 10.1 and 10.2), and then we will show the results of a sensitivity analysis of our model of the human brain (Fig. 10.4).

Each of the eight simulation parameters was varied by ±5%. Then we ran the simulation and calculated the percentage change for the perceived jump. The ratio of these two percentages is the relative sensitivity (Karnavas, Sanchez and Bahill, 1993). These sensitivity values are shown in the top of Table 10.3.

To understand these numbers, it is helpful to note from our nominal simulation results (Table 10.1) that the apparent jump of the rising fastball is 0.230 feet. When the horizontal position of the bat-ball collision point was moved back 5%, the amount of perceived jump increased to 0.231 feet, or by 0.5%, which is -0.1 (the value of the sensitivity coefficient from Table 10.3) times 5% (the perturbation size). Similarly, when the estimated pitch speed was decreased 5%, the amount of perceived jump increased to 0.472 feet, or by 105.4%, which is 21.1 times 5%. *Therefore, the simulation is most sensitive to pitch speed and the estimated pitch speed; in comparison, all other parameters are insignificant.*

| Table 10.3. Results of Sensitivity Analyses | |
|---|---|
| Sensitivity Coefficients for the Simulation | |
| Parameter | Value |
| Horizontal position of the pitcher's release point | -0.2 |
| Vertical position of the pitcher's release point | 0.0 |
| Release angle | 0.0 |
| Horizontal position of the batter's eye | 0.0 |
| Vertical position of the batter's eye | 0.0 |
| Horizontal position of the bat-ball collision | -0.1 |
| Pitch speed | 18.9 |
| Estimated pitch speed | -21.1 |
| | |
| Sensitivity Coefficients for the Model | |
| Parameter | Value |
| $\hat{D}_0$, estimated distance to release point | -2.57 |
| $\hat{t}_{\text{sinceRelease}}$, estimated time since release | 1.57 |
| $\hat{s}$, estimated pitch speed | 1.57 |
| $d\Phi / dt$, retinal image velocity | -1.00 |
| $\gamma$, retinal image size | -1.00 |
| $d\gamma / dt$, rate of change of retinal image size | 0.95 |

It is rewarding to see that the pitch speed and the estimated pitch speed are the most important parameters of the simulation. The sensitivity analysis tells us that we should not (1) try to get more accurate values for the other parameters of the simulation, (2) customize the simulation for individual players, or (3) try to get more accurate values for the thresholds of the visual parameters $\gamma$ or $d\theta / dt$, because these parameters are not that important. The bottom line of this analysis is that our simulation is a good representation for the pitched baseball.

Next, we did a sensitivity analysis of our neurophysiological model of Fig. 10.4 for a 90 mph pitch. Each of the six model parameters was varied by ±5%, and we calculated the percentage change in how far the ball dropped between 250 msec after release and the time of bat-ball contact. The ratio of these two percentages is the relative sensitivity. These sensitivity values are shown in the bottom of Table 10.3.

*The model is most sensitive to the three parameters that are used to estimate the distance to the ball.* The batter cannot get these parameters from the primary visual processes. It seems that the model is most sensitive to the things that are most difficult for the human to compute. Perhaps superior athletes are superior not because they have superior primary visual processes, but because of their subsequent processing of this information.

From these studies, we can see that estimating pitch speed is essential for the batter to achieve what Charley Lau describes as "the sweet spot in time" – that instant when the sweet area of the bat is in the right place to meet the sweet spot of the ball (Lau and Glossbrenner, 1984). It seems that a useful training technique would be to put a speech synthesizer on a radar gun behind the plate in the batting cage and announce the speed immediately after every pitch during batting practice. This might help the batter learn to estimate pitch speed better.

## 10.7 THE TWO-SEAM AND FOUR-SEAM FASTBALLS

Tom Seaver (Seaver and Lowenfish, 1984) has written that the four-seam grip is more effective in producing rising fastballs than the two-seam grip. However, wind tunnel tests have shown no difference in lift between the two-seam and the four-seam fastballs (Watts and Bahill, 2000, p. 75). Perhaps the explanation for this apparent contradiction could be based on psychology instead of physics.

As the frequency of a blinking light increases, it appears to flicker and then at a certain frequency it appears to be continuously illuminated. This transition point is called the Critical Flicker Fusion Frequency. For a person in a baseball park this frequency is probably between 40 and 50 Hz. Television screens present a new frame sixty times per second (60 Hz); therefore, their pictures do not flicker, whereas the time indicator on your VCR probably blinks once a second. This is clearly perceived as blinking.

If a fastball were rotating at 1200 rpm, it would rotate 20 times per second. For a two-seam fastball, there would be two seams crossing the field of view on each rotation (or, because the seams are close together, they might appear as one blip). So, the frequency would be between 20 and 40 Hz, which is below the critical flicker fusion frequency and perhaps the ball would appear to flicker. Therefore, the batter could detect the spin. He does not have to compute the spin rate: he just has to determine if it is top spin (a curveball) or backspin (a fastball). This would help his speed estimator of Fig. 10.4. In contrast, for a four-seam fastball, the four seams would cross the field of view symmetrically for each rotation. This would produce a frequency of 80 Hz, which is above the critical flicker fusion frequency. The ball would be seen as a blur and the batter would not be able to detect the direction of spin. Therefore, the batter would have to guess if it were a curveball or a fastball. If he guesses that it is a curveball, then he will think it is slower than it is and might perceive a rising fastball.

Almost all pitchers who use an overhand or three-quarter delivery throw four-seam fastballs and curveballs. The ball is gripped with the index and middle fingers extending across four seams. They say that each seam is a ridge that provides high impedance to the airflow around the ball when the seam is perpendicular to the direction of flight. They say that backspin with four seams perpendicular to the trajectory gives the ball maximum "lift." The four-seam pitch is probably harder to hit than a two-seam pitch. But the reason may come from Psychology, not Physics. The four seam pitches produce a flicker frequency that is too high for batters to discern the movement of the seams.

Some great hitters have stated that they do not see the direction of spin on the pitch. For example, Tony Gwynn (Gwynn and Rosenthal, 1992) has stated, "I'm not like a lot of hitters who see the ball spin a certain way on a particular pitch. I just see the ball." Frank Robinson (as quoted in Schmidt and Ellis, 1994) said, "I think the initial reaction [to the ball] coming out of the hand is an advantage over waiting to see the seams of the ball." And Mike Schmidt (Schmidt and Ellis, 1992) wrote, "So many people say that the hitter sees the spin of the ball, the seams. I don't believe it. I couldn't." Against pitchers using four-seam grips, batters are forced to determine the kind of pitch from other cues (such as the pitcher's release point and the early trajectory of the ball) rather than the direction of spin (Baker, Mercer, and Bittinger, 1993).

## 10.8 STRATEGIES

### 10.8.1 Strategies, Heuristics and Processes of the Batter

On each pitch, the batter is trying to accomplish some task. To do this he uses one or more strategies, such as "Predict the Pitch," "Protect the Plate," "Take the Pitch," "Go for the Long Ball," "Protect the Base Runner," "Hit to the Opposite Field" or "Make Contact." These strategies provide the objectives for the batter's cognitive processes (anticipatory modeling of the pitch, tracking the pitch, and consequent decision making) and kinesiological processes (position and movement during the swing – what coaches refer to as "batting mechanics" and "handling the bat"). The strategy to be used depends on such factors as the, number of outs, runners on base, score, count, coaches' signals and the speed, location and outcome of the previous pitch (Aaron and Cohen, 1974; Lau and Glossbrenner, 1984). The batters mental model will be different if he foul tipped the previous pitch or blasted it over the fence, but just foul.

For a given set of situational factors, the batter uses heuristics to help select a specific implementation of a strategy (Schmidt and Ellis, 1994). With zero or one strike, the batter would "Predict the Pitch," whereas with two strikes the batter would "Protect the Plate." Within the "Predict the Pitch" strategy, the batter uses heuristics to help with the prediction. For example, with a count of 3 and 0 the batter would expect a fastball, whereas with a 0 and 1 count the batter might look for a curveball (Williams and Underwood, 1982). If the previous pitch were a ball, the batter is more likely to expect a fastball, whereas if the previous pitch were a strike, he is more likely to predict a breaking pitch, e.g., a curve or a changeup (Gray, 2003). The basis of these heuristics is the assumption that, when a strike is needed, most pitchers have the greatest confidence in the fastball.

The batter's set of cognitive processes includes a speed estimation process (shown in Fig. 10.4) that could be modeled with a Markov process or a Kalman filter. The present pitch speed is estimated using a weighted average of the last one or possibly two pitches (Bahill and Karnavas, 1993: Gray, 2002). Pitchers believe that the batter's memory span is 20 seconds. The speed estimator probably also uses information gathered in the first third of the pitch. It is not known how this process interacts with the above strategies and heuristics to help the batter predict when and where the ball will cross the plate.

## 10.8.2 Tactics of the Pitcher Versus Batter Confrontation

The battle of pitcher versus batter is the heart of baseball. In this contest, each participant employs a time-tested tactic: the batter uses mental models of the pitcher's previous pitches in order to anticipate the kind and location of the next pitch and the pitcher attempts to vary his pitches in order to fool the batter.

The pitcher's maneuvers are based on two assumptions. First, the movement, speed and location of a pitch will influence the batter's predictive model for the following pitch. For example, a high, inside fastball sets up the batter for a slow curveball, low and outside, because the batter cannot help but remember and adjust, at least in some small way, to the speed and location of the preceding fastball.

The second assumption of the pitcher is that, with respect to preventing the batter from hitting the ball well, vertical movement of the pitch is more effective than horizontal movement. The sweet region of the bat (the best part of the barrel for hitting the ball) is perhaps four inches long or longer (Seaver and Lowenfish, 1984; Williams and Underwood, 1982), but its depth (along the transverse axis) is less than an inch. Consequently, the batter's margin of error is much smaller along the transverse axis compared with the longitudinal axis.

When the bat is swung at pitches that are mid-thigh high or higher, the longitudinal axis of the bat is almost horizontal as it passes through the strike zone. On a pitch that is knee high and on the outside corner, the angle of the bat with the horizontal plane is about ten to fifteen degrees, and this angle increases to about thirty or forty degrees on pitches that are knee high on the inside corner. For most pitches, then, vertical movement of the ball nearly parallels the transverse axis of the bat and increases the likelihood that the sweet region of the bat will pass over or under the center of the ball.

Given these assumptions, the pitcher's tactic is to select a pitch that will be different from the batter's predictive model and to provide adequate vertical movement of the ball to take advantage of the batter's vulnerability. The most important factor in pitch selection is the change of speeds. As a manager, Ted Williams exhorted his pitchers to never let a batter see consecutive pitches of the same speed. The great pitching coach, Johnny Sain, expressed the same maxim. In following this advice, pitchers exaggerate the speed differences between the overhand fastballs and the breaking pitches, thereby increasing the perceived vertical movement of the ball.

Some of baseball's fastest pitchers, such as Nolan Ryan and Rich Gossage, have used three speeds of their fastball to "go up the ladder" against the batter. This technique consists of throwing three high fastballs to a batter, each pitch faster and a little higher than the last. The batter experiences a jump on each pitch because of the speed differential. By the third pitch, the batter is swinging at an explosive fastball that is several inches out of the strike zone. The pitchers vary the speeds in this series because they recognize that this confutes the batter's predictive models, but few pitchers (if any) realize that this variation of speeds is inducing the perceptual illusion of the rising fastball.

The reverse technique, "going down the ladder," can be used on breaking pitches. Camilo Pascual had a legendary overhand curve that appeared to dive straight down just before it reached the plate. He would throw it at three speeds, enticing the batter to swing at progressively lower pitches. On one occasion, the batter actually broke his bat on home plate in trying to follow Pascual's curve.

We can describe other examples of slower speeds enhancing the illusion of the break on curveballs and on sinkers. Pitchers are advised by coaches to avoid throwing the curveball too hard (Ryan and House, 1991; Stanley and Stanley, 1988). The rationale usually given for this is that the slower curve has more time to "bite" the air - consequently, it will have a sharper break.

Relief pitchers who rely on sinkers are sometimes surprised that they are at their best when they are tired from pitching on several consecutive days. They lose some speed on their pitches and the sinker dips much more than it does when they are rested and strong (Thorn, 1979). Dusty Baker calls the sinker the toughest pitch to hit and has recommended that the batter study box scores in newspapers to determine a sinkerball pitcher's fatigue level (Baker, *et al.,* 1993).

The batter's antidote for the rising fastball and breaking pitches is to predict the next pitch and rehearse it with a mental simulation. This allows the batter to adjust expectations of the speed of the pitch. The batter can take measures to minimize the effects of the break or rise. If a breaking pitch is expected, the batter moves forward in the batter's box to "catch the ball before it breaks." Sievers (1958) and Gwynn and Rosenthal (1992) stress that the batter must concentrate on waiting as long as possible before committing the swing. If a high fastball is expected, the batter moves to the back of the box and prepares to swing at the upper half of the ball. With the count at less than two strikes, the batter has the luxury of taking the pitch if his prediction is wrong. With two strikes, the batter protects the strike zone by choking up on the bat and concentrating on hitting the ball up the middle (Baker, *et al.,* 1993).

Both pitchers and batters assume that the methods of grip and delivery a pitcher uses can make the pitches change course abruptly. Pitchers know that the speed of the pitch is important to its success, but they are unaware that success is dependent upon an illusion experienced by the batter. And the batters, too, do not realize that they are being fooled by their own expectations.

## 10.9 SUMMARY

One by one, scientists and engineers, using principles of physics, have explained most of baseball's peculiar pitches: e.g., the knuckle ball (Watts and Sawyer, 1975: Hollenberg, 1986), the scuff ball (Watts and Bahill, 2000) and the curveball (Watts and Bahill, 2000). The most mysterious remaining pitches were the rising fastball and the breaking pitches. Now, adding principles of physiological psychology, we can suggest that the rising fastball and the breaking pitches may be perceptual illusions, caused by the batter misestimating the speed of the pitch. Pitchers take advantage of these illusions when devising tactics and techniques but are generally unaware of the cause of these phenomena. However, Christy Mathewson (quoted by

Kahn, 2000) once said, "Fooling the other fellow was just as valuable as throwing hard – and a darn sight easier on the arm."

## 10.10 REFERENCES


Aaron, H., and Cohen, J., 1974, *Hitting the Aaron Way,* Prentice-Hall, Inc., Englewood Cliffs, NJ.

Adair, R. K., 1990, *The Physics of Baseball,* Harper & Row, New York.

Bahill, A. T. and Karnavas, W. J., 1993, The Perceptual Illusion of Baseball's Rising Fastball and Breaking Curveball, *Journal of Experimental Psychology, Human Perception and Performance.* **19**: 3-14.

Bahill, A. T., and LaRitz, T., 1984, Why can't batters keep their eyes on the ball, *American Scientist.* **72**: 249-253.

Baker, D., Mercer, J., and Bittinger, M. L., 1993, *You Can Teach Hitting,* Bittinger Books, Inc., Carmel, IN.

Beverley, K. I., and Regan, D., 1975, The relation between discrimination and sensitivity in the perception of motion in depth, *Journal of Physiology.* **249**: 387-398.

Flatow, I., 1988, *Rainbows, Curve Balls, and Other Wonders of the Natural World Explained,* William Morrow and Co., New York.

Frohlich, C., 1984, Aerodynamic drag crisis and its possible effect on the flight of baseballs. *American Journal of Physics.* **52**: 325-334.

Gray, R., 2002, Behavior of college baseball players in a virtual batting task, *Journal of experimental psychology: Human Performance and Perception.* **28:** 1131–1148.

Gray, R., 2003, Markov at the bat: a model of cognitive processing in baseball batters, *Psychological Science.* **13**: 542-547.

Gray, R., and Regan, D., 1998, Accuracy of estimating time to collision using binocular and monocular information, *Vision Research.* **38:** 499-512.

Gwynn, T., and Rosenthal, J., 1992, *Tony Gwynn's Total Baseball Player,* St. Martin's Press, New York.

Hollenberg, J. W., 1986, Secrets of the knuckleball, *The Bent of Tau Beta Pi.* **77**: 26-30.

Hoyle, F., 1957, *The Black Cloud,* Penguin Books, Middlesex, England.

House, T., 1994, *The Pitching Edge,* Human Kinetics, Champaign, IL

Kahn, R., 2000, *The Head Game: Baseball Seen from the Pitcher's Mound,* Harcourt Brace Inc., New York, pg. 98.

Kara, G. B., 1990, Lost vision: Babe Ruth's legend, *Argus,* from the American Academy of Ophthalmology. **13**: 6.

Karnavas, W. J., Bahill, A. T., and Regan, D. M., 1990, A model for the rising fastball and breaking curve ball, *Proceedings of the 1990 IEEE International Conference on System, Man, and Cybernetics,* pp. 925-928.

Karnavas, W. J., Sanchez, P., and Bahill A. T., 1993, Sensitivity Analysis of Continuous and Discrete Systems in the Time and Frequency Domains, *IEEE Transactions on System, Man, and Cybernetics.* **23**: 408-501.

Lau, C., and Glossbrenner, A., 1984, *The Winning Hitter,* Hearst Books, New York.

Lee, D. N., 1976, A theory of visual control of braking based on information about time-to-collision, *Perception.* **5**: 437-459.

Matin, L., and Kibler, G., 1966, Acuity of visual perception of direction in the dark for various positions of the eye in the orbit, *Perceptual and Motor Skills.* **22**: 407-420.

McBeath, M., 1990, The rising fastball: Baseball's impossible pitch, *Perception.* **90**: 545-552.

McKee, S. P., 1981, A local mechanism for differential velocity discrimination, *Vision Research.* **21**: 491-500.

Orban, G. A., De Wolf, J., and Maes, H., 1984, Factors influencing velocity coding in the human visual system, *Vision Research.* **24**: 33-39.

Palmer, J., and Cohen, J. H., 1975, *Pitching,* Atheneum, New York.

Regan, D., 1997, Visual factors in hitting and catching, *Journal of Sports Science.* **15**: 533-558.

Regan, D., 1986, The eye in ballgames: Hitting and catching. *Sport en Zien*, Haarlem, De Vrieseborcht, pp. 7-32.

Regan, D., 1992, Visual judgments and misjudgments in cricket, and the art of flight, *Perception.* **21**: 91-115.

Regan, D., and Beverley, K. I., 1978, Looming detectors in the human visual pathway, *Vision Research.* **18**: 415-21.

Regan, D., and Beverley, K. I., 1980, Visual responses to changing size and to sideways motion for different directions of motion in depth: Linearization of visual responses, *Journal of the Optical Society of America.* **11**: 1289-96.

Regan, D., Beverley, K. I., and Cynader, M., 1979, Stereoscopic subsystems for position in depth and for motion in depth, *Proceedings of the Royal Society of London* (Series B), **204**: 485-501.

Regan, D. and Cynader, M., 1979, Neurons in area 18 of cat visual cortex selectively sensitive to changing size, *Vision Research.* **19**: 699-711.

Regan, D., Erkelens, C. J., and Collewijn, H., 1986, Necessary conditions for perception of motion in depth, *Investigative Ophthalmology and Visual Science*, **27**: 584-597.

Ryan, N. and House, T., 1991, *Nolan Ryan's Pitcher's Bible,* New York: Simon and Schuster.

Saito, H., Yukie, M., Tanaka, K., Hikosaka, K., Fukada, Y. and Iwai, E., 1986, Integration of direction signals of image motion in the superior temporal sulcus of the macaque monkey, *Journal of Neuroscience.* **6**: 145-157.

Seaver, T and Lowenfish, L., 1984, *The Art of Pitching,* New York: Hearst Books.

Sievers, R., 1958, Roy Sievers on the art of hitting, *Sports Illustrated.* **8**: 35-47.

Schmidt, M. and Ellis, R., 1994, *The Mike Schmidt Study: Hitting Theory, Skills and Technique,* Atlanta: McGriff and Bell Inc.

Stanley, F. and Stanley, L., 1988, *The Complete Instructional Baseball Manual,* Scottsdale, AZ: The Stanley Co.

Steinbach, M. J., 1970, Aligning the eye to the actively or passively positioned hand, *Perception and Psychophysics.* **8**: 287-288.

Thorn, J., 1979, *The Relief Pitcher: Baseball's New Hero,* New York: E.P. Dutton.

Thrift, S. and Shapiro, B., 1990, *The Game According to Syd,* New York: Simon and Schuster.

Todd, J. T., 1981, Visual information about moving objects, *Journal of Experimental Psychology: Human Perception and Performance.* **7**: 795-810.

Watts, R. G., and Bahill, A. T., 2000, *Keep Your Eye on The Ball: Curveballs, Knuckleballs and Fallacies of Baseball,* New York: W.H. Freeman and Co.

Watts, R. G., and Sawyer, E., 1975, Aerodynamics of a knuckleball, *American Journal of Physics.* **43**: 960-963.

Will, G. F., 1990, *Men at Work*, New York: Macmillan.

Williams, T. and Underwood, J., 1982, *The Science of Hitting*, New York: Simon and Schuster.

## 10.11 BIOGRAPHICAL SKETCHES

Terry Bahill has been a Professor of Systems Engineering at the University of Arizona in Tucson since 1984. He received his Ph.D. in electrical engineering and computer science from the University of California, Berkeley, in 1975. He holds U.S. patent number 5,118,102 for the Bat Chooser, a system that computes the Ideal Bat Weight for individual baseball and softball batters. He is a Fellow of the Institute of Electrical and Electronics Engineers (IEEE) and of the International Council on Systems Engineering (INCOSE). He is chair of the INCOSE Fellows Selection Committee. His picture is in the Baseball Hall of Fame's exhibition "Baseball As America."

Dave Baldwin, as a relief pitcher for the Washington Senators, Milwaukee Brewers, and Chicago White Sox during the 1960s and 1970s, compiled a lifetime ERA of 3.08 with 6 wins, 11 loses and 22 saves in 176 games. Following a 16-year baseball career, he earned a Ph.D. in genetics and an M.S. in systems engineering from the University of Arizona. Currently, he is an information and decision analyst for the HylaPark, Inc. in Carlsbad, CA. He is also a member of the Society for American Baseball Research (SABR).

# IV

# FOOTBALL AND SOCCER

# Chapter 11

# Mechanics of the Forward Pass

William J. Rae

*Mechanical and Aerospace Engineering, University at Buffalo, The State University of New York, Buffalo, NY 14260, EM: rae@eng.buffalo.edu*

## 11.1 INTRODUCTION

The physical principles of the flight of spinning bodies are well known, but their application to the sport of American football is relatively new, and the opportunities for performance optimization are largely unexplored. This chapter contains the author's assessment of the current state of knowledge especially with regard to fundamental principles, applications, and implications for optimal athletic performance.

### 11.1.1 Physical Principles

Understanding of the physical principles must start from knowledge of the basic phenomena involved. Thus the discussion to follow begins by itemizing a number of observations of football flight, along with an introduction to the vocabulary used in describing these phenomena.

### 11.1.2 Vocabulary, Basic Phenomena

The motion of any sports ball through the atmosphere is affected by the size, shape, and speed of the ball. In the process of throwing or kicking it, an athlete imparts certain launch conditions, consisting of the speed and direction of the ball. For the case of a non-spherical object like a football, the athlete also imparts a certain orientation of the ball with respect to its flight path, i.e. it matters whether the ball is initially aligned with its flight direction, perpendicular to it, or somewhere in between. The speed and orientation of the ball plus the trajectory traced out by its center of mass define the "translational" component of the motion. In addition, one must specify the "angular" component of the motion ( the number of revolutions per minute at which the ball is spun plus the orientation of the spin axis). Thus a football requires a more detailed description of the motion than is needed in games which use a spherical ball, where orientation plays a lesser role.

## 11.2 BASIC CONCEPTS

Figures 11.1 and 11.2 show some of the terminology used for the angular degrees of freedom. The first of these figures is a computer simulation of football flight, showing a representative trajectory in isometric projection, with yard markers at five-yard intervals. Obviously, the location of the ball and its orientation change as it moves along. Figure 11.2 shows three of the angles used to define the orientation: the pitch angle $\theta$ is the angle between the ball axis and the horizontal, the angle of attack $\alpha$ is the angle from the velocity vector to the ball axis, and the flight-path angle $\gamma$ is the angle of the velocity vector above the horizontal.

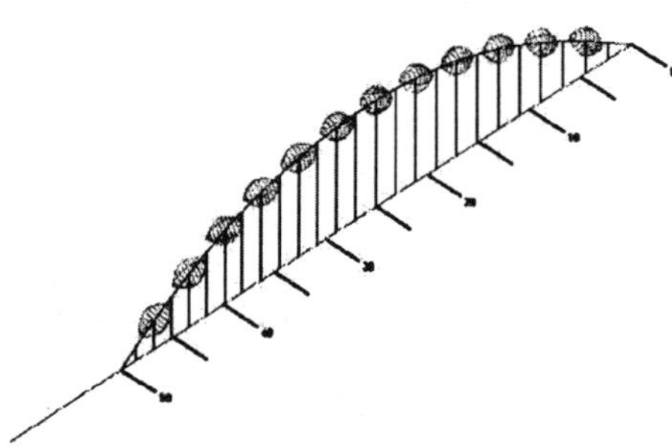

*Figure 11.1.* Typical ball position and orientation during flight.

After the ball is launched, it experiences two forces, one being gravity and the other the distribution of loads imposed by the wind flowing over the surface of the ball. This distribution of forces can be represented by a single force acting at some point on the ball, and this in turn is equivalent to the same force acting at the center of mass plus a torque about the center of mass, equal to the product of the force times the distance from the center of mass to the point at which the force acts. Simply put, flow over the ball subjects it to a force and a torque. These are *vector* quantities, i.e. they have both a magnitude and a direction, and it takes a total of six vector components to specify completely the magnitude and direction of both the force and the torque.

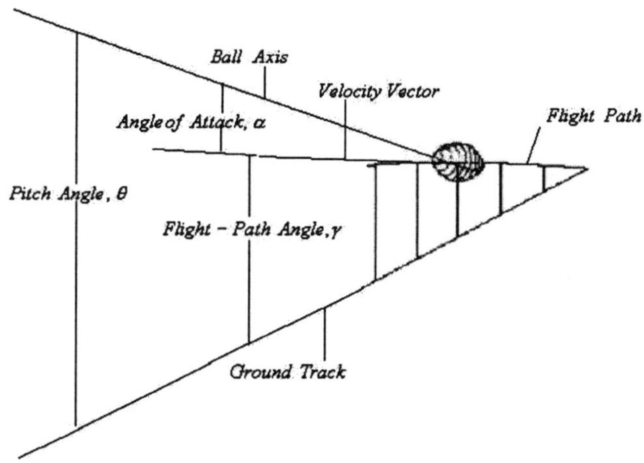


*Figure 11.2.* Definition of pitch angle, flight-path angle, and angle of attack.

In launching any sports ball, an athlete is always concerned about the interaction between the translational and angular motions imparted. For example, a baseball pitcher or soccer kicker wanting to launch a particular type of curve ball will orient the spin axis in such a way as to produce the desired trajectory. Very often the desired feature is a lateral curvature of the flight path, designed to cause a batter or goalkeeper to miss contacting the ball. However, in the case of a nonspherical ball (for example a rugby ball or an American football) a different criterion is present, namely that the ball *must* be spun about its central axis in order to overcome an inherent instability. When a football is thrown with its central axis oriented a few degrees "nose high", air pressure on the lower side near the nose will cause a torque tending to lift the nose further above the flight direction, and as the angular misalignment grows, the overturning torque also increases, leading to a tumbling motion. A similar problem affects all nonspinning bodies of elongated shape, such as aircraft fuselages and nacelles. In the absence of control surfaces, the only way to overcome this instability for a football is to spin it. This makes it into a gyroscope, which has the property that when spun about one axis and torqued about a second axis, it responds by trying to turn about the third axis, and this tendency is stronger, the larger the spin. In the case of a football the gyroscopic torque is made much larger than the overturning torque due to airflow, causing the spin axis to move to one side and to continue tracing out a cone centered on the flight direction. This motion is called a *precession* or, more colloquially, a "wobble" of the spin axis, and this means of controlling it is known as spin stabilization.

## 11.2.1 Observations

In addition to these general comments, it is useful to list a number of features that have been observed in actual flights:

1. Typical launch conditions can have speeds of up to 60 miles per hour and 600 revolutions per minute (RPM) (Brancazio, 1985, 1987).

2. As noted above, the ball is extremely unstable, and may tend toward an end-over-end tumble if it is not spun. Thus a ball that slips out of the passer's hand with relatively little spin will not follow a "tight spiral". Instead, its orientation shows large excursions, often described as the flight of a "wobbly duck". Even when the ball is spun rapidly enough to avoid the instability, it often happens that the ball will not "track" its trajectory, but will still "wobble", i.e., the ball will trace out a cone centered on the flight path. The ball spins faster than it wobbles, completing about 1.8 spins during the time it takes to make one circuit about the trajectory. The spin/wobble ratio can be seen in Fig. 11.3, which shows measurements made on a video tape from a National Football League game. The pass was thrown from around the 40-yard line of the offense, midway between the sidelines, and was caught at the ten-yard line near the left sideline. It was filmed in slow motion (90 frames per second) by a camera located on the right sideline near the ten-yard line. Thus the line of sight was approximately perpendicular to the flight path over a large portion of the flight, which lasted for about 3 seconds. Measurements were made on each frame of the angle between the horizontal and the ball centerline, and this quantity, called here the "apparent" pitch angle, is plotted versus the time inferred from the frame count. A typical frame is shown as an insert in this figure. The period of this angle is taken as a correct measure of the wobble period, even though it is not possible to measure the pitch angle exactly when the yaw angle is unknown. In Fig. 11.3, the time between revolutions is denoted by vertical lines at the instants where the laces faced the camera. Counting the number of oscillation cycles for a given number of revolutions leads to the 1.8 ratio mentioned above.

The existence of a certain spin-to-wobble ratio is not important to the flight performance of a football, but its value implies that one of the key aerodynamic parameters, namely the slope of the pitching-moment curve, is around 0.3 per radian at low angles. The fact that this agrees with the number measured in wind-tunnel tests (Rae and Streit, 2002) is an important validation of that data. The connection between the spin-to-wobble ratio and the slope of the moment curve has been derived by Rae (2003).

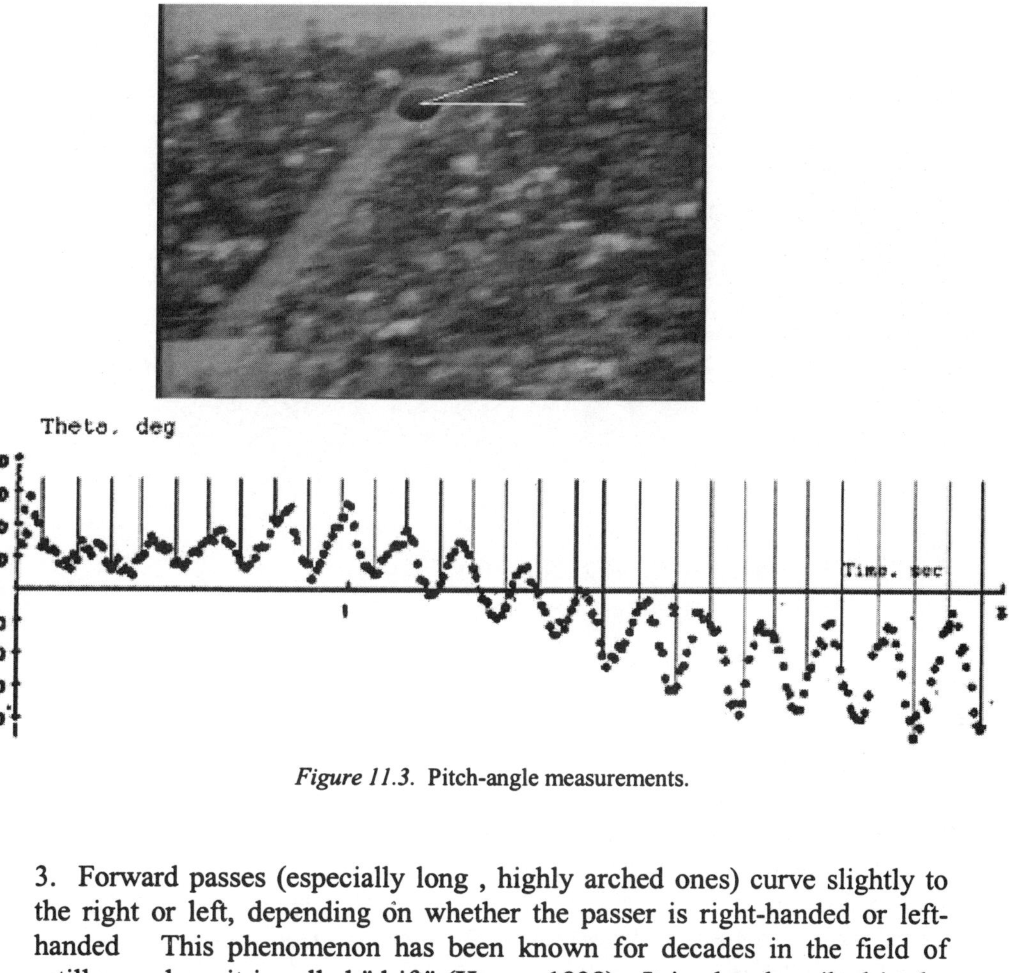


*Figure 11.3.* Pitch-angle measurements.

3. Forward passes (especially long , highly arched ones) curve slightly to the right or left, depending on whether the passer is right-handed or left-handed This phenomenon has been known for decades in the field of artillery, where it is called "drift" (Hayes, 1938). It is also described in the book by National Football League quarterback Bart Starr (Starr, 1967), and has also been observed by the present author. Figure 11.4, taken from Rae (2003), shows the lateral curvature of a pass thrown directly toward a TV camera from a point about 50 yards away. This video was made in an indoor facility, in order to eliminate any effects of wind. The location of the ball in each frame was digitized and is re-drawn on the figure as a white line. Note that the ball rises at first, moves toward the right of an observer standing behind the camera, and then reverses direction and lands to the observer's

left. No such reversal would be seen if the trajectory had lain in a vertical plane. Similar behavior is seen in computer simulations of football flight, as will be shown below.

*Figure 11.4.* End-on view of a forward pass, showing lateral curvature.

4. The spin-period measurements mentioned in item 3 show that the RPM of the ball remains essentially constant during a 2- to 3-second flight, implying that the aerodynamic torque about the spin axis is negligibly small. At present, there are no wind-tunnel measurements of this torque; the setup used by Rae and Streit (2002) used ball bearings to spin the ball about its central axis, and these did not allow roll torques to be transmitted to the strain-gage balance.

5. Wind-tunnel tests (Rae and Streit, 2002) show that footballs which are spinning and whose axis is at an angle to the wind direction experience a side force (known as a "Magnus" force), but no Magnus torque. This force is zero when the ball is pointed directly into the wind (for any value of the spin) and is also zero at any angle of attack when the ball is not spinning. The conclusion is that a side force develops when the spin axis makes a nonzero angle with the wind direction. The force results from a lateral asymmetry in the pressure distribution on the ball, caused by an interaction

of the oncoming flow with the thin layer of air adjacent to the surface of the ball, whose motion is "with the wind" on one side, and "into the wind" on the opposite side. For a ball flying at 90 degrees angle of attack, for example, one would expect the same sort of side force as that experienced by a baseball, and indeed it can be seen in the wind-tunnel data.

However, the Magnus force on a football has only a minor effect on the trajectory, and is not the quantity causing lateral curvature of the flight path, in contrast to the flight of a sphere, where it is the major cause. The reason is that a football in a well-thrown forward pass spends the majority of its time nearly aligned with the flight direction, and at these small angles to the wind the Magnus force is negligibly small.

## 11.3 Mathematical Descriptions

Most of what has been said above is well known, at least in qualitative terms. The purpose of this article is to take matters to a deeper level by describing the flight in mathematical terms. By doing so, it is possible to state the *quantitative* relationships between the launch conditions (initial speed, RPM, flight-path angle, orientation) and the flight conditions that result (such as the range, time of flight, wobble amplitude, and lateral displacement of the impact point).

The mathematical principles governing the motion of a rigid body are contained in Newton's laws. These consist of equations for the translational and angular accelerations of the body in response to the gravitational force and to the aerodynamic loads (forces and torques) which act on it. The aerodynamic loads, in turn, depend on certain features of the motion itself, such as the angles between the ball axis and the flight direction of the ball, known as the "wind" angles. This makes the problem of predicting the motion nonlinear, and requires numerical integration of the pertinent equations. Moreover, the equations are written in a coordinate system attached to the ball, and this brings into play the further complications of Coriolis and centrifugal forces as well as gyroscopic torques. It is also necessary to calculate the orientation of the ball, both with respect to the wind and with respect to the ground.

Some details of these equations are given next. Further explanation can be found in Rae (2003), and readers not interested in the mathematics can proceed to the next section. Figure 11.5 shows two coordinate systems that are used, namely a first set attached to the earth, and a second attached to the ball. The earth- fixed set (subscript F) has its origin at any convenient place on the field, such as the point from which the ball is thrown. The $x_F$ axis points downfield, the $y_F$ axis points toward the right sideline, and the $z_F$

axis points down, forming a right-handed system. The position of the center of the ball is given by its three coordinates in this system. The orientation of the ball is then defined by a second set of axes (subscript B) which are attached to the ball at its center and rotate with it. The $x_B$ axis coincides with the ball axis, the $y_B$ axis comes out of the right seam, and the $z_B$ axis out of the lower seam (the seam opposite the one containing the laces). The orientation of the body axes with respect to the space-fixed ones is defined by three angles, called Euler angles, which are shown in Fig. 11.5. Before any rotation, the ball axes are parallel to the earth, with the laces on top. Then three rotations are done, in the following order: first, the ball is rotated in yaw ($\psi$), i.e., the ball is rotated around its $z_B$-axis. Next, the ball is rotated in pitch ($\theta$), around its $y_B$-axis. Lastly, it is rotated in roll ($\phi$), about its $x_B$- axis.

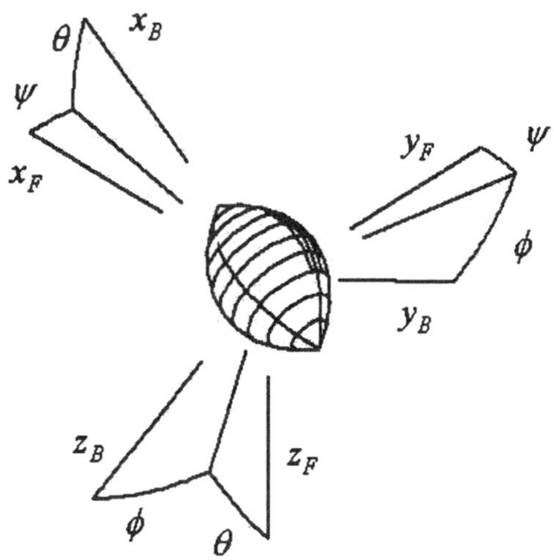


*Figure. 11.5.* Euler angles;yaw, pitch, and roll.

The vector components of the translational velocity $(u,v,w)$, the angular velocity $(p,q,r)$, the force $(X,Y,Z)$, and the torque $(L, M, N)$, resolved in the body-axis system, are shown in Fig. 11.6. These vectors have different components when resolved in the earth-fixed system. The two sets of components (of the velocity vector, for example), are related by

$$\begin{Bmatrix} \dot{x}_F \\ \dot{y}_F \\ \dot{z}_F \end{Bmatrix} = \left[ m_{ij} \right] \begin{Bmatrix} u \\ v \\ w \end{Bmatrix} \tag{11.1}$$

where

$$\left[ m_{ij} \right] = \begin{bmatrix} \cos\theta \cos\psi & \sin\phi \sin\theta \cos\psi - \cos\phi \sin\psi & \cos\phi \sin\theta \cos\psi + \sin\phi \sin\psi \\ \cos\theta \sin\psi & \sin\phi \sin\theta \sin\psi + \cos\phi \cos\psi & \cos\phi \sin\theta \sin\psi - \sin\phi \cos\psi \\ -\sin\theta & \sin\phi \cos\theta & \cos\phi \cos\theta \end{bmatrix} \tag{11.2}$$

The rates of change of the Euler angles are related to the components of the angular-velocity vector by:

$$\dot{\psi} = [q \sin\phi + r \cos\phi ] / \cos\theta \tag{11.3a}$$

$$\dot{\theta} = q \cos\phi - r \sin\phi \tag{11.3b}$$

$$\dot{\phi} = p + \tan\theta \, [ q \sin\phi + r \sin\phi ] \tag{11.3c}$$

In the body coordinates, the equations of motion are (Nelson, 1984):

$$\dot{u} = \frac{1}{m}[X - mg \sin\theta ] - qw + rv \tag{11.4a}$$

$$\dot{v} = \frac{1}{m}[Y + mg \cos\theta \sin\phi ] - ru + pw \tag{11.4b}$$

$$\dot{w} = \frac{1}{m}[Z + mg \cos\theta \cos\phi ] - pv + qu \tag{11.4c}$$

$$I_x \dot{p} = L \tag{11.4d}$$

$$I_T \dot{q} = M - pr[I_X - I_T ] \tag{11.4e}$$

$$I_T \dot{r} = N + pq[I_x - I_T] \tag{11.4f}$$

These are the full nonlinear six degree of freedom equations of motion. They will be simplified further in what follows, but are included here as a starting point for readers who might be interested in pursuing more complex aspects of football flight.

To solve these equations, it is necessary to define what the aerodynamic forces and torques depend on. Wind-tunnel tests have shown that the major factors affecting the flight of the ball, in addition to its speed, shape, and the air density, are the angles between the body axes and the velocity vector, called the angles of attack ($\alpha$) and sideslip ($\beta$). These are defined on Fig. 11.6, and are referred to as wind angles, to distinguish them from the Euler angles. The latter set affect only the gravitational force on the ball, whereas the former set affects the aerodynamic loads (forces and torques).

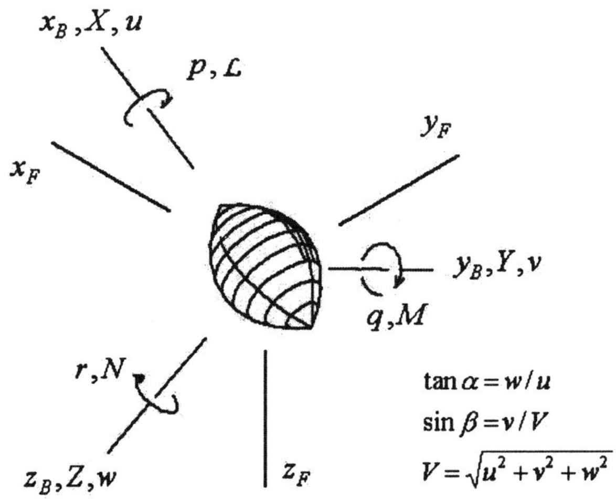


*Figure. 11.6.* Definitions of vectors: translational and angular velocities, forces, and moments.

The further simplifications made in the remainder of this article are based on the fact that wind-tunnel data suggest that the aerodynamic loads do not depend strongly on the roll angle of the ball. It behaves like a body of revolution, and consequently five degrees of freedom are sufficient to describe the motion. This amounts to using a set of non-rolling axes (denoted by $x_B$, $y$, and $\tilde{z}$) whose $x_B$-direction is still aligned with the central axis of the ball, with the $y$ and $\tilde{z}$ axes lying in vertical and horizontal planes, respectively. The geometry of these axes is shown in Fig.

11.7. Lines representing the laces and seams have been removed, in order to emphasize that the ball is regarded as a body of revolution.

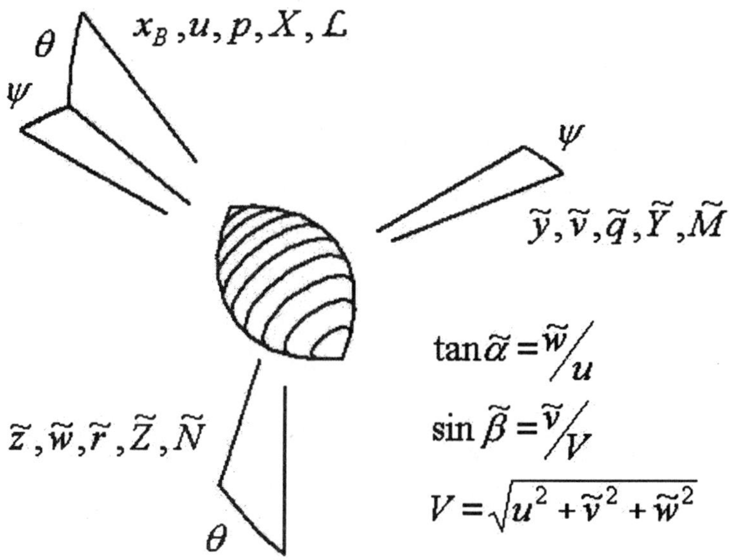


*Figure 11.7.* Notation for non-rolling coordinates, vector velocities, forces, and moments.

In these axes, the equations of motion reduce to (Rae 2003):

$$
\begin{aligned}
m(\dot{u}+\tilde{q}\tilde{w}-\tilde{r}\tilde{v}) &= X-mg\sin\theta \\
m(\dot{\tilde{v}}+\tilde{r}u+\tilde{r}\tilde{w}\tan\theta) &= \tilde{Y} \\
m(\dot{\tilde{w}}-\tilde{q}u-\tilde{r}\tilde{v}\tan\theta) &= \tilde{Z}+mg\cos\theta \\
I_x\dot{p} &= L \\
I_T\dot{\tilde{q}} &= -I_x p\tilde{r}-I_T\tilde{r}^2\tan\theta+\tilde{M} \\
I_T\dot{\tilde{r}} &= +I_x p\tilde{q}+I_T\tilde{q}\tilde{r}\tan\theta+\tilde{N}
\end{aligned}
\tag{11.5}
$$

$$
\dot{\psi}=\tilde{r}/\cos\theta,\ \dot{\theta}=\tilde{q},\ \dot{\phi}=p+\tilde{r}\tan\theta \tag{11.6}
$$

where the tildes denote vector components resolved in the non-rolling axes. A brief derivation of these equations is given in the Appendix (Section 11.8).

The motions predicted by these equations have been studied for decades, for configurations ranging from artillery shells (Hayes, 1938) to ballistic missiles (Regan, 1984) to footballs (Rae, 2003). Analytical solutions can be found for some limiting conditions, such as when the wind angles are small, but for most cases one must resort to numerical solutions.

## 11.4 SIMULATION RESULTS

### 11.4.1 Application to the Forward Pass

A number of the observations listed above can be simulated in numerical solutions of these equations. Figure 11.8a shows typical results from such a calculation.

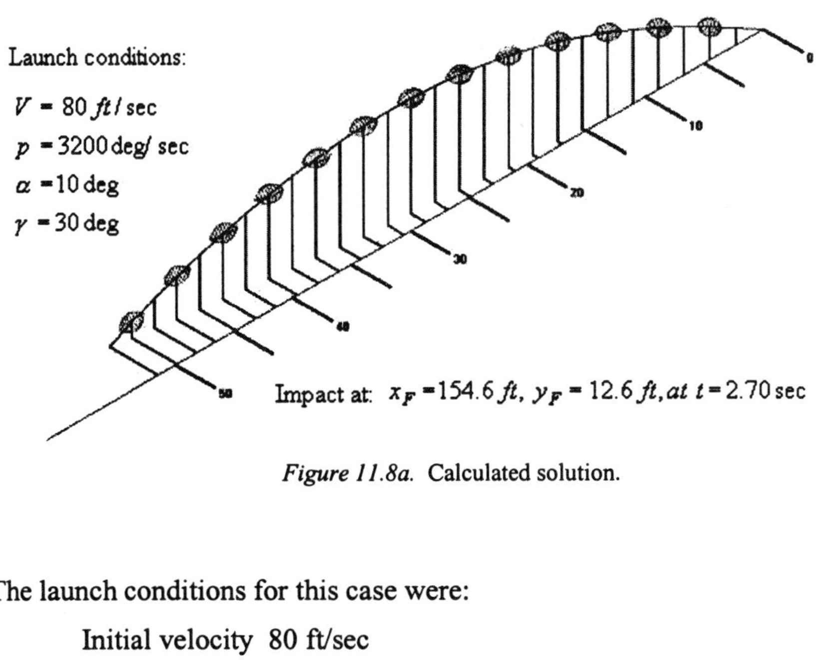


*Figure 11.8a.* Calculated solution.

The launch conditions for this case were:

Initial velocity 80 ft/sec

Initial flight-path angle gamma : 30 degrees above the horizon

Initial angle angle of attack: 10 degrees above the flight path

Initial pitch angle: 40 degrees (30 + 10) above the horizontal

Initial spin rate: 3200 degrees/sec (533 RPM)

Figure 11.8b shows the main parameters of the solution as a function of the time (in seconds) after launch. The top trace is the roll angle, taken to lie in the range –180 degrees to + 180 degrees. It shows the (approximately)

constant spin velocity, corresponding to the assumption that there is zero torque opposing the spin. The next two traces show the altitude above the launch point, and the speed. Note that the speed decreases as the ball climbs and increases as it descends, due mostly to the exchange of kinetic for potential energy, and that it returns to the launch altitude with a small deficit due to the cumulative effects of drag. The next two traces are the yaw and pitch angles, which oscillate with a period about 1.8 times longer than that of the roll angle. The last trace in Fig. 11.8b is the lateral displacement of the ball. Note that the trajectory drifts to the right ($y_F > 0$) of the original flight path.

For this case the impact point was at $x_F = 154.6$ ft downfield and $y_F = 12.6$ ft to the right of the original flight path. The time of flight was 2.7 seconds. These numbers are typical of passes thrown under game conditions. Fig. 11.8c shows the angles of attack and sideslip in the non-rolling axes ($\tilde{\alpha}$ versus $\tilde{\beta}$). Complete spins of the ball are denoted by the small circles. The angle of attack is positive when the nose of the ball is above the flight path, and the sideslip angle is positive when the nose is to the left of the flight path. Thus, Fig. 11.8c is an "end-zone" view, i.e. it is the view seen by an observer looking backward at an approaching ball. It is often the case that video taken of a ball approaching the end-zone observer shows this looping behavior, where both the spin and precession are apparent, and where the nose of the ball seems to hesitate periodically while its precession reverses direction.

For most of the flight the ball rides with its nose above the flight path and to the right of it. The negative bias of the sideslip angle means that the ball rides with its nose pointed to the right of its flight path, and this in turn leads to a side force which accounts for the rightward displacement (and hence curvature) of the flight path. This is the behavior observed in Fig. 11.8a.

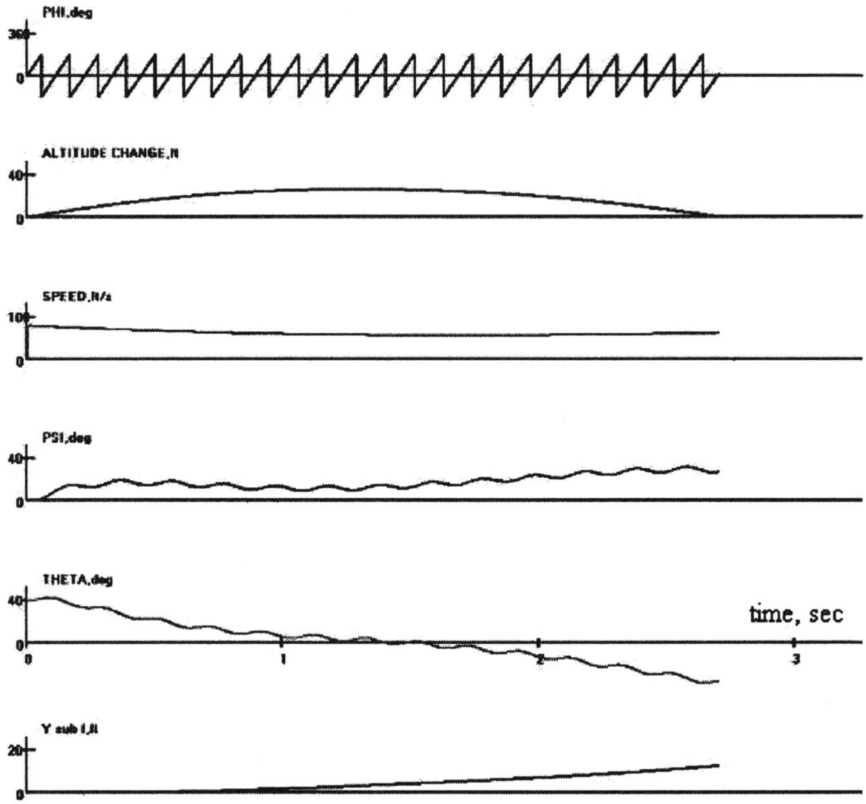


*Figure 11.8b.* Trajectory parameters.

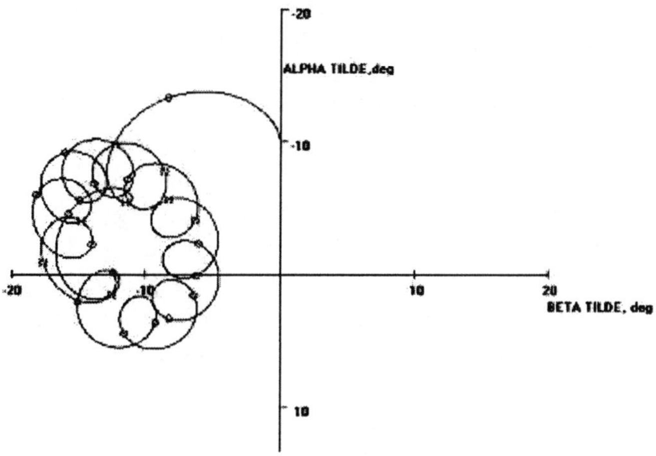


*Figure 11.8c.* End-zone view.

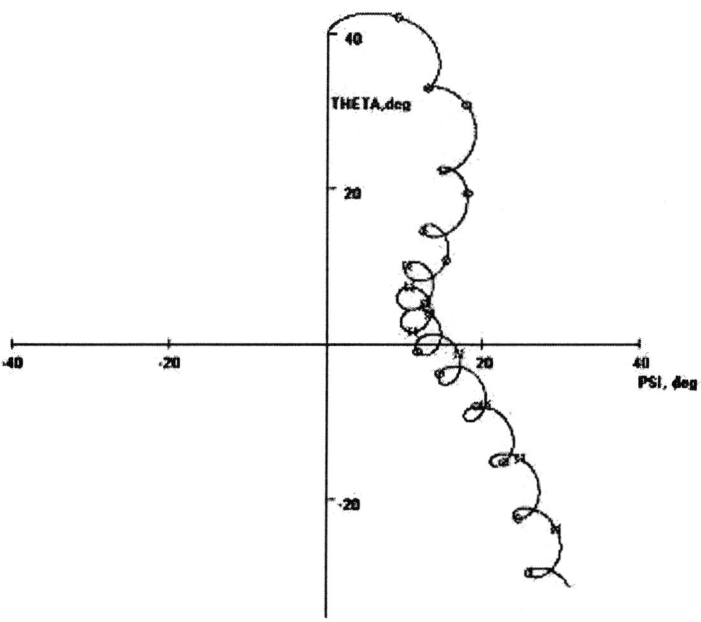


*Figure 11.8d.* Yaw and pitch angles.

The mechanism causing the bias in the orientation of the ball is the following: by spinning the ball, the unstable torque that tends to turn the ball axis away from the wind direction is converted into a precessional (or wobbling) motion, where the nose of the ball moves around the flight direction, i.e. the ball "tracks" its flight path on average. But the flight path itself curves downward. Thus a secondary result of spin stabilization is that the ball acquires, on average, a nose-down rotation rate. A gyroscope that is being rotated in a nose-down direction will experience a torque to the left or right of its spin axis, for a right- or left-handed spin, respectively. The ball compensates for this lateral gyroscopic torque by turning its nose slightly to the opposite side and slightly above the wind, and this bias in its wind angles is what is shown in the solution above.

The importance of lateral curvature in game situations is not clear. Except for the mention of it in the book by Starr (1967), the author is not aware of any further discussion of it. It may be that pass receivers compensate for it instinctively, and it is also true that it is difficult to observe when the pass receiver is moving, and viewing the ball against a background that contains no fixed vertical reference.

Figure 11.8d shows the variation of the pitch angle as a function of the yaw angle. As the pitch angle moves from positive to negative angles, the yaw angle moves, on average, to a nose-right orientation.

Figures 11.9a-d present the same variables as those of Figs. 11.8a-d, but for the case where the wind angles are zero at launch. The same lateral curvature occurs, but oscillations in the angles are minimized. The conclusion is that a "tight spiral" requires that the ball be perfectly aligned

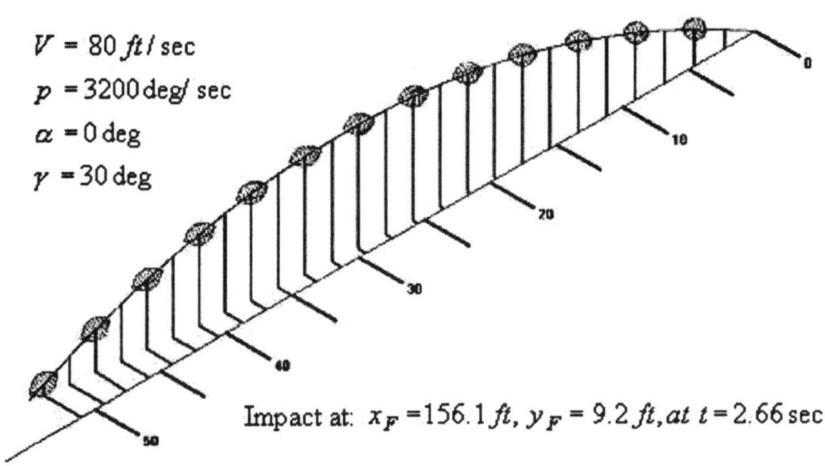


*Figure 11.9a.* Same initial conditions as Fig. 11.8a, but initial Alpha = 0.

with its flight path at launch. Even when this condition is met, however, a lateral curvature still occurs.

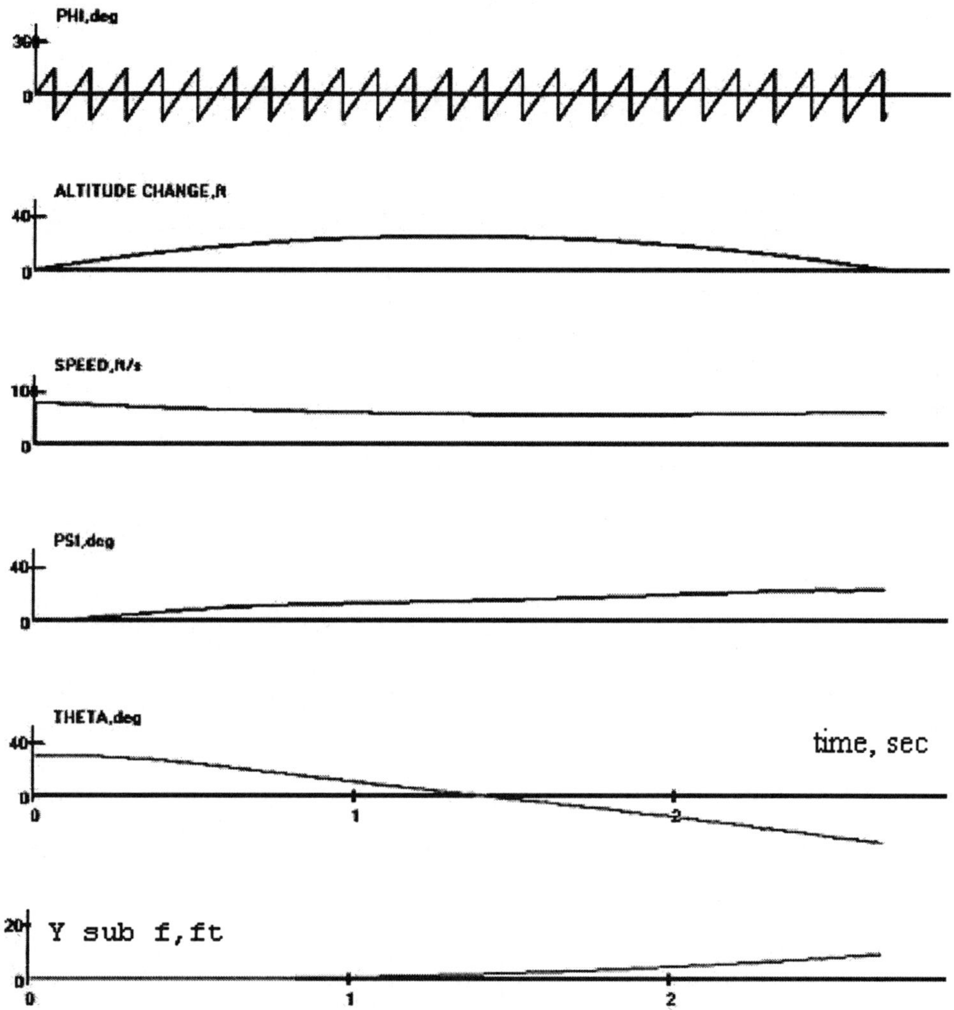


*Figure 11.9b.* Trajectory parameters.

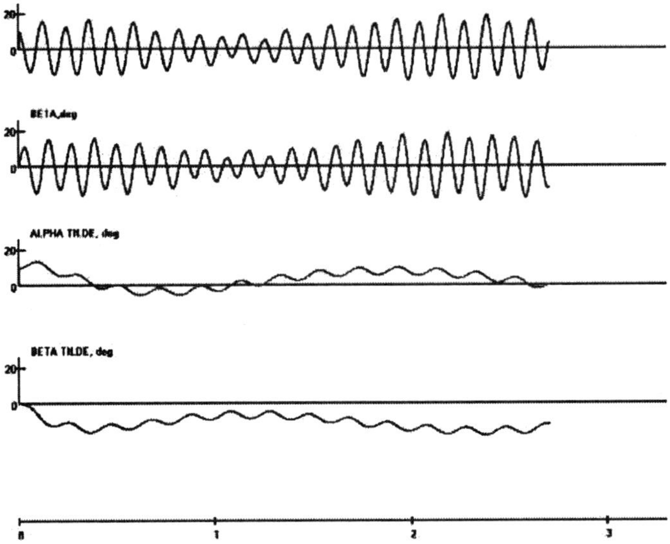


*Figure 11.9c* Angles of attack and sideslip, resolved in rolling and non-rolling coordinates.

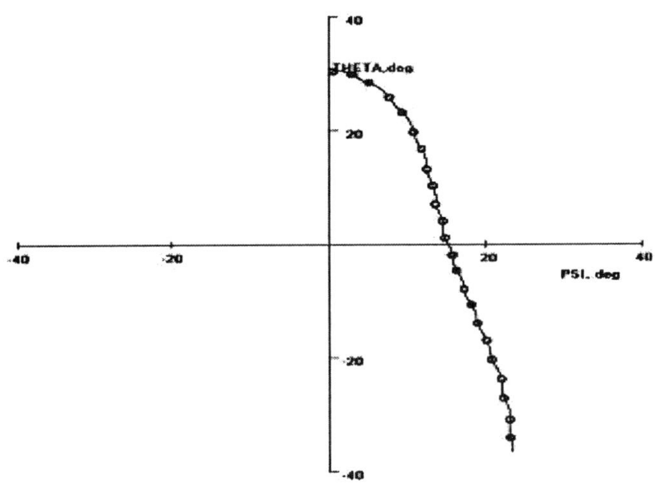


*Figure 11.9d.* Yaw and pitch angles.

## 11.4.2 Other Launch Conditions

Finally, Figs. 11.10 and 11.11 show what happens for larger initial flight-path angles and for spin rates both higher and lower than required for a tight spiral. These launch conditions can lead to wind angles which exceed 45 degrees, at which point the pitching torque reaches a maximum, after which it decreases to zero again at ninety degrees. One result of this fact is that the ball is stable when flying broadside to the wind, a condition that it can reach with a combination of minimum angle of attack at launch and maximum spin rate. On the other hand, the combination of a modest initial angle of attack with a lower spin rate produces large oscillatory wind angles, a condition often described as a wobbly pass.

Figure 11.10 shows the former behavior. For this case the initial conditions were:

Initial velocity 70 ft/sec

Initial flight-path angle gamma : 50 degrees above the horizon

Initial angle of attack: 0 degrees above the flight path

Initial pitch angle: 50 degrees (50 + 0) above the horizontal

Initial spin rate: 3600 degrees/sec (600 RPM)

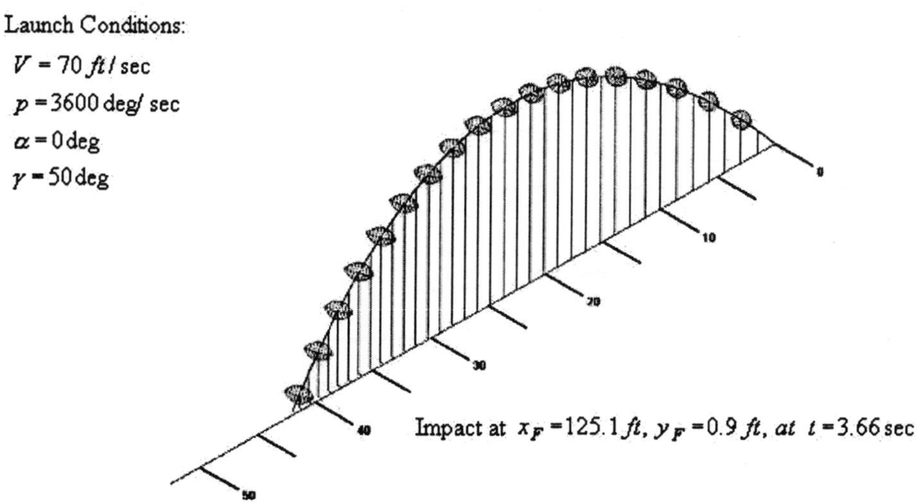


*Figure 11.10a.* Calculated solution.

Here the spin rate is much higher than required to suppress oscillations, so the ball tracks its trajectory very closely. The result is that the large nose-down rotation generates a nose-left torque, and the ball turns its nose to the right in order to achieve an equal and opposite torque due to sideslip.

The end-zone view, shown in Fig. 11.10c, indicates that the resultant wind angle moves steadily past the 45-degree circle (where the overturning moment reaches its maximum), and continues toward the 90-degree circle. Toward the end of the flight the ball is descending at a very steep flight-path angle, approaching the stable orientation broadside to the wind (Fig. 11.10d shows that the yaw angle is around 60 degrees at impact).

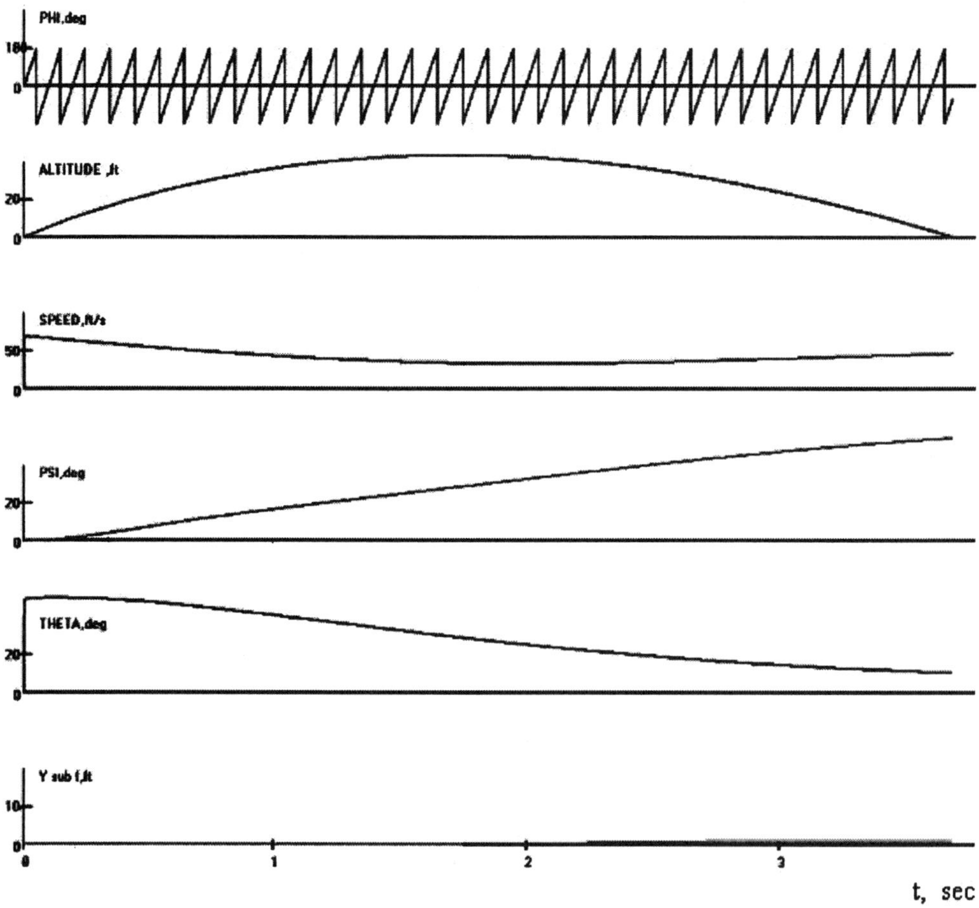


*Figure 11.10b.* Trajectory parameters.

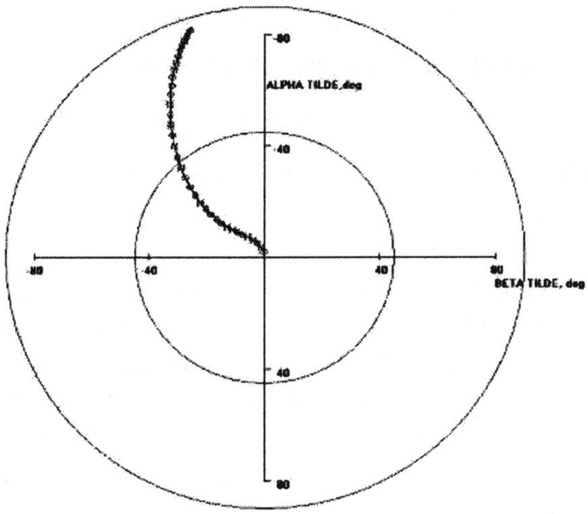


*Figure 11.10c.* End-zone view.

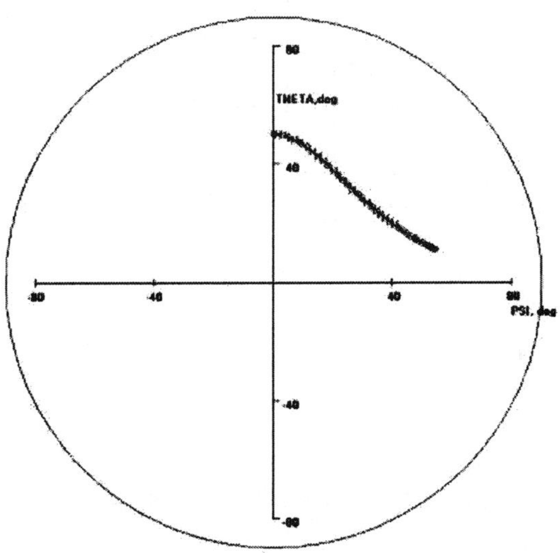


*Figure 11.10d.* Yaw and pitch angles.

Figure 11.11 shows the highly oscillatory behavior. For this case the initial conditions were:

Initial velocity 80 ft/sec

Initial flight-path angle gamma : 45 degrees above the horizon

Initial angle of attack: 10 degrees above the flight path

Initial pitch angle: 55 degrees (45 + 10) above the horizontal

Initial spin rate: 1600 degrees/sec (267 RPM)

The isometric sketch of the trajectory shows the very large angles that develop between the ball axis and the flight path, and this is confirmed by the traces in Fig. 11.11b, as well as by the end-zone view of Fig. 11.11c. In fact, the ball just narrowly escapes a 90-degree orientation to the wind. For this case, the yaw/pitch plane shown in Fig. 11.11d suggests that the ball now spends more time at positive sideslip, but still impacts the ground to the right of the original ground track.

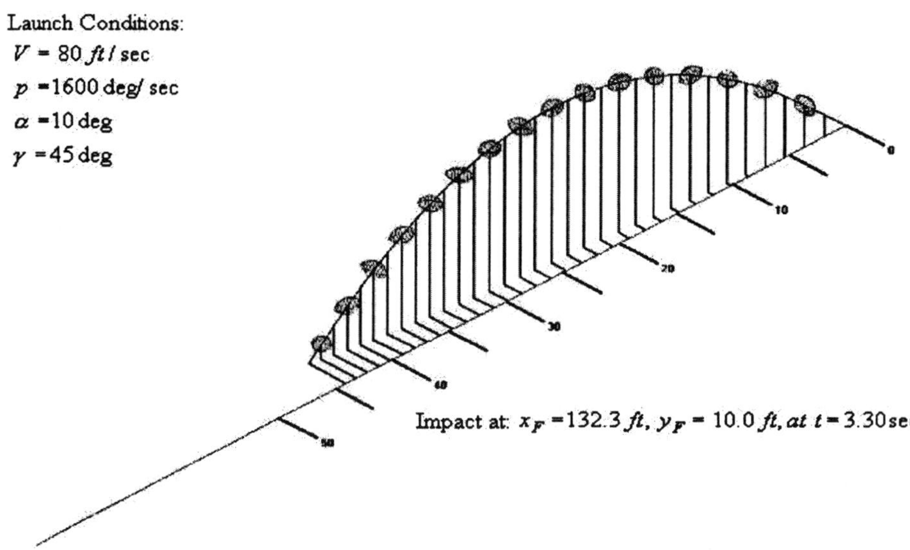


*Figure 11.11a.* Calculated solution.

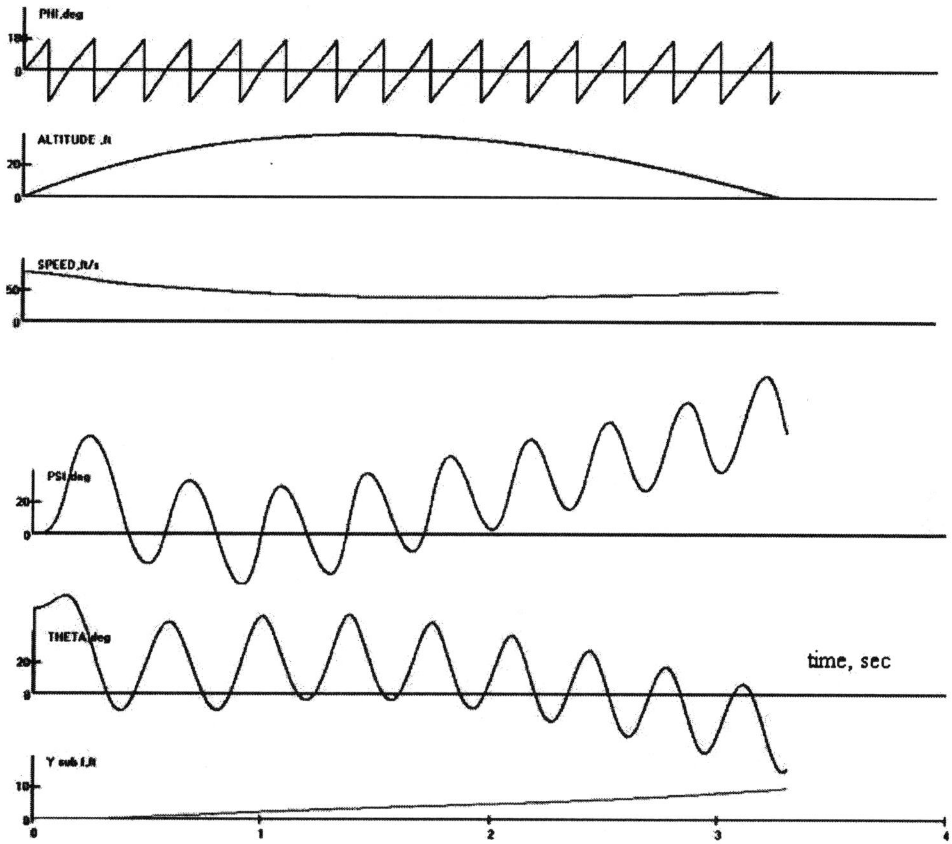


*Figure 11.11b.* Trajectory parameters.

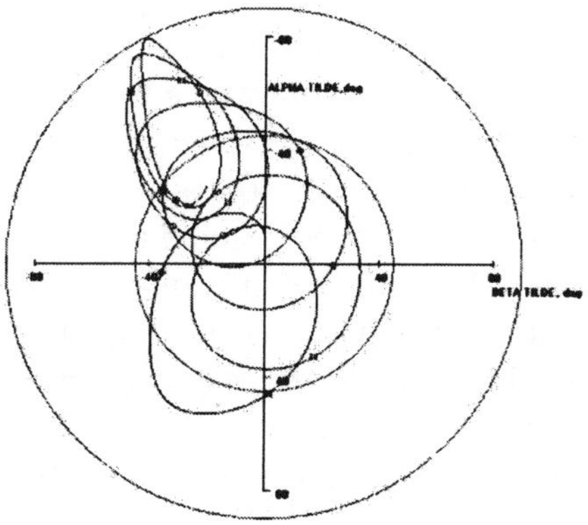


*Figure 11.11c.* End-zone view.

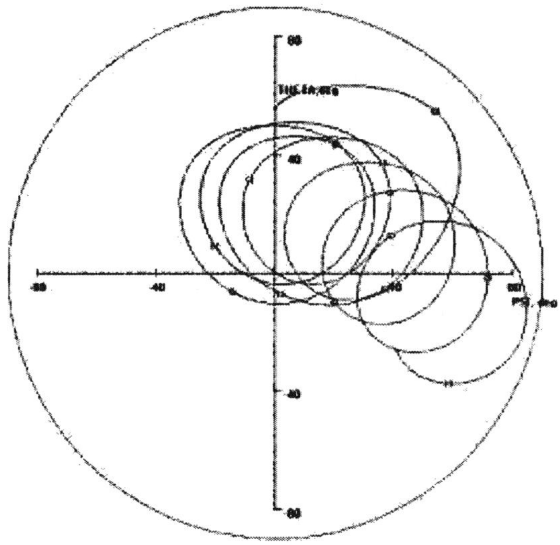

*Figure 11.11d.* Yaw and pitch angles.

## 11.5 DISCUSSION

### 11.5.1 Simulation Results

The three examples given above illustrate the wide range of motion that can take place when a football is thrown. Many of the features observed on film are also seen, at least qualitatively, in the calculated results – for example in their range, time of flight, spin/wobble ratio, and in their lateral curvature. However, much work remains to be done in refining the quantitative correspondence between the simulations and actual flights. Only the case of Fig. 11.8 contained an attempt to match calculation with field data, and its pitch-angle results showed a major difference between the prediction and the data of Fig. 11.3. However, the data itself were not taken under sufficiently controlled conditions to warrant any conclusions. Perhaps that particular flight was affected by wind, and in addition the precise launch conditions and the camera location are unknown. There is a need for more accurate and well-controlled data, and it may also be that additional aerodynamic effects need to be included in the numerical modeling.

Unless the ball is aligned at launch exactly parallel to its flight direction, it will wobble. The amplitude of the wobble is proportional to the initial misalignment - thus the "tightest spiral" is achieved when this misalignment is zero. Even when spun at a large value of the RPM, the ball is not completely stabilized, and any initial wobble will get larger, the longer the ball is in the air. Higher values of the RPM inhibit the growth rate of the wobbles, while lower ones may fail to stabilize the ball at all. It is clear that lower values produce much larger wobble amplitudes, and at very low values the ball diverges toward a 90-degree angle of attack.

When wobbles occur, the ball presents a larger projected area to the wind, leading a larger drag force which causes a reduced range and time of flight. The difference between a zero wind angle at launch and a 20 degree wind angle can lead to differences in range and time of flight on the order of ten percent (Rae, 2003).

An important conclusion revealed by computer simulations is that the aerodynamic forces on a football are small in comparison to the force of gravity (the lift-to-weight ratio is on the order of several tenths). Thus to a good approximation its trajectory is that of a stone, i.e. that of a point mass acted on only by gravity. However, the aerodynamic *torques* are not small compared to the rotational inertia of the ball. They are in fact large, and will draw the ball axis away from the wind direction very quickly unless the ball is spun. In the presence of a large spin angular velocity, the football becomes a gyroscope, and the off-axis rotation caused by the aerodynamic torques induces gyroscopic torques which act at right angles to the aerodynamic ones, resulting in the wobbling motion known as "spin stabilization".

### 11.5.2 Relation to Athletic Performance

There is no doubt that the future will see many improvements in the state of the art of physical modeling and computer simulation of football flight. As these advances are made, it will be possible to make the connection between the throwing motion and the trajectory achieved. For example, one can imagine instrumentation capable of establishing precise values for all of the velocity and orientation variables at the instant the ball is released. These can then be fed to a computer program which will calculate exactly the time of flight and the impact point. In fact, the ball can be caught in a net only a few feet from the passer and the flight can be displayed on a virtual-reality screen, thus permitting a greater rate of repetitions. It will then be possible to examine quantitatively how the trajectory parameters are affected by different throwing motions, different techniques for gripping the ball, and so forth. Most of the technologies for measuring the throwing motion and for calculating and displaying the trajectory are available, but further instrumentation development is required for measuring the launch conditions to the required accuracy. In addition, there is a great need for measurements of actual flight performance. These data may come from video images taken with calibrated cameras at known locations or from measurements made by on-board sensors mounted in thrown balls. And finally, there is a need for much more accurate wind-tunnel data, taken over a wider range of parameters than that examined by Rae and Streit (2002). Among the important quantities not yet measured in the tunnel are the aerodynamic-damping quantities, i.e. the effects of the angular-velocity components on the force and torque. These effects are also required for the kicking game, where the largest component may be about a transverse axis (for most kickoffs and field goals) and may be a combination of rotation about both axes for punts.

## 11.6 SUMMARY

In summary, it appears that continued developments in this area can have an impact on athletic performance, by correlating reliable measurements of the throwing motion with the launch conditions (and thus the trajectories) which they produce. There is no doubt that the athletes have learned from experience how to optimize their performance. But there remains a potential for developing training techniques that will facilitate the acquisition of such skills in minimum time.

## 11.7 REFERENCES

Brancazio, P. J., 1985, The Physics of Kicking a Football, *The Physics Teacher*. **23**: 403-407.

Brancazio, P. J., 1987, Rigid-Body Dynamics of a Football, *American Journal of Physics*. **55**: 415-420.

Hayes, T. J, 1938, *Elements of Ordnance*, Wiley, New York, p. 402.

Rae, W. J., 2003, Flight Dynamics of an American Football in a Forward Pass, *Sports Engineering*, to be published.

Rae, W. J., and Streit, R. J., 2002, Wind-Tunnel Measurements of the Aerodynamic Loads on an American Football, *Sports Engineering*. **5**: 165-172.

Regan, F. J., 1984, *Re-Entry Vehicle Dynamics*, AIAA, New York, Chapter X.

Starr, B, 1967, *Quarterbacking*, Prentice-Hall, Englewood Cliffs, New Jersey, pp. 168-169.

## 11.8 APPENDIX

As an example of how vector quantities are resolved in the non-rolling coordinate system, consider the velocity vector:

$$\begin{Bmatrix} u \\ \tilde{v} \\ \tilde{w} \end{Bmatrix} = \left[F_{ij}\right]\begin{Bmatrix} u \\ v \\ w \end{Bmatrix} = \begin{bmatrix} 1 & 0 & 0 \\ 0 & \cos\phi & -\sin\phi \\ 0 & \sin\phi & \cos\phi \end{bmatrix}\begin{Bmatrix} u \\ v \\ w \end{Bmatrix} \tag{11.7}$$

Similar relations apply for $(p,q,r)$, $(X,Y,Z)$, and $(L, M,N)$. The connection between the space-fixed axes and the body-fixed, non-rolling ones is (Rae 2003):

$$\begin{Bmatrix} \dot{x}_F \\ \dot{y}_F \\ \dot{z}_F \end{Bmatrix} = \left[m_{ij}\right]\begin{Bmatrix} u \\ v \\ w \end{Bmatrix} = \left[m_{ij}\right]\left[F_{ij}\right]\begin{Bmatrix} u \\ \tilde{v} \\ \tilde{w} \end{Bmatrix} \equiv \left[\tilde{m}_{ij}\right]\begin{Bmatrix} u \\ \tilde{v} \\ \tilde{w} \end{Bmatrix} \tag{11.8}$$

where

$$\left[\tilde{m}_{ij}\right] = \begin{bmatrix} \cos\theta\,\cos\psi & -\cos\phi\,\sin\psi & \sin\theta\,\cos\psi \\ \cos\theta\,\sin\psi & +\cos\phi\,\cos\psi & \sin\theta\,\sin\psi \\ -\sin\theta & 0 & \cos\theta \end{bmatrix} \tag{11.9}$$

The equations for the time rates of change of the Euler angles become:

$$\dot{\psi} = \tilde{r}/\cos\theta \;, \quad \dot{\theta} = \tilde{q}, \quad \dot{\phi} = p + \tilde{r}\tan\theta \tag{11.10}$$

In the equations of motion for non-rolling coordinates, the terms involving $\tan\theta$ are evidently contributions to the Coriolis and gyroscopic accelerations, but their physical significance is not readily apparent. The following derivation shows where these terms come from. The acceleration components in the rolling axes can be identified by taking the time derivative of the velocity vector:

$$\begin{Bmatrix} \ddot{x}_f \\ \ddot{y}_f \\ \ddot{z}_f \end{Bmatrix} = \frac{d}{dt}\left\langle \left[m_{ij}\right]\begin{Bmatrix} u \\ v \\ w \end{Bmatrix} \right\rangle = \left[m_{ij}\right]\begin{Bmatrix} \dot{u} \\ \dot{v} \\ \dot{w} \end{Bmatrix} + \left[\dot{m}_{ij}\right]\begin{Bmatrix} u \\ v \\ w \end{Bmatrix} \tag{11.11}$$

$$= \left[m_{ij}\right]\begin{Bmatrix}\dot{u}\\ \dot{v}\\ \dot{w}\end{Bmatrix} + \left[m_{ij}\right]\left[S_{ij}\right]\begin{Bmatrix}u\\ v\\ w\end{Bmatrix}, \quad \left[S_{ij}\right] = \begin{bmatrix}0 & -r & q\\ r & 0 & -p\\ -q & p & 0\end{bmatrix} \tag{11.12}$$

The term containing the matrix $\lfloor S_{ij} \rfloor$ gives the Coriolis accelerations, which arise from the time rate of change of the Euler angles. This matrix can be used as a means of updating the elements of the Euler matrix directly, rather than doing so by updating the angles and then evaluating the matrix elements from their trigonometric formulas.

A counterpart to $\lfloor S_{ij} \rfloor$ for the non-rolling case can be found by writing the time derivative of $\lfloor \widetilde{m}_{ij} \rfloor$ as follows:

$$\left[\widetilde{m}_{ij}\right] = \left[m_{ij}\right]\left[F_{ij}\right]^T \tag{11.13}$$

Thus $\left[\dot{\widetilde{m}}_{ij}\right] = \left[\dot{m}_{ij}\right]\left[F_{ij}\right]^T + \left[m_{ij}\right]\left[\dot{F}_{ij}\right]^T = \left[m_{ij}\right]\left[S_{ij}\right]\left[F_{ij}\right]^T + \left[m_{ij}\right]\left[\dot{F}_{ij}\right]^T$

(11.14)

$$= \left[\widetilde{m}_{ij}\right]\left[F_{ij}\right]\left\{\left[S_{ij}\right]\left[F_{ij}\right]^T + \left[\dot{F}_{ij}\right]^T\right\} \equiv \left[\widetilde{m}_{ij}\right]\left[F_{ij}\right]\left[H_{ij}\right] \equiv \left[\widetilde{m}_{ij}\right]\left[G_{ij}\right]$$

(11.15)

where

$$\left[H_{ij}\right] = \begin{bmatrix}0 & -r & q\\ r & 0 & -p\\ -q & p & 0\end{bmatrix}\begin{bmatrix}1 & 0 & 0\\ 0 & \cos\phi & \sin\phi\\ 0 & -\sin\phi & \cos\phi\end{bmatrix} + \dot{\phi}\begin{bmatrix}0 & 0 & 0\\ 0 & -\sin\phi & \cos\phi\\ 0 & -\cos\phi & -\sin\phi\end{bmatrix}$$

(11.16)

$$= \begin{bmatrix}0 & -\widetilde{r} & \widetilde{q}\\ r & (p-\dot{\phi})\sin\phi & -(p-\dot{\phi})\cos\phi\\ -q & (p-\dot{\phi})\cos\phi & (p-\dot{\phi})\sin\phi\end{bmatrix} \tag{11.17}$$

Thus the desired counterpart to the matrix $\lfloor S_{ij} \rfloor$ is

$$\left[G_{ij}\right]=\left[F_{ij}\right]\left[H_{ij}\right]=\begin{bmatrix} 0 & -\widetilde{r} & \widetilde{q} \\ \widetilde{r} & 0 & -(p-\dot{\phi}) \\ -\widetilde{q} & p-\dot{\phi} & 0 \end{bmatrix} \tag{11.18}$$

where $p-\dot{\phi} = -\widetilde{r}\tan\theta$ . From this it is clear that the terms involving $\tan\theta$ are genuine contributions to the Coriolis acceleration, with the effective value of the spin velocity reduced to a small quantity. The fact that this contribution is proportional to $\widetilde{r}$ is due to a peculiarity of the Euler angles, namely that it is possible for $\phi$ to change, even when $p$ is zero but $q$ and $r$ are nonzero.

# Chapter 12

# Biomechanics of Tackling

Thomas W. Journell

*Department of Health and Human Performance, University of Wisconsin at River Falls, 410 S. 3rd St, River Falls, WI 54022; PH: (715) 425-3135, FX: (715) 425-3922, EM: thomas.w.journell@uwrf.edu*

## 12.1 INTRODUCTION

The fabric of American football is blocking and tackling. There are many different types of tackling scenarios depending on position played, placement on the field (e.g., closed space or open space), type of situation (e.g., open field tackle, converge tackle, sideline tackle). Nevertheless, there are some basic, fundamental tackling principles and techniques.

To provide a perspective on the forces involved in tackling, it has been estimated that a "National Football League hit" is comparable to dropping a bowling ball from a height of 130 feet. With the offense and defense both moving at about 10 meters a second, this adds up to about 9,000 newtons of force. In other words, it gives both players an acceleration of 10 times that of gravity

Because of the limited published information available related to biomechanical considerations of tackling, this chapter is not meant to be an exhaustive review of literature. Rather, it is simply an analysis of the tackling skill.

This review begins with a discussion of the correct contact areas for safely executing a proper tackle. It is followed by an analysis of muscle anatomy, the joints involved, the joint positions, and the corresponding muscle actions. Finally, each tackling phase is reviewed during execution of a fundamental tackle.

## 12.2 SAFETY AND PROPER CONTACT AREAS

Before getting started with the fundamentals of tackling, the proper body position and contact points when executing a safe tackle needs to be examined. Figure 12.1a shows a tackler in the proper contact position. When getting into a proper preparatory tackling position, the hips, knees and ankles are in a triple flexed position (i.e. squat stance), the chest is out, shoulder blades are adducted (i.e., in towards the spine) and the head is slightly hyperextended.

Figure 12.1b illustrates and improper tackling position. This is a very weak position. Note that the legs are straight, the trunk is flexed, causing the head to turn downwards. Moreover, the player is overextended, thus exposing the head. When contact is made, force is transmitted from the head down the spine. There can be no force generated from this stance. Imagine putting a squat bar with a fair amount of weight on the back. In contrast, placing a squat bar and heavy weight on the back of the person in Fig. 12.1a could be easily done. With this body position and base, static and dynamic forces can be applied in a safe manner.

*Figure 12.1a.* Proper Tackling Position. Photo by Jens Gunelson.

*Figure 12.1b.* Improper Tackling Position. Photo by Jens Gunelson.

When contact is made, a classic "triple extension" lower body movement occurs from the ankles, knees and hips. This is why squats and Olympic lifts (e.g. cleans, snatches) are so often prescribed by the strength and conditioning coach. Contact should be made with the chest and the anterior, rounded part of the shoulder pads. A phrase that can be used to sum up tackling is the "4 ups of tackling." When performing the execution phase, the head is up, the eyes are up, the chest is up, and the arms are clubbed up in the sagittal (i.e., median) plane.

## 12.3 MUSCLE ANATOMY REVIEW

Before discussing each phase of the tackling skill, muscle anatomy associated with each phase should be understood. Figures 12.2a,b are anatomical figures displaying some of the major muscles involved in tackling.

The analysis of the tackle has been broken down into two phases: a preparation phase and an execution phase. The preparation phase includes all the eccentric actions that place body into the proper tackling position. Conversely, the execution phase includes all the concentric actions while performing the tackle. The following chart details the joints involved, the joint positions, and the corresponding muscle actions when performing a tackle.

Preparation Phase:

| Joint: | Joint Position: |
|---|---|
| Neck | Slightly Hyperextened |
| Shoulder | Hyperextended |
| Elbow | Flexed in the Semi-Supinated, Semi-Pronated position |
| Wrist | Fixed in anatomical postion |
| Fingers | Slightly Flexed |
| Trunk | Flexed |
| Hip | Flexed |
| Knee | Flexed |
| Ankle | Dorsiflexed |

Execution Phase:

| Joint: | Joint Action: | Muscles Used: |
|---|---|---|
| Neck | Slightly Hyperextended | Trapezius<br>Splenius |
| Shoulder | Flexion | Pectoralis Major<br>Coracobrachialis<br>Anterior Deltoid |
| Elbow | Flexion in the Supinated position | Biceps Brachii,<br>Brachioradialis<br>Supinator |
| Wrist | Flexion | Flexor Carpi Radialis<br>Flexor Carpi Ulnaris<br>Palmaris Longus |
| Fingers | Flexion | Flexor Digitorum |
| Trunk | Extension | Erector Spinae<br>Quadratus Lumborum |
| Hip | Extension | Gluteus Maximus |
| Knee | Extension | Rectus Femoris<br>Vastus Lateralis<br>Vastus Intermedius<br>Vastus Medialis |
| Ankle | Plantar Flexion | Gastrocnemius<br>Soleus<br>Tibialis Posterior<br>Flexor Digitorum Longus<br>Peroneus Longus |

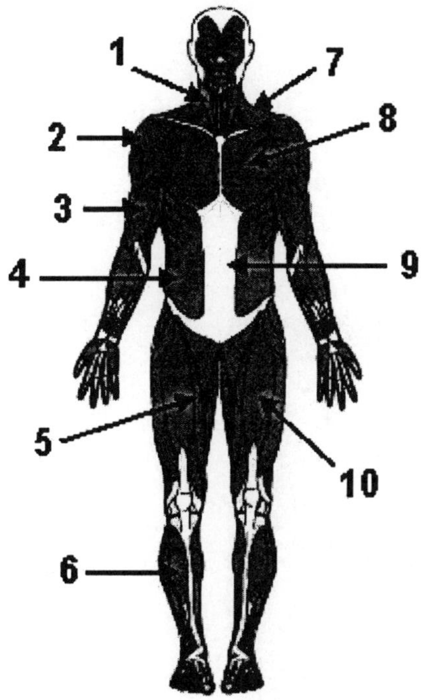


*Figure 12.2a.*

1. sternocleidomastoid
2. deltoid
3. biceps brachii
4. external oblique
5. sartorius
6. tibialis anterior
7. trapezius
8. pectoralis major
9. rectus abdominus
10. rectus femoris

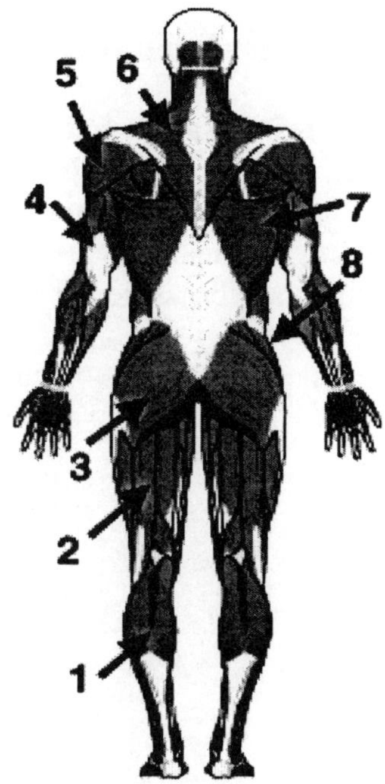


*Figure 12.2b.*

1. gastrocnemius
2. hamstrings
3. gluteus maximus
4. triceps brachii
5. deltoid
6. trapezius
7. latissimus dorsi
8. gluteus medius

*Figure 12.2.* Anterior (a) and posterior (b) views of some of the muscles involved during a tackle. Adapted from Web Anatomy, Murray Jensen General College, University of Minnesota.

## 12.4 THE TWO PHASES OF TACKLING

### 12.4.1 The Preparation Phase

Figures 12.3a and b display the position of the tackler in the preparation phase. The elbows are flexed at 90 degrees and are in a semi-supinated (i.e., pointing upwards) position. The head and chest are up. In the preparatory position, the tackler hyperextends the shoulder joint back as far as possible, keeping the elbows flexed at 90 degrees.

*Figures 12.3a (top), b (bottom).* Two stages of the preparation phase. Photo by Jens Gunelson.

When watching athletes execute a standing broad jump, they cock their shoulder joint back in a hyperextended position before jumping. Upon jumping, a simultaneous, violent flexion of the shoulder joint occurs. The elbows should remain at 90 degrees so as to create a shorter lever and thus faster arms for the tackle. This is analogous to the uppercut in boxing. When a boxer throws an uppercut, the elbow position is 90 degrees flexed. If the boxer threw this punch with a straight arm, it would have relatively little force and would be too slow.

## 12.4.2 The Execution Phase

Figures 12.4a-c illustrate the execution phase. The tackler uncoils his body and explodes with the arms and hips together (flexion of the shoulder joint along with extension of the hip joint). As mentioned previously, it is easier for the arms and hips to work together since they are working in the same sagittal plane. Furthermore, clubbing or flexing the shoulders keeps the chest and head upright. The finish should include the tackler grabbing cloth high on the back of the shoulder pads.

The execution phase demonstrates a classic triple extension movement (ankle plantar flexion, knee extension, and hip extension). An important point to remember when performing triple extension movements is that the prime mover of hip extension is the gluteus maximus. Since the hamstring muscles are antagonists to the knee extensors, they will not play a major role in extending the hip. This is commonly termed active insufficiency.

Figures 12.5a and b illustrate the conclusion of the execution phase. The emphasis here is on the wrap. The arms should be "clubbed" upward in the sagittal plane and fingers should grab cloth as high as possible on the back of ball carrier's jersey by flexing the wrists and fingers. When this is accomplished, the head is up and slightly hyperextended, the chest or trunk is up, and the eyes are up. When the tackler's arms are moved in the vertical plane, it is easier for the hips to be incorporated. Conversely, as previously illustrated in Figure 12.1b, if tacklers wrap their arms around the ball carrier in the transverse plane, the hips have a tougher time coming through, the body becomes overextended, and the trunk is flexed, thus exposing the top of the head.

*Figures 12.4a-c (top to bottom).* Initial stages of the execution phase. Photos by Jens Gunelson.

*Figures 12.5 a (top), b (bottom).* Two examples of clubbing the arm. Photos by Jens Gunelson.

Figures 12.6a and b demonstrate the final phase of the execution of the tackle. This is the finish! Emphasis is on the feet! Upon contact, the feet need to be active or churning. The tackler should "step on the toes" of the ball carrier and explode his feet, hips and arms through the ball carrier. RUN THE FEET! The tackler must continue to drive or churn the feet in order to knock back the ball carrier.

The helmet of the tackler should go across the ball carrier in order to drive the ball carrier back from where he started. Ideally, the ball carrier should be driven back 2-3 yards.

*Figures 12.6 a (top), b (bottom).* Final stages of execution phase. Emphasis is on the feet. Photos by Jens Gunelson.

## 12.5 SUMMARY

In summary, this chapter highlighted the correct contact areas for safely executing a proper tackle, the basic anatomy of the muscles, and action of the joints involved in tackling. Additionally, the preparatory and execution phases of tackling were reviewed. In the preparatory position, the tackler hyperextends the shoulder joint back as far as possible, keeping the elbows flexed at 90 degrees. The execution phase is a triple extension movement, with the arms clubbed upwards in the sagittal plane. Upon contact, the tackler explodes his feet, hips and arms through the ball carrier. A phrase that can be used to sum up tackling is the "4 ups of tackling." When performing the execution phase, the head is up, the eyes are up, the chest is up, and the arms are clubbed up in the sagittal plane.

## 12.6 REFERENCES

Bishop, P. .J., Norman, R. W., and Kozey, J. W., 1984, An evaluation of football helmets under impact conditions, *American Journal of Sports Medicine.* **12**: 233-236.

Bullis, K., and Journell, T.W., 2001, Fundamentals of Tackling, *Coaching Science Series Book,* Monterey, CA.

Bullis, K., and Journell, T.W., 2001, Fundamentals of Tackling, *Coaching Science Video Series,* Monterey, CA.

Freudenrich, C., 2003, Howstuffworks.com, http://entertainment.howstuffworks.com/physics-of-football.htm

Morris, B., Feld, F., and Junko, B., 1984, Sports Performance Series: Proper tackling technique, *National Strength & Conditioning Association Journal.* **6**: 4-7.

Journell, T. W., 1995, The Right Way to Tackle, *The Extra Point,* WHSCA Summer Manual, pp. 102-103.

Journell, T. W., 1999, Principles of Tackling at the Junior High Level, *The Extra Point.* WHSCA Summer Manual, pp. 34-36.

Journell, T. W., and Bullis, K., 2002, Positon Specific Tackling, *The Extra Point,* WHSCA Summer Manual, pp. 67-68.

Reid, S. E., and. Reid, S. E., Jr., 1978, Football, neck muscles and head impact, *Surgery, Gynecology & Obstetrics.* **147**: 513-517.

Web Anatomy, Murray Jenesn General College, University of Minnesota: http://www.gen.umn.edu/faculty_staff/jensen/1135/webanatomy/wa_muscle/wa_muscles_posterior_1.html

# Chapter 13

# Biomechanics and Aerodynamics in Soccer

Matt Carré[1] and Takeshi Asai[2]

[1]*Department of Mechanical Engineering, University of Sheffield, Mappin St., Sheffield S1 3JD, UK, PH: (44) 144-2227839, FX: (44) 144-2227855, EM: m.j.carre@shef.ac.uk*

[2]*Faculty of Education, Yamagata University, Koshirakawa Yamagata-city, Yamagata 990-8560 Japan, PH: (023) 628-4431, FX: (023) 628-4451, EM: asai@kescriv.kj.yamagata-u.ac.jp*

## 13.1 INTRODUCTION

Most of the physical activity carried out in soccer consists of either primary movements (e.g. sprinting, jumping etc.) or kicking a football (with the aim being either an accurate pass to another player or the scoring of a goal). This chapter describes a number of studies that have been carried out to examine the physical interactions that occur during these situations.

The first section examines the loading of the foot for various primary movements to give an insight into what loads are experienced commonly during play. The second section examines the impact of the foot and the ball using computer simulations to assess the amount of spin and velocity that can be imparted to a ball. Finally, the aerodynamics of a football are examined and the effects of the dynamic conditions of the ball (velocity and spin) on its ensuing flight are demonstrated using two free kick scenarios.

## 13.2 FOOT LOADING OF PRIMARY MOVEMENTS IN SOCCER

Movements of football (soccer) players consist of several complex motions and interactions. Foot loading is one of the most important factors in the analysis of the interaction of football motion with equipment and shoe design in football (Bartlett *et al.*, 1995, Hennig *et al.*, 1996, Milani *et al.*, 1995). It is also considered that the loading of the foot is related to overload type injuries in football. Therefore, it is important to analyze the loading of the foot for the improvement of skills and design of shoes in football (Eils et al., 2001).

### 13.2.1 Data Collection

Six male university football players were chosen as the subjects in this study. The playing positions of the subjects were two goalkeepers (GK), two defenders (DF) and two forwards (FW). The F-SCAN system (Nitta Inc.) was used to collect plantar pressure information (Fig. 13.1). All subjects were fitted with new football shoes (the typical 12-studs type) and the experiment was performed on an athletic field. Six football specific movements were performed: forward sprint at maximum speed, running backwards at maximum speed, side step, dive, turn and jump. The jump, turn, back step and side step all appear with high frequency in football, while diving is a specific movement of the goal keeper. The pressure scale of the foot scan system was set from 0 to 0.3 kg/cm$^2$, and was displayed as pressure contours in thirteen divisions. The data of the contact phase in this study was taken 3 ms after foot landing (Fig. 13.2). Pressure distribution and peak pressures were extracted for twelve areas such as the hallux (a-1), second and third toe (a-2), lateral toes (a-3), medial (b-1), central (b-2), lateral forefoot (b-3), medial (c-1), central (c-2), lateral mid-foot (c-3), medial (d-1), central (d-2) and lateral heel (d-3) (Fig. 13.3).

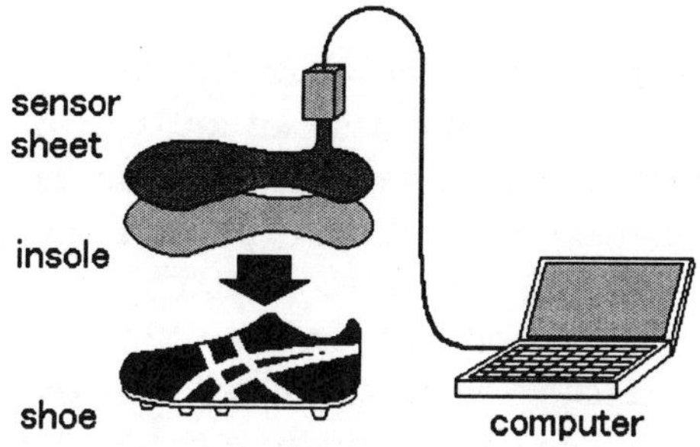


*Figure 13.1.* An insole pressure distribution measurement system (F-scan system; Nitta Inc.).

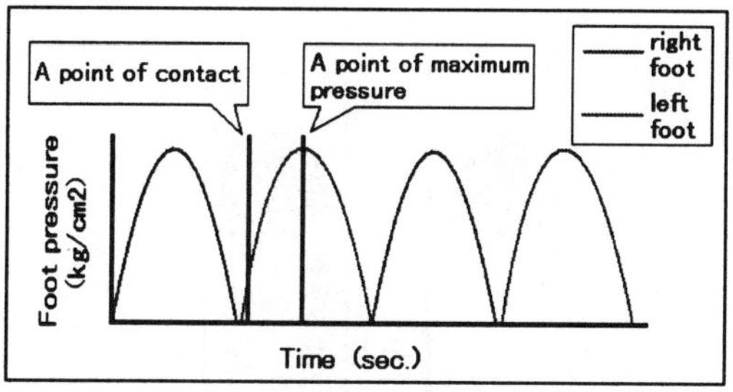


*Figure 13.2.* Definition of timing points.

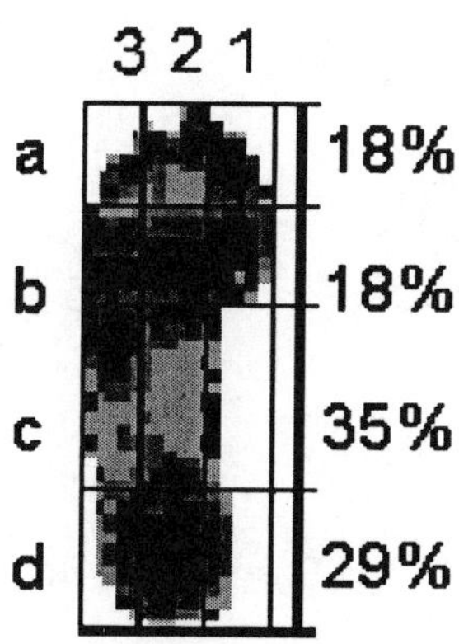


*Figure 13.3.* Definition of foot sole area.

## 13.2.2 Foot Pressure Distribution

A typical foot pressure distribution of the contact phase and the maximum pressure phase during a sprint are shown in Fig. 13.4. In the contact phase, the main loading area was b-1, and the maximum value was 3.26 kg/cm$^2$. The maximum pressure phase showed a similar pressure distribution with a maximum pressure of 5.95 kg/cm$^2$. There was almost no load in the d area. The distributions were very similar for all subjects studied and thus average pressures were calculated. In the case of the forward sprint, the highest average contact pressure occurred at b-1, with a value of 1.97 kg/cm$^2$ in the contact phase. The value of b-1 was also the highest in the maximum pressure phase, which was 3.79 kg/cm$^2$. Again, the d area at the heel showed very small pressures.

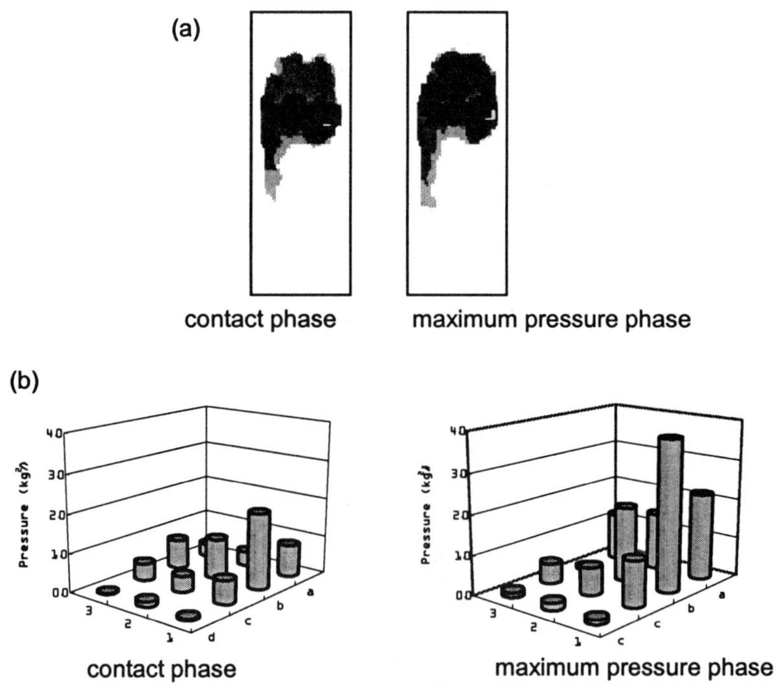


*Figure 13.4.* An example of distribution of pressure in sprint, (a) experimental data and (b) averages for six male soccer players.

The foot pressure distribution for contact and maximum pressure phases in the side step are shown in Fig. 13.5. In comparison to the sprint, there were significant differences between some of the subjects. In the contact phase on the fore foot, they were divided into two types. In the first (Type A), the d area around the heel becomes the main load area (with the largest load on d-1). In the second (Type B) b-1 and a-1 under the joints of the toes become the main load areas. In the rear foot, the main loading area was b-3. In the maximum pressure phase, both the fore foot and rear foot showed a large load at a-1 and b-1. The value of maximum loading of the rear foot is larger than that of the forefoot and the maximum difference was 4.26 kg/cm$^2$.

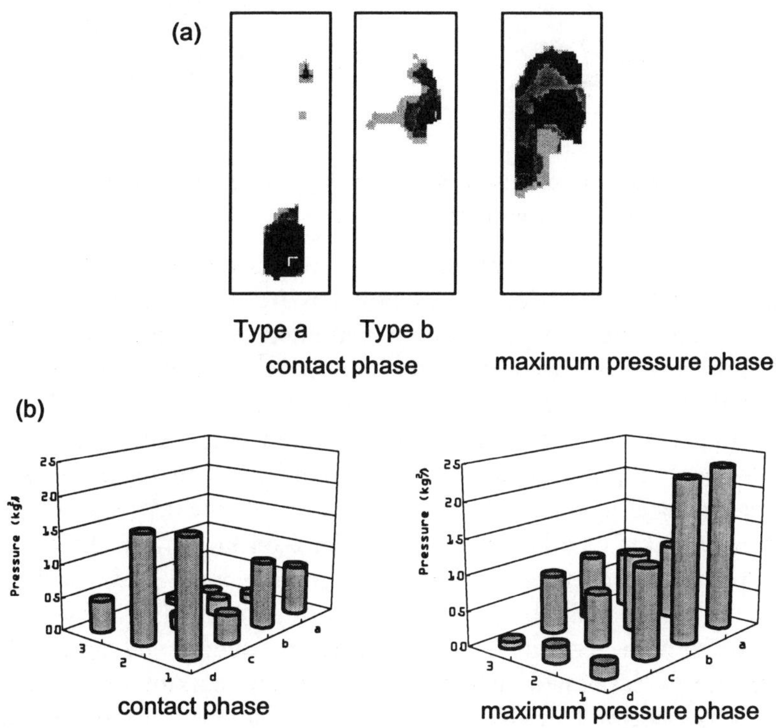


*Figure 13.5.* An example of distribution of pressure in forefoot in side step, (a) experimental data and (b) averages for six male soccer players.

In the contact phase of the fore foot in the side step, the maximum average peak pressure was observed in d-1 in the heel as 1.73 kg/cm$^2$. In the maximum pressure phase of the fore foot in the side step, the maximum average peak pressure of 4.12 kg/cm$^2$ was observed in b-1 (under the joints of the toes) and the second highest value of 3.05 kg/cm$^2$ was in a-1. In the contact phase of the rear foot in the side step, the maximum average peak pressure of 1.19 kg/cm$^2$ was observed in b-3, i.e. towards the outside of the foot (Fig. 13.6). Especially, the wide area consisting of a-1, a-2, a-3, b-1, b-2, and b-3 indicated over 0.48 kg/cm$^2$. The average peak pressure in b-1 was 2.86 kg/cm$^2$ and in a-1 was 2.29 kg/cm$^2$ (Fig. 13.6).

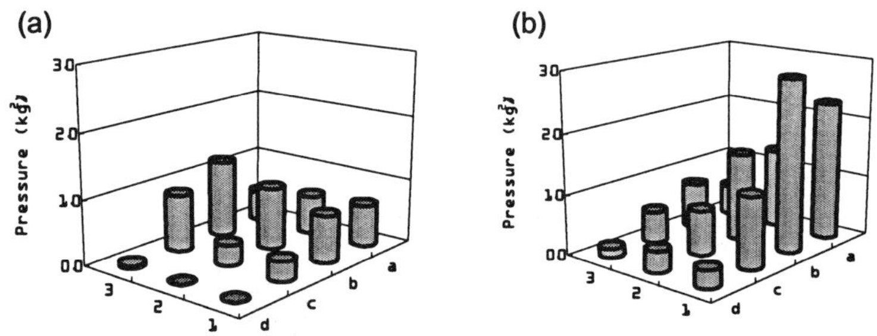


*Figure 13.6.* Average load on rear foot in side step during: (a) contact phase; and (b) maximum pressure phase, for six male soccer players.

The foot pressure distribution of the contact and maximum pressure phase in turning for the inside foot is shown in Fig. 13.7. For the inside foot, the main loading area was b-1 with a maximum value of loading of 6.54 kg/cm$^2$. For the outside foot, the main loading area was b-3, i.e. towards the outside of the foot. In the maximum pressure phase, the main loading areas were distributed across the inside of the foot and under the toes (b-1, b-2 and a-1). In the contact phase of the inside foot in the turn, the maximum average peak pressure was again observed in b-1 as 1.69 kg/cm$^2$ (Fig. 13.7). The peak pressure value of other areas was under 0.64 kg/cm$^2$. In the maximum pressure phase of the inside foot, the maximum average peak pressure was observed in b-1 as 3.58 kg/cm$^2$. The b-1 indicated high values in both phases, the similar tendency was observed in diving and sprint. In the contact phase of the outside foot during a turn, the maximum average peak pressure was observed in d-1 as 1.68 kg/cm$^2$. In the maximum pressure phase of the outside foot, the high average peak pressure was observed in a, b and c areas, and that of d areas indicated small values. In the contact phase in the backwards sprint, the main loading area was around the heel at position a-1, with a maximum value of 2.03 kg/cm$^2$. The main loading was under the toes (a-1 and b-1) in the maximum pressure phase, reducing to the heel. There was almost no load on the d-area. In the backwards run, the maximum average load in the contact phase was observed in as 1.54 kg/cm$^2$ (Fig. 13.8a) with a maximum average load of 2.45 kg/cm$^2$ (Fig. 13.8b).

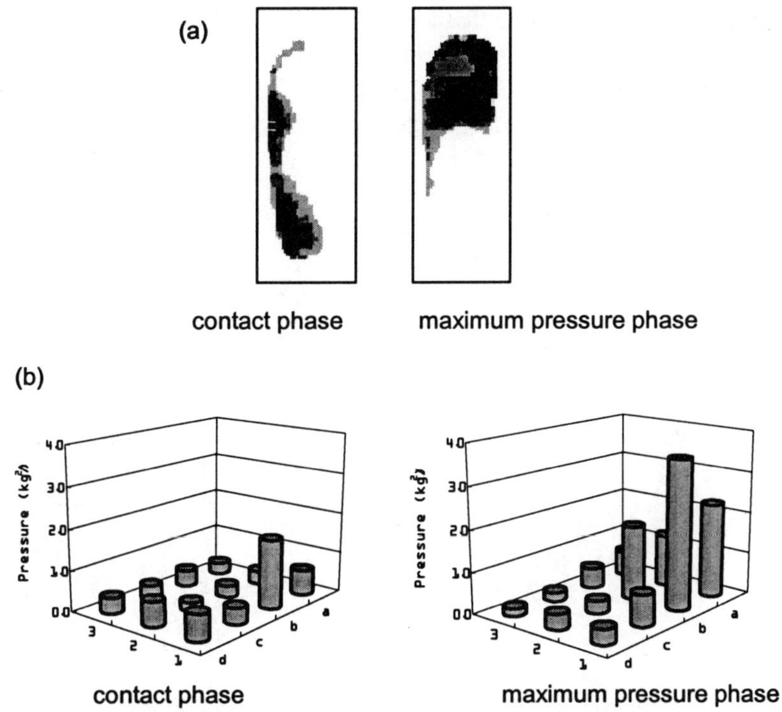


*Figure 13.7.* An example of distribution of pressure on outside foot in turn, (a) experimental data and (b) averages for six male soccer players.

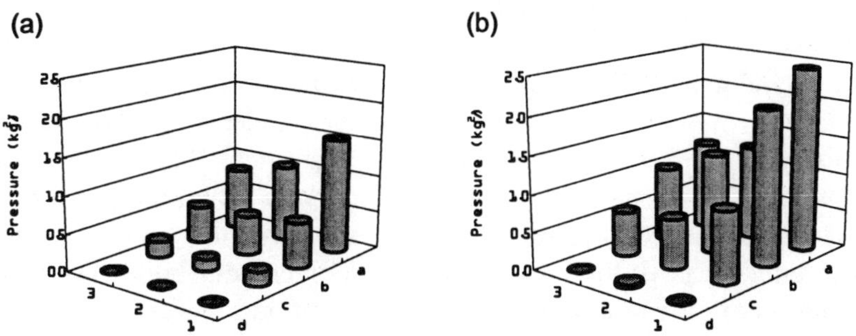


*Figure 13.8.* Average load in backwards run during: (a) contact phase; and (b) maximum pressure phase, for six male soccer players.

In the jump, the largest loading was observed on the heel area (area d) in the contact phase. The large load areas were widely distributed in areas a and b in the maximum pressure phase (Fig. 13.9). The peak pressures moved from the heel during contact to under the toes during the maximum pressure phase. During contact phase in diving, the take-off foot (the right foot) had main loading areas of b-3, d-2 and a-1. In the maximum pressure phase, the main loading areas were b-3, b-2, b-1 and a-2. In the case of the support foot (the left foot), the main loading area was b-1 for both the contacting and maximum pressure phase.

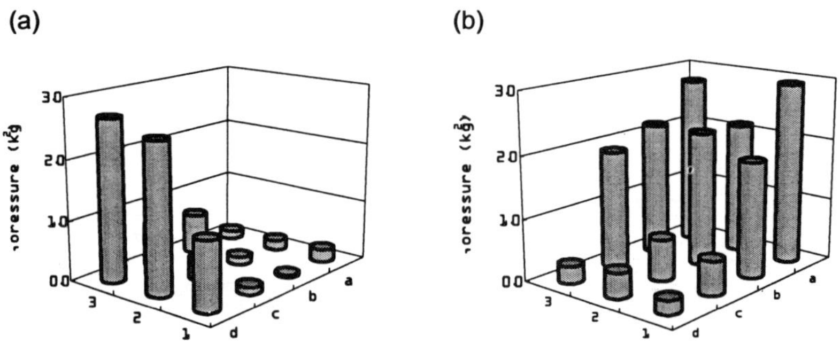


*Figure 13.9.* Average load in jump during: (a) contact phase; and (b) maximum pressure, for six male soccer players.

It is the tendency, therefore, that the value of the maximum loading point of the supported foot is greater than that of the take-off foot. In the maximum pressure phase in the jump, high average peak pressures were observed in a, b and c-3 areas and a similar tendency was observed in diving. In the forward sprint, the maximum pressures were found under the 1$^{st}$ metatarsal and generally towards the front of the foot. In the backwards run the pressures moved even further forwards under the toes in area a, presumably to get the rearwards push off the surface. The sidestep showed the highest pressures in the fore-foot with generally two different styles; in the first the peak forces started at the toes and moved towards the heel while, in the second, the peak forces moved towards the medial zone. The pressures in turning were evenly distributed between feet and peaked under area b-1 (medial), in a similar pattern to running. Higher pressures were seen around the heel in the contact phase. Not surprisingly, the jump showed the highest pressures, initially at the outside of the heel and moving forwards to the front of the foot, more or less evenly spread.

### 13.2.3 Conclusions From Foot Loading Analysis

These results show that, in the majority of movements in football, peak forces are experienced under the first metatarsal head or hallux. This has implications for football shoe design. For a comfortable impact with the ball, a stiff upper surface is required. For a comfortable impact with the ground the combination of a stiff sole with cushioning under the first metatarsal is desirable. It appears that the peak forces do not move far from these regions except in the jump. In the contact phase, a specific tendency was not observed for the foot pressure distribution. In the maximum pressure phase, the maximum vertical peak pressure was observed on the area of the first metatarsal head or the hallux in football specific movements such as sprint, backwards run, side step, turn, and dive. It is considered that the first metatarsal head and the hallux play an important role in the fundamental movements in football and boot design should have the combination of outside stiffness for comfortable impact with the ball and surface and interior cushioning, particularly around the first metatarsal.

## 13.3 FUNDAMENTAL ANALYSIS OF KICKING MOTION

The motion analysis of kicking has been studied by several investigators (Plagenhof, 1971; Roberts and Matcalfe, 1968). These studies have focused on the kinematic analysis of an instep kick. Other studies have analyzed joint force or torque in an instep kick (Roberts *et al.* 1974; Zernicke and Roberts, 1978; Luhtanen, 1988). However, there are few studies of curve ball kicking in soccer. This section discusses the fundamental characteristics of the infront curve kick, the outfront curve kick and the instep straight kick based on analysis using a 3D motion capture system (Vicon 370). Computer simulations are then used to describe various kicking strategies.

### 13.3.1 Data Collection

Three university football players were chosen as subjects for a motion analysis study. The subjects directed an infront curve kick, an outfront curve kick and an instep straight kick toward a target area 5 m away. Seven motion capture cameras (200 Hz) were set up around the subject, and eighteen digitized reflectors were attached on various joints. The ball used in this experiment was an official FIFA ball (mass 434.6 g, internal pressure 90 kPa). Stick figures from this analysis can be seen in Fig. 13.10, in three different viewing orientations. The mean of the angles of the ankle joint of the kicking leg before impact were found to be 136° for the infront curve kick, 140° for the outfront curve kick and 154° for the instep straight kick. In the same way, the mean of angles of the knee joint were 120° in the infront curve kick, 122° in the outfront curve kick and 128° in the instep straight kick. The width of swing from the top view in the kicking leg indicated the infront curve kick was 1.16 m, the outfront curve kick was 1.09 m and the instep straight kick was 1.01 m. It is the tendency that the approach angle of the infront curve kick is larger (wider) than that of the instep straight kick.

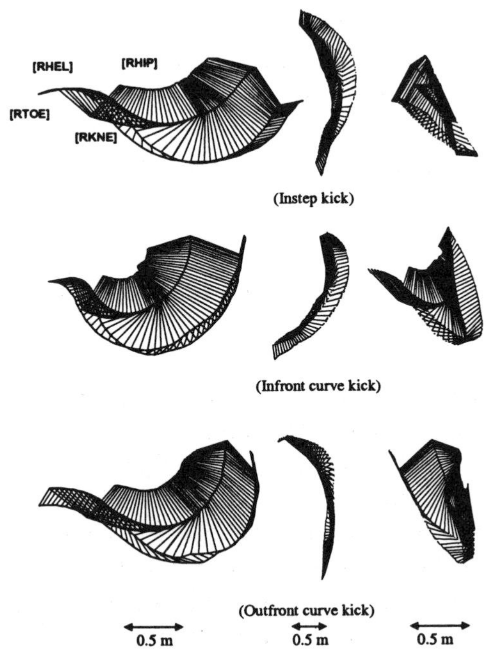


*Figure 13.10.* Stick figures in instep-kick, infront-kick and outfront-kick.

By considering the mechanics of collision between foot and ball (Leeds, A. 1996), the velocity of the ball can be stated as :

$$V_{(ball)} = V_{(foot)} \frac{M}{M+m}(1+e) \tag{13.1}$$

where $V$ is the velocity of ball and foot respectively, $M$ is the effective striking mass of the leg, $m$ is the mass of the ball and $e$ is the coefficient of restitution.

The effective striking mass is the mass equivalent of the striking object (in this case the foot and log) and relates to the rigidity of the limb. It means that the foot velocity before impact is related to the ball velocity after impact directly.

The horizontal velocity of hip joint, knee joint and ankle joint in the instep straight kick are shown in Fig. 13.11. On this graph, firstly the velocity of hip joint increased. Secondly the velocity of knee joint increased and finally the velocity of ankle joint increased, while at the same time there was a decrease in the velocity of knee joint. These appear to show an energy transfer from one segment to another respectively (flail-like action). At the same time, it seems that the muscles around knee and hip joints use the effect of a stretch-shortening cycle (Komi, P. V., 1984) by neuromuscular mechanism in the reverse phase from back swing to forward swing of kicking leg. These mechanisms produce high power (over 20000 W) at the ball impact.

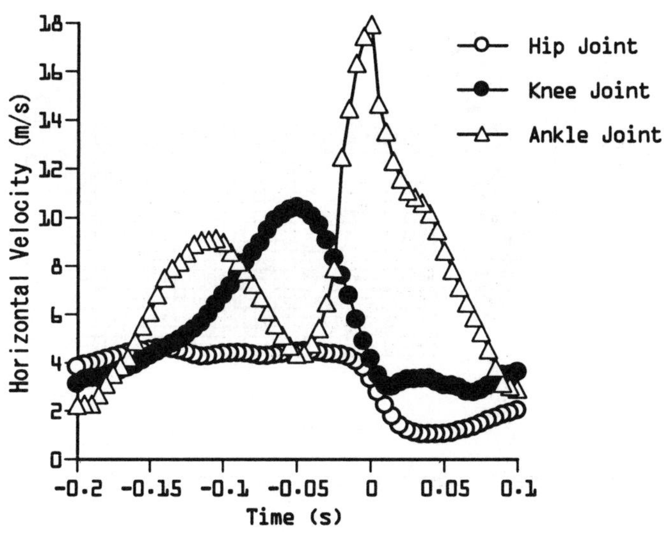


*Figure 13.11.* An example of horizontal velocity of hip, knee and ankle joints in an instep straight kick.

## 13.3.2 Computer Simulation of Ball Kicking

Finite element models of the instep kick (digitized 3D instep kick model) and the infront curve kick (digitized 3D infront kick model) were described in order to analyze the interaction between the ball and the kicking leg (Fig. 13.12). After each model was digitized the shape using the image-based digitizer based on the finite element meshes were made by MSC/PATRAN (MSC.Software Inc.) through the IGES file. Also, the infront kick model was made from the shape of an ankle with a 90° bend and 60° rotated outside. The kicking leg and the surface of the ball model were described by the Lagrangian frame of reference, and discretized by the finite element method. The air inside a ball was described with a Eulerian frame of reference and discretized by the finite volume method (Lenselink, 1991). Hexahedron solid elements (CHEXA) and shell elements (CQUAD) were used as the elements of the models to improve the calculation accuracy, and the air inside a ball was defined by the Gamma Law equation of State.

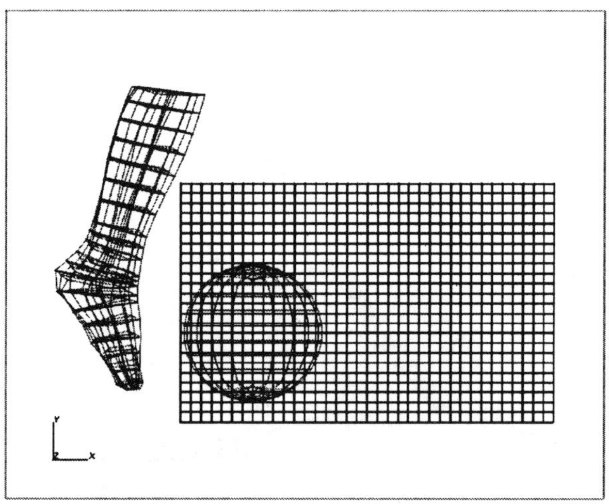

*Figure 13.12.* The digitized 3D instep kick model using coupled fluid-structure finite element model.

The foot joint of the human body has a very complex structure, which consists of bones, muscles, ligaments and so on. In this study, however, a simplified model was used which assumed that the lower leg and foot are represented by two kinds of material properties for the calf and the foot. In both models, the Young's modulus for the foot was 30 MPa (Asai *et al.*, 1998), the Young's modulus of calf was 300 MPa and the Poisson's ratio was 0.3 (Beaugonin and Haug, 1996). Each total mass of the digitized 3D instep kick model and digitized 3D infront kick model was defined to be

3.994 kg. As an initial condition of impact, the angular velocity with the rotational axis of the top end node was defined by assuming the velocity on the impact part to be 25m/s. During the analysis of an impact force, the initial speed in the horizontal direction (25 m/s) was assumed for the entire part of the kicking leg in both models.

Figure 13.13 shows the horizontal velocity of the toe (the metatarsal) during impact for the FE element model and the experiment. The model shows reasonable agreement with the experimental data, particularly in the first half of the impact. The ball deformed to 85% of its original diameter in the model compared to 86% experimentally. It was considered that the model matched the experimental data sufficiently enough to warrant use for further study.

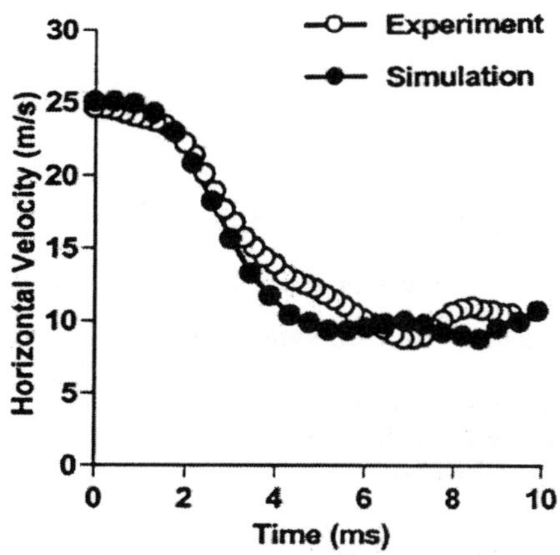


*Figure 13.13.* Comparison of simulation with experimental results.

The digitized 3D instep kick model was set up with offset (off centre) distances of 0.04m and 0.08m, and simulations were done for 12 cases of the kinetic coefficient of friction from 0 to 1.0 at intervals of 0.2, in order to analyse the relationship between a spinning ball and the kinetic coefficient of friction. To examine the relationship between the spinning ball and the offset distance at an impact point, 17 simulations was carried out which moved the impact point on both sides in directions of ±0.15 m each 0.02 m from the centre of a ball using the digitized 3D instep kick model (coefficient of friction being equal to 0.4). The static coefficient of friction and the kinetic coefficient of friction were defined by the same value to simplify the calculations in this study. In this study, MSC/DYTRAN, which was an explicit integral solver, was used for the solver of each simulation. As for the calculation result, data was visualized using MSC/PATRAN which was the pre- and post- data processor.

Figure 13.14 shows the variation of spin with coefficient of friction for impacts at 40 mm and 80 mm from the axis of the ball. It can be seen that the spin increases as the coefficient of friction increases and that doubling the offset from the axis of the ball roughly doubles the spin. These results are expected intuitively since a higher coefficient of friction stops the foot slipping across the surface of the ball and a greater offset of impact point from the axis allows the foot to apply a greater torque to the ball. It is interesting to note that, even at zero coefficient of friction, spin can still be imparted to the ball. This is due to local deformation of the ball during impact around the foot allowing forces to be transmitted to the ball around its axis. Clearly, the offset from the ball's axis has a much larger effect than a variation in the coefficient of friction.

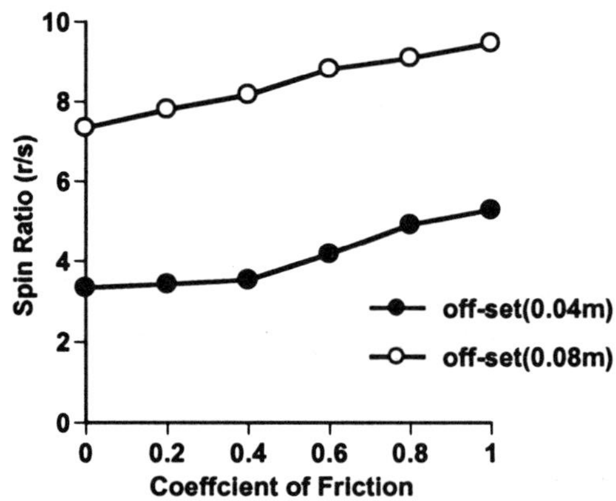


*Figure 13.14.* The variation of spin with coefficient of friction for impacts at 40 mm and 80 mm from the axis of the ball.

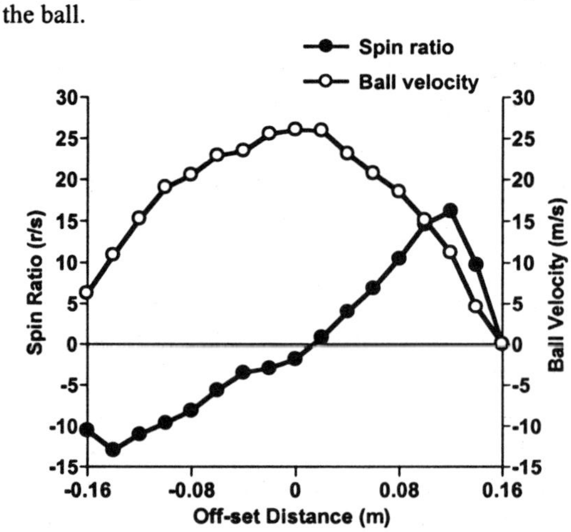


*Figure 13.15.* The relation between the offset distance of the impact point to the centre of the ball.

The relation between the offset distance of the impact point to the centre of the ball and the spin was examined using the digitized 3D instep kick model. Simulations for 17 steps which moved the impact point in each direction of ±0.16 m on both sides in 0.02 m increments from the centre of the ball were carried out (coefficient of friction is 0.4). Though the spin increases in accordance with an increase in the offset distance, it rapidly decreases after the offset distance exceeds the radius of the ball (Fig. 13.15). The ball speeds decreased with an increase in the offset distance. This suggests that in the case when the offset distance approaches the radius of the ball, the contact area and time rapidly decrease and the energy of the impact is only slightly transferred. Hence, it is suggested that the ball rotation and the ball speed are related to a trade-off of offset distance, within the range of the radius of the ball.

The simulations up to now were carried out using the digitised 3D instep kick model with the shape model extended ankle to make it easy to understand the fundamental characteristics and interactions. However, for a curve kick, not only the out front kick whose shape is similar to the shape of the instep kick, but the infront kick with an ankle bent into the shape of an "L ", is also often used. Using the digitised 3D instep kick model and the digitised 3D infront kick model, this study then analysed the pressure distribution and the impact force at impact during the instep kick and the infront kick. Also, the shape of the infront kick model was with the ankle bent rectangularly and rotated outside 45 deg., and the offset distance was 0.02 m. The pressure contours on the deformed shape at 4 ms after the impact for the simulation result using the digitised 3D instep kick model are shown in Fig. 13.16a. High compressive pressure is seen not only in the impact part but in the plantar as well, and tensile pressure is observed in the shin. The maximum pressure at this time was 126 KPa, while the minimum pressure was -543 KPa. The pressure contours on the deformed shape at 4 ms after the impact for the simulation result using the digitised 3D infront kick model are shown in Fig. 13.16b. High compressive pressure is seen not only in the impact part but also in the plantar as well, and tensile pressure is observed in the back of the ankle joint. The maximum pressure at this time was 345 KPa, while the minimum pressure was -436 KPa. Also, for this simulation, the rotation speed of the ball was 5.2 rad/s. From the analysis of the stress and pressure distributions, it is postulated that the improvement in traction on the metatarsal, navicular and cuneiform against the ball will increase the spin of the ball for the infront curve kick. The impact forces of the instep kick and the infront curve kick based on computer simulations are shown in Figs. 13.17a and b. The peak values of the horizontal impact force in the instep kick was 2439 N, that of the vertical impact force was 853 N, and that of the lateral impact force was -452 N. On the other hand, the peak values of the horizontal impact force for the infront curve kick was 2206 N, that for the vertical impact force was 1221 N, and that for the lateral impact force was -1143 N. It appears that the vertical and lateral forces are concerned with the force which increases the rotation of the ball as well as the flight of the ball in a direction different from the swing direction of the kicking leg.

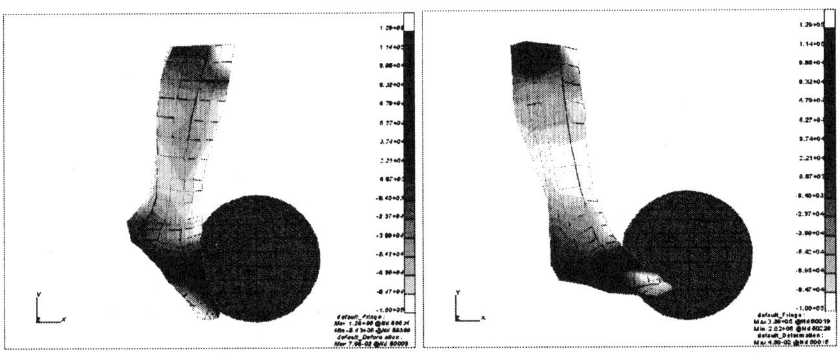

(a) (b)

*Figure 13.16.* The pressure contours on the deformed shape during: (a) the instep kick; and (b) the infront curve kick, based on computer simulations.

(a) (b)

*Figure 13.17.* The impact forces of: (a) the instep kick; and (b) the infront curve kick, based on computer simulations.

### 13.3.3 Conclusions from Kicking Motion Analysis

The computational simulation model used in this study has a simple shape and structure, and of course, the actual phenomena cannot be completely reproduced. It is considered that the results shown in this study provide a foundation to appreciate the essential mechanism of the phenomena, though these only clarify the fundamental characteristics and properties of some individual phenomenon. These will provide useful references when we will design a more precise analysis model and more accurate analysis method in the future.

## 13.4 AERODYNAMICS OF A SOCCER BALL

The aerodynamics of many sports balls (golf, cricket, tennis, baseball, volleyball) has been studied previously (Bearman & Harvey, 1976, Mehta, 1985, Haake *et al.*, 2000, Watts & Ferrer, 1987 and Mehta & Pallis, 2001) and these studies have measured the effect of velocity and spin on the drag and Magnus forces experienced by a sports ball during flight. However, little work has been published on (soccer) footballs, but this understanding is necessary to simulate the flight of a football for various launch conditions.

There are two main approaches to measuring forces on sports balls during flight. The first approach is to take measurements from controlled football trajectories using high speed video footage (Carré *et al.*, 2002). Although this is a well-recognized method that undoubtedly examines what actually happens as a ball travels through the air, it is very susceptible to errors. The trajectory must be measured very accurately, as the effect of drag and Magnus force is determined by analyzing the subtle changes in the shape of the trajectory. Monitoring the change in drag due to velocity is further complicated by the fact that the velocity changes during flight. Other effects such as turbulence, camera misalignment and inaccurate velocity measurement can all lead to further inaccuracies.

The second approach is to use a wind tunnel, where the sports ball is held in place using a 'sting' and air is blown around the ball at controlled speeds. This method has the advantage that the speed of the air can be accurately controlled and kept constant and the forces acting on the ball can be measured by mounting the ball and sting on a force balance arrangement.

## 13.4.1 Wind Tunnel Measurements

### 13.4.1.1 Scale Model Measurements

The effect of velocity on drag coefficient was examined using controlled experiments in an open circuit wind tunnel (provided by the International Tennis Federation, Roehampton, UK). Due to the limiting size of the working section, a scale plastic model of a football was created (66 mm in diameter compared to 218 mm for a full size football). The model had a generic seam pattern, commonly used by many football manufacturers (consisting of 20 hexagonal and 12 pentagonal patches). This was mounted on an 'L'-shaped sting that was attached to a two-way force balance. The effect of increased drag on the body of the sting was reduced by mounting a separate shroud in front of the sting, allowing air to flow around it. The ball model, sting and shroud can be seen in Fig. 13.18.

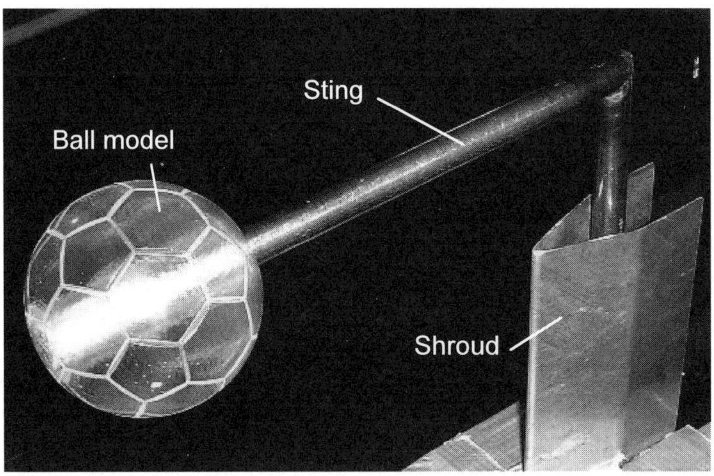


*Figure 13.18.* Scale football model (66 mm in diameter), mounted on a sting that is protected from the air flow by a shroud.

The air speed in the working section of the wind tunnel was varied from 20 m/s (the minimum consistent speed achievable) up to 70 m/s and back down again. The drag and lift forces experienced by the ball model were measured using the force balance at regular intervals. Three runs of tests were carried out, with the ball model mounted in three different orientations (the ball being rotated 90 about the sting between each run of tests).

When this testing was complete, the drag acting solely on the sting, the 'tare' drag, was measured. This was done by mounting a 66 mm diameter sphere with a hole larger than the diameter of the sting, in the same place as the football model had been, using supports from the side (see Fig. 13.19). It was mounted so that no contact was made between the sphere and the sting and the tests were run again to measure the tare drag experienced by the sting alone.

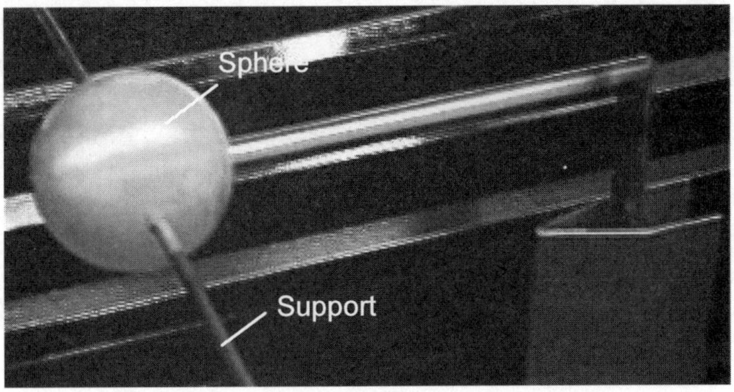


*Figure 13.19.* Apparatus for measuring tare drag.

The measured drag force for the ball model and sting combined and the tare drag for the sting alone can be seen in Fig. 13.20a for different air flow speeds. The tare drag was found to be quite high, accounting for almost half the total drag measured on the ball-sting arrangement.

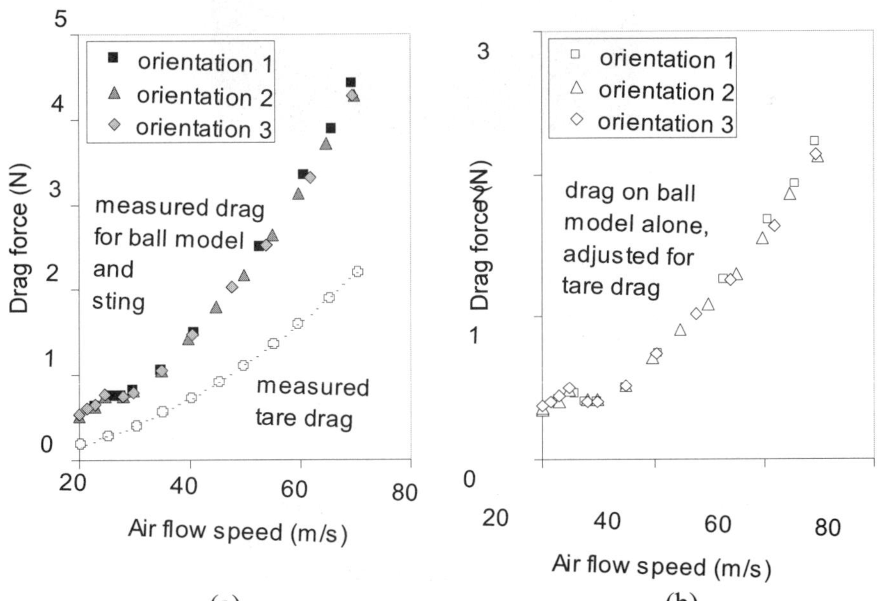


*Figure 13.20.* Drag force plotted against air flow speed: (a) measured on ball model and sting and on sting alone; and (b) adjusted for tare drag.

The measured drag forces were adjusted for tare drag and can be seen in Fig. 13.20b. Generally, the drag force increases with air flow speed, as one would expect (for simple situations drag varies with the square of wind speed). However, there is a small fluctuation in the drag force at low speeds. There was found to be no major effect on the drag due to ball orientation as each run of tests gave very similar results.

### 13.4.1.2 Mini-Football Measurements

A mini football was also mounted in the wind tunnel and tests repeated in three orientations. The football used was a smaller version of the Adidas Fernova (see Fig. 13.21) and was 140 mm in diameter (compared to 218mm for a full size ball).

*Figure 13.21.* Mini-football mounted on sting.

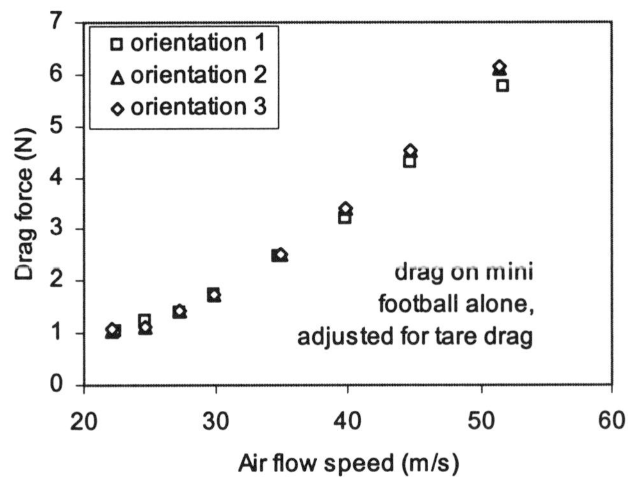


*Figure 13.22.* Drag force on the mini football, plotted against air flow speed.

The data was again adjusted for tare drag and the results can be seen in Fig. 13.22. The drag force was found to vary with the square of the air flow speed and no major effect was observed due to the orientation of the ball.

#### 13.4.1.3 Comparison with other Other Data

To predict the drag forces that a full size football would experience during flight, the drag data from the scale model and the mini football was converted to drag coefficient data and plotted against Reynolds number. Drag coefficients, $C_d$, and Reynolds numbers, Re, were calculated using the following equations,

$$F_d = \frac{1}{2} C_d \rho A v^2 \tag{13.2}$$

$$\mathrm{Re} = \frac{\rho v D}{\mu} \tag{13.3}$$

(where, $F_d$ is the drag force, ρ is density of the air, $A$ is projected area of the ball, $C_d$ is the drag coefficient, $v$ is the velocity of the air, Re is the Reynolds number, $D$ is the diameter of the ball and $\mu$ is the dynamic viscosity of the air)

In a real match situation a football has a velocity range of around 10 mph (4.6 m/s) to 70 mph (32 m/s). In normal atmospheric conditions, this equates to a range in Reynolds number of approximately 70,000 to 500,000. The drag coefficient data was plotted against Reynolds number and can be seen in Fig. 13.23.

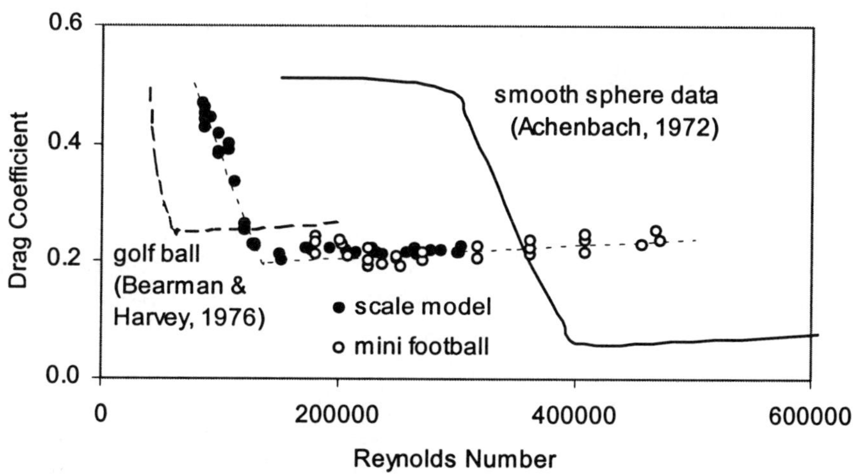


*Figure 13.23.* Wind tunnel drag coefficient data plotted against Reynolds number, compared with data for other spheres.

The data show a sharp drop in drag coefficient from about 0.5 to 0.2 over a range in Reynolds number from about 90,000 to 130,000 and then a slight increase over the rest of the range. This is due to a transition in the behaviour of the air in the boundary layer, surrounding the ball. At low Reynolds number (and therefore, low speeds) the boundary layer contains *laminar* flow, an ordered type of flow where the air can be thought of as travelling in stream tubes, which do not mix with the surrounding flow. An example of this type of flow can be seen in Fig. 13.24a, which shows a jet of smoke travelling around the ball model, when the Reynolds number is 90,000 (equivalent to 6.1m/s for a full size football). The boundary layer separates quite early and there is a large wake behind the ball and consequently, a large drag force acting on the ball. As the speed of the air is increased, the flow in the boundary layer becomes *turbulent* and mixing occurs with the surrounding air. This gives the boundary layer sufficient energy to travel further round the ball and separate later. This effect can be seen in Fig. 13.25b, which shows a jet of smoke travelling around the ball model for a Reynolds number of 130,000 (equivalent to 9.1m/s for a full size football). There is less wake behind the ball and the drag coefficient is reduced.

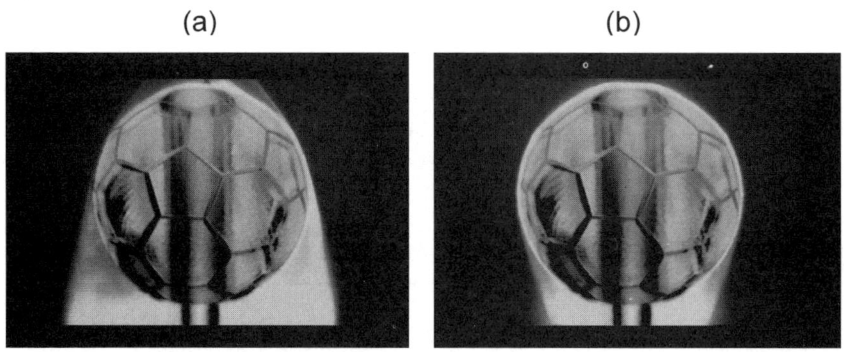


*Figure 13.24.* Smoke flow around the scale football model showing: (a) laminar flow at Re = 90,000; and (b) turbulent flow at Re = 130,000.

Also shown in Fig. 13.23 are the results of other aerodynamic studies on spheres. Achenbach (1974) found that an increase in surface roughness had the effect of moving the transition from laminar to turbulent behaviour to a lower Reynolds number. A similar effect has been found for sports balls that have surface roughness (Mehta and Pallis, 2001) It is well known that dimples on golf balls help to reduce drag by moving the transition point to lower Reynolds number so that the flow around a golf ball is turbulent for more of its flight. Judging by Fig. 13.23, the seam pattern on a football, seems to have a similar effect, but not to such a great extent. If a football were completely smooth, according to Achenbach's data, it would have a higher drag coefficient for most of the range of Reynolds numbers, seen in play.

## 13.4.2 Trajectory Measurements

As spin could not be achieved simply in the wind tunnel, data was taken from a previous study that utilized trajectory analysis to measure the Magnus force experienced by a ball due to spin (Carré *et al.* 2002).

### 13.4.2.1 Data Collection

The football used for the trajectory analysis was a Mitre *Max* (218 mm diameter, 430 g, when inflated to 75.8 kPa). The ball was launched with a near constant velocity of 18 m/s at approximately 15° to the ground using a specially built ball projection machine. Spin was imparted about a horizontal axis and was varied from 211 rad/s of top spin to 224 rad/s of back spin. As it was very difficult to keep the orientation of the ball seams constant for all the tests, a random seam orientation was used.

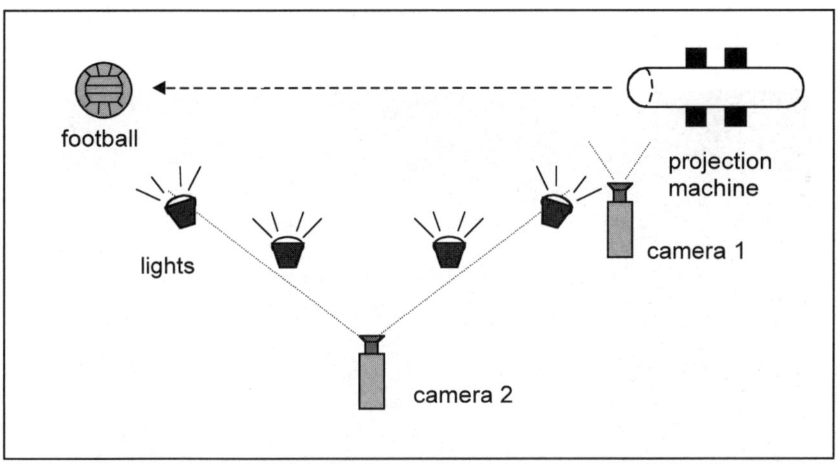


*Figure 13.25.* Diagram of experimental set-up for trajectory analysis.

A diagram of the experimental setup, as seen from above is shown in Fig. 13.25. The ball trajectories were filmed using two high speed video cameras (Kodak motioncorders). The first camera (camera 1) was focused close to the exit of the projection machine to measure the launch speed and angle of trajectory. It was set-up to record at a rate of 240 frames/s with a capture time of 1/1,000 s. A typical set of images can be seen in Fig. 13.26 (they have been superimposed). A pattern was drawn on the ball with indelible ink, to allow any displacements of the ball to be measured from the images.

The second camera (camera 2) filmed the progress of each ball's flight over a horizontal distance of 10m. The purpose of this camera was to monitor the shape of the trajectory for each ball. The camera was set-up to record at a rate of 120 frames/s with a capture time of 1/500 s. A typical set of superimposed images from camera 2 can be seen in Fig. 13.27.

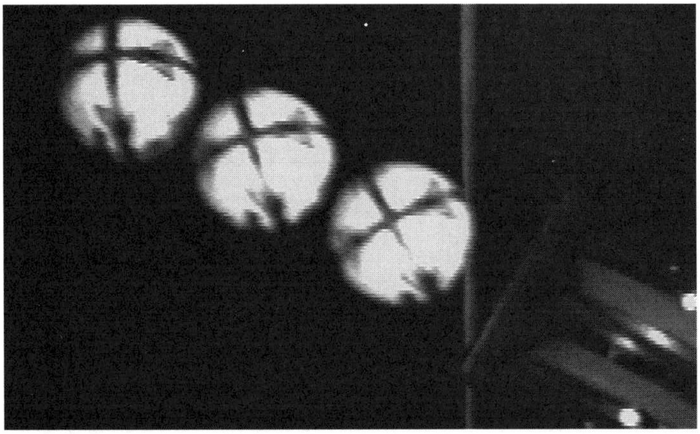

*Figure 13.26.* Typical view from Camera 1. Reprinted from Carre et al. (2002) with permission of Blackwell Publishing Ltd.

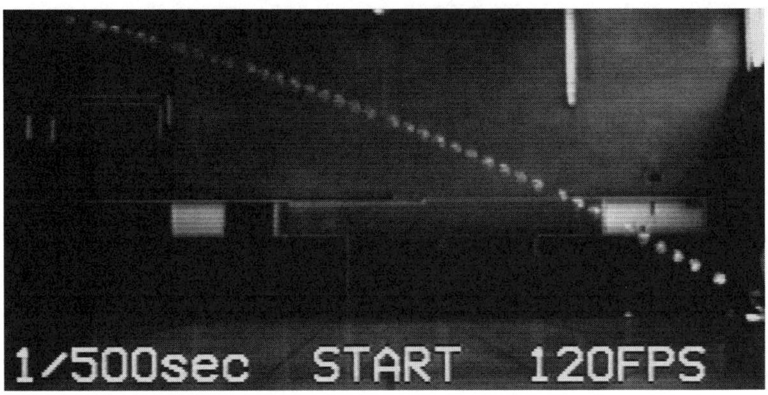


*Figure 13.27.* Typical view from Camera 2. Reprinted from Carre et al. (2002) with permission of Blackwell Publishing Ltd.

A dedicated image analysis program (Richimas v1.0) was then used to collect co-ordinate data from both cameras, which could be used to calculate the launch conditions and trajectory for each test.

#### 13.4.2.2 Trajectory Simulation

A step-by-step model, similar to that used by Mehta (1985) was designed for the simulation of a football with spin imparted about a horizontal axis, in flight. The model assumed that the ball is subjected to two forces, $F_d$ and $F_l$, due to drag and lift as well as *mg* due to its weight. The force, $F_l$ in this case is due to the Magnus effect. The original explanation of this effect was that the flow on one side of a spinning ball is faster than that on the other due to the spin. This causes a pressure differential, resulting in a transverse force. Since the concept of boundary layers was introduced by Prandtl in 1904, it has been found that boundary layer separation is delayed on the retreating side of a spinning ball and enhanced on the advancing side (Mehta, 1985). It is this asymmetry that causes the pressure difference, resulting in a transverse force (upwards if back spin is imparted, downwards if top spin is imparted). A force diagram of a ball traveling with a velocity $v$ at an angle θ to the ground under these conditions is shown in Fig. 13.28.

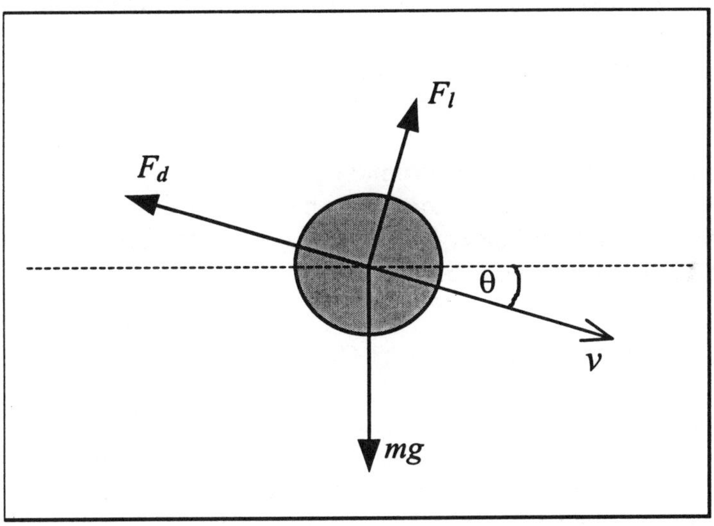


*Figure 13.28.* Forces acting on a football during flight.

The equations of motion for this situation can be derived by resolving the resulting accelerations vertically and horizontally to give

$$m\frac{\mathrm{d}}{\mathrm{d}t}(v\sin\theta) = mg - F_d \sin\theta - F_l \cos\theta \tag{13.4}$$

$$m\frac{\mathrm{d}}{\mathrm{d}t}(v\cos\theta) = -F_d \cos\theta - F_l \sin\theta \tag{13.5}$$

These equations can then be re-written to deal with finite differences using

$$m\delta(v\sin\theta) = \delta t\left[mg - F_d \sin\theta - F_l \cos\theta\right] \tag{13.6}$$

$$m\delta(v\cos\theta) = \delta t\left[-F_d \cos\theta - F_l \sin\theta\right] \tag{13.7}$$

The forces $F_d$ and $F_l$ were assumed to vary with velocity squared according to the following equations

$$F_d = \frac{1}{2} C_d \rho A v^2 \tag{13.8}$$

$$F_l = \frac{1}{2} C_l \rho A v^2 \tag{13.9}$$

(where, $\rho$ is density of air, $A$ is projected area of ball, $C_d$ is the drag coefficient and $C_l$ is the lift coefficient)

For the purposes of this model, it was assumed that the drag and lift coefficients remained constant throughout the trajectory. Given values of $\rho$, $A$, $C_d$ and $C_l$ and using initial values of $v$ and $\theta$, the ball trajectory was calculated using a step-by-step process with a small time period, $\delta t$.

For each set of trajectory co-ordinate data, an accompanying trajectory simulation was carried out using a time step of 2.5 ms and an initial guess for $C_d$ being equal to 0.3 and $C_l$ being equal to 0.0. An iterative solving algorithm was then used to determine $C_d$ and $C_l$ by minimising the difference between the simulated trajectory and the measured values. The algorithm was Microsoft Excel Solver, which uses a generalised, reduced gradient (GRG2) non-linear optimisation code.

#### 13.4.2.3 Resulting Drag And Lift Coefficient Data

Figure 13.29 shows the variation in drag and lift coefficient with increasing spin. In order to collate the data, only the moduli have been plotted. In other words a negative lift coefficient, as found with top spin, has been given the same positive value as a lift coefficient found with back spin.

It can be seen that both the drag and lift coefficients increased non-linearly with spin. The lift coefficient rapidly increased from zero as the spin was increased to 100 rad/s, but then became constant at around 0.26. The increase in lift coefficient is due to an increase in Magnus force (lift increases with back spin and becomes negative with top spin). The drag coefficient appeared to increase steadily from 0.2 at 0 rad/s of spin to 0.5 at 240 rad/s. Curves were fitted to the data (see dotted lines in Fig. 13.29) which allowed a smoother data set to be used to predict how the trajectory of a ball would be affected by varying the imparted spin.

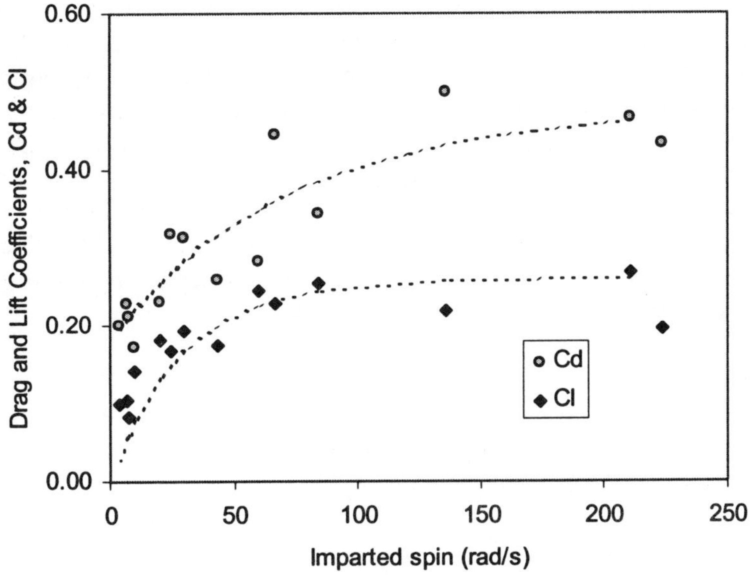


*Figure 13.29.* Drag and lift coefficient versus imparted spin for balls projected at 18m/s. Reprinted from Carre et al. (2002) with permission of Blackwell Publishing Ltd.

## 13.5 FLIGHT OF A SOCCER BALL IN PLAY

Data from the aerodynamic studies could then be used to predict the flight of a football for different kicking scenarios. As discussed in Section 13.3, if a ball is struck in its centre, with a foot velocity of 25 m/s and a coefficient of friction of 0.4, it will have an effective launch velocity of 26 m/s and no spin (see Fig. 13.15). However, if the impact location of the foot striking the ball was off-set by 0.08m to the right, this would cause the ball velocity to drop to 18.5 m/s and a spin of 10.2 rev/s (64 rad/s) to be imparted to the ball, with a vertical spin axis (i.e., in an anti-clockwise direction when viewed from above). These two kicking situations will be described as 'Kick a' and 'Kick b', respectively.

### 13.5.1 Combining Aerodynamic Data

Data from the wind tunnel and trajectory studies was combined to give drag coefficient data for the two situations, Kick a and Kick b. The trajectory study showed that for a particular launch velocity (18 m/s), the drag coefficient increased with spin. Therefore, it was assumed that the 'effective Reynolds number', as experienced by the ball, increased with spin and this was the cause of the increase in drag coefficient. Figure 13.30 shows the wind tunnel data for the scale model and mini football tests, where no spin was applied. Also plotted is the drag coefficient data from the trajectory analysis of a spinning ball launched at 18 m/s (equivalent to Re = 270,000), with spin varying from 0 to 200 rad/s. The drag coefficient curve for balls with no spin was then shifted to intersect with the measured drag coefficient for a ball spinning at 64 rad/s, at a Reynolds number of 270,000. These two curves could then be used to predict the change in drag during the flight of a football launched with either Kick a or Kick b.

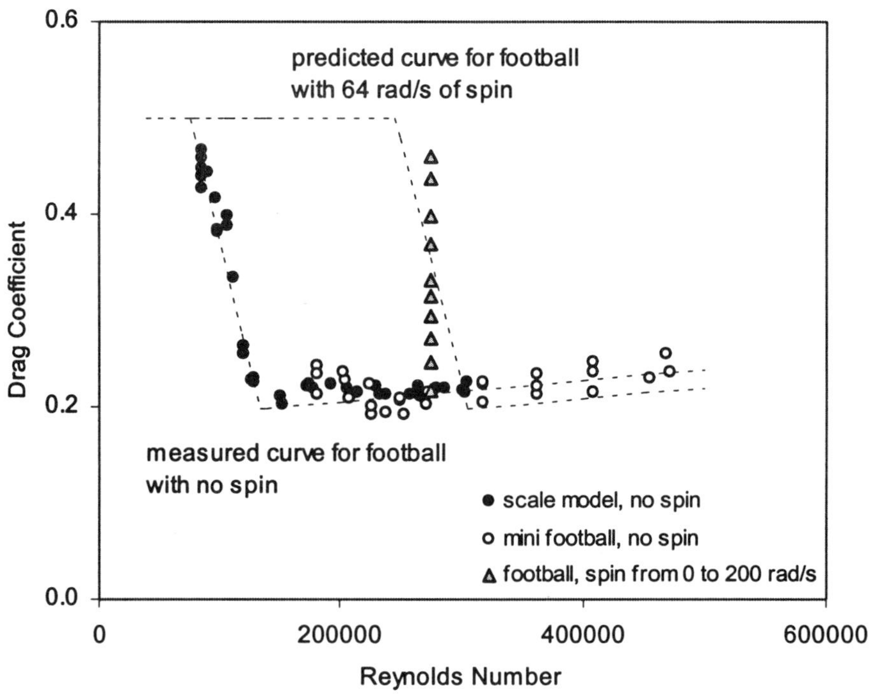


*Figure 13.30.* Wind tunnel drag coefficient data plotted against Reynolds number, compared with data on other spheres.

## 13.5.2 Trajectory Simulations

Trajectory simulations were carried out for both kicks, taken just outside the penalty area, 18 m from the centre of the goal. Kick a (struck in the centre of the ball) had an effective launch velocity of 26 m/s and no spin. According to the wind tunnel data in Fig. 13.30 this would lead to a

Reynolds number of 390,000 at launch, giving a drag coefficient of 0.226, with the flow around the ball being turbulent. A three-dimensional trajectory model was used to simulate the flight of the ball through the air and as the ball slowed down, the model adjusted the drag coefficient to account for the change in Reynolds number. As no spin had been applied at launch, there was assumed to be no Magnus force acting on the ball. The launch direction was adjusted manually to allow a goal to be scored in the top right hand corner of the goal (just inside the right post). This kick can be seen in Fig. 13.31.

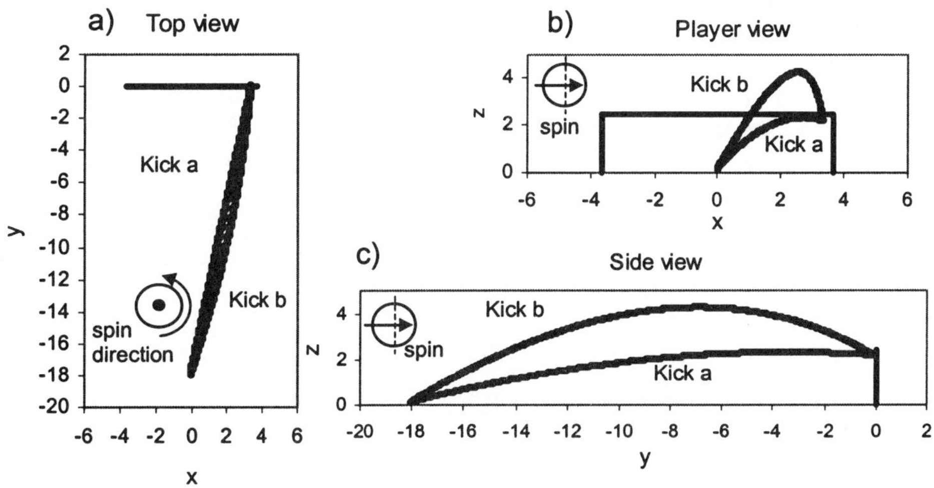


*Figure 13.31.* Prediction for flight of two free kicks in 3 dimensions, based on varying foot impact location. Kick a launched at 26 m/s, no spin; Kick b launched at 18.5 m/s with 64 rad/s of spin.

Kick b (ball struck off-centre by 0.08 m to the right), had a launch velocity of 18.5 m/s and a spin of 64 rad/s imparted to the ball, with a vertical spin axis. For spin in this direction, a sideways Magnus force is applied to the ball, rather than 'lift'. It was assumed that the lift coefficients found from the trajectory analysis, where spin was applied about a horizontal axis, could be used as 'Magnus' coefficients for a situation where spin is applied about a vertical axis, in order to calculate the sideways force acting on the ball. This gave a Magnus coefficient of 0.22 for a spin of 64 rad/s. For the purposes of this simulation the spin axis was assumed to remain vertical throughout the flight and to be in an anti-clockwise direction when viewed from above. The Magnus coefficient was also assumed to remain constant throughout flight. According to the predicted curve in Fig. 13.31 a launch velocity of 18.5 m/s, equivalent to a Reynolds number of 270,000, would give a drag coefficient of 0.35, with the flow around the ball being under transition from turbulent to laminar. As the ball slowed down in this simulation, the ball was assumed to be in the laminar phase when the drag coefficient reached 0.5 (based on previous studies, Mehta, 1985) and this value was set as the maximum. The launch direction was adjusted so that a goal was scored in exactly the same position as for Kick a (see Fig. 13.31). In other words, for both free kicks to be successful, the launch angle had to be altered in the same way that a player would do instinctively for a choice of kick.

The trajectory plots in Fig. 13.31 show how Kick b curves to a much greater extent than Kick a, due to the sideways Magnus force, caused by the ball spinning. However, because the ball is traveling more slowly for Kick b it must be launched higher into the air to reach the same point as Kick a. This effect, combined with the variation in drag means that the total time required to score is 0.8s for Kick a, whereas it is 1.6 s for Kick b. This comparison demonstrates the choice of strategy available to a player when taking such a kick. He can either strike the ball centrally to gain as much velocity as possible and therefore give the goalkeeper less time to react, or he can strike it off-set from the centre to put spin on the ball, which may allow him to bend it round a defensive wall of players.

## 13.6 CONCLUDING REMARKS

The research activity described in this chapter demonstrates how scientists and engineers can use measured observations and computer simulations to provide information that is useful to football coaches, players, equipment manufacturers and interested spectators.

The analysis of foot loading during typical movements has led to suggestions for football boot design. As peak forces are experienced under the first metatarsal head for the majority of movements, it is suggested that a boot should include a stiff sole with cushioning supplied specifically in this area.

The analysis of kicking motion, combined with an understanding of the aerodynamics of a football has led to a better understanding of free kick scoring strategy. This analysis demonstrates the choice available to players when taking a free kick. They can either strike the ball centrally to make the ball travel as fast as possible, or they can strike the ball off-centre, putting spin on the ball, which will cause it to bend round a defensive wall of players. It is thought that an experienced, professional player would be able to strike the ball with sufficient force to gain high velocity *and* spin.

## 13.7 REFERENCES


Achenbach, E., 1972, Experiments on the flow past spheres at very high Reynolds numbers. *J. Fluid Mech.* **54**: 565-575.

Achenbach, E., 1974, The effects of surface roughness and tunnel blockage on the flow past spheres. *J. Fluid Mech.* **65**: 113-125.

Asai, T., Akatsuka, T., Nasako, M., and Murakami, 0., 1998, Computer simulation of curve-ball kicking in soccer, in: *The Engineering of Sport*, S. J. Haake, ed., Blackwell Science, Oxford, U. K., pp. 433-440,

Asai, T., Carré, M. .J., Akatsuka, T., and Haake, S. J., 2002, The Curve Kick of a Football I: Impact with the Foot, *Sports Engineering.* **5**: 183-192.

Bartlett, R. M., Muller, E., Raschner, C., Lindinger, S., and Jordan, C., 1995, Pressure distributions on the plantar surface of the foot during the javelin throw. *J. Biomech.* **11**: 163-176.

Bearman, P. W., and Harvey, J. K., 1976, Golf ball aerodynamics, *Aeronautical Quarterly.* **27**: 112-122.

Beaugonin, M., and Haug, E., 1996, A numerical model of the human ankle/foot under impact loading in inversion and eversion, *Society of Automotive Engineers 962428*, pp. 239-249.

Carré, M. J., Asai, T., Akatsuka, T., and Haake, S. J., 2002, The Curve Kick of a Football II: Flight Through the Air, *Sports Engineering.* **5**: 183-192.

Eils, E., Streyl, S., Linnenbecker, S., Thonrwesten, L., Völker, K., and Rosenbaum, D., 2001, Shoe pressure distribution analysis of soccer specific movement, *Abst. of XVIIIth Congress of the International Society of Biomechanics*, p. 364.

Haake, S..J., Chadwick, S. G., Dignall, R. J., Goodwill, S., and Rose, P., 2000, Engineering tennis – slowing the game down, *Sports Engineering.* **3:** 131-144.

Hennig, E. M., Valiant, G. A., and Liu, Q., 1996, Biomechanical variables and the perception of footwear, *J. Applied Biomech.* **12**: 143-150.

Komi, P. V., 1984, Physiological and biomechanical correlates of muscle function: Effects of muscle structure and stretch-shortening cycle on force and speed, *Exerc. Sports Sci. Rev.* **12**: 81-117.

Luhtanen, P., 1988, Kinematics and kinetics of maximal instep kicking in junior soccer players, in: *Science and Football*, T. Reilly, A. Leeds, K. Davids, and W. J. Murphy, eds., E and FN Spon, London, pp. 441-448.

Lees, A., 1996, Biomechanics applied to soccer skills, *Science and Soccer*, T. Reilly, ed, E and FN Spon, London, pp. 123-133.

Lenselink, H., 1991, *Oblique penetration in ductile plates*, The MacNeal-Schwendler Corporation, paper 91-03, The Mac-Neal Schwendler Corporation.

Mehta, R. D., 1985, Aerodynamics of sports balls. *Ann. Rev.Fluid Mech.* **17:** 151-189.

Mehta, R. D., and Pallis, J.M., 2001, Sports ball aerodynamics: Effects of velocity, spin and surface roughness, in: *Materials and Science in Sports, Proceedings of Symposium sponsored by TMS (The Minerals, Metals and Minerals Society), San Diego, CA, April, 2001*, F.H. Froes and S.J. Haake, eds, TMS, pp. 185-197.

Milani, T. L., Schnabel, G., and Hennig, E. M., 1995, Rearfoot motion and pressure distribution patterns during running in shoes with varus and valgus wedges, *J. Applied Biomech.* **11**: 177-187.

Plagenhoef, S., 1971, *Patterns of Human Motion*, Prentice Hall, Englewood Cliffs, New Jersey, pp. 98-116

Roberts, E. M., and Metcalfe, A., 1968, in: *Biomechanics I*, J. Wartenweiler, E. Jokl, and M. Hebbelinck, eds., University Park Press, Baltimore, pp. 315-319.

Roberts, E. M., Zernicker, R. F., Youm, Y., and Hung, T. C., 1974, Kinetic parameters of kicking, in: *Biomechanics IV*, R. Nelson and C. Morehouse, eds., University Park Press, Baltimore, pp. 157-162.

Watts, R. G., and Ferrer, R., 1987, The lateral force on a spinning sphere: Aerodynamics of a curveball, *Amer. J. Physics*. **55**: 40-44.

Zernicker, R. F. and Roberts, E. M., 1978, Lower extremity forced and torques during systematic variation of non-weight bearing motion, *Med. Sci. Sports*. **10**: 21-26.

# V

# BASKETBALL

# Chapter 14

# Aerodynamics and Biomechanics of the Free Throw

George K. Hung[1], Brian Johnson[2], and Adrienne Coppa[3]

[1]*Dept. of Biomedical Engineering, Rutgers University, 617 Bowser Rd., Piscataway, NJ 08854-8014, PH: (732) 445-4137, FX: (732) 445-3753, EM: shoane@rci.rutgers.edu*

[2]*Kinematic Consultants,Functional Capacity Evaluation Dept., 645 Ocean Rd., Point Pleasant, NJ 08742, PH: (732) 714-1907, EM: jl2ball@hotmail.com*

[3]*635 Raymond St., Westfield, NJ 07090, PH: (908)233-9466, EM: adrienne723@hotmail.com*

## 14.1 INTRODUCTION

Free throw shooting should be one of the simplest actions in sports; there is no physical contact, jumping, or twisting involved. It is a consistent action in that the free throw is always taken 15 feet from the face of the backboard, which is 10 feet above the floor, and there is never anything between the player and the basket. The basketball is standardized and has a weight of 21 ounces, a circumference of 30 inches, and a diameter of 9.7 inches. The rim is standardized and has an inner diameter of 18 inches.

Indeed, in the early history of basketball, all free throws were taken by the same specialist player each time a foul was committed (Sharman, 1965). This is analogous to football in which a field goal specialist kicks the extra point after each touchdown. However, in the 1920s, the rules were changed so that the player who was fouled shot his own free throw. This changed the entire complexion of the game because now each player needed to be proficient in shooting the free throw.

Players found, however, that free throw shooting was not as simple as it appeared. Statistics obtained for U.S. men's college basketball since the 1970s showed a consistent average free throw shooting percentage of 69%

(Krause and Hayes, 1994). Also, 20 to 25 % of a team's scoring was from the free-throw line. Moreover, in about 20 to 30 % of all games, the winning team made more free throws than the losing team (Barnes, 1972). This may mean the difference between winning or losing several game in a season, or even between having a winning or losing season. Thus, the free throw, despite appearing relatively simple and innocuous, is an extremely important part of the game.

There have been numerous studies on the trajectory and biomechanics of the free throw. These are reviewed in some detail below. We expanded on these earlier studies by examining free throw trajectories using video recording and computer simulations, with simultaneous quantitative measurements of accelerations of the finger, wrist, and elbow during the free throw using a newly developed device. These findings provide some new insights into the aerodynamics and biomechanics of the free throw.

## 14.2 BACKGROUND

### 14.2.1 Free Throw Trajectory

There are two main styles of free throw shooting: the underhand and overhand shots. At one time in the early days of basketball, the underhand free throw was the only recommended shot at the foul line, and there have been arguments in favor of this technique, such as its stability and smaller release angle error (Bunn, 1964; Tan and Miller, 1981). In a study of games played in New York City, Hobson (1955) found that the percentage of successful free throws for underhand, two-handed, and one-handed shots were 56.5, 53.1, and 50.4, respectively. Indeed, Rick Barry, a Hall of Famer who used the underhand technique, had a lifetime free throw percentage of 90%. Generally, however, in modern basketball, the one-handed overhand shot has become the accepted standard. There are many factors in its favor. It decrease the distance to the hoop, minimizes the angle of release and hence minimizes the velocity of release, and increases the margin for error for entry into the hoop (Cooper, 1969; Cousy, 1970; Maugh, 1987, and Mortimer, 1951).

Earlier subjective assessments of the trajectory for the free throw were based on experience and observations, with advocates for low (Veeneker, 1937), medium (Lambert, 1932; Wooden, 1966), and high (Meanwell, 1924; Bunn, 1972; Miller and Coffey, 1992) angles of release. More recent quantitative analyses of the free throw (discussed in more detail below) have led to similar recommendations of low (35 to 45 deg, Sharman, 1965), medium (49 to 55 deg, Hay, 1978), and high (56 deg, Mullaney, 1957; 55 to

60 deg, Hartley and Fulton, 1971; and 56 to 62 deg, Hamilton and Reinschmidt, 1996) angles of release.

Mortimer (1951) noted that for release angles between 56 and 59 deg, the initial velocity was about the same (24.1 ft/sec) for making the basket. That is, slight errors in release angle at this velocity should not have an adverse effect on ball trajectory. However, beyond 59 deg, as release angle increased, the required initial velocity also increased. Thus, errors in release angle would have a relatively greater adverse effect on ball trajectory. Hence, she concluded that release angles between 56 and 59 deg provided the greatest margin for error.

Mullaney (1957) analyzed free throw trajectory using an assumption of an optimal entry angle into the hoop of 45 deg, as well as a ballistic trajectory, i.e., without air resistance and spin. He provided a simple formula for calculating the maximum height and distance to the hoop. Based on the formula, Hamilton and Reinschmidt (1997) estimated that the corresponding release angle would be 56 deg.

Hartley and Fulton (1971) noted that the low trajectory arch requires the most accuracy for a successful shot, but takes the least amount of strength and effort. The high trajectory arch requires the least accuracy, but takes the most amount of strength and effort. They argued that a medium arch would provide the advantages of both the low and high arch trajectories. They recommended, however, an above average arch of 55 to 60 deg to provide a relatively optimal tradeoff between accuracy and required power.

Hay (1978) investigated the kinematic (description of motion of bodies) and kinetic (analysis of the forces that move the bodies) properties in shooting the basketball. He found that at a set release height of 7 ft and distance of 15 ft, there is a unique value of release speed for a given release angle that will result in the basketball hitting the center of the hoop. In turn, there is a corresponding unique entry angle. Figure 14.1 shows a set of curves with different release angles, with the appropriate release speeds, and the corresponding entry angles. Hay also noted that as the ball approaches the hoop at a particular entry angle, there is a corresponding effective viewing shape of the hoop. For a high entry angle (90 deg, Fig. 14.2a), the effective hoop shape is circular. But as the entry angle decreases (60 deg and 32.5 deg, Figs. 14.2 b,c, respectively), the shape become increasingly elliptical. Until at 20 deg entry angle, the effective hoop shape is highly elliptical (Fig. 14.2d).

It would appear at first that a high entry angle would provide the best error margin for entry into the hoop. However, this is not correct for three reasons. First, the release speed required for an entry angle of 90 deg would be 44 mph, far beyond the capability of an athlete using any orthodox shooting technique. Second, even if one can throw the ball at that speed, the

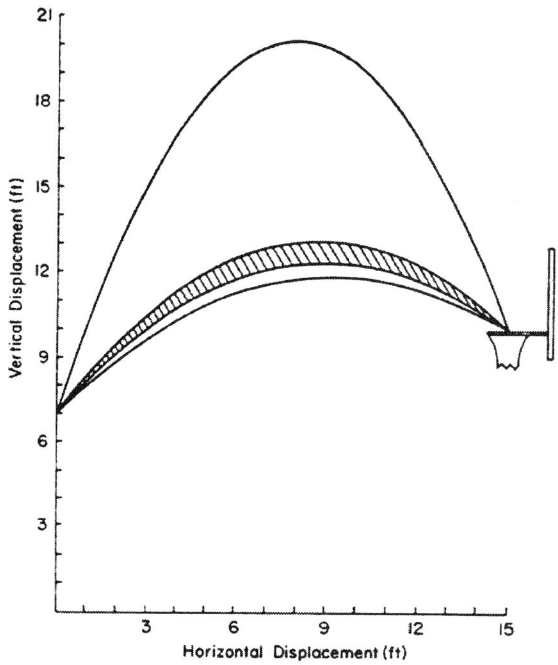


*Figure 14.1.* Curves showing higher release result in higher entry angles, when the release speed is adjusted to result in entry at the center of the hoop. Release height is set at 7 ft and the distance is 15 ft. The error in the length of the shot due to a ±1 deg error in release angle was found to be larger at the lower (46 deg) and higher (73 deg) release angles, but smaller in the range of release angles between 49 and 55 deg, corresponding to entry angles between 38 and 45 deg. Adapted from Hay (1978), Fig. 114, pg. 220.

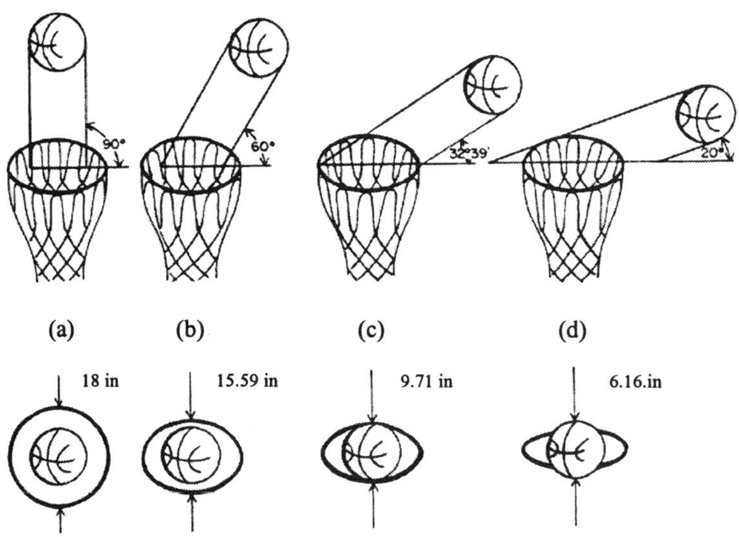


*Figure 14.2.* Top half of figure shows the geometry of entry opening for the ball for different entry angles. Bottom half of figure shows the corresponding views from directly behind the basketball as it approaches the hoop. The effective hoop shape is seen to changes from circular (a) to highly elliptical (d) as the entry angle is decreased. Adapted from Hay (1978), Fig. 112, pg. 216.

height reached would be 72 ft above the level of the court, which would correspond to the height of a six-story building. Third, at such high velocities and large distances, the error in the length of the shot would be very large for any given error in release angle.

It turns out that this error in distance is large for high as well as low entry angles, but smaller for intermediate entry angles. Hay defined the optimal entry angle as that which provides a distance error of less than ± 1.4 inches and an acceptable effective hoop shape for ball clearance. The optimal entry angles were found to be between 38 and 45 deg, which corresponded to release angles of between 49 and 55 deg.

While Hay's study has provided quantitative guidelines for optimal release angles, there are three aspects that could have improved the accuracy of these values. The first is air resistance, and the second is the spin of the basketball. These factors, which were not included in the calculations, could have an effect on the shape of the trajectories, and in turn the optimal release angles. This is because the drag due to air resistance retards the forward motion of the ball. In addition, the Magnus force (Swanson, 1961), which acts in a direction perpendicular to the plane formed by vector of ball motion and its axis of rotation, introduces an additional component to the trajectory. In the case of the free throw, the underspin adds lift to the ball, thereby increasing the range, and in turn the entry angle. The third aspect is the distance to the basket. Hay (1978) had assumed a distance of 15 ft from the foul line to the center of the hoop, rather than the actual distance of 13 ft 8 in. Also, the arm of the shooter would extend beyond the foul line rather than ending at the foul line. This overestimation of distance would introduce a small error in the optimal release angle.

Toyoshima et al. (1979) filmed free throws by two varsity basketball players and two students. The average initial velocities for the skilled and unskilled players were 23.92 and 24.02 ft/sec, respectively. Also, the release angles were 53.29 and 53.61deg, respectively. Their analysis indicated that 65 to 85% of the scatter of the ball at the hoop was due to control of the initial ball velocity regardless of the subject. This result contrasted with that of Mortimer (1951), who reported that the control of release angle was the more important factor.

Hamilton and Reinschmidt (1996) also used a kinematic analysis to determine the optimal trajectory for the free throw. The parameters included release angle, release speed, as well as spin. They set the release height at 6.72 ft, the distance from the release point to the center of the hoop at 13.12 ft, and the backspin rate at 916.7 deg/sec. For a particular combination of release angle, release rate, and spin, they calculated the deviation of the resultant trajectory from a selected reference trajectory. This was done for a range of combinations of release angle, release rate, and spin. They defined the optimal trajectory as a combination for which the accumulated error between the calculated and reference trajectory is less than two standard deviations. They concluded that, when air resistance was neglected, the optimal release angle ranged from 58.00 to 63.25 deg, and the release speed ranged from 23.69 to 24.93 ft/sec. They then made the assumption that air resistance was a constant factor in the distance-direction, so that its effect could be represented by a gain factor for matching the shape of the trajectory with that calculated without air resistance. They concluded that the optimal release angle, with the inclusion of air resistance, ranged from 55.80 to 61.38 deg. It should be noted, however, that their use of a gain factor to match the shape of trajectories with and without air resistance is not entirely correct since, as discussed above, the combination of the effects of air resistance and underspin results in a Magnus force, which would cause the ball to rise higher and thereby alter the shape of the trajectory. Thus, the corresponding optimal release angles would be slightly different than those predicted by Hamilton and Reinschmidt (1996).

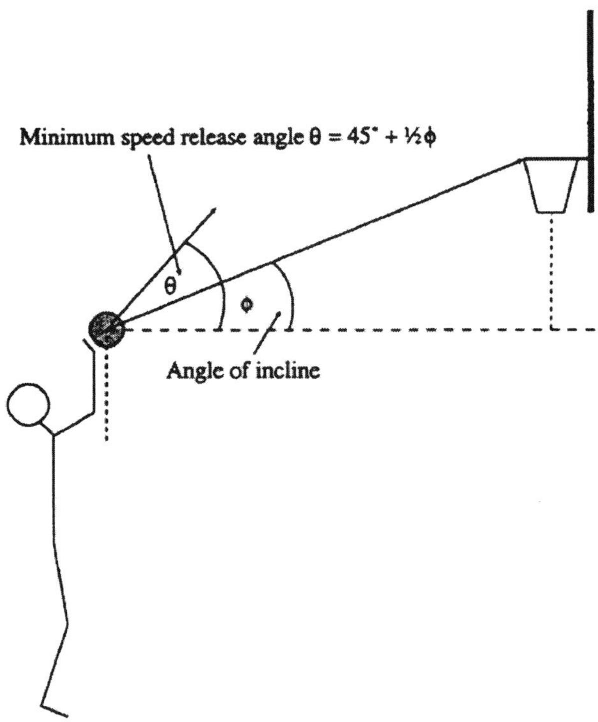


*Figure 14.3.* Angle of release for minimum release speed. Adapted from Miller and Bartlett (1986), Fig. 3, pg. 244, which was based on Brancazio (1984).

Brancazio (1984) quantitatively analyzed free throw kinematics under the condition of no air resistance and determined that the angle of release (θ) for minimum force, or release speed, is given by 45 deg plus one-half the angle of incline to the basket (ϕ) (Fig. 14.3). Moreover, he found that this condition provided the largest margin for angle error.

### 14.2.2 Biomechanics of the Free Throw

Coaches have emphasized the importance of proper technique in shooting a free throw. This included: bending the knees to provide leg drive; keeping the elbow directly below the ball in line with the basket and bent in an L shape; using the tips of all five fingers rather than the palm; keeping the eyes on the hoop throughout the shot; releasing the ball using the elbow as a hinge, springing the forearm forward, and snapping the wrist to provide backspin; shooting a high arching shot; and keeping the hands up during the follow-through (Brancazio, 1984; Miller and Coffey, 1992). Nordland (1998) described the proper technique as one in which the leg drive provided the major power and the relaxed wrist, hand, and fingers just snapped to completion to result in a soft high-arching shot.

There have been some studies that quantified these biomechanical motions during the free throw. Toyoshima et al. (1979) filmed free throws by two varsity basketball players and two students under both visual and non-visual conditions. They found that, under the non-visual condition, the scatter of the ball at the hoop was smaller for the skilled than the non-skilled players. This indicated that the skilled players exhibited greater kinesthetic control of the ball than non-skilled players.

Hudson (1982) investigated biomechanical parameters during the free throw in college women. They were divided into three groups. The high-skill group consisted of 9 members on the United States team in the World University Games. The moderate-skill group consisted of 7 varsity players. And the low-skill group consisted of 9 members on an instructional class. Film records were taken of the players as they attempted the free throws, and the records were analyzed frame-by-frame to assess body position and angle of release. As expected, accuracy was highest for the high-skill players (78%), intermediate for the moderate-skill players (69%), and lowest for the low-skill players (47%). The height of release of the high-skill group was 10.6 in. higher than the low-skill group. In addition, the center of gravity of the high and moderate-skill groups were significantly lower than the low-skill group, indicating better stability. Perhaps surprisingly, the angle of release (~ 52.5 deg) and velocity of release (~ 23.5 ft/sec) were not statistically significantly different among the three groups. Spin was not measured in this study, but could have exhibited differences among the groups.

Hayes (1987) filmed members of the Men's Canadian National Basketball Team as they attempted free throws under simulated game conditions. He divided the free throw motion into the preparation and propulsion phases, and found that the duration of the preparation phase was longer and more variable (duration = 0.38 sec, s.d. = 0.127 sec) than the propulsion phase (duration = 0.19 sec, s.d. = 0.043 sec). He digitized the positions of body segments frame-by-frame to determine their relative contributions to the velocity of the basketball (see Fig. 14.4a,b). Hayes found that the relative contributions of the hip, trunk, upper arm, forearm, and hand during the preparation phase were 47, 24, 27, -3, 5 percent, respectively, and those during the propulsion phase were 4, 6, -2, 15, 77 percent, respectively. The velocity of the hand and basketball at the moment of release were 16.1 and 21.0 ft/sec, respectively. Thus, during the early part of the preparation phase, the forearm, upper arm, and trunk were the main contributors, whereas during the late part, the hip was the main contributor. On the other hand, during the early part of the propulsion phase, the lower body was the main contributor, whereas during the late part, the contribution of the forearm increased and then dropped off, whereas the contribution by the hand continued to increase and became dominant at the end of the propulsion phase.

Tsarouchas et al. (1988) analyzed video recording of free throws by 22 players who participated in the 25th European Championship to investigate the trajectory of the ball and the movements of the players' body segments. The players were divided into two groups: Group One (n = 10) had a success rate of > 80%, whereas Group Two (n = 12) had a success rage of < 65%. They found that: the lower extremities contributed mainly to the vertical displacement of the ball; the trunk had negligible contribution to either vertical or horizontal components; the arm contributed in the vertical direction; the forearm contributed in the horizontal direction; and the hand contributed evenly to both components. They also found, similar to Hudson's (1982), that the angle and height of ball release, the ball's displacement prior to release, as well as the initial and final angles of the body's segments with respect to the horizontal, were not correlated with the success rate in free throw shooting.

Miller and Bartlett (1996) filmed 15 male basketball players who were on a college team or otherwise played at a high level. They were divided into three equal-size groups into guards, forward, and centers, and were instructed to shoot the basketball at three different distances. For the present chapter, only their 15 ft shot results will be discussed. They digitized the

(a)

(b)

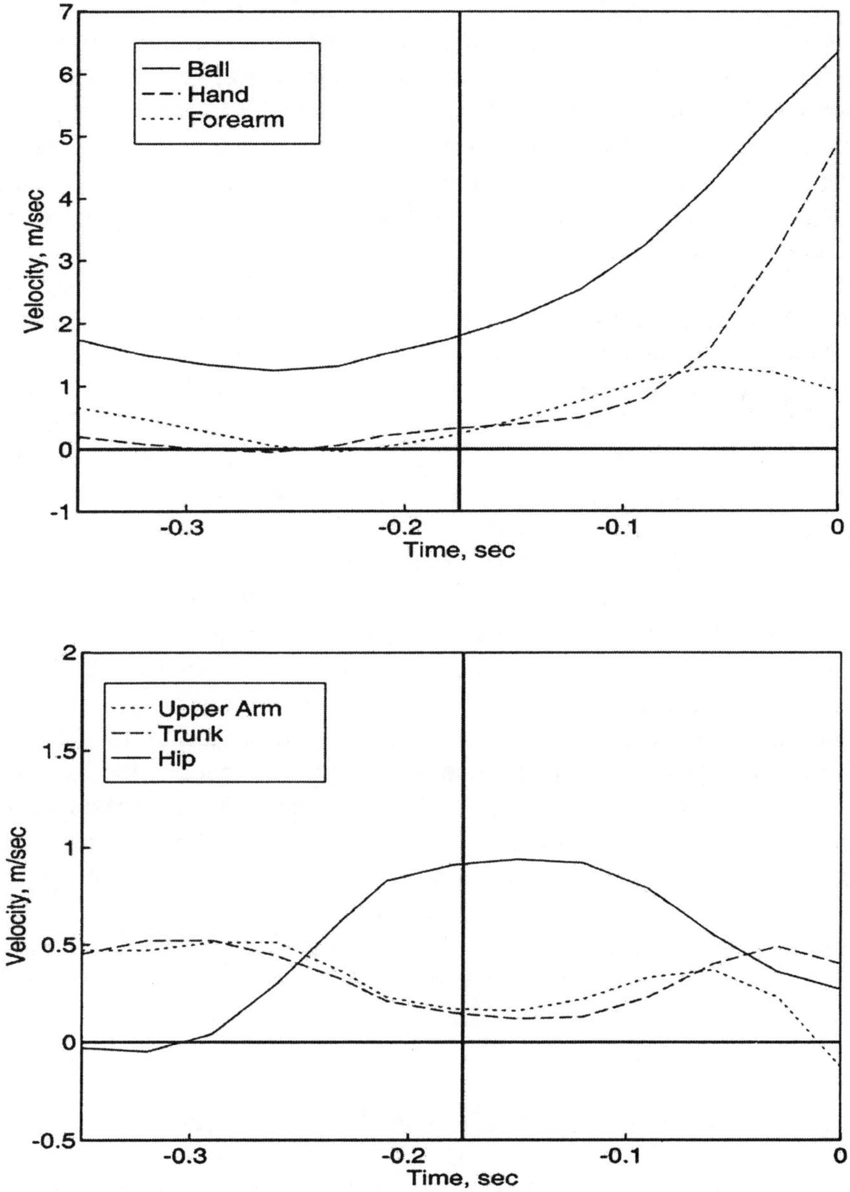


*Figure 14.4.* Average velocity of limb segments from the preparation phase (left of vertical line) through the propulsion phase (right of vertical line) to the point of release (time = 0). Note different ordinate scales for (a) and (b). Adapted from Hayes (1987), Fig. 2, pg. 208.

film record and marked various joint positions during the shot. They found that the shoulder rotational velocities for the guards, forwards, and centers were 338.1, 229.2, and 257.8 deg/sec, respectively. Also, the elbow rotational velocities were 790.7, 624.5, and 676.1 deg/sec, respectively. They did not measure the wrist rotational velocities because they considered the wrist to provide only fine control of the shot. The release heights for the jump shots were 7 ft 3 in, 8 ft, and 9 ft 3 in, respectively. Moreover, they found, in agreement with Elliott (1992), that for all groups there was a forward lean of the trunk. The release angle was found to be 52 deg for the guards and forwards, and 54 deg for the centers. Further, the release speed was 20.6 ft/sec for the guards and forwards, and 21.0 ft/sec for the centers.

Skoglund et al. (2001) used a videotape technique to record motions of the limbs during the free throw in a novice and an expert player. They found that even though the novice was occasionally successful, he did not have the proper form. One element that made the novice inconsistent was the angle of the elbow, which was less bent than the expert and was also slightly misaligned with respect to the plane of the wrist and shoulder. In addition, the novice's knee angle was 11 deg less than that of the expert. This caused the novice to jump, or lunge, when attempting the shot to make up for the lack of knee bend. Finally, the novice did not release the ball rapidly, so that the hand was in contact with the ball a brief moment after the initial release. All these factors resulted in less force being generated by the novice, and consequently inconsistency in repeated attempts in shooting the free throw.

Amberry (Amberry and Reed, 1997) provided a 7-step procedure for the proper execution of a free throw. He emphasized that besides the mechanics of the shot, there should be a positive mental attitude. For example, instead of dreading the free throw, one should consider it one of the most vital scoring opportunities in a game. This is detailed in Chapter 15 of this book. He holds the world record for the most consecutive free throws made in one hour (307) and in a single day (2,750).

## 14.2.3 Eye Movements During the Free Throw

Harle and Vickers (2001) used a Vision-in-Action device to train university women basketball players. The device consisted of an Applied Sciences Laboratory 501 eye tracker and a scene camera (Vickers, 1996). The eye tracker measured the eye gaze direction relative the helmet worn by the subject. The scene camera viewed the hoop and backboard as seen from the perspective of the subject. A monitor displayed the scene, and a small square cursor was superimposed to show the subject's eye fixation. The system has an accuracy of ± 1 deg visual angle. A 100 ft cable, which was attached at the subject's waist, relayed the eye and scene images to a nearby computer.

Women basketball teams from 3 universities participated in the study. Team A received visual fixation training during their practice in which they were provided quantitative feedback regarding their eye positions during free throws. Teams B and C served as controls, and had a similar level of win/loss record as Team A. In Season 1, the Free Throws Made (FTM%) accuracy of Team A was low (54.1%) as compared to 67.7% and 61.3% for Teams B and C, respectively. However, by the end of Season 2, Team A had improved their FTM% to an impressive 76.7% while Team B declined to 66.2, and Team C improved to 74.1%. Indeed, the training in Team A appeared to have led to a more economical visual-motor routine, as indicated by a longer duration of eye fixation on the hoop, a slower preparation phase, and a faster shot motion. The improved visual-motor routine in turn led to improved free throw performance in real game situations.

## 14.3 OUR RECENT STUDY

Our recent study expanded on previous research to provide a comprehensive analysis of free throw dynamics using video, modeling, and accelerometer techniques. There were several novel aspects to our research. First, spin and drag were both included in the model simulations. Second, vertical eye movements were recorded during the free throw. Third, accelerometers were employed to study the movement of upper body joints. And fourth, wireless transmission of these signals was used for data recording and display.

Our study consisted of two parts, aerodynamics and biomechanics. The aerodynamics portion compared model and experimental free throw trajectories. The model simulation used two second-order differential equations, which were based on a previously conducted tennis study (Rüde and Kowarschik, M., 2000). These equations were modified for the basketball free throw. The two 2nd-order equations were first converted to four 1st-order equations. Then, a Matlab program was used to simultaneously solve the equations. These equations provided realistic simulations of basketball free throw motion through the incorporation of spin, lift, and drag and could accurately measure the x (horizontal) and y (vertical) components of motion. Experimental free throw trajectories were obtained from studies conducted in a gym that used a video camera, display grid, and digital timer. The indoor court optimized the shooting situation by providing a standard and controlled environment. Through distance and close-up filming, experimental trajectory and kinematic data were collected.

The biomechanical portion of the study used accelerometers to measure fingertip, wrist, and elbow acceleration during the free throw. Also, a visor-mounted infrared eye movement detector measures the vertical eye movements. The joint acceleration and eye movement signals were input to a controller which multiplexed the data for serial wireless transmission using one radio frequency. A Matlab program was used for analysis and display of the data records.

Ten subjects were recruited for the aerodynamic and biomehanics study. They consisted of eight male and two female Rutgers University students. Six of the male subject and one female subject were considered experienced players at the intramural college basketball level, whereas the remainder were novice players.

Each of the subjects shot two sets of ten free throws while being videotaped. In addition, in two of the subjects, one male experienced player and one female novice player, joint accelerations and vertical eye movement were measured simultaneously.

### 14.3.1 Theoretical Aerodynamic Analysis

Our initial calculations used a set of simple ballistic equations, which neglected the backspin, lift, and drag: where Range is the total horizontal distance that the ball travels and $t_{peak}$ represents the time is seconds until the peak height of the shot, which is noted as the $Height_{peak}$. $V_y$ is velocity of the ball in the vertical direction and $V_x$ is the horizontal velocity of the ball. The angle θ represents the initial angle of release for the ball. Simulation of Eqs. 14.1a-e provided a basis for comparison with a more realistic simulation (see below) that included spin, lift, and drag.

$$\text{Range} = \frac{v_0 \sin 2\theta}{g} \tag{14.1a}$$

$$t_{peak} = \frac{v_{oy}}{g} \tag{14.1b}$$

$$\text{Height}_{peak} = \frac{v_{0y}^2}{2g} = \frac{v_0^2}{2g}\sin^2\theta \tag{14.1c}$$

$$v_y = v_0 \sin\theta \tag{14.1d}$$

$$v_x = v_0 \cos\theta \tag{14.1e}$$

The more realistic model was derived from a previous quantitative analysis of the trajectory of tennis balls (Štěpánek, 1988; Rüde and

Kowarschik, M., 2000). The parameters were appropriately modified for the quantitative analysis of basketball trajectories. The two 2nd-order equations were given by:

$$\frac{d^2x}{dt^2} = -C_D \alpha v v_x + \eta C_M \alpha v v_y \tag{14.2a}$$

$$\frac{d^2y}{dt^2} = -g - C_D \alpha v v_y - \eta C_M \alpha v v_x \tag{14.2b}$$

The two 2nd-order differential equations were converted to four 1st-order differential equations, and a Matlab analysis program was used to solve these equations simultaneously.

$$\frac{dx}{dt} = v_x \tag{14.3a}$$

$$\frac{dv_x}{dt} = -C_D \alpha v v_x + \eta C_M \alpha v v_y \tag{14.3b}$$

$$\frac{dy}{dt} = v_y \tag{14.3c}$$

$$\frac{dv_y}{dt} = -g - C_D \alpha v v_y - \eta C_M \alpha v v_x \tag{14.3d}$$

where

| | |
|---|---|
| $C_D = f(v/w)$ | Drag coefficient (Štěpánek, 1988; Rüde and Kowarschik, 2000). |
| $C_M = h(v/w)$ | Lift coefficient associated with the Magnus force (Štěpánek, 1988; Rüde and Kowarschik, 2000). |
| $\alpha = \frac{\rho \pi d^2}{8 m}$ | Ball dimension coefficient |
| $v = \sqrt{v_x^2 + v_y^2}$ | Net velocity |

| | |
|---|---|
| $g = 9.81$ N/kg | Gravitational acceleration |
| $\rho = 1.29$ kg/m$^3$ | Density of air |
| $d = 0.247$ m | Diameter of the basketball |
| $m = 0.595$ kg | Mass of the basketball |
| w | Spin rate of basketball - subject dependent |
| $\eta$ | Direction of spin. $\eta = +1$ denotes topspin, and $\eta = -1$ denotes backspin) |

## 14.3.2 Measurement Techniques

### 14.3.2.1 Video-Based Method

The aerodynamics and biomechanics of the free throw was studied by means of a video-based technique in indoor basketball court. The camera (JVC GR-SX850U, 30 frames/sec) was set up on a tripod a few feet to the right of the subject as he stands on the free throw line (for simplicity, he/she will be designated by he). Just in front of the camera, a digital clock was mounted on a second tripod, so that its image occupied a small lower right hand portion of the frame. A large poster-board, consisting of a repeated 10 cm x 10 cm grid pattern, was mounted to the left of the subject. In this way, the camera captured the subject's free throw motion with a fix grid pattern in the background. Thus, frame-by-frame displacements of the limbs and the basketball could be measured. In addition, the digital clock readout provided the temporal change needed for the calculation of velocity of motion.

### 14.3.2.2 Acceleration Measurements

The Analog Devices ADXL202 is an accelerometer which can be used to measure both static and dynamic acceleration in two-axes. It is mounted on a small circuit board about ½ in by 1 in. The accelerometer has a bandwidth of 5 KHz and a range of either ±2 g or ±10 g depending on the sensitivity required. Its operation is based on the capacitance changes associated with the relative motion of a movable central mass-driven plate and two fixed plates. An output circuit in the accelerometer converts the analog capacitance change to either an analog voltage output or a change in the duty cycle of a rectangular waveform with a known preset frequency. The use of the duty cycle output provided a robust means to transmit the signals wirelessly.

A specially-designed lightweight "sleeve" was used for the measurement of finger, wrist and elbow acceleration (Fig. 14.5). It consists of a long ribbon cable which is interrupted by connectors at the finger, wrist, and elbow. At each junction, a small accelerometer board is attached to the connector to provide electrical connection to the ribbon cable. In this way, the ribbon cable provides both the power supplied to the accelerometers and the return signals from the accelerometers. A modified athletic wristband was used for fitting the assembly to the fingertips. Also, a Velcro wristwatch band was used for attachment at the wrist, and an athletic wristband was used for attachment at the elbow. The distances between accelerometers were determined by fitting it to the tallest of our subjects. This enabled a proper fit for the subject with the longest arms, and also allowed for a good fit for shorter subjects. The extra slack in fitting the shorter subjects was found to be not a hindrance to the motion of the subjects.

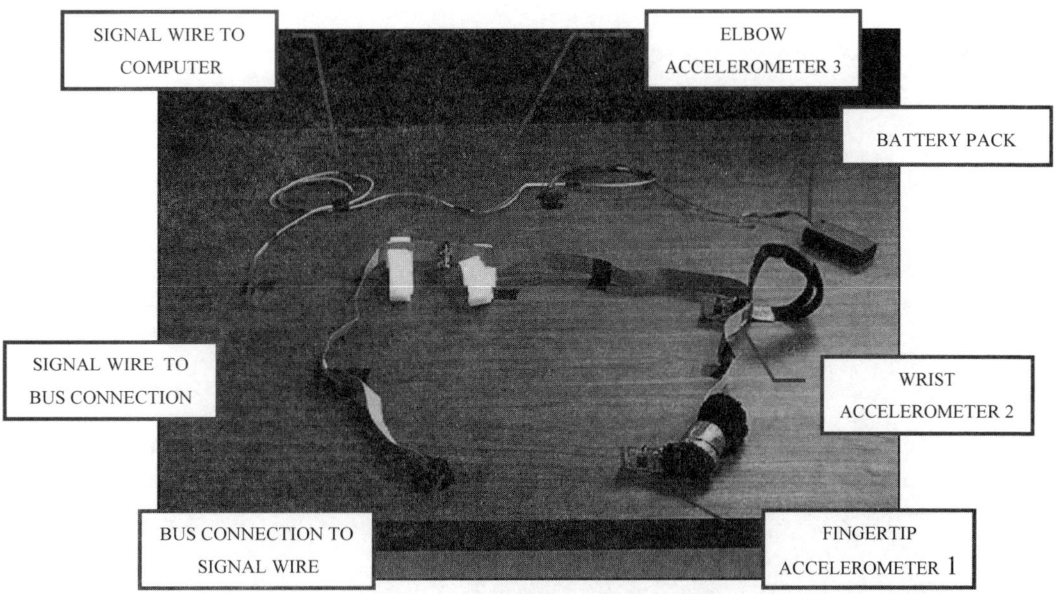


*Figure 14.5.* Photo layout of accelerometers on a cable bus to be worn on the arm and hand, along with a signal wire for connection to the computer.

### 14.3.2.3 Eye Movement Measurements

In addition to the acceleration measurements, we had an interest in the vertical eye motion during the free throw shot. Vertical eye movements are an important factor in the overall technique for shooting free throws. Not only does eye motion affect the trajectory and flight of the ball, but the mental focus of the shooter as well. Indeed, coaches have stressed that keeping the eyes focused on the rim is essential to shooting accuracy. Thus, quantitative measures of vertical eye movements can provide an assessment of the focusing state of the shooter, and thus can serve as a useful training tool for both novices and experts.

We employed a Skalar IRIS eye movement monitor. This device uses the principle of infrared reflection from the limbus region, or the boundary between the white sclera and dark iris of the eye. Infrared emitters provide an infrared source to the eye. Vertical eye movements result in changes in the amount of white sclera that is exposed, and in turn changes in the amount that is reflected to the infrared sensor. Thus, the sensor output provides a signal which is proportional to the vertical eye movement amplitude.

The eye movement monitor was attached via a custom-fabricated mount (Fig. 14.6), placed on a visor. The mount was made of nylon and was attached to the visor using stainless steel screws. This custom piece allowed for adjustment in the vertical and horizontal directions to accommodate different facial dimensions of the subjects.

Figure 14.6. Photo of eye movement monitor attached via a custom-fitted mount and placed on a visor. Also shown is the accelerometer bus assembly worn on the elbow and the hand. Further, note the grid pattern in the background used for calibration of joint and ball positions.

#### 14.3.2.4 Microcontroller and Wireless Transmission

For ease of use and effective data collection, a newly developed wireless transmission system was employed in the study. The key to the wireless system is a four-signal controller. The controller, Atmel T89C51AC2, acts as a link between the accelerometers and the transceiver. In the free throw study, the data from the three accelerometers and the single eye motion detector were fed into the controller. The controller organized the data into a serial data stream for transmission through the wireless transceiver. In the case of the free throw study, these four signals were recorded and transmitted. Significantly, this system allowed information from multiple data sources to be wirelessly transmitted over one radio frequency.

Two transceivers (TR1000 916.50 MHz Hybrid Transceiver, manufactured by RFM) were utilized in this study (Fig. 14.7). One transceiver was connected to the programmable controller. It sends the gathered data wirelessly to the second transceiver. The second transceiver receives the transmitted data and feeds the signal into a computer. After the data are received at the second transceiver, an RS232 computer cable carries the data to the computer for recording and analysis using a Matlab program.

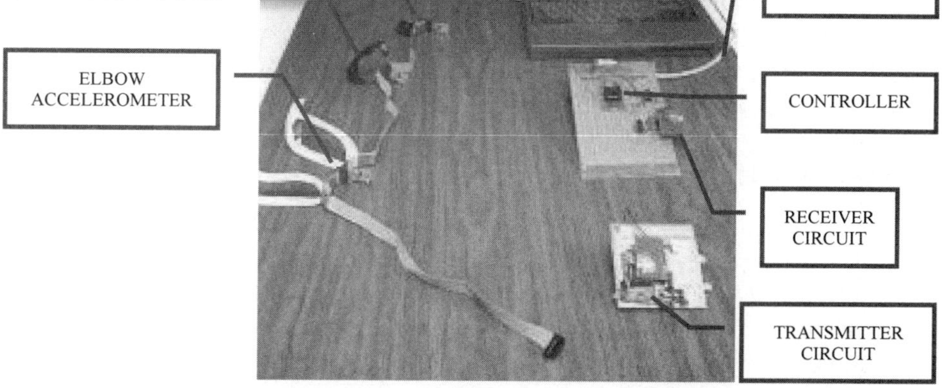


*Figure 14.7.* Photo of complete equipment layout for wireless transmission of acceleration signals consisting of: measurement device, controller, transmitter, receiver, RS232 interface, and personal computer.

## 14.3.3 RESULTS

### 14.3.3.1 Experimental and Theoretical Trajectories

Typical experimental free throw trajectories (* symbols curve) are shown for a novice (Fig. 14.8a) and an experienced player (Fig. 14.8b). The novice player exhibited a release angle of 61 deg and velocity of 24.4 ft/sec, while the experienced player exhibited a release angle of 54.8 deg and velocity of 24.5 ft/sec. For both players, model parameters were adjusted to provide a reasonable fit of the simulation trajectory (solid curve) to the experimental trajectory. In addition, simulated trajectories at the same release angle and velocity, but without lift, drag and spin (dashed curve), are shown for comparison.

The similar initial velocity values for the novice and experienced players is consistent with the finding by Hudson (1982). The novice player in this example exhibited a somewhat higher release angle. It appears, however, that the main difference is the amount of spin, with a higher simulated basketball surface spin rate (6.6 ft/sec) for the experienced than the novice player (3.3 ft/sec). Indeed, the spin is an important component that may determine whether the basketball falls short (Fig. 14.8a) or goes through the hoop (Fig. 14.8b). This is due to the Magnus force, which as stated above, introduces lift in the trajectory. In the case of the free throw, the increased lift provided by the greater underspin imparted by the experienced player helps the ball rise higher so that the peak height is reached further downstream, by about 2 ft more than that for the novice player. Moreover, the entry angle is larger, thus, improving the chance of success. In contrast, despite similar release angle and velocity, the smaller lift due to lower spin imparted by the novice player resulted in the peak height being reached earlier and thus further away from the basket, thereby reducing the entry angle. Indeed, the ball does not quite reach the basket. Moreover, in this example, even if the release height of the basketball for the novice player were artificially increased by a foot, thereby raising the curve, the projected trajectory would still just hit the front rim.

(a)

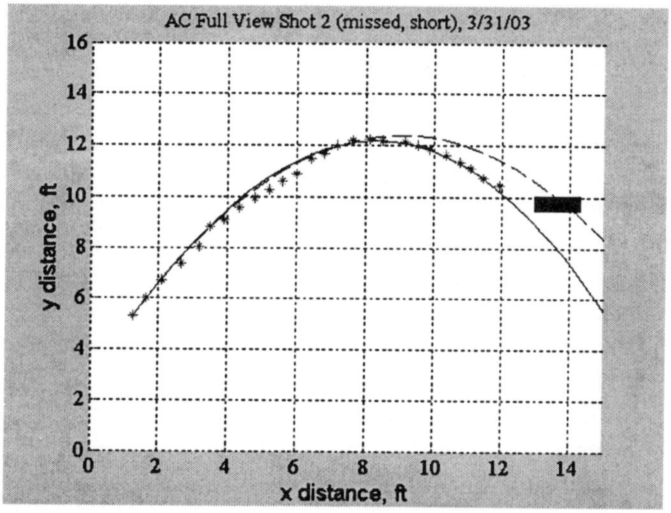


(b)

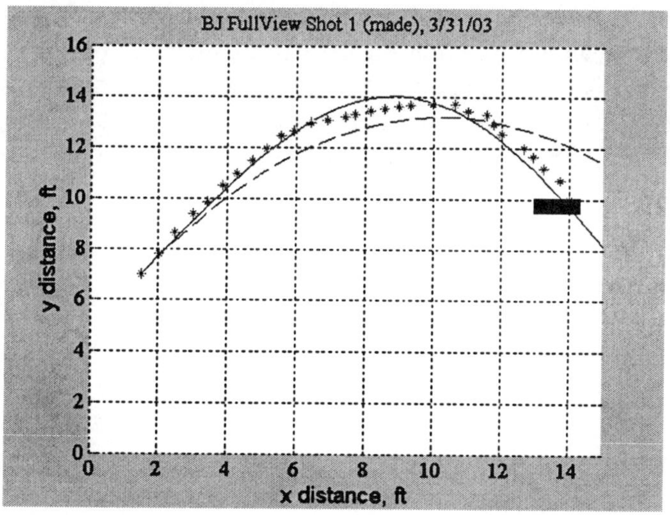


*Figure 14.8.* Trajectories of experimental free throw (*) and model simulation with (—) and without (---) lift, drag and spin for the (a) novice and (b) experience player.

#### 14.3.3.2 Experimental Finger, Wrist, and Elbow Acceleration, and Eye Movements

The main difference in joint accelerations between the novice and experienced player can be seen in Figs. 14.9a, b. For the novice player, the magnitude of acceleration is similar for the finger and wrist (Fig. 14.9a). This indicates that the novice player tends to push rather than spin the ball. As mentioned above, the smaller spin results in reaching the peak height earlier, thereby reducing range as well as entry angle, and providing a smaller chance for making the free throw. In contrast, the experienced player exhibits a large differential in the magnitude of acceleration between the finger and the wrist. This corresponds to a "snapping" action, which imparts substantially more spin. In turn, this results in reaching the peak height later, thereby increasing the range as well as entry angle, and providing a greater chance for making the free throw.

The main difference in vertical eye movements between the novice and experienced player can also be seen in Figs. 14.9a,b. The novice player has a tendency to look upwards even before releasing the ball and try to follow the trajectory of the basketball (Fig. 14.9a). This will lead to reduced accuracy in executing the free throw. On the other hand, the experienced player tends to maintain fixation on the rim as much as possible, until the moment the ball is released and the change in momentum causes the head to jerk back in a natural manner, and in turn causes the eyes to look up slightly (Fig. 14.9b). This fixation technique increases the accuracy in executing the free throw.

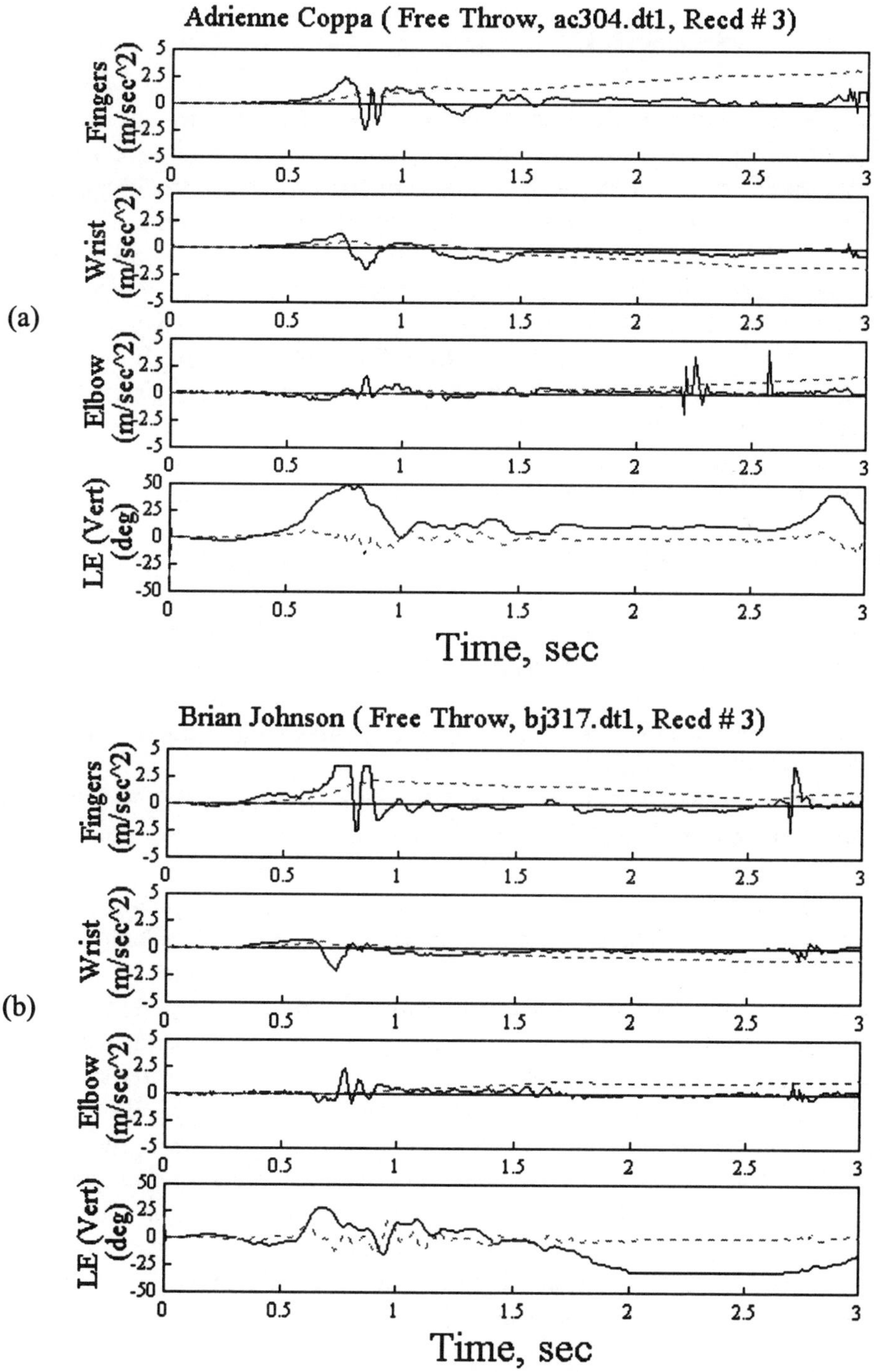


*Figure 14.9.* Typical acceleration time courses for finger, wrist and elbow, and vertical left eye position for the (a) novice and (b) experienced player. The dotted lines are the corresponding velocity traces, with units of (x 2) m/sec for the joint motions and deg/sec for eye motion.

## 14.3.4 DISCUSSION

The similarity in release angle and velocity between the novice and experienced player indicates that both players attempt to achieve a trajectory that would provide a reasonable chance of success. However, it appears that the limb segments used by the players may be rather different. As suggested by Skoglund et al., the novice uses less elbow and knee-bend, and hence requires more shoulder, arm and leg motion than the expert. The accuracy that can be attained with these large muscle groups is less than that with the wrist and finger tips, which can perform fine motor control. Hence, this can explain the lower accuracy and consistency of the novice players as compared to the experienced players.

We found that snapping the wrist is important because it generates increased spin. During the rising phase of the trajectory, the combined effect of underspin and air drag increases the amount of lift due to the Magnus force. This results in the ball reaching the peak height further downstream, and in turn attaining a greater range and increased entry angle. The overall effect of these factors is to increase the chance of making the free throw. The wrist snap can also provide increased release velocity. Indeed, Miller and Bartlett (in press) noted that wrist flexion contributed to 59% of the ball release speed, whereas the elbow extension and shoulder flexion contributed 8% and 14%, respectively.

Our study has shown that the device we developed can be effectively used for motion analysis in sports applications. Indeed, this study was the first to use this device, consisting of accelerometers and eye movement sensor, to measure these important biomechanical components during the free throw. The wireless feature provides a quick and reliable means to study the biomechanical parameters without the hindrance of connecting wires. The device can also serve as a training tool at all experience levels to provide immediate quantitative feedback regarding limb and eye motions during the free throw. The data can be compared to reference curves obtained on excellent free throw shooters to assess specific areas for improvement. Moreover, it can serve as a prototype for wireless measurements under other dynamic conditions such as during the jump shot, or even in a mock-up game which simulates the real life contest environment. Further, the ideas behind the device can be applied to other sports as well, such as the training the mechanics of arm motion in the pitch, the football throw, or the slapshot.

## 14.4 CONCLUDING REMARKS

Aerodynamics analysis has provided an optimal release angle in the range of 51 to 56 deg, and velocity in the range of 20.5 to 24 ft/sec. Biomechanical analysis has shown the importance of bent elbows and knees, and a snap of the wrist at the moment of release. This imparts the spin and consequent Magnus force that adds lift to the trajectory, increasing the range and the angle of entry. Also, eye fixation on the rim is important for accuracy and consistency. Our video, accelerometer, and eye movement measurements have demonstrated and verified these effects. Moreover, the wireless device that has been developed can serve as a useful training tool to provide immediate feedback regarding the biomechanics of limb motions and eye fixation for optimal free throw performance. Continued development of efficient wireless devices can provide detailed understanding of biomechanics and aerodyamics of the free throw under real-game conditions. Further, this measurement technique can be applied to the study of athletic motions in other sports.

## 14.5 REFERENCES

Amberry, T., with Reed, P., 1997, *Free Throw; 7 Steps to Success at the Free Throw Line*, HarperCollins.

Barnes, M. J., 1980, *Women's Basketball*, 2nd Edition, Allyn and Bacon, Boston.

Brancazio, P. J., 1984, *Sports Science - Physical Laws and Optimal Performance*, Simon and Schuster, New York, NY, pp. 281-289, 306-314.

Bunn, J. W., 1964, Basketball Techniques and Team Play, Prentice-Hall, Englewood Cliffs, NJ, pp. 89-95.

Bunn, J. W., 1972, *Scientific Principles of Coaching*, 2nd Edition, Prentice-Hall, Englewood Cliffs, NJ.

Cooper, J. M., and Siedentop, D., 1969, *The Theory and Science of Basketball*, Lea and Febiger, Philadelphia, PA.

Cousy, B., and Power, F. G., 1970, *Basketball Concepts and Techniques*, Allyn and Bacon, Boston, MA.

Elliott, B., 1992, A kinematic comparison of the male and female two-point and three-point jump shots in basketball, *Australian J. Science and Medicine in Sports.* **24**: 111-117.

Hamilton, G. R., and Reinschmidt, C., 1997, Optimal trajectory for the basketball free throw, *J. Sports Sciences.* **15**: 491-504.

Harle, S. K., and Vickers, J. N., 2001, Training Quiet Eye improves accuracy in the basketball free throw, *The Sports Psychologist*, **15**: 289-305.

Hartley, J., and Fulton, C., 1971, Mechanical analysis of the jump shot, *Athletic J.*, **51**: 92.

Hay, J. G., 1978, *The Biomechanics of Sports Techniques*, 2nd. Edition, Prentice-Hall, Englewood Cliffs, NJ, pp. 211-234.

Hobson, H. A., 1955, *Scientific Basketball for Coaches, Players, Officials, Spectators, and Sportswriters,* Prentice-Hall, Inc., New York, pp. 75-85.

Hudson, J. L., 1982, A biomechanical analysis by skill level of free throw shooting in basketball, in: *Biomechanics in Sports*, J. Terauds, ed., Academic Publishers, Del Mar, CA, pp. 95-102.

Krause J., and Hayes, D., 1994, Score on the throw, in: *Coaching Basketball*, J. Krause, ed., Master Press, Indianapolis, IN, pp. 138–141.

Lambert, W. L., 1932, *Practical Basketball*, Athletic Journal, Chicago.

Maugh, T. H., 1981, Physics of basketball: those golden arches, *Science 81*, **2**: 106-107.

Meanwell, W. E., 1924, *The Science of Basketball for Men*, Democrat, Madison, WI.

Miller, F. Y., and Coffey, W., 1992, *Winning Basketball for Girls*, Facts on File, New York, pp. 29-33.

Miller, S., and Bartlett, R., 1996, The relationship between basketball shooting kinematics, distance and playing position, *J. Sports Sciences*. **14**: 243-253.

Miller, S., and Bartlett, R., in press, The biomechanics of intra-subject variability in basketball shooting: kinematic considerations. J. Sports Science.

Mortimer, E. M., 1951, Basketball shooting, *Research Quarterly*. **22**: 234-243.

Nordland, T., 1998, Interview of Tom Nordland by Alan Lambert of the Basketball Highway, <http:/www/swish22.com/BBHwyInt.html>

Rüde, U., and Kowarschik, M., 2000, Numerical simulation of spinning tennis balls. University of Erlangen-Nürnberg. < http://www10.informatik.uni-erlangen.de/~markus/ss00/sheet07.pdf>

Skoglund, J., Syverson, J., and Melius, L., 2001, Biomechanical analysis of the basketball free throw. <http://www.cord.edu/homepages/lbmelius/scipap.html>

Sharman, B., 1965, *Sharman on Basketball Shooting*, Prentice-Hall, Englewood Cliffs, NJ.

Štěpánek, 1988, The aerodynamics of tennis balls - the topspin lob, *Am. J. Physics*, **56**: 138-142.

Swanson, W. M., 1961, The Magnus effect : a summary of investigations to data, *J. Basic Engin.*, Trans. ASME, Series D, **83**: 461-470.

Tan, A., and Miller, G., 1981, Kinematics of the free throw in basketball, *Am. J. Physics*. **49**: 542-544.

Toyashima, S., Hoshikawa, T., and Ikegami, Y., Effects of initial ball velocity and angle of projection on accuracy in basketball shooting, 1979, *Biomechanics, VII-B*, Morecki, A., Fidelus, K., Kedzior, K., and Wit, A., PWN - Polish Scientific Publishers, Warszawa, Poland, and University Park Press, Baltimore, MD, pp. 525-530.

Tsarouchas, E., Kalamaras, K., Giavroglou, A., and Prassas, S., 1988, Biomechanical analysis of free throw shooting in basketball, in: *Proceedings of the 6th International Society of Biomechanics in Sports Symposium*, E. Kreighbaum and A. McNeill, eds., Montana State University and the International Society of Biomechanics in Sports, Bozeman, MT, pp. 551-560.

Veeneker, G. F., 1937, *Basketball for Coaches and Players*, A. S. Barnes, New York.

Vickers, J. N., 1996, Visual control when aiming at a far target, *J. Experimental Psych.: Human Perception and Performance*. **2**: 324-354.

Wooden, J, 1966, *Practical Modern Basketball*, Ronald Press, New York.

# Chapter 15

# Make Every Free Throw

Thomas Amberry[1], with Philip Reed[2]

[1]*Make Every Free Throw, P. O. Box 14664, Long Beach, CA 90803, PH: (800) 378-7777, FX: (310) 420-1893, EM: DrAmberry@aol.com*

[2]*Editor, Edmunds.com, Inc., 2401 Colorado Ave., Suite 250, Santa Monica, CA 90404, PH: (310)-309-6331, FX: (310) 309-6400, EM: Philreed@aol.com*

## 15.1 INTRODUCTION

When I give clinics for basketball players, I tell them, "I'm going to show you how to make all your free throws. I'm going to teach you how to make two baskets, under pressure, when you really need to."

You may think that this is an outrageous claim, that no one can be perfect from the stripe. But I find it necessary to be completely optimistic for several reasons. First, the free throw is one of the few situations in sports when the player is in complete control. The basket is always 15 feet away, the hoop is always ten feet off the ground. Secondly, it is essential that the player have a positive attitude when preparing for the shot. It would be disaster if the player was thinking, "I'm a 75 percent free throw shooter so I have a three out of four chance of making this free throw."

I've been a positive thinker all my life (see Fig. 15.1). But it was only in my so-called retirement, that I found how the free throw line is where the *cans* are separated from the *cannots* or *maybes*. When I was preparing to set the world's record, back in 1993, I spent three weeks conditioning my mind for the event. I barred all negative thoughts from my consciousness. And it paid off. With ten witnesses watching me, at the age of 72, I shot for 12 hours without missing a free throw and made 2,750 in a row. I could have kept shooting, but they were closing the gym, so they kicked me out. My accomplishment is now listed in the *Guinness Book of World Records.*

*Figure 15.1.* Dr. Tom Amberry was a former All-American basketball player who took up free throw shooting as a hobby after his retirement from practicing podiatry. A year later he set the all-time record of 2,750 consecutive free throws.

## 15.1.1 My Second Career: Free Throw Champion

Ever since I set the *Guinness* record, I've continued to compete in free throw tournaments, setting new records and teaching my free throw shooting technique to other players. I have set the world's record for the most free throws in one hour during three different national television appearances: on Fox television I made 274 in a row in an hour with one ball. Then I beat my own record on "The Crook and Chase Show" by making 305 consecutive baskets. In 2002, on "Ripley's Believe It or Not!," I made 307 baskets in a row. I've won all the free throw tournaments I've entered, including the World Master's Game in 1994, 1998 and 2002. On television, I've shot against some of the best players in the NBA, including Kobe Bryant. And I've always won.

I have coached many men and women high school and college basketball players. Mostly, people contact me because of my book *Free Throw; 7 Steps To Success at the Free Throw Line,* written with Philip Reed, and published by HarperCollins. My dream of coaching on the pro level was realized in 2001 when Chicago Bulls coach Bill Cartwright hired me as a free throw shooting coach for some of his players. The Bulls flew me to the windy city to work with Eddy Curry, Tyson Chandler, and Charles Oakley. On February 28, 2002, after I'd returned to my home in California, I picked up the paper and saw that the Bulls had been perfect from the line making 14 of 14 free throws against the Detroit Pistons.

## 15.2 PSYCHOLOGICAL AND BIOMECHANICAL ASPECTS OF FREE THROW SHOOTING

You might be wondering if I'm suggesting that my success at the free throw line is purely my positive attitude. Absolutely not. My positive approach is the final ingredient to a process that combines biomechanics and psychology. But I want you to open your mind to a new paradigm before I describe my method.

### 15.2.1 Psychological Aspects

You see, many players have convinced themselves the free throw is a tough shot and there is nothing that can make it easier. But these same players have probably never had instruction on how to shoot free throws. Their style comes from a variety of sources: tips from other players, hand-me-down information and copying the shooting style of the pros. There might even be elements of superstition in the way they shoot. What this boils down to is that everyone knows how to *shoot* free throws -- but not many players know how to successfully *make* free throws.

#### 15.2.1.1 Free Throws - A Pressure Shot

Players tell me that it's the pressure that makes the free throw so hard. They say they can make them all day in practice, but in a game the ball just won't go through the hoop. These players have what I call "performance anxiety." I tell them that of course they will feel nervous if they are shooting with the game on the line. But if they have a process to focus on, and their concentration is strong, the process replaces the pressure. Focus on the process, not the result and you will be perfect from the line. I call this being "process oriented" rather than, "result oriented."

Before I even begin talking about free throw mechanics, I try to change the mindset of players by making the target larger in their minds. Players have spent their careers looking at the hoop from the side. This makes the hoop appear like a narrow opening rather than a large circle. So I carry a basketball rim into clinics, lay it on the gym floor and show them that there is almost enough room to fit two basketballs through the hoop at once (see Fig. 15.2). If the basketball goes through the center of the hoop, there will be almost a nine-inch clearance on either side. Many players say, "I didn't know that. I thought the ball just kind of squeaked through the hoop."

### 15.2.1.2 Practice for Improvement

*Figure 15.2.* The hoop is wider than most players realize, some 18 inches across. This means that you can almost fit two regulation basketballs through it at the same time. Knowing there is a greater margin of error than you thought will make the task seem easier.

I also tell players that the results they get from my method will depend on how much they practice. I recommend 100 free throws a day, every day of practice. To maximize improvement, they should keep track of their results. Players should set goals and continually strive for perfection. Coaches can organize free throw competitions and give rewards for the best performers.

At my clinics the players always want to see me shoot. They can watch me on a video, or see me on TV shooting free throws. But there is something about seeing me, in person, with their own eyes, that finally convinces them. So I usually shoot 25 free throws. If the clinic is for kids, I invite them to try to distract me while I'm shooting. They love yelling and screaming and jumping up and down behind the basket. But I always make my free throws anyway. Then we settle back down and I tell them my seven steps to successful free throw shooting.

#### 15.2.1.3 Are You a Believer or a Skeptic?

When people hear about my record, they sometimes can't believe it. They want to know if I used a regulation ball or a larger hoop. Sometimes, for fun, I just tell people, "I have a homing device in the ball."

At a coaches' clinic I made 100 baskets in a row. Afterwards, most of the coaches tell me, "you know, I've been coaching 20 years and I've never seen anyone do that. That's great shooting." But four or five of them will come up and say, "you're a pretty good shooter. But I like my way of shooting better."

Another time, I made 50 in a row while being watched by a panel of "evaluators" who were coaches from across the country. Afterward, one coach said I didn't have enough follow through. Another said I didn't have enough backspin on the ball. And one other guy said I made them all because I was "very, very lucky." Still another coach said my free throw shooting would improve if I had a higher arc. My answer to that is, in the final analysis, free throw shooting is about putting the ball in the basket. If they all go in, how can they be "improved."

#### 15.2.1.4 Documented Success

I've never had anyone who accepted my process and practiced who didn't get better. Players have improved a minimum of 10 percent. After I did a clinic in Palm Springs, this beautiful blond girl came up to me and said, "I'm Nicole Quinn. Remember me? You came to my school and told me that if I practiced I could be the best free throw shooter in Woodbridge High. Well I practiced. But I wasn't only the best free throw shooter in Woodbridge High, but I became the best woman's high school free throw shooter in the United States."

### 15.2.2 Biomechanical Aspects

My seven-step method has been used in high schools, colleges and now it is being used by some of the NBA pros. If you learn it, and practice it, you will improve. How much? It depends on how much you practice. And how high you set your goals. As I said in the beginning of this chapter, my recommendation is that you try to make every free throw.

#### 15.2.2.1 Seven-Step Method

Here are the seven steps I follow every time I shoot a free throw:

### Step 1. Feet square to the line.

Many players, and even some that get good results, shoot with a staggered stance. However, if you are right handed the staggered stance aims you to the left of the basket because your shoulders are turned. I put both my feet square to the free throw line while straddling the "painter's hole." On wooden gym floors they pound a nail into the floor and draw the top of the key with a pencil on a string. This leaves a small hole dead center on the free throw line. Use this hole to center yourself by putting each foot nine inches to either side of it – 18 inches apart (the width of the basketball hoop). See Fig. 15.3. You will want to shoot from the painter's hole on the free throw line to the painter's hole under the basket. That's not hard to do if you follow the rest of my steps.

*Figure 15.3.* When shooting free throws find the "painter's hole" and straddle it with your feet about 18 inches apart or 9 inches from the centerpoint.

### Step 2. Bounce the ball three times with the inflation hole up.

Bouncing the ball starts the process and clears your mind. Putting your eyes on the inflation hole helps to focus your attention. Why use the inflation hole? Because, as you play in different arenas, maybe in different countries, the inflation hole is the only universal thing on a regulation basketball (some cheapies and knockoffs put the inflation hole to the side). Look at the inflation hole as you bounce the ball *slowly* three times (see Fig. 15.4).

Some players might feel that dribbling the ball quickly or spinning it in their hands will give them rhythm and help them make the shot. After all, they see professional basketball players on television do this sort of thing. But most pros aren't particularly good as role models since many of them make less than 65 percent of their free throws.

*Figure 15.4.* As you bounce the ball, look at the inflation hole. It is the center of the basketball and, focusing on it, will help clear your mind. Furthermore, it is the only universal trait on all basketballs.

## Step 3. Thumb in the channel, third finger pointing at the inflation hole.

Every time I shoot a free throw I put my hand on the ball the exact same way. After the third bounce, I fit my hand on the ball with my thumb in the channel (or groove) and my middle finger in line with the inflation hole. You can use any groove you want as long as your middle finger is in line with the inflation hole (see Fig. 15. 5)

All the hand surgeons I've talked to say the middle finger is the center of your hand. And the center of the hand should be in the center of the ball right in line with the inflation hole. When I put my thumb in the channel, that's my stabilizer or guidance counselor. If you put your thumb in the channel you don't have to worry about spin or rotation – it will happen automatically.

*Figure 15.5.* This is the proper position of the hand while shooting free throws: the thumb in the groove, the middle finger in line with the inflation hole. Players with larger hands can use any groove that fits their hands as long as their middle finger points at the inflation hole.

## Step 4. Elbow in.

This is one of the most important steps. If a player successfully shoots with the elbow in there is a much better chance of the ball traveling from the center of the free throw line to the center of the hoop. Let me put it another way. If you hold the ball correctly, positioned in the center of the body, all you have to do is shoot it straight ahead. If you do that there is a very good chance of the ball going into the basket.

Two coaches came to the gym where I shoot and studied me as I shot 10,000 free throws. Those that didn't swish had 74 % chance of going into the basket providing they were on line. So if they hit the backboard or the rim they would still go in. You can call it "shooter's touch," or a charitable bounce or flat out luck, but a huge part of free throw shooting is putting the ball down the middle. If you shoot the ball straight ahead, down the middle above the rim 74% of the balls will go into the basket.

Some players are like contortionists in every other way, but when I tell them to bring their elbow in they say it's impossible. This is because the current shooting style is with a staggered stance and the elbow stuck out in the wind. Today's player feels unstylish with the elbow in and therefore won't do it. But when Leonardo Da Vinci drew the universal man his painting had a line down the center of the body (see Fig. 15.6). If you don't think you can bring your elbow in, let me ask you this: what do you do when you sit in the driver's seat of your car and reach for the seat belt? See? Your elbow just touched your navel. That's where the elbow should be when the shot begins.

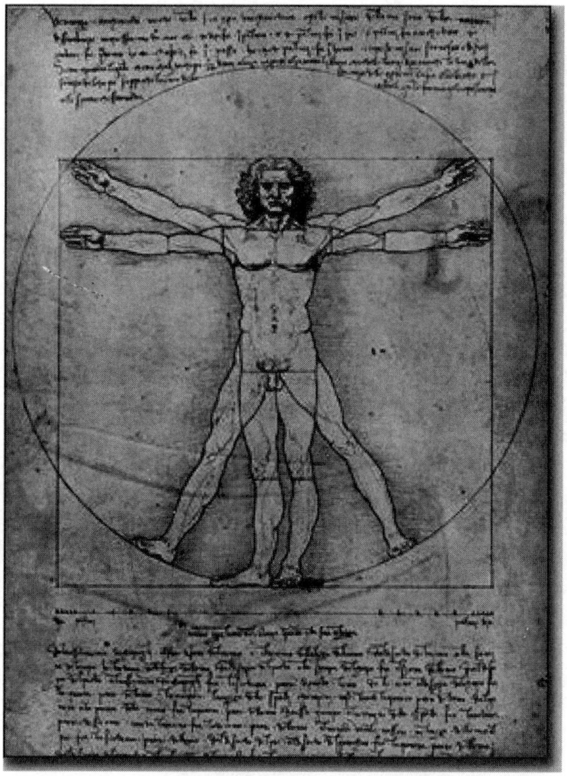

*Figure 15.6.* Leonardo Da Vinci drew the "Universal Man" to show the symmetry of the human body. Using this balanced concept of the body will help players center themselves on the free throw line and keep the ball on target as it travels into the basket.

Just remember, there are only four ways to miss a free throw: right or left, short or long (see Fig. 15.7). If you bring the elbow in (see Fig. 15.8), and go straight at the basket, you've just eliminated missing to the right or left.

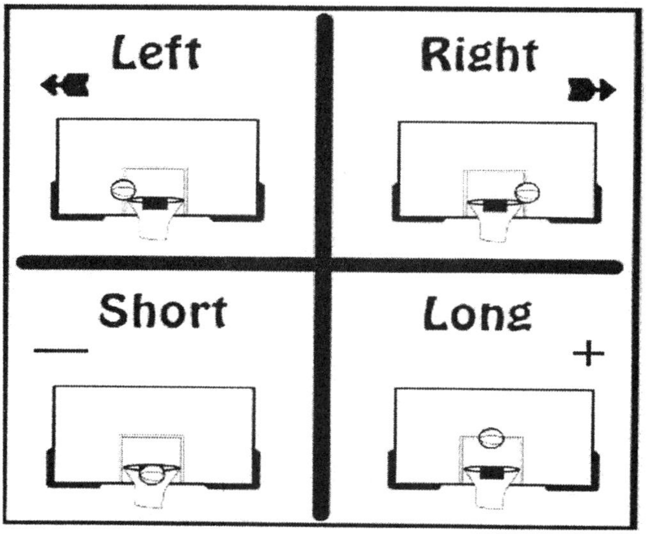


*Figure 15.7.* There are only four ways to miss a basket: left, right, short and long. Bringing your elbow in will eliminate missing to the right and left. Making a good knee bend will give you the proper distance.

*Figure 15.8.* This is the correct position just prior to shooting. Notice that the elbow is tucked in so that when the arm is extended the middle finger of the shooting hand goes directly at the center of the basket.

### Step 5. Bend your knees.

Some coaches say the free throw is an upper body shot. They seem to be afraid of extra motion that might throw off the shooter's aim. But I shoot with my legs contributing a good push to the ball. In fact, a middle school-aged player might not be easily able to reach the hoop without some help from the legs.

Occasionally, players will bend the legs, straighten, then shoot. This leaves all the work to the arm and shoulder. Really, the motion of the legs and arm should be combined. The shot starts in the legs and moves up through the body. This makes it sound a little complicated; it's not. Try it and you see how natural it feels.

### Step 6. Eye on the target.

If you look at nothing, you will hit it. So where do you look when you shoot a free throw? There are several points here that are important. First of all, don't look at the target any earlier than at this point in the process, just before you are ready to shoot. If you stare at the target you risk getting "focal dystonia," which means what you are looking at loses its vividness. Besides, there's no reason to stare at the hoop because there is nothing new to see. You know where the target is. It hasn't changed. To stand frozen at the free throw line, staring at the basket, is likely to amplify your anxiety.

Instead, look at the target just before you shoot, just after you have bounced and positioned your hand on the basketball, and bent your legs. Then, move your eyes up to the target like a lens on a camera. And look at the space above the back rim of the basket. This is the area through which you want to see the ball drop on its way into the hoop.

### Step 7. Shoot and follow through.

All the steps up to this point are preparation. Everything I've told you to do has been to position the body and prepare the mind for this moment. Do you think you can do the preceding steps I have described successfully? If you do these steps, the result is virtually predetermined. The ball is in the center of the body, the legs are ready to give it just the right push. The ball almost has to go where you are pointed.

Some players will want to know if they should snap the wrist. Others will want to know how long to hold the follow through. All I can say is that, once the ball has left your hand, no amount of motion will influence the flight of the ball. Furthermore, if you hand was positioned correctly on the ball, backspin or rotation will automatically be added.

On the other hand, I have to say that if a fancy follow through makes a player do the preceding step correctly, then I'm not against it. They can do anything they want to afterwards if they want to.

## 15.3 Other Factors Influencing the Free Throw

My seven steps integrate all the necessary motions for a successful free throw. However, there are a few related questions that frequently come up. For example, many players want to know about the position of the "off hand" the hand that holds the basketball while the shooting hand provides the push. The off hand is like a golf tee that holds the ball in the right position. One check point, however, is to make sure that the off hand doesn't move too much during the shot. "Shooting straps" are on the market to hold the off hand from moving. I don't think you really have to tie yourself up. But keep an eye on the off hand and make sure it doesn't try to get into the act.

Something I see from many young players is that the off hand creeps up behind the ball, rather than under it. Now they are in a position that looks like they are actually going to block their own free throw. This doesn't make sense; reposition the hand under the ball and don't let it interfere.

Many players want to know how high the arch should be. If I said, 74 percent or 68 percent would you be able to duplicate that? Who's going to come out and measure it for you? True, it's not good to have a flat trajectory. But you want to have just enough loft on the ball to comfortably clear the rim. Shoot it up in the air, out in front of you.

## 15.4 Free Throw Shooting in Game Conditions

Now let's put my seven steps into a real basketball game. You've just been fouled and the referee gives you two shots. You're heading for the charity stripe when you hear a whistle – time out. The thought flashes into your head that the other team is trying to "ice" you, to let you cool off so you will miss the shot. Instead of letting this psyche you out, consider this. A study was done to see how many well players shot free throws after a break from the game action. The result was that free throw performance *improves* as time goes by. So put these thoughts of being "iced" out of your mind.

One of the first thoughts you should have after getting fouled and facing a foul shot is that "free throws are a gift" or "great, here's my chance for an easy couple of points."  It's time to make all your practice pay off. Remain focused and positive as you move to the line. There's nothing in particular you need to think about since the seven steps take care of all that for you.

Yes, you may be nervous. However, as you stand square on the line, and slowly bounce the ball three times, your mind will clear and muscle memory will be triggered. Remember, if you have practiced free throw shooting faithfully, you will have seen the success of your efforts. Now, it's time to put them into action.

## 15.5 CONCLUDING REMARKS

Shooting free throws has been a wonderful thing for me. It's taken me all around the world, opened doors and given me a second career. Free throws can do a lot of you too. It is something you can practice on your own and use at clutch moments of a game. It is one of the few aspects of basketball that you can completely control. So consider the steps I've outlined here, commit them to memory and practice every day. When your time comes, and you find yourself standing on the free throw line, with the game in the balance, every moment you've spent practicing and improving will pay off for you, and for your team.

## 15.6 ACKNOWLEDGMENT

The authors with to thank John Doyle for providing the photographs and graphics.

## 15.7 REFERENCES

Amberry, T., with Reed, P., 1997, *Free Throw; 7 Steps to Success at the Free Throw Line*, HarperCollins, New York.

Gallwey, W. T., 1982, *The Inner Game of Tennis*, Bantam Books, New York.

Herrigel, E., 1999, Zen and the Art of Archery; Random House, London, U.K.

Penick, H., and Shrake, B., 1999, *Harvey Penick's Little Red Book; Lessons and Teachings from a Lifetime of Golf*, Fireside, Wichita, KS.

## 15.8 APPENDIX - BIOGRAPHICAL SKETCHES

Dr. Tom Amberry was born and raised in North Dakota. He served in the United States Navy during World War II and was on a destroyer at Omaha Beach during the D-day invasion. He played basketball while attending the University of North Dakota and, later, was a two-time All-American while playing for Long Beach City College. He was offered a two-year no-cut contract with the then Minneapolis Lakers but turned it down to become a podiatrist. After his retirement in 1992 he began shooting free throws as a hobby. He set the all-time record of 2,750 consecutive free throws on November 15, 1993 at the age of 72. Now, at the age of 81, Dr. Amberry still shoots 500 free throws a day.

Philip Reed studied journalism and creative writing in college while playing on the soccer and tennis teams. After graduating from the University of North Carolina with a B. A. in English, he became a police reporter in Chicago and Denver. He then moved to Los Angeles and began to write plays, screenplays, and books. After reading a newspaper story about Dr. Amberry's Guinness Record, Phil contacted him hoping to learn something that could improve his golf game, since he believed that the skills used for free throw shooting could be successfully applied to golf and other sports. He and Dr. Amberry became friends and then worked together to write *Free Throw; 7 Steps to Success at the Free Throw Line.* In the meantime, Phil applied Dr. Amberry's technique to his own free throw shooting and improved it to the point where his accuracy is currently 93 percent -- and he is still improving.

# VI

# PERFORMANCE AND REHABILTATION

# Chapter 16

# Vision Training and Sports

Kenneth J. Ciuffreda[1] and Bin Wang[2]

[1] *Dept. of Vision Sciences, State University of New York, State College of Optometry, 33 West 42nd St. New York, NY 10036; PH: (212) 780-5132, FX: (212) 780-5124; EM: kciuffreda@sunyopt.edu*

[2] *Dept. of Vision Sciences, State University of New York, State College of Optometry, 33 West 42nd St. New York, NY 10036; PH: (212) 780-5139, FX: (212) 780-5124; EM: bwang@sunyopt.edu*

## 16.1 INTRODUCTION

### 16.1.1 Vision Contribution to Sensory Perception and Sports

The five senses provide us with continual information regarding their on-going status. If all information from each of the senses were processed fully, "sensory overload" would result due to channel capacity limitations. Thus, much of the input information both within and between each sense must be neurologically "filtered" to rid the brain of unwanted and unnecessary aspects via attentional and related/other mechanisms (Welford, 1960; Pashler, 1997) (Figure 16.1). Then, this information is "weighted" with respect to the immediate relevance in relation to the task on hand (Abernethy et al, 1997; Dosher and Lu, 1999), such as batting a baseball (Watts and Bahill, 1990).

What is the relative contribution of each sense, in particular vision, to sports? Is vision dominant? Probably so. Some relevant examples will demonstrate the point. Imagine one is attempting to catch a fly ball, hit a hockey puck, or throw a football. Certainly *gustation* and *olfaction* are not essential, as one does not have to either taste or smell the object of interest to

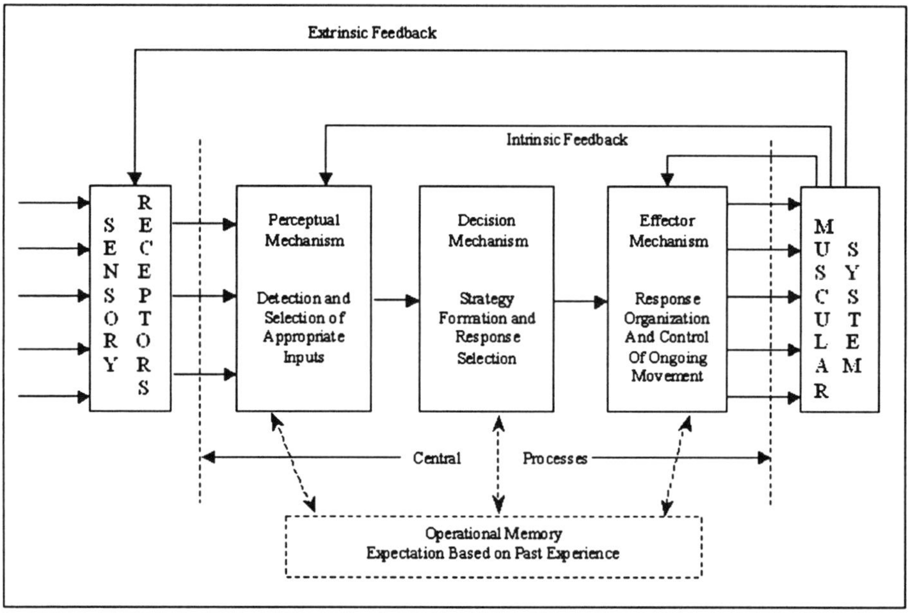


*Figure 16.1.* An information processing model of skilled performance.

identify it, and then to elicit a proper motor response. *Audition* may be helpful to hear instructions from either the coach or one's teammates, detect movement of opposition players beyond one's visual field, etc. *Taction* may also be assistive, for example in gripping a bat more tightly or feeling the exact location of a baseball as it lands in one's glove. However, *none* of these common sports' tasks could be accomplished, at least with any degree of skill, without the sense of *vision*. Thus, as appears to be true for much of our daily experiences and activities (Colavita, 1974; Posner and Nissan, 1976), vision must also be dominant for the wide range of sports-related tasks.

## 16.1.2 Vision-Related Tasks in Sports

The general categories of tasks believed to be important for specific visual demands are presented in Table 16.1 (Getz, 1978; Sherman, 1980; Stine et al, 1982; Seiderman and Schneider, 1983; American Optometric Association, 1984; Martin, 1984; Gregg, 1987; Pizzarello and Haik, 1987; Classe, 1993; Hitzeman and Beckerman, 1993; Kirscher, 1993; Reichow and Stoner, 1993; Berman, 1995; Coffey and Reichow, 1995; Gardner and Sherman, 1995; Loran and MacEwen, 1995; Zagelbaum, 1996; Leslie, 1997a,b; Elliott, 1998). The relative weights are based on our own subjective assessment, as well as that of others (Seiderman and Schneider, 1983; Gardner and Sherman, 1995).

*Table 16.1*

**Relative importance of selected vision functions for common sports in the U.S.**

| | SVA | DVA | CS | VA | SA | EM | ACC | E-H/E-F | VF |
|---|---|---|---|---|---|---|---|---|---|
| Golf | 3 | 1 | 1 | 3 | 5 | 2 | 3 | 5 | 2 |
| Football | 3 | 4 | 1 | 1 | 5 | 5 | 3 | 5 | 5 |
| Baseball | 4 | 5 | 2 | 1 | 5 | 5 | 4 | 5 | 5 |
| Basketball | 4 | 5 | 1 | 1 | 5 | 5 | 3 | 5 | 5 |
| Tennis | 4 | 5 | 2 | 1 | 5 | 5 | 3 | 5 | 3 |
| Soccer | 3 | 5 | 1 | 1 | 4 | 5 | 3 | 5 | 3 |

Symbols: SVA = Static Visual Acuity, DVA = Dynamic Visual Acuity, CS = Contrast Sensitivity, VA = Vernier Acuity, SA = Stereo Acuity, EM = Eye Movement, ACC = Accommodation, E-H/E-F = Eye-Hand coordination/Eye-Foot coordination, and VF = Visual Field. Scoring: 1 = not very important and 5 = very important

There are five major categories directly related to vision (See Appendix 16.1 for vision glossary):

1. Resolving detail: The athlete must be able to detect and resolve the overall form, and sometimes fine details such as surface texture, of a baseball, puck, playing field, etc., to produce a proper response (Welford, 1960). This involves directing the high-resolution fovea of the eye to the object of interest, and then activating the accommodative system to focus accurately for that distance. This would be true for both suprathreshold and threshold target contrasts.

2. Estimating depth: One must be able to determine the absolute depth of people or objects in the playing field, for example to gauge the force necessary to throw a football to a receiver forty yards away. The athlete must also have an accurate sense of relative depth based on the multitude of both binocular (e.g., stereopsis) and monocular (e.g., object overlap) cues to depth in the playing field (Ciuffreda, 2002; Ciuffreda and Engber, 2002).

3. Tracking moving objects: The object of interest, such as a football or baseball, must be accurately tracked throughout its trajectory by the eye movement system (Ciuffreda and Tannen, 1995), frequently with concurrent head and body movements (Leigh and Zee, 1999). This involves both oculomotor subsystems, namely version and vergence (Tables 16.2 and 16.3; Figure 16.2). The process also involves the accommodative system to attain and maintain accurate focus on the target as it is being followed along its trajectory.

| *Table 16.2* Versional eye movements | | |
|---|---|---|
| **Subsystem** | **Stimulus** | **Function** |
| Fixational | Stationary target | To stabilize a target onto the fovea |
| Saccadic | Step of target displacement | To acquire an eccentric target onto the fovea |
| Pursuit | Target velocity | To match eye velocity with target velocity to stabilize the retinal image |
| Optokinetic | Target or field velocity | To maintain a stable retinal image during sustained head movement |
| Vestibular | Head acceleration | To maintain a stable retinal image with the target on the fovea during transient head movement |

| *Table 16.3* Vergence eye movements | |
|---|---|
| **Subsystem** | **Stimulus** |
| Disparity (or fusional) | Target disparity |
| Accommodative | Target blur |
| Proximal | Apparent nearness or perceived distance of target |
| Tonic | Baseline neural innervation (midbrain) |

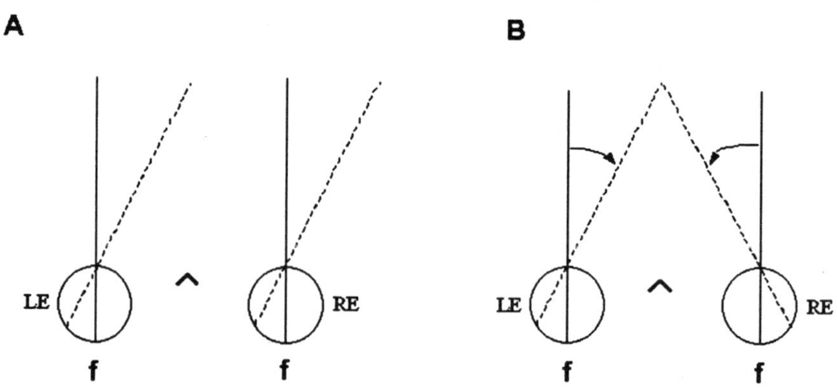


*Figure 16.2.* (A) Schematic drawing of versional eye movements used to track objects moving laterally (e.g., left or right), and (B) vergence eye movements used to track objects moving in depth (e.g., inward). LE, left eye; RE, right eye; f, fovea.

4. Visuomotor integration: Related to the above, the athlete must employ complex visuomotor commands involving synchronization and coordination of various body parts, such as the eye and hand in baseball, eye and foot in soccer, etc., with respect to the task of catching or hitting the ball, or other targeted objects (Abernethy et al, 1997). Frequently, this is performed while concurrently running, jumping, etc., which further complicates the dynamic task demands.

5. Visual information processing: The athlete must be able to detect, differentiate, and process the relevant stimuli across the entire visual field. This involves the processes of selective and simultaneous attention (Pashler, 1997), for example looking at the basketball hoop in preparation for a shot to be made, while concurrently attending in the periphery to the opposing team's players who may be preparing to block the ball. However, other higher-level cognitive factors are also important in sports (e.g., prediction, recall, visual imagery, etc.), as will be discussed in Section V.

Since vision is the dominant sensory modality in sports and is integral in many specific sports-related tasks, understanding the visual system and its functional capabilities and plasticity in athletes is of particular relevance. Hence, in this chapter, we will answer three important questions:

- Can normal vision function be enhanced through specific laboratory-based vision training paradigms?
- Do experienced athletes possess unique visual attributes?
- Following successful training, can the athlete's enhanced vision function translate/transfer into improved specific sports-related skills and performance in the playing field?

## 16.2 TRAINING NORMAL VISION FUNCTION

Over the past several decades, there have been numerous studies involving the training, or enhancement, of normal vision function. The results are nearly unanimous --- most vision functions can be improved by specific laboratory-based vision training paradigms (i.e., repeated practice of some highly specific task). Much of this has revolved around sensory/perceptual aspects (e.g., visual acuity, stereoacuity, etc.), where typically thousands of trials are required to produce a significant and sustained improvement (Goldstone, 1998; Rosenbaum et al, 2001). However, some have also involved motor aspects (e.g., saccadic adaptation, Semmlow, et al., 1989), where typically only hundreds of trials are required to produce a significant and sustained effect.

In Table 16.4, pertinent results are summarized for selected studies in each of the above two main areas. In general, positive training effects ranged from

***Table 16.4***
**Training effect of various vision functions for visually-normal individuals in selected studies**

| Vision Function | | Training Effect ( % relative improvement) | Amount of Training |
|---|---|---|---|
| Visual Acuity | Static Visual Acuity [1] *[1-6]* | 4% at the fovea; 40% in the periphery (20° to 60°) | 11 test sessions (1.5hr/session) |
| | Dynamic Visual Acuity [2] *[7,8]* | Average 30%; effect greater for stimuli with higher velocities and shorter durations | 4 training sessions (30min/session) |
| Contrast Perception | Threshold (contrast sensitivity) [3] | 50%, 45%, and 80% for low, medium, and high range spatial frequencies, respectively | 18 months of nearly daily practice |
| | Suprathreshold (contrast discrimination) [4] | 100% and 70% for reference contrasts of 50% and 25%, respectively; training effect lasting at least 8 months | 3 phase training cycle (3 days /cycle, 500-1,000 trials /day) |
| Vernier Acuity [5] *[1,9]* | | 60% | Over 10,000 trials |
| Stereoacuity [6] *[10,11]* | | 50% at the fovea; 60-80% in the periphery (2.5° and 5°) | 20,000-25,000 trials |
| Eye Movement | Saccade (adaptation) [7] *[12,13]* | Compensation: 25% for modifications that increase amplitude; 70% for modifications that decrease amplitude | 50-100 trials |
| | Pursuit [8] *[14-17]* | Velocity gain: 20% for 30 deg/sec; 30% for 40 deg/sec; training effect: greater for higher speed target, with effect lasting several months | 20-30 min |
| | Vergence [9] *[19-21]* | Convergence: 130-195% Divergence: 10-15%; with training effect lasting at least 21 weeks | 150 min total over 3 weeks |
| Accommodation [10] *[21-27]* | | Velocity 18%, gain 30%, phase lag 28% | 7 days (140-210 min/day) |

Numbers in square brackets refer to specific citations in the References used to construct the Table. Numbers in italicized brackets refer to additional citations in the area.

10-200%, with a mean of approximately 40%. These values also reflect the general trends of related studies in much of the literature in the field (see Table 16.4). The results for two sensory/perceptual functions and two motor functions are presented in Figure 16.3. In Figure 16.3A, the results are shown for static visual acuity in the central fovea (0°) and in the retinal periphery (20-60°) (Johnson and Leibowitz, 1979). In the periphery, improvements were

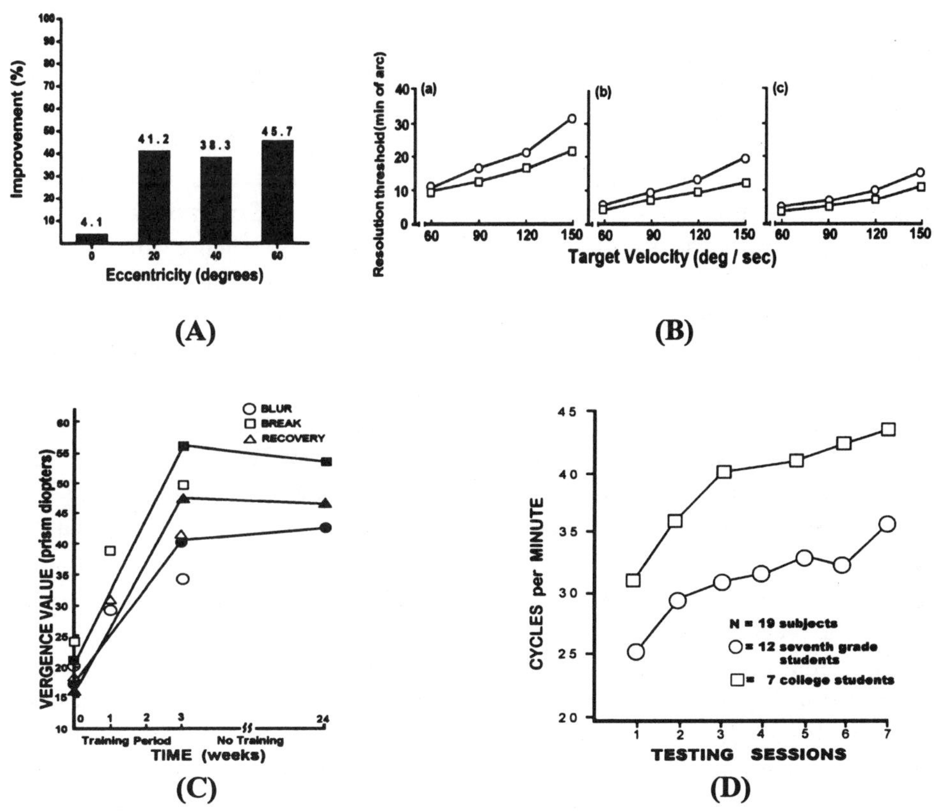


*Figure 16.3.* (A) Percent improvement in static visual acuity as a function of retinal eccentricity. (B) Mean resolution thresholds in pretraining (circles) and posttraining (squares) sessions as a function of target velocity in the (a) 200ms, (b) 400ms, and (c) 600ms duration conditions. In all cases, the standard error for a particular data point is not larger than the symbol defining that point. (C) Changes in the relative convergence vergence values at near. The open symbols represent the mean for 35 subjects. The changes after 1 week are computed from a subpopulation of 23 subjects. The filled symbols represent the mean vergence range values for a subpopulation of 12 subjects. (D) The result of fourteen training sessions on proximal-distal response time in cycles per minute. Mean cycles per minute for the standard distance rock test are plotted on the vertical axis. Numbers 1 through 7 on the horizontal axis indicate the testing sessions which were interspersed between daily training sessions. Pre- and post-training responses are shown by items marked 1 and 7.

approximately 40%. This probably reflected learning to shift one's visual attentional focus from the fovea to the retinal periphery, as well as some degree of perceptual learning. In contrast, at the central foveal region, which is that retinal region involved in the specification of one's visual acuity (e.g., 20/20 vision), there was a notable exception. The training effect was only 4%. This might reflect the very high level of visual attention and perceptual learning habitually occurring in the central foveal region. Hence, the small improvement may reflect a saturation effect. In Figure 16.3B, results are presented for dynamic visual acuity (Long and Riggs, 1991), another sensory/perceptual task believed to be very important in many sports. Improvements were found for all velocities of target movement, but more so at the highest target velocity and shortest stimulus duration. This result is important, as it demonstrates that practice/training effects may only be evident, or at least largest, for the most challenging stimuli/tasks used in vision training and subsequent vision testing protocol. For nonchallenging stimuli/tasks, response saturation may be present, and hence little if any training effect will be evident. These two sensory-based results suggest that resolution of detail in the visual field for both stationary and moving targets can be improved following systematic practice/training. In Figure 16.3C, motor vergence range findings are presented (Daum, 1982). The range of improvements was very large and rapid, and furthermore these effects were sustained for many weeks. And, in Figure 16.3D, results for the accommodative motor system are presented (Haynes and McWilliams, 1979). There was a rapid and progressive improvement in overall dynamic responsivity over the 7 day training period. These two motor-based results suggest that tracking of objects moving in depth can be improved following systematic practice/training.

The clinical results in one laboratory-based study involving cricket are presented in Figure 16.4 for eye movements and accommodation (West and Bressan, 1996). Improvements ranged from approximately 25-45% for these two basic oculomotor skills. These clinical findings were in general agreement with the more stringent laboratory results described above (Table 16.4). Furthermore, these positive clinical findings were in agreement with the post-training improvements found for the specific sports-related vision task used in cricket.

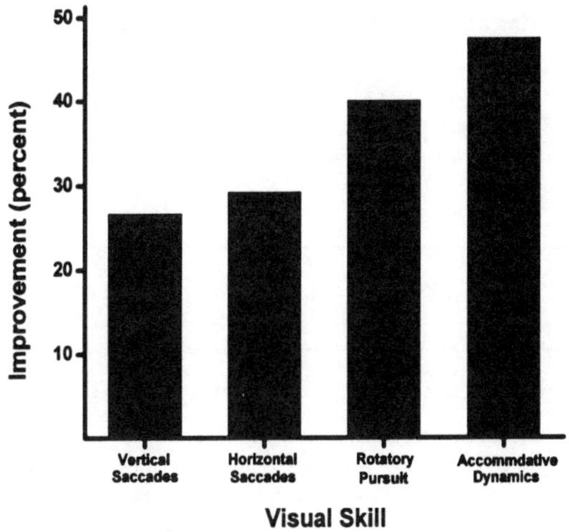


*Figure 16.4.* Category of visual skill versus percent improvement level for clinical vision training.

## 16.3 UNIQUE VISUAL ATTRIBUTES OF ATHLETES

Are there unique visual attributes that discriminate between athletes and non-athletes, or experienced/skilled athletes from inexperienced/unskilled athletes? Are their basic vision skills well above average? Probably so, based on several clinical and laboratory-based studies. Some of the positive findings are summarized in Table 16.5. These enhanced attributes encompassed the range of sensory, motor, and perceptual aspects of basic vision and visual information processing. While negative results have also been found (see Abernethy, 1986 for a review; McLeod, 1987; Abernethy et al, 1998; Abernethy and Neal, 1999), they are in the minority.

Some examples of visual attribute differences between highly experienced and novice athletes are presented in Figures 16.5 and 16.6. In Figure 16.5A, the two-dimensional (i.e., horizontal and vertical) eye positions of elite and novice precision shooters are presented (Di Russo et al, 2003). The task was to fixate carefully upon a target through a rifle sight, while eye position was recorded objectively. The better the fixational ability, the smaller the scatter of points about the x, y target intersection. In the simple, non-distraction condition, the eye position scatter was only slightly greater and biased upward (i.e., less accurate) in the novice. However, with visual distractions present, the novice's response became considerably more variable and inaccurate, whereas that of the elite shooter remained essentially unchanged.

| *Table 16.5*<br>**Enhanced visual attributes of athletes** | |
|---|---|
| Visual Resolution | Dynamic Visual Acuity [11-14], Static Visual Acuity [15, 16], and Contrast Sensitivity [16, 17-19] |
| Depth Perception | Stereopsis [20-26] |
| Visual Tracking | Vergence [26, 28], Pursuit [27], Saccades [28-30], and Fixation [30] |
| Visuomotor Integration | Eye-Hand Coordination [29, 32], Visual Reaction Time [25, 28, 29, 31, 33] |
| Visual Information Processing | Visual Field [33], Simultaneous Vision [20], Peripheral Awareness [28, 29], Peripheral Field [25, 32], Peripheral Vision [35], Speed of Recognition [36], Field of Motion Perception [34], and Movement Peripheral Limits [37] |

Numbers in square brackets refer to specific citations in the References.

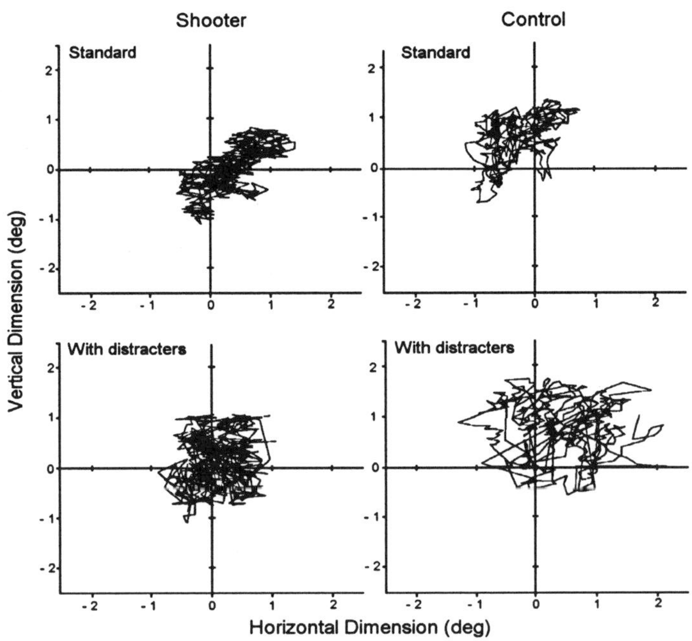


*Figure 16.5.* (A) Eye position during 60-s fixation task in one shooter (left) and in one control subject (right) in the standard condition (top) and when distracters were flashed in the parafovea (bottom).

These results are further quantified in Figure 16.5B where, in addition, the increased positional error over time found only in the novices suggested fatigue and/or attentional decay. This result is important, as it demonstrates that such differences may be very subtle and highly specific in nature. Furthermore, it confirms and reinforces the important role of visual attention in sports. Similar results are presented in Figure 16.6 for dynamic visual acuity (Ishigaki and Miyao, 1993). Only for the smallest targets and hence more difficult stimulus conditions do these two groups exhibit divergent responses. Again, this suggests presence of subtle differences that require more demanding test paradigms to become manifest.

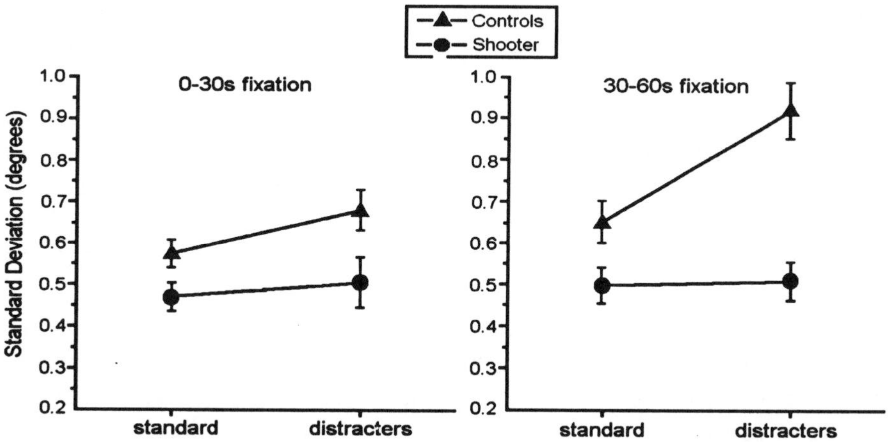


*Figure 16.5.* (B) Data showing eye variability shift around the fixation point (fixation stability) for both subject groups during the first and second 30-s interval. The vertical bars indicate the standard deviations.

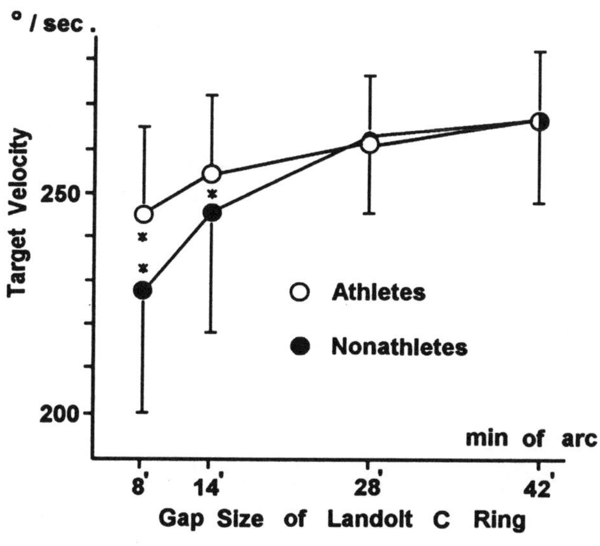


*Figure 16.6.* Comparison of the dynamic visual acuities of the athletes and the nonathletes (minutes of arc gap in the target versus velocity). * = a statistically significant effect ($p<0.05$).

## 16.4 TRAINING TRANSFER TO SPECIFIC SPORTS SKILLS AND COMPETITIVE FIELD PERFORMANCE

### 16.4.1 Why Train Vision Enhancement in Athletes?

There are at least three reasons why one might consider specific and targeted vision training in athletes. First, it has been demonstrated that many aspects of vision can be trained in the laboratory environment. Thus, one might hope that any improvement would transfer to the athlete's specific sports-related skills, as well as actual performance in the playing field, at least in those aspects where such a correlation might be expected. For example, since stereoscopically-based binocular depth perception can be improved, then theoretically it seems plausible to predict that one's ability to execute a long pass successfully with a football might also be improved, as stereopsis operates in a rangefinder manner to enhance one's fine binocular depth discrimination, as well as global relative and absolute depth assessment. Even if there is a genetic basis for their stellar vision, this fact does not preclude further enhancement by specific vision training paradigms. Second, over the years, the athlete may have been unknowingly performing "self-training" by executing thousands of repetitions of some specific related task that effectively conditions a visual attribute, such as smooth pursuit eye movements or accommodation. Assuming absence of a saturation phenomenon, subsequent formal and targeted training of specific components would simply be an extension of their earlier self-training. Such improvement occurs via the processes of perceptual (Goldstone, 1998; Dosher and Lu, 1999) and motor (Newell, 1991; Abernethy et al, 1997; Sanes and Donaghue, 2000) learning, as well as other aspects of learning, as will be described later. And, third, since the visual attributes of elite/professional athletes typically cluster well above the norm as described earlier in Section III, even a small improvement might translate into a relatively large effect or gain in one's sport's performance and effectiveness. For example, a 50 msec reduction in reaction time, heightened peripheral awareness, or time optimization of the arm's trajectory, might be sufficient to block the winning basketball at the end of the fourth quarter in a championship game.

### 16.4.2 Skilled Task and Performance Transfer

Thus far, we have answered two of the three critical questions posed at the beginning of this chapter. First, a full range of static and dynamic, sensory/perceptual and motor, and monocular and binocular vision functions can be trained. This process involves perceptual and motor learning, which may require up to several thousands of trials for enhancement to occur. Second, many studies demonstrate that skilled athletes possess better (i.e., more acute and accurate) vision functions over a range of conventional tests/tasks than do either non-athletes or novice athletes.

Given the positive responses to the first two questions, the answer to the third question becomes even more important --- do the positive effects of either laboratory-based or clinically-based vision training transfer to specific sports-related, visually-based skills, and more importantly to actual field performance? The results are reasonably positive.

### *Positive Findings*

The positive findings relate to a range of sports (Table 16.6). Since there have been relatively few studies conducted, each will be considered briefly.

***Table 16.6***

**Vision training sports-related task transfer**

| Sport | Task Transfer | |
|---|---|---|
| | **Yes** | **No** |
| Shooting | (elite) [38] | (novice) [39] |
| Basketball | [40] | |
| Cricket | [41] | |
| Soccer | [42] | |
| Tennis | | [43,44] |

Numbers in square brackets refer to specific citations in the References.

The first sport is precision shooting, where acute vision is surely critical. Before and after an unspecified vision training program (approximately 20 hours over a 3 month period in 11 subjects) to enhance the visual skills presumably involved in precision shooting in highly experienced Olympic team members, vision function and shooting performance were tested (Quevedo i Junyent and Sole i Forto, 1995). Most basic vision functions showed statistically significant improvements. More importantly, shooting performance also exhibited statistically significant improvement immediately after training. Performance progressed steadily during the course of training, and the effects were greatest in those with the initially lowest scores. However, shooting performance regressed at the 3 month follow-up, therefore suggesting the need for a longer program for full training-related consolidation effects to occur. Use of a larger sample size and incorporation of both placebo and control groups would have strengthened the study.

Later, this same group investigated novices comprised of first-year university students just beginning to learn precision shooting and the related skills (Quevedo et al, 1999). The experimental group (n = 37) received both shooting training (~1700 hours) and vision training (~7.5 hours total over 9 weeks) incorporating visual acuity, saccadic tracking, and dynamic accommodative tracking exercises. The control group (n = 34) received both shooting training (~1700 hrs) and lectures on psychological training techniques (~7.5 hours); no vision training was performed. Pre-/post testing included visual acuity, combined saccadic/accommodative tracking, and visual search. Following training, improvements in shooting scores were statistically equivalent in both groups, as were the saccadic/accommodative and visual search measures; only visual acuity improved more in the experimental group. When taken in conjunction with their earlier positive results, these findings suggested that vision training would be beneficial in elite versus novice athletes, in agreement with Wood and Abernethy (1997), in which case improvement in the finer visual aspects and technique components may now predominate.

Similar positive results have been found for the sports of cricket, soccer, basketball, and volleyball. In the cricket study (West and Bressan, 1996), there were two experimental groups of cricket players: one (n = 12) received general vision training incorporating seven basic visual skills, and one (n = 12) received selected vision training involving only three basic visual skills, with each skill assumed to be most relevant to the sport. In both cases, the individuals received daily training sessions for three weeks. Training in the first group involved eye movements (version and vergence) and accommodative dynamics, whereas the second group did not include version. The control group did not receive any vision training. Testing involved pursuit and saccadic eye movements, and accommodative dynamics, as well as the specific sports-related skill of "judging length-of-ball" (i.e., predicting the zone in which the cricket ball would land). Following training, both experimental groups showed significant improvement in the specific sports-related skill (Figure 16.7), thus demonstrating a positive training effect. The control group did not. Regarding vision function, the first experimental group improved in all areas, whereas the second only improved their accommodative dynamics. Since both groups improved equally well in task transfer to the specific sports-related skill, the results suggested that the improvement in accommodation may have been the critical visual factor. Use of a larger sample size in all groups would have strengthened the study. In a basketball

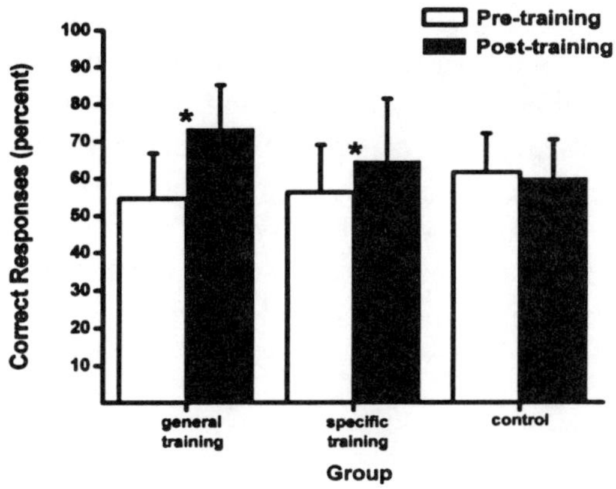


*Figure 16.7.* Experimental group versus percent correct responses before and after training. * = a statistically significant effect ($p<0.05$).

study (Kofsky and Starfield, 1988), six experienced college level players received general vision training (static and dynamic visual acuity, visual reaction time, peripheral awareness, eye-hand coordination, and visualization/visual imagery). Twenty sessions of twenty minutes duration over a five week period were given. There was a control group (n = 6), which did not receive any vision training. Actual game performance was assessed over a range of parameters (i.e., basic shots, steals, rebounds, foul shots, fouls, assists, and blocks). Most vision functions improved in the experimental but not the control group. More importantly, actual game performance improved more so across parameters (e.g., number of rebounds) in the experimental versus control group. Thus, both basic visual skills and actual field performance exhibited larger improvements in the group receiving vision training. Use of a larger sample size and more detailed statistical analysis would have strengthened the study. And, in a group (n = 18) of varsity soccer players (McLeod, 1991), a commercially-available vision training program ("Eyerobics") (Revien, 1987) was used (12 sessions of 18-38 minutes each), which incorporated basic versional, vergence, accommodative, and visualization skills. Two specific soccer-related transfer skills were tested (i.e., the Dribble Test and the Wall-Volley Test). Significant improvement was found in the former, which demanded continuous visual feedback, but not in the latter, which only demanded intermittent visual feedback. These findings suggested that vision training would be most beneficial for sports-related tasks involving a dominate visual component. Inclusion of a control/placebo group would have strengthened the study. Lastly, in a group (n = 8) of female varsity volleyball players, eye-hand coordination training was performed with the

Wayne Saccadic Fixator (Sexton-Radek et al, 1998). Training was conducted over a 14 week period. It was performed once per week with the Wayne Saccadic Fixator, which conditioned eye-hand coordination and saccadic timing/accuracy (Leslie 1996b), and six times with an on-court task, which involved prediction of ball-landing. Testing was conducted before and after the training period. There were statistically significant improvements in the various subcomponent measures with the Wayne Saccadic Fixator, and marginal improvements in the on-court task. Use of a larger sample size, more test parameters, and inclusion of a control/placebo group would have strengthened the study.

### *Negative Findings*

The primary negative findings have been by the Abernethy group (Abernethy, 1986; Wood and Abernethy, 1997; Abernethy et al, 1998; Abernethy and Wood, 2001) (Table 16.6). This involved two primary investigations, with emphasis on racket sports (e.g., tennis). In the first (Wood and Abernethy, 1997), thirty inexperienced subjects ages 16-35 years were tested. There was an extensive test battery of general optometric tests (Benjamin, 1998; Grosvenor, 2002) (e.g., static and dynamic visual acuity, eye movements, stereopsis, etc.), sports-specific perceptual tests (i.e., coincidence timing, rapid ball detection, and anticipation), and sports-specific motor tasks (i.e., tennis forehand drive accuracy), with all being tested before and after training. There were three groups with 10 subjects in each: (1) vision training experimental group (four 20 minute sessions per week for four weeks; 5.5 hours total), with the training mirroring the test areas described above using conventional optometric training paradigms (i.e., eye movements, accommodation, stereopsis, and peripheral awareness) (Scheiman and Wick, 2002), (2) placebo group which performed the reading of tennis instructional manuals and watched instructional tennis videos for the same amount of time, and (3) control group, which only practiced the weekly 20 minute motor task. For virtually all tasks across the three groups, the improvements were statistically similar, thus suggesting lack of any unique effect resulting from the vision training per se. They attributed lack of any true training effect to the training tasks not having "program situation specificity", a basic training tenet for motor skill acquisition (Henry, 1961). Thus, the vision training was too general. Other factors such as prediction/anticipation, recall, and cognitive strategy were suggested to also be of importance (Abernethy, 1986). And, in a second study (Abernethy and Wood, 2001), the overall paradigm was similar to their earlier study (Wood and Abernethy, 1997), except that each of the two experimental groups received vision training in conventional areas using commercially-available, video-based programs with reported positive clinical results, namely either "Eyerobics" (Revien, 1987) or "Sports Vision" (Revien and Gabor, 1981). Once again, the findings suggested lack of a true and specific vision training effect in all assessed areas. Use of a larger sample size (about 30) in each group, testing of a wider range of specific sports-related tasks, testing of both novice and elite athletes, and assessment of actual field performance would have strengthened these two studies.

## 16.5 CONCLUSIONS AND FUTURE DIRECTIONS

Knowledge in this most interesting and important area of "sports vision" remains incomplete. Absence of critical information in part contributes to the lack of closure related to the efficacy of sports-related vision training (Stein et al, 1989; Wood and Abernethy, 1997; Abernethy and Wood, 2001). How can one begin to understand the area better, and therefore have optometrists and others render the appropriate level of sports vision care? At one end of the spectrum is the conventional comprehensive vision examination to assure that there are no readily correctable impediments either to clarity of vision (e.g., the need for a more accurate spectacle correction) (Benjamin, 1998) or binocular vision (e.g., vergence/accommodative dysfunctions remediated by conventional vision therapy) (Ciuffreda, 2002). No one would argue with this level of vision care, as it is most basic. At the next level, one might use contact lenses versus spectacles for safety reasons, as well as for increasing one's clear field of vision (Sherman, 1981; Coffey and Reichow, 1990; Benjamin, 1998). Other suggestions might include the use of a nylon flexible "sports" spectacle frame to prevent injury upon impact (Gregg, 1987; Benjamin, 1998), or even refractive surgery to remove the potential need for optical correction (Grosvenor, 2002). This too would be met with little argument. At the other end of the spectrum is sports-related vision training, the main topic of this chapter. Based on the above, the following comments and suggestions are both important and relevant.

### 16.5.1 Learning: Perceptual, Motor, and Other Types

Learning how to perform a specific task, such as hitting a rapidly-moving hockey puck, involves perceptual and motor learning (Welford, 1960; Abernethy et al, 1997, 1998; Ciuffreda, 2002), as well as other forms of learning (Gould and Damarjian, 1998; Abernethy et al, 1998) (Figure 16.1). These will be described below.

Perceptual learning refers to the ability to improve on a specific sensory/perceptual task with practice (beyond simple task familiarity), such as judging when two very fine objects are aligned (i.e., vernier acuity) (Fahle and Edelman, 1993). Typically, this requires thousands of trials distributed over a period of days, weeks, or even months to obtain consistent, sustained, and significant levels of improvement. The mechanism is frequently believed to involve a reweighting of sensory channel inputs biased in favor of the more relevant information (Dosher and Lu, 1999). Such training is highly specific with respect to stimulus aspects such as direction, orientation, visual field location, etc. (Goldstone, 1998; Dosher and Lu, 1999). Thus, by its nature it demands repeated trials under a full range of conditions for global attribute enhancement and the potential generalization to the specific sports-related skill, as well as actual sports performance in the field.

Motor learning refers to "the process that improves motor performance through practice" (Lisberger, 1988). This is believed to occur by increasing synaptic efficiency in a Hebbian neural-network manner (Lisberger, 1988). This practice effect follows a power function, which is optimized using distributed rather than massed training trials (Keele, 1986; Abernethy et al, 1998). Motor learning involves three stages (Abernethy et al, 1997; Ciuffreda, 2002): determining by a trial-and-error process the optimal motor program to accomplish a particular task, repeated practicing of the optimal motor program/movement for more rapid and precise execution, and finally attaining a level of automaticity such that the movement essentially becomes "pre-programmed" or "open-loop" in its execution, thereby allowing attention to be allocated to other related tasks. Error-related feedback specific for the motor task and optimization of performance is crucial to the learning process (Lisberger, 1988; Gottlieb et al, 1988). The proposed oculomotor research training paradigm follows the basic tenets of neuroplasticity, and more specifically adaptive or learning-based brain plasticity (Gottlieb et al, 1998; Abernethy et al, 1997):

- it incorporates attended to, highly specific, goal-directed, repetitive, behaviorally-based training activities with feedback related to performance.
- it involves unimodal, as well as intermodal, conditioning and strengthening of responses.
- it conceptually employs Hebbian-like neural adaptive network mechanisms of strengthening, and synchronizing, input coincidence-based connections (i.e., synapses).
- the above three factors allow for residual portions of the different pathways to functionally "self-organize" via neural shaping and remodelling of the underlying cortical cell clusters.

Any clinical or laboratory-based training of either basic vision functions or specific sports-related tasks, with the hope for transfer to the playing field, should carefully follow and subscribe to the above basic tenets of perceptual and motor learning, especially with respect to task specificity and use of a large number of trials for the greatest possible effect. Furthermore, such training might be of more benefit to the elite athlete, for whom basic aspects of the specific sports-skill are well established, and therefore attentional allocation can be shifted to other related domains.

Other forms of learning should also be considered. This would include recall function, visualization/visual imagery, anticipation/prediction, and a range of cognitive strategies/mental skills (Gray, 1990; Abernethy et al, 1998; Gould and Damarjian, 1998; Williams, 2000). In fact, some of these functions have been found to be better in athletes (Abernethy, 1986; Abernethy et al, 1998). In addition, visual attention clearly plays a fundamental role. For example, elite shooters have a narrower attentional focus (Tremayne and Barry, 2001) and are less affected by distractions in the visual field (Di Russo et al, 2003) than novices. In contrast, volleyball players have a broader attentional focus (Pesce-Anzender and Bosel, 1998). Hence, modulation of attention is presumably conditioned based on the visual demands of the specific sport, and hence suggests a highly specific sports-related practice effect (Di Russo et al, 2003). Therefore, visual attentional training (e.g., dynamically shifting or weighting one's visual attentional focus from one region of the visual field to another) should be incorporated into any sports vision training paradigm. Optometrists have been doing this for years in their "peripheral awareness" training (Marrone et al, 1991; Birnbaum et al, 1994). Similar arguments hold true for visual imagery, a common technique used in optometric vision training (Forrest, 1981).

## 16.5.2 Future Research

We have adopted a relatively broad and positive position on the role and application of optometric vision care, in particular vision training, in the area of sports. We have viewed the research results using a global, meta-analysis approach (Glass, 1976). Hence, when the findings and general trends of all studies are considered together, as opposed to that of any single study, the evidence in support of the use of vision training in sports is reasonably good. Thus, we believe that vision training should be continued with strict adherence to the scientific principles underlying the various types of "learning" involved as described in this chapter.

However, further studies are needed. In particular, we suggest that a comprehensive randomized, double-blind clinical trial be conducted, or perhaps even multiple trials for various sports having substantially different basic visual and specific sport task demands (e.g., precision shooting versus hockey). This should include testing and training in a range of simple and more challenging vision functions, attentional aspects, and cognitive/higher-level functions and strategies, as well as more specific sports-related, visually-based tasks with emphasis on those that are most engaging and challenging. This might involve a virtual reality environment to improve potential transfer to the actual field environment. Most importantly, actual competitive field performance in trained athletes must be assessed across a range of parameters. Whenever possible, testing should incorporate objective measures of selected and perhaps more subtle aspects of vision function. These might include accommodative peak velocity, saccadic gain, vergence dynamics (e.g., time constant), etc., using a range of sophisticated modes of quantitative analysis, such as Bode plots and Nyquist functions (Milsum, 1966; Stark, 1968; Bahill, 1981) to assess response dynamics (i.e., gain/phase, and stability, respectively), which would likely not be detected using either basic gross visual analysis or a very simple performance task. And, lastly, in selected subjects, dynamic brain responsivity (e.g., visually-evoked cortical response) and brain imaging (e.g., fMRI) should be assessed. This would be beneficial to gain insight into the neural events and underlying sensorimotor plasticity, as well as the precise neural loci, related to vision and sports performance enhancement.

## 16.6 ACKNOWLEDGMENTS

We thank Drs. Carl Gruning, Leonard J. Press, and Gary Williams, and Marcelline A. Ciuffreda, for their helpful comments on an earlier version of the manuscript.

## 16.7 REFERENCES*

Abernethy, B., 1986, Enhancing sports performance through clinical and experimental optometry, *Clin. Exp. Optom.* **69**: 189-196.

Abernethy, B., Kipper, V., and Mackinnon, L. T., 1997, *The Biophysical Foundations of Human Movement,* Human Kinetics Pub, Champaign, Ill.

Abernethy, B., and Neal, R. J., 1999, Visual characteristics of clay target shooters, *J. Sci. Med. Sport.* **2**: 1-19.

Abernethy, B., Wann, J., and Parks, S., 1998, Training perceptual-motor skills for sport, in *Training in Sport: Applying Sport Science*, Elliott, B. ed., John Wiley & Sons, New York, pp. 1-68.

Abernethy, B., and Wood, J. M., 2001, Do generalized visual training programs for sport really work? An experimental investigation, *J. Sports Sci.* **19**: 203-222. [43]

Adini, Y., Sagi, D., and Tsodyks, M., 2002, Context-enabled learning in the human visual system, *Nature.* **415**: 790-793. [4]

American Optometric Association, 1984,1985,1988,1993, Sports Vision Section, *Sports Vision Guidebook*, St. Louis.

Bahill, A. T., 1981, *Bioengineering: Biomedical, Medical and Clinical Engineering*, Prentice Hall, Englewood Cliffs, NJ.

Bahill, A. T., and LaRitz, T., 1984, Why can't batters keep their eyes on the ball?, *Am. Scientist.* **73**: 249-253. [27]

Bailey, I. L., Sheedy, J. E., and Fleming, D. P., 1988, Dynamic visual acuity: rotational measurement (abstract), *Invest. Ophthal. Vis. Sci.* **28**(suppl): 78. [11]

Barnes, G. R., and Schmid, A. M., 2002, Sequence learning in human ocular smooth pursuit, *Exp. Brain Res.* **144**: 322-335. *[14]*

Beard, B. L., Levi, D. M., and Reich, L. N., 1995, Perceptual learning in parafoveal vision, *Vision Res.* **35**: 1679-1690. *[1]*

Benjamin, W. J., 1998, *Borish's Clinical Refraction*, W.B. Saunders Company, Philadel., PA.

Berman, A., 1995, Setting up a sports vision practice, in *Sports Vision*, Loran, D.F.C. and MacEwen, C. J., eds, Butterworth-Heinemann, Boston, pp. 178-188.

Birnbaum, M. H., Ciuffreda, K. J., and Tannen, B., 1994, Peripheral awareness in the treatment of an unusual neuromotor vision disorder, *J. Optom. Vis. Dev.* **25**: 253-258.

Boman, D. K., and Hotson, J. R., 1987, Smooth pursuit training and disruption, *Neuroophthalmology.* **7**: 185-194. [8]

Buckellew, W. F., 1954, Peripheral perception and reaction time of athletes and non-athletes, *Thesis*, University of Illinois. [31]

Christenson, G. N., and Winkelstein, A. M., 1988, Visual skills of athletes versus nonathletes: development of a sports vision testing battery, *J. Am. Optom. Assoc.* **59**: 666-675. [28]

Ciuffreda, K. J., 2002, The scientific basis for and efficacy of optometric vision therapy in nonstrabismic accommodative and vergence disorders, *Optometry.* **73**: 735-762.

Ciuffreda, K. J., 2002, Why two eyes?, J. Behav. Optom. 13: 35-37.

Ciuffreda, K. J., and Engber, K., 2002, Is one eye better than two when viewing pictorial art?, *Leonardo.* **35**: 37-40.

Ciuffreda, K. J., Levi, D. M., and Selenow, A., 1991, *Amblyopia: Basic and Clinical Aspects*, Butterworth-Heinemann, Boston.

Ciuffreda, K. J., and Tannen, B., 1995, *Eye Movement Basics for the Clinician*, Mosby Yearbook, St. Louis.

Coffey, B., and Reichow, A. W., 1989, Athletes vs non-athletes: Static visual acuity, contrast sensitivity, dynamic visual acuity (abstract), *Invest. Ophthal. Vis. Sci.* **30** (suppl): 517. [16]

** Numbers in square and italicized brackets correspond to items listed in Tables 16.4 to 16.6.*

Coffey, B., and Reichow, A. W., 1990, Athletes vs non-athletes: 6m vergence ranges, accommodative-vergence facility, and 6m speed of stereopsis(abstract), *Optom. Vis. Sci.* **67**(suppl): 81. [26]

Coffey, B., and Reichow, A., 1995, Visual performance enhancement in sports optometry, in *Sports Vision*, Loran, D. F. C. and MacEwen, C. J., eds, Butterworth-Heinemann, Boston.

Colavita, F. B., 1974, Human sensory dominance, *Percept. Psychophys.* **16**: 409-412.

Collins, F. L. Jr., Epstein, L. H., and Hannay, H. J., 1981, A component analysis of an operant training program for improving visual acuity in myopic students, *Behav. Ther.* **12**: 692-701. *[2]*

Cornsweet, T. N., and Crane, H. D., 1973, Training the visual accommodation system, *Vision Res.* **13**: 713-715. *[21]*

Daum, K. M., 1982, The course and effect of visual training on the vergence system, *Am. J. Optom. Physiol. Opt.* **59**: 223-227. [9]

Daum, K. M., 1983, A comparison of the results of tonic and phasic vergence training, *Am. J. Optom. Physiol. Opt.* **60**: 769-775. *[19]*

Daum, K. M., 1986, Negative vergence training in humans, *Am. J. Optom. Physiol. Opt.* **63**: 487-496. *[20]*

Daum, K. M., 1986, Double-blind placebo-controlled examination of timing effects in the training of positive vergences, *Am. J. Optom. Physiol. Opt.* **63**: 807-812. *[21]*

De Valois, K. K., 1977, Spatial frequency adaptation can enhance contrast sensitivity, *Vision Res.* **17**: 1057-1065. [3]

Di Russo, F., Pitzalis, S., and Spinelli, D., 2003, Fixation stability and saccadic latency in elite shooters, *Vision Res.* **43**: 1837-1845. [30]

Elliot, B. (ed.), 1998, *Training in Sport: Applying Sports Science*, John Wiley and Sons, New York.

Fahle, M., and Edelman, S., 1993, Long-term learning in vernier acuity: effects of stimulus orientation, range and of feedback, *Vision Res.* **33**: 397-412. [5]

Fendick, M., and Westheimer, G., 1983, Effects of practice and the separation of test targets on foveal and peripheral stereoacuity, *Vision Res.* **23**: 145-150. [6]

Forrest, E. B., 1981, *Visual Imagery: An Optometric Approach*, OEP Foundation Press, Duncan, OK.

Fujita, M., Amagai, A., Minakawa, F., and Aoki, M., 2002, Selective and delay adaptation of human saccades, *Cogn. Brain Res.* **13**: 41-52. *[12]*

Fukushima, K., Tanaka, M., Suzuki, Y., Fukushima, J., and Yoshida, T., 1996, Adaptive changes in human smooth pursuit eye movement, *Neurosci. Res.* **25**: 391-398. *[15]*

Gardner, J. J., and Sherman, A., 1995, Vision requirements in sport, in *Sports Vision*, Loran, D. F.C. and MacEwen, C. J., eds, Butterworth-Heinemann, Boston.

Getz, D. J., 1978, Vision and sports, *J. Am. Optom. Assoc.* **49**: 385-388.

Glass, G. V., 1976, Primary, secondary, and meta-analysis of research, *Educ. Res.* **5**: 3-8.

Goldstone, R. L., 1998, Perceptual learning, *Annu. Rev. Psychol.* **49**: 585-612.

Gottlieb, G. L., Corcos, D. M., Jaric, S., and Agarwal, G. C., 1988, Practice improves even the simplest movements, *Exp. Brain Res.* **73**: 436-440.

Gould, D., and Damarjian, N., 1998, Mental skills training in sport, in *Training in Sport*, Elliott, B. ed., John Wiley & Sons, New York.

Graybiel, A., Jokl, E., and Trapp, C., 1955, Russian studies in vision related activity and sports, *Res. Q. Am. Assoc. Health. Phys. Educ.* **26**: 212-223. [22]

Gregg, J. R., 1987, *Vision and Sports: An Introduction*, Butterworth-Heinemann, Boston, MA.

Grosvenor, T. P., 2002, *Primary Care Optometry*, Butterworth-Heinemann, 4th ed., Boston.

Haynes, H. M., and McWilliams, L. G., 1979, Effects of training on near-far response time as measured by the distance rock test, *J. Am. Optom. Assoc.* **50**: 715-718. *[23]*

Hazel, C. A., 1995, The efficacy of sports vision practice and its role in clinical optometry, *Clin. Exp. Optom.* **78**:98-105.

Henry, F. M., 1961, Specificity vs generality in learning motor skill, in *Classical Studies in Physical Activity*, Brown, R. C., and Kenyon, G. S., eds., Prentice-Hall, Englewood Cliffs, NJ.

Hitzeman, S. A., and Beckerman, S. A., 1993, What the literature says about sports vision, *Optom. Clin.* **3**: 145-169.

Hobson, R., and Henderson, M. T., 1941, A preliminary study of the visual field in athletics, *Proc. Iowa Acad. Sci.* **48**:331-337. [33]

Hoffman, L. G., Polan, G., and Powell, J., 1984, The relationship of contrast sensitivity functions to sports vision, *J. Am. Optom. Assoc.* **55**: 747-752. [17]

Hoffman, L. G., Rouse, M., and Ryan, J. B., 1981, Dynamic visual acuity: a review, *J. Am. Optom. Assoc.* **52**: 883-887. *[7]*

Isaacs, L. D., and Finch, A. E., 1983, Anticipatory timing of beginning and intermediate tennis players, *Percept. Mot. Skills* **57**: 451-454. [36]

Ishigaki, H.,and Miyao, M., 1993, Differences in dynamic visual acuity between athletes and nonathletes, *Percept. Mot. Skills* **77**: 835-839. [13]

Johnson, C. A., and Leibowitz, H. W., 1979, Practice effects for visual resolution in the periphery, *Percept. Psychophys.* **25**: 439-442. [1]

Johnson, W. G., 1952, Peripheral perception of athletes and non-athletes and the effect of practice, *Thesis*, University of Illinois. [15]

Keele, S. W., 1986, Motor control, in *Handbook of Perception and Human Performance: Cognitive Processes and Performance*, Boff, K. R. et al, eds., John Wiley & Sons, New York: pp.1-60.

Kirscher, D. W., 1988, A comparison of reaction time in athletes and non-athletes, *Highlights of the National Academy of Sports Vision.* **4**: 8-9. [33]

Kirscher, D. W., 1993, Sports vision training procedures, *Optom. Clin.* **3**: 171-182.

Kluka, D., Love, P., and Allen, S., 1989, Contrast sensitivity functions of selected collegiate female athletes, *Sportsvision.* **5**: 18-24. [19]

Kofsky, M., and Starfield, B., 1989, Sports vision visual training and experimental program with Australian Institute of Sport Basketball players, *Aust. J. Optom.* **6**: 15-17. [40]

Kumar, T., and Glaser, D. A., 1994, Some temporal aspects of stereoacuity, *Vision Res.* **34**: 913-925.

Leigh, R. J., and Zee, D. S., 1999, *The Neurology of Eye Movements*, 3rd ed., Oxford University Press, New York .

Levine, S., Ciuffreda, K. J., Selenow, A., and Flax, N., 1985, Clinical assessment of accommodative facility in symptomatic and asymptomatic individuals, *J. Am. Optom. Assoc.* **56**: 286-90. *[24]*

Lisberger, S. G., 1988, The neural basis for learning of simple motor skills, *Science.* **242**: 728-735.

Long, G. M., and Riggs, C. A., 1991, Training effects on dynamic visual acuity with free-head viewing, *Perception.* **20**: 363-371. [2]

Long, G. M., and Rourke, D. A., 1989, Training effects on the resolution of moving targets--dynamic visual acuity, *Hum. Factors* **31**: 443-451. *[8]*

Loran, D. F. C., and MacEwen, C. J., (eds),1995, *Sports Vision*, Butterworth-Heinemann, Boston.

Low, F. N., 1943, The peripheral visual acuity of 100 subjects, *Am. J. Physiol.* **140**: 83-88. *[3]*

Low, F. N., 1946, Some characteristics of peripheral visual performance, *Am. J. Physiol.* **146**: 573-584. *[4]*

Low, F. N., 1951, Peripheral visual acuity, *Arch. Ophthal.* **45**: 80-99. *[5]*

Martin, W. F., 1984, *An Insight to Sports: Featuring Trapshooting and Golf*, SportsVision Inc., Seattle.

McHugh, D. E., and Bahill, A. T., 1985, Learning to track predictable target waveforms without a time delay, *Invest. Ophthal. Vis. Sci.* **26**: 932-937. *[16]*

McKee, S. P., and Westheimer, G., 1978, Improvement in vernier acuity with practice, *Percept. Psychophys.* **24**: 258-262. *[9]*

McLeod, B., 1991, Effects of eyerobics visual skills training on selected performance measures of female varsity soccer players, *Percept. Mot. Skills.* **72**: 863-866. [42]

Miller, D. M., 1960, The relationship between some visual-perceptual factors and the degree of success realized by sports performers, *Dissertation*, University of Southern California. [24]

Millslagle, D. G., 2000, Dynamic visual acuity and coincidence anticipation timing by experienced and inexperienced women players of fast pitch softball, *Percept. Mot. Skills.* **90**: 498-504. [14]

Milsum, J. H., 1966, *Biological Control Analysis*, McGraw-Hill Book Company, New York.

Mizusawa, K., Sweeting, R. L., and Knouse, S. B., 1983, Comparative studies of color fields, visual acuity fields, and movement perception limits among varsity athletes and non-varsity groups, *Percept. Mot. Skills.* **56**: 887-892. [37]

Montebellow, R. A., 1953, The role of stereoscopic vision in some aspects of baseball playing ability, *Thesis*, Ohio State University. [21]

Newell, K. M., 1991, Motor skill acquisition, *Annu. Rev. Psychol.* **42**: 213-237.

Olson, E. A., 1956, Relationship between psychological capacities and success in college athletics, *Res. Q. Am. Assoc. Health Phys. Educ.* **27**: 79-89. [23]

Parker, R. L., 1986, Optometry's role in the 1980 Olympics, *Am. J. Optom. Physiol. Opt.* **51**: 655. [35]

Pashler, H. E., 1997, *The Psychology of Attention*, MIT Press, Cambridge, Mass.

Paterno, J. V., 1986, Importance of vision in college athletes, *J. Am. Optom. Assoc.* **51**: 654. [32]

Pesce-Anzender, C., and Bosel, R., 1995, Modulation of the spatial extent of the attentional focus in high-level volley ball players, *Eur. J. Cog. Psychol.* **10**: 247-267.

Pizzarello, L. D., and Haik, B. G., 1987, *Sports Ophthalmology*, Charles C. Thomas, Springfield, Ill.

Posner, M. I., Nissen, M. J., and Klein, R. M., 1976, Visual dominance: an information-processing account of its origins and significance, *Psychol. Rev.* **83**: 157-171.

Provine, R. R.,and Enoch, J. M., 1975, On voluntary ocular accommodation, *Percept. Psychophys.* **17**: 209-212. *[22]*

Quevedo i Junyent, L., and Sole i Forto, J., 1995, Visual training programmme applied to precision shooting, *Ophthalmic Physiol. Opt.* **15**: 519-23. [38]

Quevedo, L., Sole, J., Palmi, J., Planas, A., and Soana, C., 1999, Experimental study of visual training effects in shooting initiation, *Clin. Exp. Optom.* **82**: 23-28. [39]

Randle, R. J., and Murphy, M. R., 1974, The dynamic response of visual accommodation over a seven-day period, *Am. J. Optom. Physiol. Opt.* **51**: 530-544. [10]

Reading, R. W., 1983, *Binocular Vision: Foundations and Applications*, Butterworth-Heinemann, Boston.

Reichow, A. W., and Coffey, B., 1986, A comparison on contrast sensitivity in elite athletes vs normal population (abstract), *Am. J. Optom. Physiol. Opt.* **63**: 82P. [18]

Revien, L., 1987, *Eyerobics (videotape),* Visual Skills Inc., Great Neck, NY.

Revien, L., and Gabor, M., 1981, *Sports Vision: Dr. Revein's Eye Exercises for Athletes*, Workman Publishing Co., New York: 20-21.

Ridini, L. M., 1968, Relationship between psychological functions test and selected sport skills of boys in junior high school, Res. Q. *Am. Assoc. Health. Phys. Educ.* **39**: 674-683. [25]

Roscoe, S. N., and Couchman, D. H., 1987, Improving visual performance through volitional focus control, *Hum. Factors.* **29**: 311-325. *[25]*

Rosenbaum, D. A., Carlson, R. A., and Gilmore, R. O., 2001, Acquisition of intellectual and perceptual-motor skills, *Annu. Rev. Psychol.* **52**: 453-470.

Rouse, M. W., Deland, P., Christian, R., and Hawley, J., 1988, A comparison study of dynamic visual acuity between athletes and nonathletes, *J. Am. Optom. Assoc.* **59**: 946-950. [12]

Saarinen, J., and Levi, D. M., 1995, Perceptual learning in vernier acuity: what is learned?, *Vision Res.* **35**: 519-527.

Sanes, J. N., and Donoghue, J. P., 2000, Plasticity and primary motor cortex, *Annu. Rev. Neurosci.* **23**: 393-415.
Saugstad, L., and Lie, I., 1964, Training of peripheral visual acuity, *Scand. J. Psychol.* **5**: 218-224. *[6]*
Scheiman, and M.,Wick, B., 2002, *Clinical Management of Binocular Vision: Heterophoric, Accommodative, and Eye Movement Disorders*, 2nd ed., Lippincott Williams, New York.
Seiderman, A., and Schneider, S., 1983, *The Athletic Eye: Improved Sports Performance Through Visual Training*, Hearst Books, New York.
Semmlow, J. L., Gauthier, G. M., and Vercher, J. L., 1989, Mechanisms of short-term saccadic adaptation, *J. Exp. Psychol. Hum. Percept. Perform.* **15**: 249-258. [7]
Sherman, A., 1980, Overview of research information regarding vision and sports, *J. Am. Optom. Assoc.* **51**: 661-665.
Shih, Y. F., Lin, L. L., Hwang, C. Y., Huang, J. K., Hung, P. T., and Hou, P. K., 1995, The effects of Qi-Qong ocular exercise on accommodation, *Chin. J. Physiol*, **38**: 35-42. *[26]*
Sisson, E. D., 1938, Voluntary control of accommodation, *J. Gen. Psychol.* **18**: 195-198. *[27]*
Sowden, P., Davies, I., Rose, D., and Kaye, M., 1996, Perceptual learning of stereoacuity, *Perception* **25**: 1043-1052. *[10]*
Stark, L., 1968, *Neurological Control Systems*, Plenum Press, New York.
Stein, R., Squires, G., Pashby, T., and Easterbrook, M., 1989, Can vision training improve athletic performance?, *Can. J. Ophthalmol.* **24**: 105-106.
Stine, C. D., Arberburn, M. R., and Stern, N. S., 1982, Vision and sports: A review of the literature, *J. Am. Optom. Assoc.* **53**: 627-633.
Stroup, F., 1957, Relationship between measurements of field of motion perception and basketball ability in college men, *Res. Q. Am. Assoc. Health. Phys. Educ.* **28**: 72-76. [34]
Tremayne, P., and Barry, R. J., 2001, Elite pistol shooters: physiological patterning of best vs. worst shots, *Int. J. Psychophysiol.* **41**: 19-29.
Vaegan, 1979, Convergence and divergence show large and sustained improvement after short isometric exercise, *Am. J. Optom. Physiol. Opt.* **56**: 23-33.
Vogel, G. L., and Hale, R. E., 1992, Does participation in organized athletics increase a child's scoring ability on the Wayne Saccadic Fixator?, *J. Behav. Optom.* **3**: 66-69. [29]
Watanabe, S., Ogino, S., Nakamura, T., and Koizuka, I., 2003, Saccadic adaptation in the horizontal and vertical directions in normal subjects, *Auris. Nasus. Larynx.* **30** (Suppl): S41-45. *[13]*
Watts, R. G., and Bahill, A. T., 1990, *Keep Your Eye on the Ball*, W H Freeman & Co, New York
West, K. L., and Bressan, E. S., 1996, The effect of a general versus specific visual skills training program on accuracy in judging length-of-ball in cricket, *Int. J. Sport Vis.* **3**: 41-45. [41]
Whittaker, S. G., and Eaholtz, G., 1982, Learning patterns of eye motion for foveal pursuit, *Invest. Ophthal. Vis. Sci.* **23**: 393-397. *[17]*
Winograd, S., 1942, The relationship of timing and vision to baseball performance, *Res. Q. Am. Assoc. Health. Phys. Educ.* **13**: 481-493. [20]
Wittenberg, S., Brock, F. W., and Folsom, W. C., 1969, Effect of training on stereoscopic acuity, *Am. J. Optom. Arch. Am. Acad. Optom.* **46**: 645-653. *[11]*
Wood, J. M., and Abernethy, B., 1997, An assessment of the efficacy of sports vision training programs, *Optom. Vis. Sci.* **74**: 646-659. [44]
Zagelbaum, B. M., 1996, *Sports Ophthalmology*, Blackwell Science, Cambridge Mass.

# APPENDIX 16.1: Glossary of Vision Terminology

**Accommodation:** The eye's lens-based focusing mechanism used to obtain and maintain a clear retinal image of an object of interest on the high-resolution fovea.

**Accommodative dynamics/facility:** The ability to change focus rapidly to objects at different distances; in the clinic, this is specified as the maximum number of focusing changes/shifts possible within a specified period of time (e.g., cycles/min).

**Contrast discrimination:** The ability to detect contrast changes in a readily visible target.

**Contrast sensitivity:** The ability to first detect very small amounts of contrast in an initially invisible (i.e., extremely low contrast) target.

**Convergence:** The inward-directed turning of the eyes' lines-of-sight toward each other for tracking of objects moving nearer in depth.

**Diopter:** A unit of ophthalmic lens power; one diopter focuses light from infinity at a distance of one meter.

**Divergence:** The outward-directed turning of the eyes' lines-of-sight away from each other for tracking of objects moving farther in depth.

**Dynamic accommodation:** Focusing response changes of the eye during the actual shifting from one steady-state level to another to see targets at different distances clearly.

**Dynamic visual acuity:** The ability to resolve details in a moving target with eye (and sometimes concurrent head) tracking.

**Fixation:** The ability to maintain the visual axes accurately upon an object of interest.

**Fovea:** The central retinal region of maximum visual acuity/visual resolution, and principal visual direction, used to fixate upon an object of interest.

**Fusion:** Single vision under binocular viewing conditions.

**Peripheral awareness training:** The attentional weighting of the peripheral visual field over the central visual field.

**Prism diopter:** A unit of ophthalmic prism power; one prism diopter deviates light from infinity 1 centimeter at 1 meter; 1.745 prism diopters equal 1 degree of displacement.

**Pursuit:** The smooth and relatively slow tracking movement of the eyes (as when following a moving car or a bird in flight).

**Retinal disparity:** The angular difference in retinal images between the fixated object and any other object in the visual field.

**Saccades:** The discrete and very rapid tracking movement of the eyes (as when reading or when scanning a picture).

**Static accommodation:** Steady-state focusing response of the eye following completion of dynamic changes.

**Static visual acuity:** The ability to resolve fine detail in a stationary target (such as an eye chart); 20/20 clinical based specification.

**Stereoacuity:** The ability to detect very small relative depth differences (i.e., seconds of arc; 1 degree = 60 minutes of arc = 3600 seconds of arc) between objects under binocular viewing conditions based on the presence of horizontal retinal-image disparity.

**Stereopsis:** Binocular depth discrimination based on horizontal retinal-image disparity.

**Vergence:** The coordinated and disjunctive inward/outward movement of the eyes when binocularly tracking an object moving in depth.

**Vergence dynamics/facility:** The ability to change binocular eye position rapidly to objects at different distances; in the clinic, this is specified as the maximum number of eye position changes/shifts in depth possible within a specified period of time (e.g., cycles/min).

**Vernier acuity:** The ability to resolve very fine (i.e., seconds of arc) target misalignments.

**Version:** The coordinated and parallel movement of the eyes (laterally, vertically, or obliquely) when tracking a moving object at a fixed distance.

**Vision therapy:** Highly specific, sequential, sensory-motor-perceptual stimulation paradigms and regimens aimed at normalizing binocular vision. It incorporates the use of blur (via lenses and target distance), disparity (via prisms and target distance), and proximity (via perceived target distance) stimuli; it is also referred to as v.t., vision training, visual training, visual therapy, eye exercises, and orthoptics.

**Visual acuity:** The ability to resolve very fine details (i.e., minutes of arc) in a form.

# Chapter 17

# Application of Biomedical Principles to the Maturation of Skill in Children

Jaci L. VanHeest [1] and Jane M. Cappaert [2]

[1]*Department of Kinesiology, University of Connecticut, 2095 Hillside Road U-1110, Storrs, CT 06269, PH: (860) 486-5123;* ***FX: (860) 486-1123****; EM: Jaci.vanheest@uconn.edu*

[2] *Reebok International Ltd., Human Performance Engineering Lab., Canton, MA 02021,PH: (781) 401-4074; FX: (781) 401-7222; EM: Jane.cappaert@reebok.com*

## 17.1 INTRODUCTION

Participation of children in youth sport is rapidly growing in the United States and around the world. As children become involved in sport, the risks and benefits of participation must be evaluated. Activity promotes physical fitness, reduces risk factors for cardiovascular disease, and helps combat the growing obesity epidemic in children and youth. Children are becoming exposed to a larger variety of activities such as triathlons, weight lifting, long distance running, soccer, and basketball. These varied experiences stress the child's body in differing ways and demand a wide range of skills for success. It is critical that scientists and practitioners approach the issues of sport participation for children in a manner that does not assume that children are miniature versions of adult athletes. Training programs for adults cannot be diluted and then given to children with the hopes of sporting success. The biomedical issues important to children and youth are very specific. This chapter focuses on issues of physical and motor skill growth and development. In addition, the integration of these concepts into sport training will be addressed.

## 17.2 PHYSICAL AND PHYSIOLOGICAL DEVELOPMENT

The progression of physical development in children follows a typical pattern that is driven by alterations in reproductive and growth hormones. The linear growth and body weight increases occur from birth and reach their peak between the ages of 12-14 years for girls and 13-15 years for boys (Weber, Kartodihardjo, and Skissouras, 1976). Puberty is the window of time whereby the adolescent becomes reproductively mature (American Academy of Pediatrics, 2000; Bar-Or, 1983; Gladish, Washington and Bull, 2002; Weber, Kartodihardjo, and Skissouras, 1976). This period of puberty spans approximately 1.5-2 years on average. During this window of time, adolescents are rapidly growing and changing in stature, in body weight and in body composition. Gender plays an important role in these changes with young girls maturing approximately two years earlier than young boys (Weber, Kartodihardjo, and Skissouras, 1976; Gladish, Washington, and Bull, 2002; Malina and Bouchard, 1994). Within this framework, physiological and metabolic development is occurring. The following section will delineate the changes that occur in several aspects of the child's physiological function.

### 17.2.1 Aerobic Development

Aerobic endurance is the foundation of performing physical work. Aerobic metabolism allows the body to generate energy from carbohydrates, fats, and proteins, fuels while using oxygen in the processes. The aerobic pathways can produce energy for extended periods of time (large capacity) when the intensity of the exercise is relatively low. Maximal aerobic power ($VO_2max$) or endurance capacity is the ability to maximally take in, transport and utilize oxygen. Maximal aerobic power has traditionally been described as a marker of endurance performance capacity and cardiovascular fitness (Bar-Or, 1983; Krahenbuhl, Skinner, and Kohrt, 1981; Mirwald and Bailey, 1986).

Developmental rate, body composition and habitual physical activity all influence aerobic capacity. Maximal aerobic power is achieved through a rather complex series of reactions within the human body that involve the lungs, heart, circulatory system, and skeletal muscle. Maximal oxygen consumption is used to represent aerobic power. Cardiac output (heart rate times stroke volume) and a-v $O_2$ difference are mathematically summed to produce $VO_2max$ (Astrand, 1952). Cardiac output has been shown to relate to body surface area in children (4.1 to 4.3 L/min per square meter)

(Krovetz, 1966; Locke, 1973). The factors that affect either cardiac output or a-v $O_2$ difference will ultimately dictate $VO_2max$.

Aerobic capacity is expressed in either absolute terms (L/min) or relative terms (ml/kg/min). Absolute $VO_2max$ increases with increasing age in healthy untrained children (Andersen et al., 1974; Baxter-Jones, Goldstein, and Helms, 1993; Binkhorst et al., 1984; Mirwald and Bailey, 1986; Murase et al., 1981; Rutenfranz et al., 1981). A 200 ml/min increase up to the age of 12 years for both boys and girls is typically cited in the literature (Baxter-Jones, Goldstein and Helms, 1993; Hennig, 2002; Johnson, 1962; Krahenbuhl, Skinner and Kohrt, 1981; Malina, 1994). Following puberty, a gender difference becomes clearly evident. The increase seen in young girls flattens into a plateau while boys continue to increase. By the age of 16 years, there is a 50% difference between the genders (Mirwald et al., 1981; Rutenfranz et al., 1981). The difference seen in young women is due primarily to the alterations in body composition during this time period with an increase in body fat relative to lean tissue. Table 17.1 illustrates the 50$^{th}$ percentile times for a one mile run/walk test that is frequently used to evaluate the ability of the aerobic system.

***Table 17.1.*** **Performance scores for the 50$^{th}$ percentile in the one mile run/walk***

| Age (years) | Boys | Girls |
|---|---|---|
| 6 | 12:36 | 13:12 |
| 7 | 11:40 | 12:56 |
| 8 | 11:05 | 12:30 |
| 9 | 10:30 | 11:52 |
| 10 | 9:48 | 11:22 |
| 11 | 9:20 | 11:17 |
| 12 | 8:40 | 11:05 |
| 13 | 8:06 | 10:23 |
| 14 | 7:44 | 10:06 |
| 15 | 7:30 | 9:58 |
| 16 | 7:10 | 10:31 |
| 17+ | 7:04 | 10:22 |

* Time is in minutes and seconds. (President's Counsel on Physical Fitness).

### 17.2.2 Anaerobic Development

Production of energy is the goal of metabolism. Energy is needed to enable muscles to contract. The demand for energy is related to the intensity of the work placed on the body. If the energy demand is high, the body must generate energy very rapidly. To accomplish this task, muscle cells contain immediately accessible fuel stores—ATP (adenosine triphosphate) and CP (creatine phosphate) (Mercier et al., 1992; Williams and Armstrong, 1991). ATP and CP are energy-rich fuel stores that are contained within the muscle cells. These fuels are limited in supply. High-intensity work, therefore, can be maintained for relatively short periods of time (less than 10 seconds) and at the start of exercise (Falk and Bar-Or, 1993; Mercier et al., 1992). After ATP and are reduced to a low level another energy-producing pathway must contribute for muscular contraction to continue. Studies involving children and adolescents report values such as 5.5 up to 6.5 W/kg and 5.6 up to 8.0 W/kg in 8-14 year old girls and boys, respectively (Rowland, 1996).

Another source of energy production in the muscle that can be maintained for longer periods of time is muscle glycogen. If the energy demand is relatively high, the muscle breaks down glycogen through anaerobic pathways. A by-product is produced during the breakdown of glycogen called lactic acid. The lactic acid accumulates in the muscle cells and can begin to inhibit further energy production and muscle contraction (Williams and Armstrong, 1991a; Williams and Armstrong, 1991b). The glycolytic pathway can produce relatively high amounts of energy for up to about two minutes. If continued work is necessary, energy must be generated through aerobic pathways.

Peak and mean anaerobic power increase with increasing chronological age (Gabriel and Boucher, 2000; Mirwald and Bailey, 1986). In addition, evaluation of blood lactate in children and youth indicates a linear rise in both maximal and submaximal exercise during childhood (Williams and Armstrong, 1991a; Williams and Armstrong, 1991b). The point of interplay between the aerobic and anaerobic systems (ventilatory threshold) declines as a percentage of maximal oxygen consumption as the child ages and is typically higher in child athletes than non-athletes (Gabriel and Boucher, 2000). It is clear that the development of anaerobic potential occurs throughout childhood into adulthood. The rapid expansion of these capacities is closely aligned with the progression of puberty in both young boys and girls.

### 17.2.3 Muscular Development

Development of muscular strength has long been thought to be a parameter left only for adulthood. However, as more children become involved in competitive sport earlier in their lives, the rush to resistance train has grown. The incorporation of heavy resistance training during the pre-pubescent period has resulted in many injuries both in soft tissue and in the bony growth plates (American Academy of Pediatrics, 1983; Strickler and VanHeest, 2001). It is important to understand the factors involved in the development of muscular strength, endurance and power.

Muscular endurance is the ability to lift a given amount of weight for a number of repetitions. Muscular strength can be defined as the maximum amount of weight that can be moved by a muscle or muscle group. Muscular power incorporates the component of time and is the ability to move a given weight rapidly or explosively (Faigenbaum, 2000). These parameters are developed both in pre and post pubescent individuals. The early gains in muscular strength are due to alterations in the neuromuscular patterning and firing (American Academy of Pedicatrics, 2000; Falk and Tenenbaum, 1996; Faigenbaum, 2000; Strickler and VanHeest, 2001). During puberty, reproductive hormones (ie. testosterone) provide the impetus for muscular development. The additional strength and power gains are defined as muscular hypertrophy gains (Faigenbaum, 2000; Falk and Tenenbaum, 1996). The time course for muscular strength, power and endurance enhancements follows the aerobic development in children. Typically, the period beginning at puberty is reported as the window of time whereby large gains in muscular parameters are evident. The increase in strength and power continue into early and middle adulthood (20-60 years) (Strickler and VanHeest, 2001). Although muscular strength, endurance and power vary greatly during development, pulls-ups, push-ups and flexed arm hang are often used to estimate these parameters. Table 17.2 illustrates the 50$^{th}$ percentile data for these markers.

Training young children to become successful athletes is a challenge. Care must be taken to carefully construct training programs that optimally load the various systems without overloading their capacities. Too high a load and the youngster will show signs of overtraining and/or overuse. Loading the athlete less than desired will result in poor adaptation and performances. The critical element is timing the stimulation of the various physical components (American Academy of Pediatrics, 2000; Astrand, 1952). It is very clear that aerobic capacity develops early in development followed by both anaerobic abilities and muscular strength and power (Baxter-Jones, Goldstein, and Helms, 1993; Krahenbuhl, Skinner, and Kohrt, 1985; Mirwald and Bailey, 1986; Mirwald et al., 1981). Stressing anaerobic systems before developing the aerobic capacity can result in an inability to maximize the potential of either system (Urhausen, 1995; Uusitulo, 1998). The further integration of these developing systems with the maturation of motor capacities is essential in the development of sport performance in children and youth.

***Table 17.2.*** **50th percentile scores for pull-ups, push-ups, and flexed arm hang.**

| Age | Boys | | | Girls | | |
|---|---|---|---|---|---|---|
| | Pull-up | Push-up | Flexed Hang * | Pull-up | Push-up | Flexed Hang * |
| 6 | 0 | 7 | 6 | 0 | 6 | 5 |
| 7 | 1 | 8 | 8 | 0 | 8 | 6 |
| 8 | 1 | 9 | 10 | 0 | 9 | 8 |
| 9 | 2 | 12 | 10 | 0 | 12 | 8 |
| 10 | 2 | 14 | 12 | 0 | 13 | 8 |
| 11 | 2 | 15 | 11 | 0 | 11 | 7 |
| 12 | 2 | 18 | 12 | 0 | 10 | 7 |
| 13 | 3 | 24 | 14 | 0 | 11 | 8 |
| 14 | 5 | 24 | 20 | 0 | 10 | 9 |
| 15 | 6 | 30 | 30 | 0 | 15 | 7 |
| 16 | 7 | 30 | 28 | 0 | 12 | 7 |

* Score in seconds, (President's Counsel on Physical Fitness).

## 17.3 BIOMECHANICAL CONSIDERATIONS IN THE DEVELOPMENT OF MOTOR SKILLS

For children, motor development progresses from large muscles (earlier years of life) to smaller muscles (grade school years of life). Similarly, the acquisition of movement skills progresses from gross (using larger muscles) motor skills to fine (using smaller muscles) motor skills. This plays out in the way skills should be learned. The gross motor phase is the time to develop the basic skills such as running, jumping, kicking, throwing and striking (approximately ages 2-5+ years), and the fine motor phase provides the opportunity to master these basic skills by adding elements of accuracy, placement and speed (approximately ages 6-10 years) (Roncesvalles, Woolacot and Jensen, 2001).

### 17.3.1 Foot Development

The feet of infants grow rapidly and reach almost half their adult size during the first year. Many of the bones of infant feet are still in cartilage form and are therefore undeveloped, soft and flexible. Extra fat under the arch and lack of development of the muscles supporting the arch makes an infant's foot appear flat and broad. The infant/toddler foot needs to exercise to develop the muscles of the arch. Barefoot walking is recommended during toddler years to strengthen the developing foot.

### 17.3.2 Walking

Children begin crawling and standing between 6-9 months of age. Walking will typically begin when the infant begins to feel stable on two feet (usually between 10-18 months of age). A toddler's gait begins with short choppy steps with the hands held high for balance, and progresses to a heel to toe foot strike pattern with opposing arm motions by about 25 months (Gallahue and Ozmun, 2002). Footwear for toddlers is mainly for protection and to provide some warmth for the foot. When toddlers wear shoes, the shoes should be soft and flexible with a wide base so that they do not interfere with the development of the foot.

As a child develops from 6-10 years of age, an adult gait pattern emerges. Research has shown that girls and boys (6-10 years old) have similar walking patterns (Hennig, 2002). Hennig (2002) has also shown that the underfoot foot pressures of children during walking differ from adults. Children have similar pressures across the entire foot, whereas adults have more pressure under the forefoot and heel than in the arch area. By about age 6 years, the longitudinal arch has fully developed and the pressures have progressed away from the arch to the 3rd -5th metatarsal heads (Hennig, 2002; Figure 17.1).

*Figure 17.1.* Underfoot pressure differences between a child (left) and an adult foot. Children's feet are flatter and wider due to extra arch fat and under-developed arch muscles.

## 17.3.3 Running

Running is a basic gross motor skill that develops naturally from walking. Running can also be considered a sport as well as a skill. A running motion that includes a true non-support phase develops between 2-3 years of age. The running pattern progresses through an advanced pattern to a mature pattern by age 5 years. The mature pattern will include opposing arm and leg motions and the speed of running becomes faster.

Mastery of the running skill is developed during the grade school years. Mastery of this skill includes the ability to change running speeds, efficiently change running directions, and the ability to use different speeds for different distances of running (pacing). All of these variations of running (forward, backward and sideways running, as well as the ability to change directions while running) are used in many different sports (basketball, tennis, soccer, baseball, softball, field hockey, football, etc.). Mastery of this skill is critical in the grade school years as a prerequisite to achieving excellence in sport performance in teenage years. Typical running speeds for boys and girls ages 6-12 years are presented above in Table 17.1.

Footwear for children should be comfortable and flexible in the forefoot area so that the shoes do not interfere with foot development. Children's footwear should also be durable since growing children are usually very active.

## 17.3.4 Jumping

The development of jumping begins at about 18 months as children learn to step down from low objects. By approximately 2 years of age, children can jump down from an object with one foot leading the other. Shortly thereafter, (age 2.5 years) children can jump off the floor with both feet. Children continue to improve their jumping ability for height and distance until about age 5 years, and a mature jumping pattern is usually in place by age 6 years.

Once the mature pattern is in place, mastery of the jumping motion includes perfecting the coordination and timing of the arm swing, and increasing the height and distance that can be jumped. During this mastery phase, children will also begin to combine skills such as running and jumping with a one foot take off.

## 17.3.5 Kicking

At about 18 months of age, children are able to begin developing the skill of kicking by pushing against a ball with their foot. As children develop to about age 2-3 years, they kick at a ball with a straight leg and use very little body movement. As the skill develops, flexion of the lower leg only is seen just before kicking (age 3-4 years). About a year later, children have greater backward and forward swinging motion of leg with correct arm opposition. Finally, a mature kicking pattern emerges at about age 5-6 years.

Refinement and mastery of the kicking skill involves kicking for accuracy, kicking for distance, kicking for placement, which is a combination of accuracy and distance. This mastery of placement is necessary later on for sports such as soccer and kicking in football. Another element of mastery includes placing different spins on the ball during the kick.

### 17.3.6 Eye-Hand Coordination

Eye-hand coordination can begin to develop as early as 3-7 months as children learn to use their fingers for grasping and poking. Whole arm coordination develops 1-2 years later as a child learns to roll, then bounce and finally underarm toss a ball. These precursors are necessary to prepare a child for the basic eye-hand motor skills of throwing, catching and striking.

### 17.3.7 Throwing

Children are ready to learn the skill of throwing at about age 2-3 years. Throwing at this stage is accomplished with the feet stationary and elbow extension of the forearm only to throw at a target. During the next stage, children usually throw with stationary feet, but body rotation is used as the elbow extends the forearm to throw at a target. By about 4-5 years, children throw using a step with the foot on the same side as the throwing arm. Finally a mature pattern in which the foot opposite the throwing arm steps forward and the body rotations is achieved by about age 5+ years.

Once a mature pattern has been developed, mastery of the throwing skill includes throwing for speed, throwing for accuracy, throwing for distance, placing different spins on a ball and throwing other objects besides a small ball (such as football). Research has shown that the best progression for working on both speed and accuracy is to increase the speed of the movement first and then work on accuracy (Gabriel and Boucher, 2000). Mastery of these refinements to the throwing skill should be achieved throughout grade school.

However, research has shown that even by age 13 years, baseball pitchers have not perfected their throwing pattern. In a study of youth baseball pitchers (13 years), significant differences were found in trunk rotation velocities and timing of maximum trunk rotation (Sabick, Torry and

Hawkins, 2002). Sabick (2002) found that the angular velocity of the trunk segment was significantly higher in youth pitchers than professional pitchers. Peak pelvic rotation velocity occurred near the time of stride foot contact in professionals and much later in the youth pitchers. Increased trunk rotation velocity may be a compensation for improper timing of that rotation. Improper energy transfer from the trunk to the throwing arm may have implications to shoulder injuries in youth pitchers.

The above implies that mastery of the more complex throwing skill takes a longer time to accomplish than simpler skills such as jumping or catching. It is possible that for some children, the mature throwing pattern is never fully developed.

### 17.3.8 Striking

At about age 2-3 years, children begin learning the skill of striking by swinging at a target in vertical plane while facing forward. Turning sideways to object and swinging in a horizontal plane has developed with practice by about age 4-5 years. Body rotation and a forward shifting of body weight during the striking motion is developed by about age 5 years. Finally a mature two-handed striking pattern at a stationary ball is developed by age 6-7 years. Once a mature two-handed pattern has developed, the striking motion can usually be transferred to one-handed striking easily.

Mastery of the striking skill progresses from a stationary ball to striking moving balls of different speed (baseball, tennis), as well as balls moving with different arcs (softball, tennis). Overhead striking such as those motions used in tennis and volleyball serves combines the skill of throwing and the principles of striking.

Mastery of the striking skill can be enhanced with the use of appropriately sized striking equipment. Tennis rackets and baseball/softball bats should be lighter in weight than their adult counterparts to adjust for the lack of muscular strength. Striking equipment should also be shorter to reduce the moment of inertia, and make it easier to swing. Appropriately sized equipment will reduce the risk of muscle injury and it will allow the skills to be developed properly.

### 17.3.9 Catching

Initially, children do not respond to a ball lightly tossed to them. By about age 2-3 years, they respond to a tossed ball, but with delayed arm movements (age 2-3 years). The next progression in the skill includes using the body to help catch (age 3 years). The skill develops until by about age 5 years, children can catch a small ball using only their hands (age 5 years). A mature two handed catching pattern is developed by age 6+ years.

Mastery of catching includes learning one handed catching as well as two handed catching. One handed catching usually involves the use of a baseball or softball glove. Refinement of catching includes combining running and judgement of depth (or depth perception) to catch fly balls in such activities as baseball, softball or catch balls in kick ball.

### 17.3.10 Swimming

Front crawl or freestyle swimming is a complex motion that requires a sense of stability and balance in the water. Feeling balanced in the water is usually achieved with the body in a vertical position because the natural floating position of the body in water is closer to vertical than horizontal for most children (and adults). Since swimming occurs with the body in a horizontal position, excellence in swimming is difficult to achieve as the body must somehow compensate for the tendency to rotate vertically.

Early motions in the water that are precursors to the swimming motion are mostly up and down splashes in the water with the body in a vertical position. This can happen as early as age 2-3 years, but developmental timelines in swimming are dependent on the age in which a child is exposed to water. At ages 3-4 years, children will put their faces in water while they hold on to the side of the pool and they begin to move their hands vertically underwater to tread water for a few seconds when they let go of the side of the pool. Progression to front crawl swimming occurs with a straight arm windmill type of motion with the head lifted forward in the water (about age 4 years). Next, children will perform a dog-paddle type of stroke putting their faces in the water for about 5 feet (about age 5 years). Children then begin to swim across the length of a pool (usually 25 yds.) using a front crawl type pattern with straight arms and a sideways turning of the head to breathe. A mature front crawl or freestyle pattern with sweeping pulling patterns and proper breathing usually does not occur until puberty when muscular strength develops. Freestyle swimming speed for 100 yards would be approximately 58.6s for 9-10 year old boys and 58.9s for 9-10 year old girls. Eleven and twelve year olds should complete the 100 yards freestyle swim in 51.9s and 53.7s for boys and girls, respectively.

Motor skill development clearly corresponds with the physical development of children. As various physical components develop (ie. foot development, hand-eye coordination), the fundamental motors skills follow. Table 17.3 summarizes the motor skills and activities appropriate for children as they develop.

***TABLE 17.3.* Activities during each stage of motor development**

**Chronological Age Range:** 5 - 6 years
**Stage:** Basic movement skills and coordination (running, jumping, throwing, catching, striking, kicking).
**Activities:** Activities should focus on mastering the basic motor skills in a fun, non-competitive environment. The activities should be simple and repetitive that allow the child to practice and develop the different skills.

**Chronological Age Range:** 7-9 years
**Stage:** Transitional period between the basic skills to mastery of the skills (throwing for speed, kicking for accuracy).
**Activities:** Activities should continue to focus on having fun and mastering motor skills. Competition is not recommended at this stage. Activities should be simple with flexible rules (entry-level tee-ball, youth soccer).

**Chronological Age Range:** 10-11 years
**Stage:** Ready for entry level sport.
**Activities:** The activities at this stage will be more complex as children learn the different sports and how to combine basic motor and cognitive skills to create sport specific skills (playing basketball, tennis, etc.).

## 17.4 SUMMARY

Basic motor skills (running, jumping, throwing, etc.) are prerequisites to learning the sport specific movements that are found in baseball, basketball, soccer, tennis and gymnastics. These basic motor skills must first be learned and developed into a mature pattern by about the age of 6 years. Then, they must be mastered during the grade school years. Research has shown that the performance level of basic skills improves throughout grade school until about grade 5 (Johnson, 1962) suggesting that mastery of basic skills does not occur until approximately 10-11 years of age. The integration of physical/physiological development with the development of motor skills is critical. Figure 17.2 illustrates the relationship between these parameters.

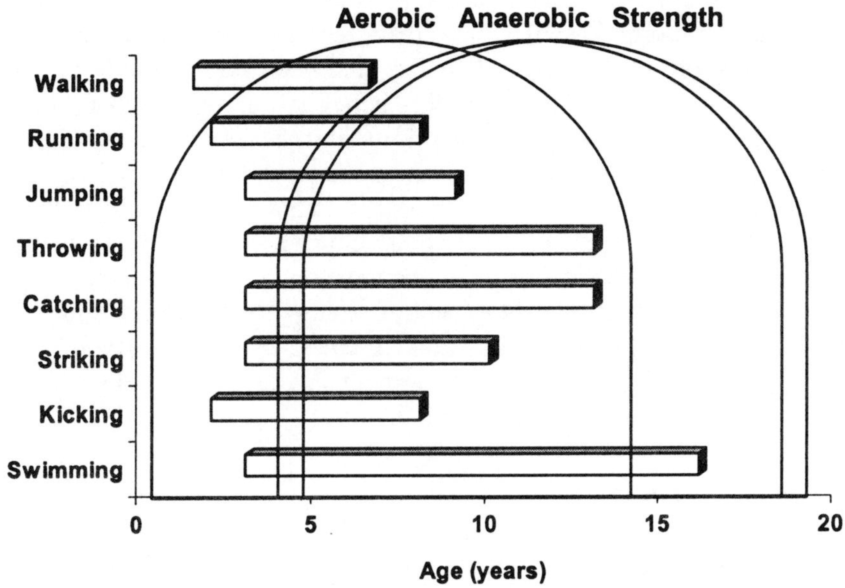


*Figure 17.2.* Horizontal bars represent the time segments (in years) associated with the development of each fundamental motor skill listed in the column on the left. The curved lines represent the time windows during which each physiological parameter (labeled above) is developed.

If the basic motor skills are not mastered in the developmental years (up to age 10-11 years) it will be difficult to master the sport specific skills (Berger, 1995). Therefore, children should be encouraged to develop basic skills and coordination patterns early in life (up to 10 years) rather than participating in organized sports during these years (Baxter-Jones, 1995). Furthermore, both coordination and aerobic capacity should be the developmental focus during this timeframe. Without the development of these systems, young children will not possess the foundation upon which to build anaerobic capacity, muscular strength and power. The competitive atmosphere found in organized sports will only add stress to the child making it more difficult to perform the motor skills involved with the specific sport. If the skills have been mastered by the child over years of practice and development, the child will be better able to perform the skills at a top level. Placing the child in a competitive atmosphere of game play before the skills have been mastered (for example, playing competitive baseball at age 7 years) will only hinder further development and mastery of the necessary skills.

It is widely believed that in order to attain a high level of success at playing specific sports (baseball, basketball, tennis, soccer, etc.), competitive levels of that sport must begin early in life, and that sport should be played at the exclusion of others. However, this belief is not supported by current research. Instead, a non-competitive, fun atmosphere is more successful at encouraging participation, and therefore practice and mastery of skills, for pre-pubescent children (Baxter-Jones, 1995). This practice and mastery of skills and the development of whole body coordination is a more balanced precursor to specific sports and gives the developing athlete a better chance to meet his/her potential. Development of young athletes and/or sport skills in children is an important task. It is essential that coaches and parents follow appropriate timelines related to growth and development to enable the child to maximize their sporting potential.

## 17.5 REFERENCES

American Academy of Pediatrics, Committee on Sports Medicine and Fitness, 2000, Intensive training and sports specialization in young athletes, *Pediatrics.* **106**: 154-157.

American Academy of Pediatrics, 1982, Weight training and weight lifting: information for the pediatrician, *The Physician and Sports Medicine.* **11**: 157-161.

Andersen, K. L., Seliger, V., Rutenfranz, J., Mocellin, R., 1974, Physical performance capacity of children in Norway, *Eur. J. Appl. Physiol.* **33**: 177-195.

Astrand, P. O., 1952, *Experimental studies of physical working capacity in relationship to sex and age,* Munksgaard, Copenhagen, Demark.

Bar-Or, O., 1983, *Pediatric sports medicine for the practitioner,* Springer-Verlag, New York.

Baxter-Jones, A. D., 1995, Growth and development of young athletes, *Sports Med.* **20**: 59-64.

Baxter-Jones, A., Goldstein, H., Helms, P., 1993, The development of aerobic power in young athletes, *J. Appl. Physiol.* **75**: 1160-1167.

Berger, K. S., 1995, *The Developing Person Through Childhood and Adolescence (4th Edition),* Worth Publishers, New York.

Binkhorst, R. A., de Jong-va de Ker, M. C., Vissers, A. C .A., 1984, Growth and aerobic power of boys aged 11-19 years. In: Ilmarinen, J, Valimaki, I, eds., *Children and sport,* Heidelberg, Springfield, IL., pp. 99-105.

Faigenbaum A., 2000, Strength training for children and adolescents, *Clin Sports Med.* **19**: 593-619.

Falk, B., Tenenbaum G., 1996, The effectiveness of resistance training in children: a meta-analysis, *Sports Med.* **22**: 176-186.

Falk, B., Bar-Or, O., 1993, Longitudinal changes in peak aerobic and anaerobic mechanical power of circumpubertal boys, *Pediatr.Exerc. Sci.* **5**: 318-331.

Gabriel, D. A., Boucher, J. P., 2000, Practicing a maximal performance task: A cooperative strategy for muscle activity, *Research Quarterly for Exercise and Sport.* **71**: 217-228.

Gallahue, D. L., Ozmun, J. C., 2002, *Understanding Motor Development: Infants, children, adolescents, adults (5th Edition),* McGraw-Hill, New York.

Gladish K., Washington, R. L., Bull, M. J., 2002, Swimming programs for infants and toddlers, *Pediatrics.* **109**: 168-169.

Hennig, E., 2002, Pressure patterns under the feet of children, adults and over weight persons - the influence of gender, *Proceedings of the IV World Congress on Biomechanics*, Calgary, CA.

Johnson, R., 1962, Measurement of achievement in fundamental skills of elementary school children, *Research Quarterly.* **33**: 94-103.

Krahenbuhl, G. S., Skinner, J. S., Kohrt, W. M., 1985, Developmental aspects of maximal aerobic power in children, *Exerc. Sport Sci. Rev.* **13**: 503-538.

Krovetz, L. J., McLoughlin, T. G., Mitchell, M. B., Schiebler, G. L., 1967, Hemodynamic findings in normal children, *Pediatr. Res.* **1**: 122-130.

Lehmann, M., Keul, J., Korstein-Reck, U., 1981, The influence of graduated treadmill exercise on plasma catecholamines, aerobic and anaerobic capacity in boys and adults, *Eur, J. Appl. Physiol.* **47**: 301-311.

Locke, J. E., Einzig, S., Moller, J. H., 1978, Hemodynamic responses to exercise in normal children, *Am. J. Cardiol.* **41**: 1278-1285.

Malina, R. M., 1994, Physical growth and biological maturation of young athletes, *Exerc. Sports Sci. Rev.* **22**: 389-434.

Malina, R. M., Bouchard, C., 1991, *Growth, maturation, and physical activity*, Human Kinetics Pub., Champaign, IL.

Mercier, J., Mercier, B., Granier, P., LeGallais, D., Prefaut, C., 1992, Maximal anaerobic power: relationship to anthropometric characteristics during growth, *Int. J. Sports Med.* **13**: 21-26.

Mirwald, R. L., Bailey, D. A., 1986, *Maximal aerobic power, A longitudinal analysis*, Sport Dynamics, London, Ontario, Canada.

Mirwald, R. L., Bailey, D. A., Cameron, N., Rasmussen, R. L., 1981, Longitudinal comparison of aerobic power in active and inactive boys aged 7 to 17 years, *Ann. Hum. Biol.* **8**: 404-414.

Murase, Y., Kobayashi, K., Kamei, S., Matsui, H., 1981, Longitudinal study of aerobic power in superior junior athletes, *Med. Sci. Sports Exerc.* **13**: 180-184.

President's Council on Physical Fitness and Sports, 2003, Normative data. *http://fitness.gov.*

Roncesvalles, M. N., Woollacot, M. H., Jensen, J. L., 2001, Development of lower extremity kinetics for balance control in infants and young children, *J. Motor Behavior.* **33**:180-92.

Rowland, T. W., 1996, *Developmental Exercise Physiology*, Human Kinetics Pub., Champaigne, IL.

Rutenfranz, F., Andersen, K.L., Seliger, V., Klimmer, F., Berndt, I., Ruppel, M., 1981, Maximum aerobic power and body composition during the pubertal growth period: similarities and differences between children of two European countries, *Eur. J. Pediatr.* **136**: 123-133.

Sabick, M. B., Torry, M. R., Hawkins, R. J., 2002, Differences in kinematics between professional and youth baseball pitchers, *Proceedings of the IV World Congress on Biomechanics*, Calgary, Canada.

Strickler, P. R., VanHeest, J. L., 2001, Strength training and endurance training in the young athlete. In: Eds: Birrer, R., Griesemer, B., and Cataletto, M. *Pediatric Sports Medicine in Primary Care,* Lippincott-Williams & Wilkins Publishers, Philadelphia, PA, pp. 83-94.

Urhausen, A., 1995, Blood hormones as markers of training stress and overtraining, *Sports Med.* **20**: 251-276.

Uusitalo, A. L., 1998, Hormonal responses to endurance training and overtraining in female athletes, *Clin. J. Sport Med.* **8**: 178-186.

Weber, G., Kartodihardjo, W., Skissouras, V., 1976, Growth and physical training with reference to heredity, *J. Appl. Physiol.* **40**: 211-215.

Williams, J. R., Armstrong, N., 1991, The influence of age and sexual maturation on children's blood lactate response to exercise, *Pediatr. Exerc. Sci.* **3**: 111-120.

Williams, J. R., Armstrong, N., 1991, Relationship of maximal lactate steady state to performance at fixed blood lactate reference values in children, *Pediatr. Exerc. Sci.* **3**: 333-341.

# Chapter 18

# Medical Advances in the Treatment of Sports Injuries

Anthony C. Luke[1], Marc R. Safran[2], Lyle J. Micheli[3]

[1]*Department of Orthopaedics, University of California, San Francisco, 1701 Divisadero St., Suite, 240, San Francisco, CA 94114, PH: (415)353-7590,* FX: 415-353-9675; *EM: antluke@yahoo.com*

[2]*Department of Orthopaedic Surgery, Sports Medicine Center, University of California, San Francisco, 1701 Divisadero, Suite 240, San Francisco, CA 94115, PH: (415)353-7586, FX: 415-353-7593, EM: safranm@orthosurg.ucsf.edu*

[3]*Division of Sports Medicine, Department of Orthopaedic Surgery, Children's Hospital of Boston, Harvard University, 300 Longwood Ave., Boston, MA 02115, PH: 617-355-6028, FX: (617)731-5298, EM: Sports.Medicine@tch.harvard.edu*

## 18.1 INTRODUCTION

### 18.1.1 The "Balance" of the Skeletal System in Sports

In the most simplistic biomechanical model, nerves stimulate muscles, muscles produce forces, tendons transmit forces from muscle to bone, bones act as levers, and joints act as fulcrums to create movement. To achieve the intricate movements of a gymnast or the power and speed of a sprinter, a series of complex interactions within the components of the musculoskeletal system must be orchestrated. Neurophysiological, intracellular and biomechanical processes are all involved in creating movement.

To perform a specific task, a coordinated sequence of signals and movements are required to accomplish a purposeful, efficient action. The body naturally maintains an equilibrium in its systems called homeostasis. During movement, the interaction of each joint of the body influences and is dependent on the others, producing a kinetic chain of activity. Joints require stability offered by static structures, such as ligaments and the anatomic shape of the joint itself, as well as, dynamic tissues, such as the surrounding muscles, which are under neurological control. When a muscle contracts, there are usually one or more antagonist muscles that relax but continue to function to help control motion. Several muscles can contract synergistically to allow more complex movements. For example, when throwing a baseball, the wrist and hand utilize the force generated by the elbow and shoulder. Further force comes from the rotation at the waist and the transfer of weight forward through the legs. Anatomic limitations and excessive stress from thc high angular velocities achieved during the pitching motion apply stress to the joints and tissue restraints (Zheng and Barrentine, 2000). Dysfunction in any link of the kinetic chain can result in abnormal biomechanics, which can ultimately lead to injury

Despite the health benefits, in many cases, sport is still a risk-taking behavior. When the balance in the musculoskeletal system is disturbed, injuries occur. Unexpected trauma or repetitive overload from training can lead to stresses that overcome the capacity of the musculoskeletal system to withstand the stress, leading to breakdown of tissues and injury.

Nature has evolved such that the healing process of humans is for the most part effective and efficient. However, in light of the extreme demands of sports to push the human body to the limit, healing may occur in less than ideal circumstances in conjunction with modified training. Many of the principles and theories of treatment for sports injuries focus on restoring the balance and function of the injured joint or tissue. Advances in medical technology and the science of tissue healing enable more accurate diagnoses to be made and improve recovery by assisting and promoting the healing process.

## 18.1.2 Etiology of Sports Injuries

The etiology of sports injuries is complex. It is often classified into two categories based on the mechanism of injury, namely macrotrauma and microtrauma. Macrotrauma typically involves a relatively high velocity acute trauma, which injures the joint or soft tissues, such as an ankle sprain. In comparison, microtrauma develops from repetitive stress and overuse, usually presenting as a chronic complaint, such as Achilles tendinopathy. The mechanism of trauma can also bc dircct or indirect. Direct trauma results from contact or unexpected movement such as a fall. Indirect trauma can cause injury to tissues secondarily, for examples from over-stretching, excessive exercise, or ischemia-reperfusion injury (Best and Hunter, 2000). Fatigue during exercise is a serious complicating factor to the development of injury. Fatigue results in a decreased ability for affected muscles to maintain force production through complex effects on the intracellular mechanisms responsible for muscle contraction (MacIntosh, 2000).

Ultimately, the development of sports injuries is often multifactorial in nature.

| Contact or Collision | Limited Contact | *Noncontact* |
|---|---|---|
| *Basketball* | *Baseball* | *Archery* |
| *Boxing* | *Bicycling* | *Badminton* |
| *Diving* | *Cheerleading* | *Body building* |
| *Field hockey* | *Canoeing or kayaking* | *Bowling* |
| *Football* | (white water) | *Canoeing or kayaking* |
| Tackle | *Fencing* | (flat water) |
| *Ice hockey* | *Field events* | *Crew or rowing* |
| *Lacrosse* | High jump | *Curling* |
| *Martial arts* | Pole vault | *Dancing* |
| *Rodeo* | *Floor hockey* | Ballet |
| *Rugby* | *Football* | Modern |
| *Ski jumping* | Flag | Jazz |
| *Soccer* | *Gymnastics* | *Field events* |
| *Team handball* | *Handball* | Discus |
| *Water polo* | *Horseback riding* | Javelin |
| *Wrestling* | *Racquetball* | Shot put |
| | *Skating* | *Golf* |
| | Ice | *Orienteering* |
| | In-line | *Power lifting* |
| | Roller | *Race walking* |
| | *Skiing* | *Riflery* |
| | Cross-country | *Rope jumping* |
| | Downhill | *Running* |
| | Water | *Sailing* |
| | *Skateboarding* | *Scuba diving* |
| | *Snowboarding* | *Swimming* |
| | *Softball* | *Table tennis* |
| | *Squash* | *Tennis* |
| | *Ultimate frisbee* | *Track* |
| | *Volleyball* | *Weight lifting* |
| | *Windsurfing or surfing* | |

*Table 18.1.* Classification of sports by contact. Reprinted from Committee on Sports Medicine (2001), pg. 1206, Table 1, with permission from the American Academy of Pediatrics.

Individual athletes are subject to both intrinsic and extrinsic risk factors for sports injuries. Intrinsic risk factors are those that are inherent to the athlete and are usually slow to change, such as, anatomical malalignment, muscle imbalances, pre-existing medical conditions, nutritional concerns, and psychological issues. Extrinsic risk factors involve variables associated with the athlete that can change from moment to moment and may be altered or avoided. Examples of extrinsic risk factors include training errors (excessive frequency, duration, and intensity), the environment (temperature, weather hazards, humidity, etc.), the activity surface, equipment issues, and mental stresses. Each sport has its own risk of injury. The American Academy of Pediatrics has categorized the sports into: full contact, limited contact and no contact activities, suggesting a declining level of risk of macrotraumatic injury (Table 18.1).

### 18.1.3 Medical Advances in Treating Sports Injuries

Medical advances in the field of sports medicine have been made through basic science and clinical research, the experiences of pioneering health professionals, and sometimes the trials and errors of dedicated athletes in training. The sports fields and arenas are practical laboratories for comprehending exercise physiology and injury management. Better understanding of tissue repair and developments in technology have impacted the traditional methods of treatment and rehabilitation of sports injuries.

Medical advances have enhanced the ability of the sports medicine specialist to return the athletes to play sooner, while minimizing the risk for re-injury. Imaging technologies have improved our ability to diagnose problems earlier and more accurately. Symptoms such as pain and swelling can be controlled using constantly improving medications that influence the biochemical mediators of inflammation. Specific exercise programs help control load and stress on tissues in order to stimulate proper healing during rehabilitation. Arthroscopy has made "minimally-invasive" surgery possible, which has revolutionized the way orthopaedic sports medicine is practiced.

Sports medicine is a young, growing field involving multiple specialties and professions, for examples, primary care, orthopaedics, physical medicine and rehabilitation, athletic and physical therapy, psychology, nutrition, chiropractic, podiatry, and other health professionals. In managing the injured athlete, a multidisciplinary team approach is often required. Treatments of injuries can be classified as conservative and surgical. In the majority of athletes, a conservative approach including medical management and rehabilitation is sufficient. However, in cases where pain, instability or loss of function cannot be improved with non-operative treatment, surgical management may offer a solution with a more predictable prognosis. In sports medicine, the greatest challenge for the physician is often deciding when and with what injuries players can return to play with acceptable safety. Understanding the pathophysiology of the injury, the psychology of the athlete, and the demands of the sport are critical in managing the risks in return to play decisions.

### 18.1.4 Overview of the Chapter

This chapter is intended to provide an overview of how advancements in the understanding of tissue healing and changes in technology have improved the management of sports injuries. A number of topics will be introduced in order to highlight diverse concepts and issues in the field of sports medicine. The reader is directed to the references for more in depth information. By demonstrating highlights in the evolution of sports medicine, the reader can appreciate the importance of the pathophysiological, biochemical, and physical processes responsible for injury, which can help explain the theories behind medical treatments. The first section focuses on the science of tissue repair and the body's typical inflammatory response to injury. The structure, function, and fundamental principles of injury and healing for different tissues of the body are discussed, specifically muscle, tendon, ligament, bone and cartilage. The processes of injury and healing lay the foundation for understanding the theories and principles behind various medical treatments.

The healing process can be enhanced through major advancements in the fields of medicine, rehabilitation and surgery. Some technological advances in diagnostic imaging are presented, in particular magnetic resonance imaging and ultrasound. The use of medications to affect the inflammatory response, nutritional issues in sports and management of concussion are discussed. The importance of mobilization, "proprioception" (sensation of joint position and detection of motion), core stabilization, and eccentric loading are popular concepts leading to a more "active" approach to the rehabilitation of sports injuries. Arthroscopy has developed increased indications and is now the most commonly performed surgical procedure worldwide. Reconstructive anterior cruciate ligament surgery is now performed successfully for previously career ending injuries, and is an example demonstrating how technology, bioengineering principles and rehabilitation have advanced the field of orthopaedic surgery. Transplantation techniques, extracorporeal shock wave therapy, and gene therapy are emerging technologies that will also be reviewed. By discussing approaches to problems and the principles of managing sports injuries, the chapter will hopefully stimulate ideas for future clinical innovations.

## 18.2 SCIENCE OF TISSUE INJURY AND REPAIR

### 18.2.1 The Plight of Early Athlete

During human evolution, the body has adapted so that there is maximum potential for recovery without relying on medical treatment. Take for example a prehistoric hunter who is chased by a sabertooth tiger. Though our caveman eludes his predator, unfortunately, he plants his foot and with a twisting motion, injures his right knee, a classic mechanism for an anterior cruciate ligament tear. He develops immediate pain and swelling in the knee due to intra-articular bleeding, which splints the knee at 30° of flexion which helps prevent further movement and damage to knee in the short term. Over the next few days, he has difficulty bearing weight on the leg, which enables his innate healing processes to begin to stabilize the torn ligaments through the development of scar tissue. The improvement in symptoms coincide with the healing process such that his swelling and pain resolve, he recovers his range of motion, and he starts to function again. However, the damaged tissue in the torn ligament no longer provides the same biomechanical functions as uninjured tissue, which may compromise the stability of the knee. Also, there most likely has been associated injury to the cartilage, and surrounding soft tissues, as well as muscle atrophy from disuse. When our caveman returns to the hunt, will his knee have the strength and stability necessary to function? Will he be able to elude his predator in his next encounter?

### 18.2.2 Musculoskeletal Injury

A basic question in the development of sports injuries is: "Can the tissue take the stress applied?" Muscle, tendon, ligaments, bone and cartilage, known as "connective tissues" are subjected to stress and strain, and have capacity to withstand these forces. However, when the threshold limit of force is reached failure of the tissue occurs. (See Stress-strain curve Fig. 18.5). Each tissue is biological and dynamic in nature. It is in a constant remodeling process, particularly as the tissue adapts to training and stress. In the musculoskeletal system, the connective tissues are made up mainly of combinations of collagen, water, and glycoproteins.

Collagen is the greek word for "glue" (Kielty et al., 1993). There are at least 13 types of collagen and they are classified into three groups based on size and physicochemical properties (Kielty et al., 1993) (Table 18.2). Group 1 is characterized by an uninterrupted helical domain of approximately 300 nm (Type I, II, III, V, XI), while Group 2 has a large molecular weight (>160000 Daltons), extended triple helices (>350 nm)

(Types IV, VII) and finally Group 3 consist of short triple helices (short-chain collagens). Type I collagen has high tensile strength and stiffness properties, while type III collagen is thinner and more elastic (Best and Hunter, 2000).

Connective tissues often have a ground matrix component, including proteoglycan and glycosaminoglycan proteins, as well as other glycoproteins. A proteoglycan monomer has a central protein core usually hyaluronic acid. Several glycosaminoglycan chains, such as dermatan sulfate or chondroitin sulfate, are covalently bound to the central core (Newman, 1998). They have a high anionic negative charge responsible for extension of the molecules and attraction of water, which give these macromolecules tensile properties and resistance to pressure.

| Collagen Type | Chain Composition | Molecular Weight | Tissue Distribution |
|---|---|---|---|
| Interstitial collagens | | | |
| I | $\alpha_1(I)_2$, $\alpha_2(I)$ | 300,000 | Skin, bone, tendon, synovium |
| I trimer | $\alpha_1(I)_3$ | 300,000 | Tumors, fetal skin, liver |
| II | $\alpha_1(II)_3$ | 300,000 | Hyaline cartilage, vitreous, nucleus pulposus |
| III | $\alpha_1(III)_3$ | 300,000 | Fetal skin, blood vessels, intestine |
| V | $\alpha_1(V)_2$, $\alpha_2(V)$ | 300,000 | Same as type I collagen |
| XI | $1\alpha$, $2\alpha$, $3\alpha$ | 450,000 | Hyaline cartilage |
| FACIT collagens | | | |
| IX | $\alpha_1(IX)$, $\alpha_2(IX)$, $\alpha_3(IX)$ | 500,000 | Hyaline cartilage, vitreous, cornea |
| XII | $\alpha_1(XII)_3$ | 600,000 | Same as type I collagen |
| XIV | $\alpha_1(XIV)_3$ | ? | Skin, tendon |
| Basement membrane collagen | | | |
| IV | $\alpha_1(IV)$, $\alpha_2(IV)$ | 450,000 | Lamina densa of the basement membrane |
| Other non-fibrillar collagens | | | |
| VI | $\alpha_1(VI)$, $\alpha_2(VI)$, $\alpha_3(V)$ | 570,000 | Aortic intima, placenta, skin, kidney, muscle |
| VII | $\alpha_1(VII)_3$ | 960,000 | Amnion, dermo-epidermal anchoring fibrils |
| XIII | ? | ? | Endothelial Cells |
| Short-chain collagens | | | |
| VIII | $\alpha_1(VIII)_3$ | 500,000 | Endothelial cells, Descemet's membrane |
| X | $\alpha_1(X)_3$ | 180,000 | Growth plate cartilage |

*Table 18.2.* Collagen types. Reprinted from Schumacher et al. (1993), pg. 17, Table 3A-1, with permission of the Arthritis Foundation.

### 18.2.3 Inflammation

When an injury occurs to connective tissues, there is a typical inflammatory response by the body, though there are specific differences in healing response depending on the tissue type. In general, connective tissue responses to injury include inflammation, hemorrhage, edema, and hypoxia, which can result in cell death. The healing process usually involves a proliferation of cells and increased vascularity, triggered by chemical mediators. If the tissue has poor vasculature, there is typically a poorer healing response.

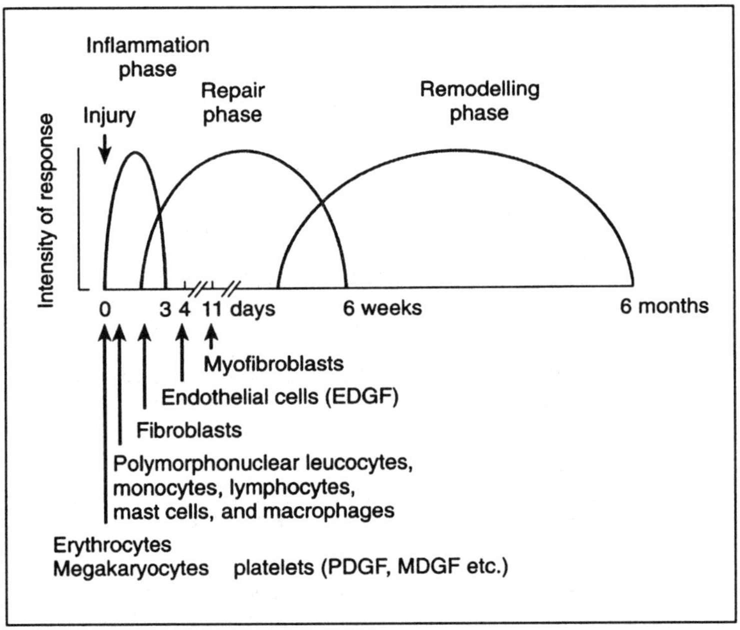


*Figure 18.1.* Stages of inflammation. Reprinted from Oakes (1998), pg. 606, Fig. 36, with permission of Oxford University Press.

The traditional sequence of tissue healing involves three stages: 1) substrate / inflammatory phase (3-4 days after injury), 2) the proliferative phase (around 3-6 weeks), and 3) the remodelling phase (lasting 6-12 months) (Fig. 18.1). During this time, an athlete may experience involuntary muscle spasm, since the body's spontaneous reflex is to splint and protect the injured area. Reduced mobility can lead to disuse atrophy of muscles and other connective tissues.

## 18.2.4 Muscle

### 18.2.4.1 Structure and Function

Muscle has a unique structure, in order for contraction to occur. Skeletal muscle tissue contains types I, III, IV, and V collagen (Best and Hunter, 2000). Muscle fibers are composed of sarcomere units, which contain myosin protein chains that are shaped like golf clubs between actin protein filaments, which are attached to the Z-line that serves as an anchor (Fig. 18.2). The myosin and actin are attached via "cross-bridges" involving the myosin heads (Ait-Haddou and Herzog, 2000). Muscles are organized into motor units, comprising up to 1000 muscle fibers. The smallest myofibrils are grouped to become muscle fibers, which are arranged into muscle bundles. The number of muscle fibers that are activated determines the power of the muscle. If fine motor control is required, there are more muscle fibers and more nerve fibers.

When the muscle unit is signalled, an influx of calcium is released by the sarcolemmal membrane, which surrounds the muscle fibers, triggering contraction of the myosin heads along the actin filaments. This results in shorting of the sarcomere unit. With the coordinated shortening of sarcomere units, the muscle fiber shortens, causing force and movement. Adenosine triphosphate (ATP) then releases the myosin proteins allowing relaxation of the contraction. These mechanisms are based on numerous scientific contributions particularly of A.V. Hill and A.F. Huxley (Huxley, 2000).

There are two main types of muscle fibers that serve different purposes. Slow twitch fibers (Type 1) compose the large motor units required for endurance activities and posture. They typically are innervated by small neurons, use aerobic metabolism, and contain greater amounts of mitochondria and stored triglycerides for energy (Clanton and Coupe, 1998). The fast twitch fibers (Type 2) provide explosive power and are innervated by large neurons. These fibers are designed to function more anaerobically and become fatigued more easily.

A concentric contraction produces tension while the muscle is being shortened, while an eccentric contraction generates tension while the muscle is stretched (Garrett, Safran et al., 1987). Also, fewer motor units are activated maximally during eccentric versus concentric muscle contraction (Komi et al, 2000), which may result in more stress with eccentric exercises. Muscles are important in providing dynamic stability to joints and allowing energy absorption during motion, which decreases shock to the other structures and enables more efficient movement. During rest, a few fibers continue firing producing a resting muscle tone.

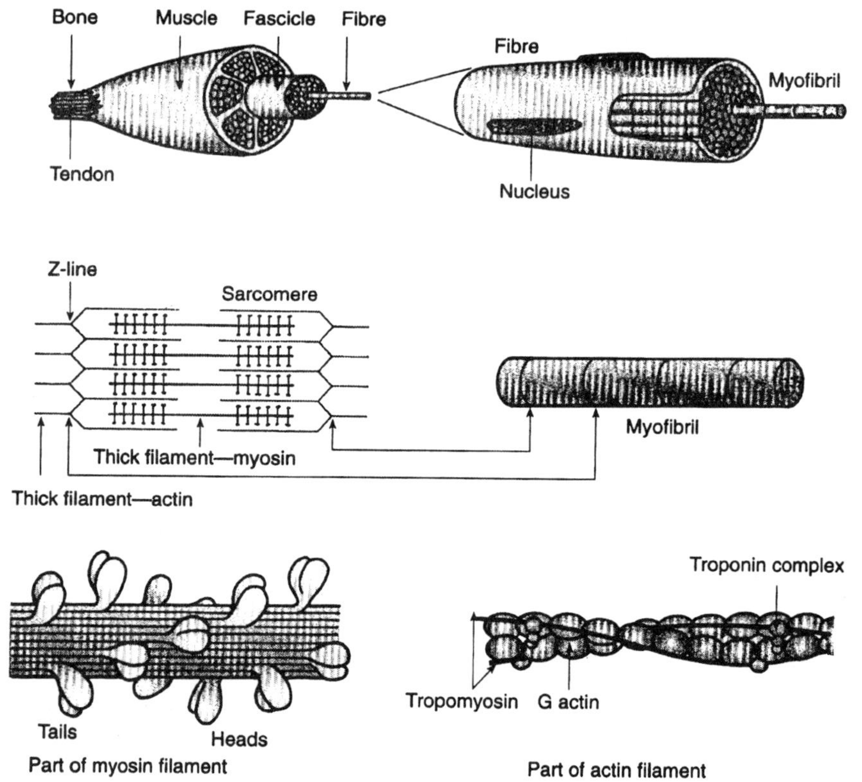


*Figure 18.2.* The structure of skeletal muscle. Reprinted from Lakomy (1998), pg. 172, Fig.1, with permission of Oxford University Press.

## 18.2.4.2 Injury and Healing

Muscle can be injured through direct and indirect trauma (Best and Hunter, 2000). Direct trauma may cause contusion or laceration of the muscle units, producing hemorrhage and edema. Indirect trauma results from stretch injury to the muscle, usually through eccentric contraction, where the muscle is shortening from activation but is also being lengthening from stretch (Garett, Safran et al, 1987). The sarcomeres are overstretched beyond the lengths of their actin and myosin overlaps (Best and Hunter, 2000; Best et al. 1995; Lieber and Friden, 1993). These stretch injuries disrupt the contractile apparatus and cause focal disorganization of the sarcomeres (Mishra et al., 1995). Under increasing stress, partial tearing of muscle fibers can occur, and under excessive force, complete muscle rupture may result. Muscle rupture almost always occurs at the musculotendinous junction (Garrett, Safran et al., 1987). Animal studies suggest that an early sign of muscle injury is decreased contractility and that muscle fibers may become partially disrupted when a specific threshold is surpassed. In the case of rabbits, partial tearing occurred from stretch with greater than 30% of failure force (Noonan et al., 1994).

Muscle repair involves initial inflammatory process in response to fiber damage, hemorrhage and metabolic free radical generation. The muscle injury results in the influx of calcium, cytokines and growth factors leading to an inflammatory process. Neutrophil invasion begins 1-2 hours after injury and peaks at 24 hours and the macrophage response begins at 24 hours and is maximal at 72 hours (Best and Hunter, 2000). These cells remove the damaged myofibers through digestion. IL-1 and TNF stimulate fibroblasts to proliferate and synthesize collagen (Best and Hunter, 2000). Fibroblasts and satellite cells are recruited into the area by cytokine mediators. Collagen production by fibroblasts increases by day 3 after injury, synthesizing higher proportions of type III collagen and also some type V collagen (Best and Hunter, 2000). The type III collagen scar helps preserve the tensile strength of muscle, however, the area is susceptible to reinjury while the muscle fibers are being regenerated. Satellite cells act as reserve stem cells to regenerate the damaged myofibers (Best and Hunter, 2000). They differentiate into myoblasts, which can fuse into multinucleated myotubes, and finally mature into myofibers.

Injuries to the nerves can result in damage to the muscle. The innervation of the muscle units must be restored before proper contraction can occur. Myofibers typically are innervated at a neuromuscular junction, located within the middle-third of the unit (Jarvinen et al., 1998). Damaged axon units can grow at a rate of approximately 1 mm per day, and are capable of sprouting additional terminal fibers to create new neuromuscular junctions across short distances of injury (Jarvinen et al., 1998). Otherwise, myofibers losing innervation atrophy and become dysfunctional.

Chronic exertional compartment syndrome is a muscle problem which usually presents in the athletic population. Muscle groups are divided into compartments separated by fibrous fascia. Compartments exist throughout the musculoskeletal system. Chronic exertional compartment syndrome presents most commonly with leg pain below the knee associated with "tightness" in the muscles, usually in the area of the anterior tibialis and foot extensor muscles, which is reproducible with activity. The underlying pathophysiology is still unclear. The most popular theories on etiology subscribe that symptoms are due to increased pressure in the compartment, which possibly compromises regional blood flow causing mild ischemia (Mohler et al., 1997), decreases metabolic waste product removal (Balduini et al., 1993), stretches the surrounding fascia (Rorabeck et al., 1988;

Schepsis et al., 1993) or results in a combination of mechanisms. The contractility of the muscles and function of nerves may become affected transiently. Symptom improvement has been shown with compartment release performed by surgical fasciotomies (Micheli et al., 1999; Schepsis et al., 1993).

## 18.2.5 Tendon

### 18.2.5.1 Structure and function

Tendons consist mainly of Type 1 collagen organized into large, highly, orientated fibers (Fig. 18.3). The collagen fibrils made of tropocollagen (86%) align and form regular compact bundles in layers of collagen to form the endotenon, along with tenocytes (fibroblasts), some elastin, and a proteoglycan matrix (1%) (Arangio, 2001) The endotenon is surrounded by the epitenon, a tendon sheath of loosely interwoven type 1 and 3 collagen, forming a layer of connective tissue rich in blood vessels. Finally, the epitenon is covered by the paratenon, a loose areolar connective tissue of type 1 and 3 collagen, elastic fibrils and an inner lining of synovial cells (Khan et al., 1999). There are few nerve endings in tendon, however, the paratenon and the bone-tendon interface (osteotendinous junction) are innervated. Meanwhile, the musculotendinous junction is the location most commonly injured along the muscle-tendon unit.

The role of tendons is to transfer force from muscles to bone and improve locomotor efficiency. Viscoelasticity is an important property of tendon, since it allows the tendon to resist large loads as well as absorb shock with minimal elongation and energy loss. Thus, energy stored in the tendon during movement is often recovered to improve kinetic efficiency (Alexander, 1984). The tensile strength of tendons increases during childhood, peaks from 25 to 35 years of age and then decreases slowly afterwards (Józsa and Kannus, 1997). As a result, tendons become stiffer and less compliant, and have a lower metabolic activity in the later decades (Almekinders and Deol, 1999). As a result, tendon injury is more common in older populations.

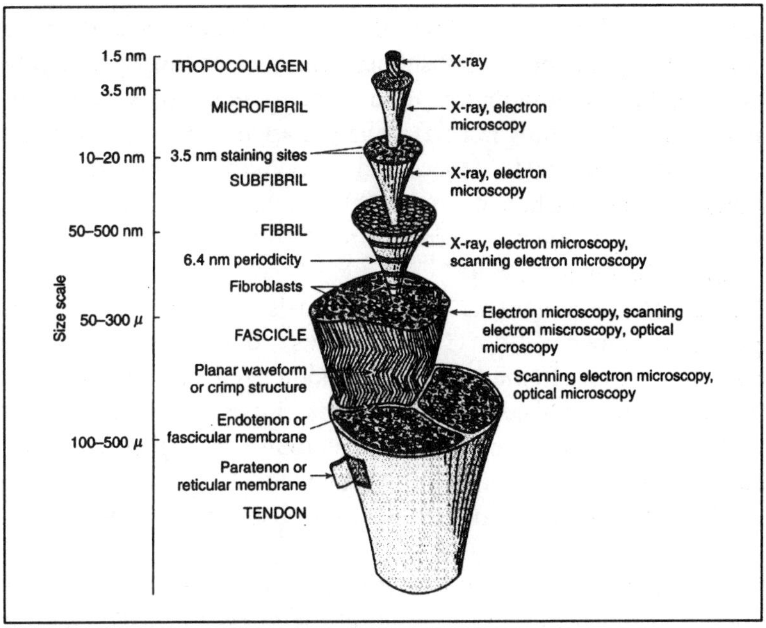


*Figure 18.3.* The structure of tendon and ligament. Reprinted from Oakes (1998), pg. 586, Fig. 4, with permission of Oxford University Press.

### 18.2.5.2 Injury and Healing

Stress in a tendon is limited by the size of its associated muscle as well as the total motion of the joint involved. Tendons located around joints with increased mobility, such as the shoulder and hip, or cross two joints are at higher risk for injury, since there is greater need for dynamic muscle stability. High eccentric stresses can lead to tendon injury, especially in muscles with large percentages of fast-twitch fibers or if there are misfiring muscle patterns.

Mechanical stresses can lead to degeneration of collagen fibers, which is referred to as "tendinosis". Histopathological assessments of areas of tendon injury reveal degeneration and an absence of inflammatory cells, making the description "tendonitis" inappropriate. Microscopic examination of symptomatic and asymptomatic Achilles tendons by Åström et al. demonstrated various pathologies (Åström and Rausing, 1995). Areas of tendinopathy showed blurring of individual bundles and loss of the normal parallel alignment of collagen bundles, with increased number of cells which were not inflammatory in nature. Irregular hypervascularity was noted demonstrating new blood vessel formation or "angiogenesis". Partial tears were seen in some cases with an obvious discontinuity in the tendon and fibrin deposition but otherwise changes were similar to those found in tendinosis. Finally, only a few of the specimens showed inflammatory reactions in the paratenon. Subsequent analyses of biopsies of patients with chronic patellar tendinopathy and elbow tendinosis confirm mucoid degeneration with loss of normal collagen appearance, increased cellularity identified as fibroblasts, increased vascularity and the absence of inflammatory cells (Khan et al., 1999; Nirschl, 1992). These areas have characteristic high signal intensity on MRI and hypoechogenic areas on ultrasound (Khan et al., 1999).

The exact pathogenesis of tendinosis is still being discovered. Temperatures above 42-43°C could cause degenerative changes in the central core of some human Achilles tendon (Wilson and Goodship, 1994). Recent work by Alfredson et al. has investigated the roles of the neurotransmitter glutamate, which were found in higher concentrations in symptomatic extensor carpi radialis brevis and Achilles tendons (Alfredson et al., 2000, Alfredson et al., 1999), as well as, angiogenesis in mediating pain.

The treatment of tendon injuries often involves rest, modification of activities, and physical therapy in an effort to eccentrically load the tendon without causing re-injury to stimulate the tenocytes and allow the area to heal (Khan et al., 1999) (see Section 18.5.2). Persistent cases may require surgical debridement of the degenerative tendinosis. If there is significant disruption of the tendons with displacement of the tendon ends, surgical repair may be necessary to restore function to the tendon unit.

## 18.2.6 Ligament

### 18.2.6.1 Structure and Function

The structure of ligaments consists of fascicles of collagen oriented in parallel or branched and interwoven arrangement with a characteristic "crimp" pattern (Fig. 18.3). The major constituent is type 1 collagen, which is similar to tendon and bone (Woo et al., 1993) and the associated proteoglycan molecules with glycosaminoglycan side chains enhance the viscoelastic properties of the ligaments.

Ligaments play an important role in providing stability to joints, minimizing stress and load on the joints, maintaining proper alignment of the bones, guiding joint motion and possibly enhancing proprioception. Ligaments can provide static and dynamic stability to a joint, via direct osseous insertions and/or indirect insertions, where superficial fibers blend with the outer layer of bone called the periosteum (Woo et al., 1993). The direct insertions have 4 zones over a short distance sometimes 1 mm across, consisting of: 1) ligament midsubstance, 2) fibrocartilage, 3) mineralized fibrocartilage, 4) and bone surface. Neural elements have been discovered in the ligaments and the joint capsule that appear to generate neurologic impulses on stretching and may be involved in proprioception and neuromuscular feedback (Frank, 1996). Ligaments also demonstrate evidence of microvasculature, suggesting healing potential.

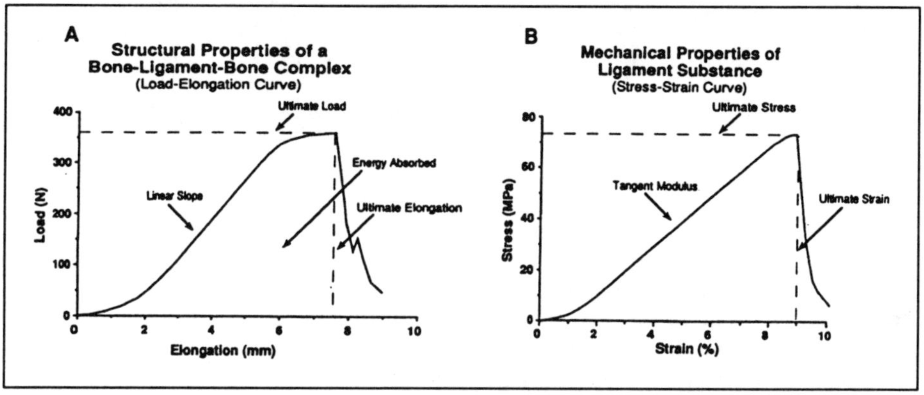


*Figure 18.4.* Load elongation curve and stress strain curve. (A) Typical load-elongation curve for determining the structural properties of a bone-ligament-bone complex. (B) Typical stress-strain curve for determining the mechanical properties of a ligament. Reprinted from Woo (1993), pg. 76, Fig. 4, with permission from Elsevier Science.

Like tendons, the structural, mechanical and viscoelastic properties are important in order for ligaments to fulfill their roles. Under tension, the stiffness and the number of recruited fibers help determine how much stress the ligament can withstand. Figure 18.4 demonstrates an example of the load-elongation curve and stress-strain curve for a ligament. Stiffness of the tissue refers to "the amount of load required to displace the bones to which a ligament is attached in a particular direction" (Frank, 1996) and is represented by the slope of the load-elongation curve. The slope of the stress-strain curve refers to the tangent modulus or Young's modulus of the ligament midsubstance, rather than including the bone insertions, describing a mechanical property of the ligament (Frank, 1996).

The viscoelastic properties of creep and stress relaxation are important to understand the forces required to tear a ligament and in ligament repair (Fig. 18.5). Creep refers to the increase in length of a material when a constant load is applied (Woo et al., 1993). Meanwhile, stress relaxation explains that less force required to keep visco-elastic material stretched to a constant length.

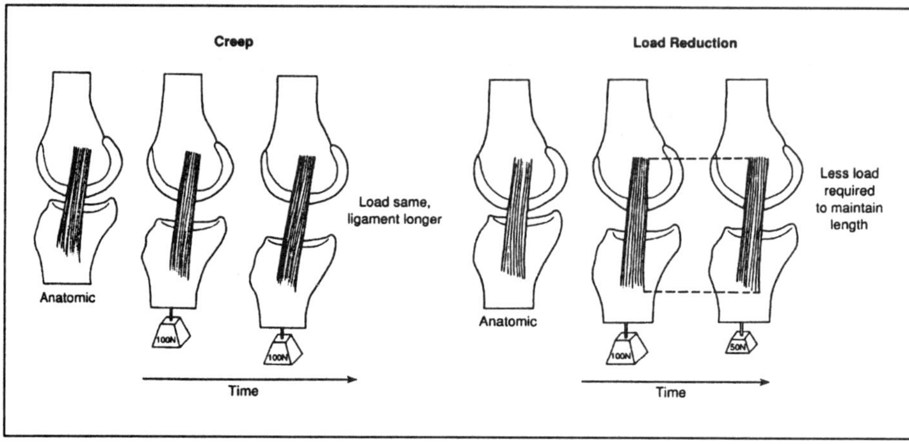


*Figure 18.5.* Creep and stress relaxation phenomenon. Left, Creep is the increase in length of a ligament that occurs over time when a ligament is subjected to a constant load. Right, Stress relaxation/Load reduction is the decrease in load that a ligament experiences over time when it is held at a specific. Reprinted from Frank CB (1996), pg. 75, Fig. 1, with permission from the American Academy of Orthopaedic Surgeons.

## 18.2.6.2 Injury and Healing

The most serious concerns for ligaments are ruptures and tears. When an applied force surpasses the ultimate stress or load, mechanical failure can occur. When fibers are disrupted, the inflammatory response occurs, followed by formation of scar tissue then remodeling as discussed in Section 18.2.3. However, the scar tissue that results only has approximately one third to half the tensile strength of healthy tissue, despite remodeling (Hildebrand and Frank, 1998). Intraarticular ligaments, such as the anterior cruciate ligament (ACL), degenerate at the ends due to increased collagenase activity affecting repair (Amiel et al., 1989). Also, the surrounding synovial fluid, the ligament ultrastructure, and response of the ACL fibroblasts are factors accounting for poor healing of the ACL (Hyman and Rodeo, 2000).

Ligament injuries have an effect on the load sharing principles of the involved joint (Frank, 1996). If one structure is injured then the balance is disrupted and there is a shift of the load onto the remaining structures (Fig. 18.6). The static stabilizers particularly the anatomic architecture of the joint, the joint capsule, and the ligaments must be able to maintain stability, otherwise the function and sharing of the load of the damaged ligament needs to be replaced, usually through surgery or bracing. The dynamic

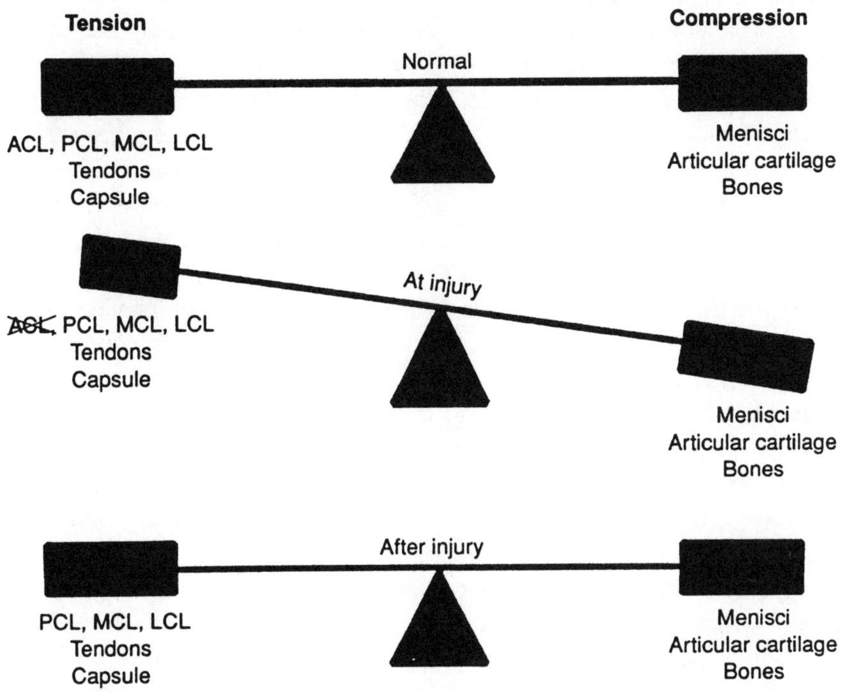


The balance between the tension-bearing elements and the compression-bearing elements of a knee joint. **Top,** In the uninjured knee joint, the tensile and compressive elements are in equilibrium. **Center,** Immediately after failure of one or more structures, imbalance occurs instantly. **Bottom,** To restore joint function, the forces must be redistributed among the remaining structures, thus increasing the stresses on the remaining structures and predisposing some of them to subsequent failure. ACL = anterior cruciate ligament; LCL = lateral collateral ligament; MCL = medial collateral ligament; PCL = posterior collateral ligament.

*Figure 18.6.* Model of Knee in Equilibrium. Reprinted from Frank CB (1996), pg. 77, Fig. 3, with permission from the American Academy of Orthopaedic Surgeons.

stabilization of the joint is also important, which is improved with physical therapy. Proprioception is a critical component of the rehabilitation program during recovery (see Section 18.5.3).

## 18.2.7 Bone

### 18.7.1 Structure and Function

Bone is composed of an organic collagen matrix (40%) and an inorganic component of calcium hydroxyapatite. Bone is in a constant process of turnover. Osteoblast cells secrete the matrix collagen, while osteoclasts provide an antagonistic function, by releasing enzymes that resorb bone. Some osteoblasts that are trapped within the structure of the bone become osteocytes and serve to maintain bone by secreting calcium and phosphate resulting in calcium hydroxyapatite formation, which mineralizes the bone giving bone its strength. Vascular supply is a necessity for bone growth.

There are 206 bones in the human body. They can be classified by shape and function (Nigg, 1998). Long bones provide appropriate levers for movements in the extremities, for examples the humerus and the femur. Short bones are compact and help to provide strength in joints such as the wrist and the foot. Flat bones are broad providing wider protection and allowing a large surface area for tendon and ligament attachments, for examples the scapula and the skull. Finally, irregular bones, such as the vertebrae of the spine and the pelvic bones, are uniquely shaped and each has special functions and characteristics. Therefore, the functions of skeleton allow the bony framework to support the body against gravity, act as a lever system to transfer muscular forces, and protect the internal organs from injury (Nigg, 1998).

Bone has a complex architecture. Cortical or compact bone has strong tensile and compressive strengths and is appropriately found in the shafts of the long bones. Trabecular, cancellous or spongy bone is softer and found more in the axial skeleton, such as the spine. The covering of bone is referred to as the periosteum, which contains osteoprogenitor cells, which can differentiate into osteoblasts to create new bone.

The development of the skeleton in children is unique in that the bones enlarge with age and the shape and size of bones are genetically inherited. Most bones develop from primary and secondary centers of ossification, which mineralizes bone from cartilage. The growth plate or "physis" contains areas of reserve cells, which proliferate, hypertrophy and become calcified. Under gene control, the physis ceases growth at a specific time, completing calcification and attaining the adult form.

#### 18.2.7.2 Injury and Healing

In sports, macrotrauma to the bone can lead to fracture while microtrauma can result in overuse stress injury to the bone. Sufficient force can overcome the structural properties of the bone resulting in frank disruption of the bone and fracture. When a bone breaks, a hematoma forms and local and systematic messengers are released. Surviving local cells are sensitized, then proliferate, differentiate and organize to promote healing. Fibroblasts and new capillaries use the loose fibrin framework to organize and secrete a procallus. This is slowly remodeled into a fibrocartilaginous callus, composed of matrices of cartilage and woven bone, and ultimately becomes an osseous callus, as it mineralizes. Over time, the remodeling process improves the strength and shape of the callus. This process typically can take from four to more than sixteen weeks (Frost, 1989), requiring limited stress loading to the fracture site while healing.

In certain cases, displacement, shortening, rotation and angulation of the fracture site leads to healing in a non-functional position. Therefore, it is often crucial for the orthopaedic surgeon to restore the anatomy of the fracture, by closed reduction and cast immobilization, or open reduction and fixation of the fracture site. Repair may require use of hardware, such as screws, plates, and implants, to maintain alignment and help absorb stress at the fracture site.

Repetitive loading of a bone can result in declining osseous stiffness and mechanical fatigue even though the forces are submaximal to create a fracture (Schaffler, et al., 1990). Overuse stress injuries to bone progress through a spectrum of severity starting with subtle microfractures in the cortex and ending in overuse mechanical failure and a clinical stress fracture.

The removal of the inciting repetitive trauma especially from sports-specific training is often necessary to allow for proper bone healing. With limited stress, the remodeling process begins five days after stimulation. Bone remodeling can continue for approximately 180 days (Beck, 1998). Osteoclastic tunneling occurs over the next 30 to 45 days, and after 30 days, osteoblasts begin to fill the resorbed space with new bone.

The growth plate or "physis" represent an area of potential injury in children. The growth plate is usually the weakest link in the musculoskeletal system in young people. Whereas adults with the same mechanism of injury would have a ligament tear, children who are growing experience fractures to the growth plate, classified as "Salter fractures". The physis is particularly susceptible to repetitive microtrauma, resulting in a group of injuries referred to as "apophysitises" (Micheli, 1987). These injuries are the result of traction and repetitive load often through sports activities, possibly in association with rapid growth and muscle-tendon unit flexibility imbalances (Micheli et al., 1986).

## 18.2.8 Cartilage

### 18.2.8.1 Structure and Function

The articular cartilage refers to the hard, smooth areas covering bones at the articulating surfaces of the joints. The ultimate stress for articular cartilage (dry) is approximately 5 MPa or 500 N/cm$^2$, and its friction coefficient is 0.0025, which is the lowest for any solid material (Nigg, 1998). It is composed of hyaline cartilage, which consists predominantly of type II collagen, but also types VI, X, IX and XI collagens. The latter two are unique to cartilage (Ghivizzani et al., 2000). Large hydrophilic proteoglycan molecules and type II collagen form a tight lattice structure referred to as the matrix or "solid phase" of cartilage (Ghivizzani et al., 2000). The water associated with the matrix or "liquid phase" gives cartilage its resilient mechanical properties.

Articular cartilage is divided into four zones, which vary in collagen architecture and cellular organization of chondrocytes (Newman, 1998) (Fig. 18.6). The superficial zone, which lies on the articular surface, has thin collagen fibrils parallel to the joint surface. The middle and deep zones have progressively greater degrees of vertically aligned collagen fibers with more active chondrocytes. The deepest zone consists of calcified cartilage. The specific orientation and mechanical properties must remain intact in order to meet the functional demands of the joint. There is a small population of chondrocytes responsible for maintaining and replacing the cartilage. Chondrocytes are continually involved in the degradation, digestion, and synthesis of the matrix (Buckwalter and Mankin, 1998). Proteoglycans can be turned over in days or weeks, while fibrillar collagens are replaced over months or years (Evans, 1999).

The fluid dynamics through the matrix are crucial for maintaining the health of the articular cartilage and for determining the mechanical properties. Articular cartilage is avascular and has no nerve supply. Nutrients arrive to chondrocytes via the capillaries of the synovium and diffusion into the synovial fluid, permeating the cartilage matrix through the fluid dynamics created by normal joint loading (Ghivizzani et al., 2000). The fluid movement also removes cellular metabolic waste products. In addition, water flow through the matrix of cartilage causes a cushioning effect (Fig. 18.7). For example, under initial loading of a joint, up to 75% of the compressive stress is absorbed by the fluid in the matrix. However, after sustained loading, the fluid is squeezed out of the matrix, and the solid portion of the matrix bears the load (Mow et al., 1984; Mow et al., 1980). Under load for greater than 5 minutes, the deformation that occurs in the cartilage takes several minutes to recover (Nigg, 1998). The arrangement of type II collagen restrains the large proteoglycans and limits their ability to swell, providing a pressure that contributes to the stiffness of cartilage enabling it to withstand heavy loads. If the proteoglycans are less compact, they can "imbibe additional water", reducing the stiffness of cartilage (Ghivizzani et al., 2000).

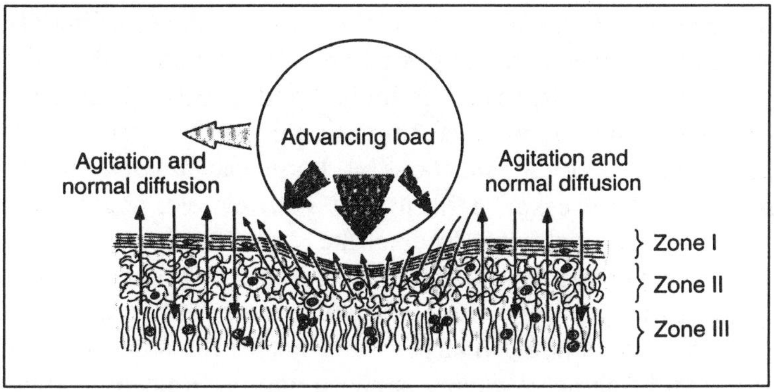


*Figure 18.7.* Effect of load on articular cartilage. Reprinted from Stanish and Wood (1998), pg. 682, Fig. 6, with permission of Oxford University Press.

In addition to articular cartilage, the knee specifically has meniscal cartilage structures. The menisci are thick disks of fibrocartilage that are located in the medial and lateral compartments of the knee between the tibia and femur. They are congruently shaped to help support and stabilize the tibia and femur, to absorb and handle load placed on the knee, and to protect the articular cartilage. Almost 85 per cent of the load on the knee is transmitted through the meniscus during knee flexion, while 50 per cent is transmitted through the meniscus during extension (DeHaven, 1994).

### 18.2.8.2 Injury and Healing

Excessive loading of the articular cartilage can lead to degeneration of the matrix and injury. Direct blunt trauma or torsional loading can cause acute injury to the joint surface, while repetitive impact or joint malalignment can lead to abnormal loading of the articular cartilage (Ghivizzani et al., 2000). Damage to the joint can fracture the matrix and cause chondrocyte necrosis. Pain and dysfunction may lead to an altered load distribution pattern causing excess stress to surrounding areas of cartilage. The process of mineralization may be activated, which causes thickening of the subchondral bones, and thinning of the articular cartilage leading to early osteoarthritis (Ghivizzani et al., 2000). In some cases, sufficient trauma may cause defects in the surface of the cartilage. Partial thickness defects involve the matrix only, while full thickness injuries may involve the subchondral bone.

Usually with minor injury, the chondrocytes respond through the mediation of cytokines and can restore the matrix. The avascular nature of cartilage gives it poor healing potential. Also, immobilization due to pain may affect the nutrient flow to the chondrocytes, leading to further chondrocyte death. Therefore, movement and controlled loading of the joint are important to resume as soon as possible. Unfortunately, the ability of the chondrocytes to repair the matrix is limited and in cases of partial cartilage defects, the lesion may remain causing catching in the joint. Full thickness tears that extend into the subchondral bone can involve injury to the vasculature. In these cases, a fibrin clot with platelet aggregation occurs from bleeding, drawing chondrocyte-like cells to the area. A typical inflammatory response may occur, however, the collagen which is laid down forms a fibrocartilage scar, which has weaker strength and less durability than that of the surrounding matrix. As a result, the area of fibrocartilage may fracture and fragment, leaving exposed bone. Clinically, exposed bone, which is innervated, can cause pain, swelling and catching symptoms due to the irregular surface of the joint. Damage to the articular surface and subchondral bone may lead to osteoarthritis in some athletes. This appears to be particularly more prevalent in athletes with significant malignment, previous injury or surgery. Degenerative joint disease is common in the hip, ankle and knees.

Meniscal tears in young athletes are usually traumatic from a twisting knee injury, while they are more commonly due to degenerative wear and tear in adults. Unfortunately, the menisci have blood supply limited to their peripheral portions, which gradually declines with age. Therefore, the healing potential for meniscal tears improves in patients, who are younger and sustain stable tears in the periphery of the cartilage. If possible, a large meniscus tear is repaired with suture techniques or mechanical stabilization devices during arthroscopy. If a significant mechanical tear cannot be repaired, it is usually removed and debrided through the arthroscope. It is always desirable to preserve as much of the meniscus as possible to protect the articular surface. Increased loss of the meniscus results in decreased contact area and increased pressure in the involved compartment (Baratz and Fu, 1986).

## 18.3 MANAGEMENT OF SPORTS INJURIES

Understanding the pathophysiology of sports injuries allows the sports medicine practitioner to determine the appropriate management plan, including any necessary modifications of training to avoid jeopardizing the athlete's health. Technology has improved our ability to manage damaged tissue and shorten recovery time. Surgery may be necessary to restore appropriate function. Principles of tissue healing, surgery and rehabilitation must be considered to devise a treatment plan and determine safe return to play. To simplify the treatment goals of sports medicine, one may consider resolving pain, restoring function, and reducing risk of complications, including re-injury.

### 18.3.1 Imaging

The ability to visualize the structures of the human body has revolutionized medicine. Diagnostic imaging has enabled earlier, more accurate diagnoses to be made, and thus proper, earlier treatment may be implemented. Two-dimensional projectional X-ray images have improved to three-dimensional images and reconstructions in multiple planes through computed tomography (CT scan). Magnetic resonance imaging (MRI) creates images in multiple planes or slices through the body in different directions using a strong magnet and radio waves to detect the alignment and motion of protons in the body's water molecules. Imaging technologies and techniques, such as fluoroscopy, ultrasound and dynamic MRI, can assist in treatment by visualizing the area of concern, in the operating room during surgery or in the radiology suite through interventional procedures by the radiologist. Functional testing is becoming available. Ultrasound can evaluate a structure as it moves through its positions of function. Dynamic MRI allows assessment through a range of motion. Function at the cellular level is evaluated in the research setting using very high field strength (3-4 Tesla) MRI scanners. The appropriate selection of an imaging test can help a physician identify injuries affecting bones to soft tissues, providing valuable information regarding treatment and prognosis.

### 18.3.2 X-Ray and CT

Basic radiographs are still the baseline test for any potential bone injury. X-rays are directed toward a film through a part of the body. The bones absorb most of the X-rays, allowing fewer to expose the film. Since an X-ray is a photographic negative, areas that get less exposure to the X-ray beam are more white on an X-ray film. An X-ray is a two-dimensional projection of a three-dimensional object. Despite newer elaborate imaging modalities, X-ray is still the best preliminary test to check for fractures, bone tumors, infections and other bony pathology. Bone mineral density can also be evaluated to help estimate the strength of bone. Digital films are becoming standard, removing the need for hard copy film, and allowing better evaluation of both soft tissues and bone on a single exposure.

Using a pencil thin beam of X-rays, a CT scanner can produce images of axial slices through the body. These are two-dimensional slices rather than a projectional image so there is no overlap of structures. In addition, the CT scanner can use coordinates on two-dimensional slices to recreate a three-dimensional structural projection. This allows for volumetric, geometric and biomechanical assessments (Genant et al., 1999) Ultra-high resolution CT can improve spatial resolution, assess the trabecular and cortical structure, and assess bone mineral density (Genant et al, 1999).

## 18.3.3 Bone Scan

Nuclear medicine uses the administration of a radiopharmaceutical agent for the diagnosis and treatment of disease. The common bone scan uses a radioactive metastable technetium-99 (Tc-99m) molecule coupled to a diphosphonate molecule, which has an affinity for sites of osteoblastic activity (Tung and Brody, 1997). A two to three hour period of delay after injection allows uptake of tracer by the skeleton and renal excretion of excess chelate. The patient is then positioned next to a gamma camera that detects emissions from Tc-99m. Bone scans are useful for identifying pathology that results in increased bone turnover, such as stress injuries to bone, stress fractures, osteoarthritis, and benign and malignant bone tumors, as well as demonstrating bone metabolism and activity. The test is extremely sensitive for bone pathology but not specific. Bone scan may be the earliest test to demonstrate pathology, even before MRI shows signs of edema (Van der Wall et al., 2001). Single photon emission computed tomography (SPECT) can help localize a lesion in three dimensional space by taking "slices" in three planes of the body and is particularly useful for assessing the spine. The radiopharmaceutical is retained in particular tissues in the body for minutes to days with no reported late deleterious effects.

## 18.3.4 MRI

Magnetic Resonance Imaging (MRI) uses the varying proton content and physical constraint of the protons in different tissues to produce contrast. The patient is placed into the main magnetic field, ranging from 0.2 to 4 Tesla. A stronger magnetic field typically enables better resolution. There are investigational devices as high as 8 Tesla with more advances in development. Hydrogen protons found in the water molecules in tissues align in the main magnetic field. A radiofrequency pulse is applied to excite the protons. The protons absorb the energy then relax, emitting the MR signal. The net vector of MR signal is calculated and localized in space by the scanner with the use of a gradient field. Through the Fourier transformation, the results of all the points of information produced by the repeated sequences of excitations and relaxations registered during the scan create the images.

The appearance of the images depend on the field strength, the echo time (TE), the repetition time (TR) and the signal-to-noise (SNR) ratio achieved (Genant et al., 1999). Different sequences are used to identify specific pathologies (Figs. 18.8 and 18.9). Common sequences include T1 weighted views and proton density sequences, which are useful to define anatomy. T2 weighted images are valuable to identify pathology associated with edema and fat suppression sequences can help assess articular cartilage (Peterfy et al., 1994).

Technical parameters that affect performance include signal-to-noise ratio (SNR), contrast to noise ratio (CNR), spectral separation (between protons in fat and those in water) and acquisition speed (Peterfy et al., 1997). These will also have effects on the diagnostic capabilities of MRI. CNR is more important clinically since it determines the ability to discriminate one structure from an adjacent structure (Peterfy et al., 1997).

Magnetic resonance imaging has become a common non-invasive diagnostic procedure in sports medicine injuries. Diagnostic accuracy for meniscal tears is commonly reported around 90% (sensitivity ranges from 83-100%) while accuracy for anterior cruciate ligament tears is approximately 93% (sensitivity ranges from 71-100%) (Rappaport et al., 1996). In adults (Gelb et al., 1996) and children (Kocher et al., 2001), MRI has been found to be specific for ligament and meniscal pathology, however, judicious use is indicated as clinical history and physical exam is often sufficient for diagnosis. MR arthrography uses an intrarticular injection of gadolinium contrast in order to improve visualization of lesions in the labrum or capsule of joints, especially the shoulder, hip or wrist (Tung and Brody, 1997).

Changes in hardware technology facilitate the use of MRI for musculoskeletal problems. Dedicated extremity MRI uses a circumferential small-bore magnet in which only the patient's extremity is inserted, thereby saving on cost and space, as well as improving patient comfort and safety. However, only the distal two thirds of an extremity can be investigated (Peterfy et al., 1997). The disadvantages are the lower field strength (0.2 to 0.3 Tesla (T)) which leads to lower SNR. On the other hand, open MRI avoids the circumferential enclosure of the patient, allowing large athletes to be positioned properly. It also enables joint motion as well as patient access, which allows dynamic MRI studies of a joint to be performed as well as some interventional procedures (Seeger, 1998). Open magnet systems do have decreased SNR compared to the standard tunnel magnet and thus produce images with relatively poorer quality and resolution.

Functional MRI is being performed in research to assess the cellular processes involved in training and disease. Nuclear magnetic resonance spectroscopy allows study of muscle metabolism by measuring phosphorus metabolites, such as phosphocreatine (PCr) and phosphorus ($P_i$), which are involved in energy production. This method can detect decreases in oxygenated blood flow to specific muscle groups (Balduini et al., 1993). Functional MRI may also be used in assessing cerebral blood flow and oxygenation following concussion, by identifying different magnetic properties of oxyhemoglobin and deoxyhemoglobin (McAllister et al., 1999).

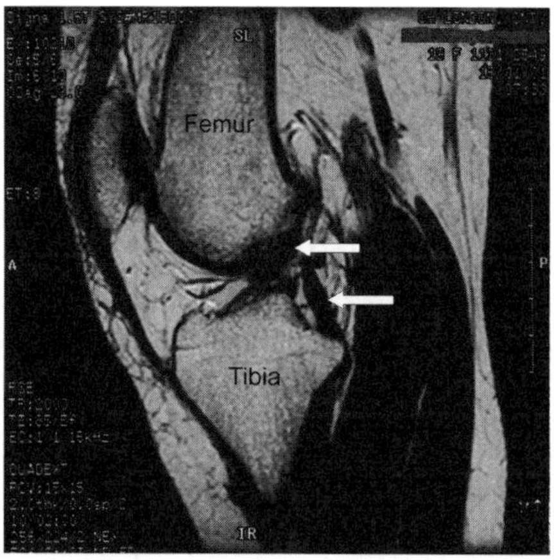


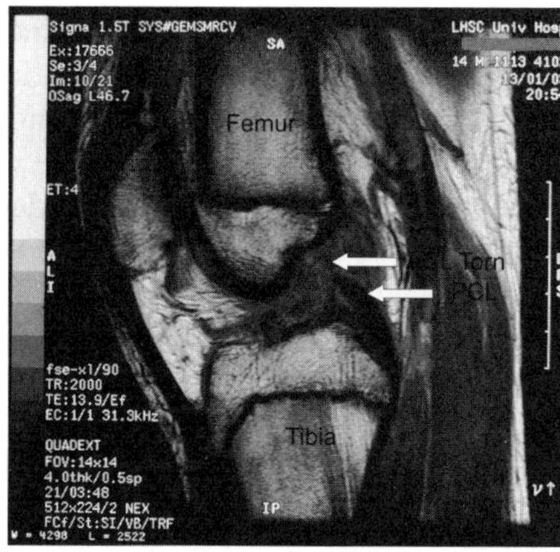


*Figure 18.8.* Left: Sagittal MRI demonstrating ACL in a 19 year old female.
*Figure 18.9.* Right: Sagittal MRI showing torn ACL in a 14 year old male (Note: Growth plates are still open).

Improvements in sequencing and hardware continue to make MRI more efficient and economical. MRI is a safe and extremely valuable imaging modality because of its absence of radiation, superior definition, and high sensitivities for pathology. The applications of MRI will continue to expand as proper sequencing, radiologic interpretation and clinical significances of findings are further understood.

## 18.3.5 Ultrasound

Ultrasound has been developing into a very useful diagnostic tool for musculoskeletal injuries. Though the examination is highly operator dependent, ultrasound provides a safe, economical alternative to X-ray and MRI technology. The junction of two media of different acoustic properties is the acoustic impedance. An ultrasound beam gets reflected, refracted or diffracted at interfaces between different media. Greater returned energy is reflected back at an interface along the incident path, hitting the transducer and forming a signal by the reverse piezoelectric effect (Lopes et al., 1994).

Signals in one plane are stored in a computer as a "brightness" resolved on a scale of grey tones.

Improvements in ultrasound involve the uses of high-resolution linear array probes, tissue harmonic imaging, two-dimensional matrix probe technology, extended field-of view imaging and power Doppler sonography (Winter III et al., 2001) . Linear array transducers using frequencies up to 15 MHz can produce images of superficial structures with axial resolution in the range of 250 to 500 microns (Merritt, 2001).

Ultrasound is useful for looking at superficial musculoskeletal structures such as shoulder pathology particularly the rotator cuff, ankle ligament tears, the compartments of the lower leg (posterior tibialis, Achilles, peroneals) (Thain et al., 2001), superficial knee ligaments (MCL, LCL) and tendons (patellar and quadriceps), and some hip pathology. Ultrasound can distinguish if a mass is cystic or solid. Power Doppler can be used to assess blood flow to tissues and identify areas of inflammation or active healing (Neuman et al., 1994) . A major advantage of ultrasound is that it can be used to study structures dynamically, so it can assess subluxation of tendons, view impingement of structures such as rotator cuff tendons between the humerus and the acromion in the shoulder. Ultrasound-guided procedures enable direct visualization of the intended area of treatment. These include aspiration of cysts, acute hematomas, joint effusions, and bursae (Cardinal et al., 1998). Biopsies and removal of calcifications can also be peformed. Bone-mineral density can also be determined using portable ultrasonography of the calcaneus (Taaffe et al., 1999) although the accurate correlation with central (spine and hip) bone mineral density assessments has not yet been proven and testing may be operator dependent.

## 18.4 MEDICAL

Today's athletes come in all shapes and sizes. Special populations participate with medical diagnoses, including physical or mental challenges, diabetes, and even post-organ transplants. Medical advice regarding sports activities is individualized based on the athlete's interests and medical history. Pre-participation physicals can be performed to identify risk factors for injury and for medical-legal purposes. Because of the health benefits of exercise, the ideal goal is to maximize fitness in all individuals.

### 18.4.1 Anti-inflammatory medications

Pain is the body's signal that there is a problem. Unfortunately, nature's healing process does not always accommodate an athlete's training and competition schedule. Medical treatment of sports injuries often involves influencing the pathways of inflammation and pain.

The biochemical pathways of inflammation and pain have been well-studied. The effect of anti-inflammatory medications such as aspirin, were discovered to inhibit arachidonic acid production and consequently formation of prostaglandins and thromboxanes by blocking the function of the cyclo-oxygenase (COX) enzymes (Fig. 18.10). There are two types of COX enzymes. Type 1 COX enzymes produce homeostatic prostaglandins that maintain gastrointestinal mucosal integrity and renal blood flow (Mukherjee et al., 2001). They are also expressed in platelets mediating thromboxane A2 production, and perform house-cleaning cellular type functions. Cyclo-oxygenase type 2 is involved in inflammation.

Non-steroidal anti-inflammatory drugs, known as NSAID's have been commonly used in the treatment of musculoskeletal injuries. The analgesic and mild anti-inflammatory effects make it a popular medication for pain. They are usually taken orally but can be administered intramuscularly, intravenously and topically. However, there can be side effects from long-term NSAID use, particularly gastrointestinal bleeding, acute renal failure,

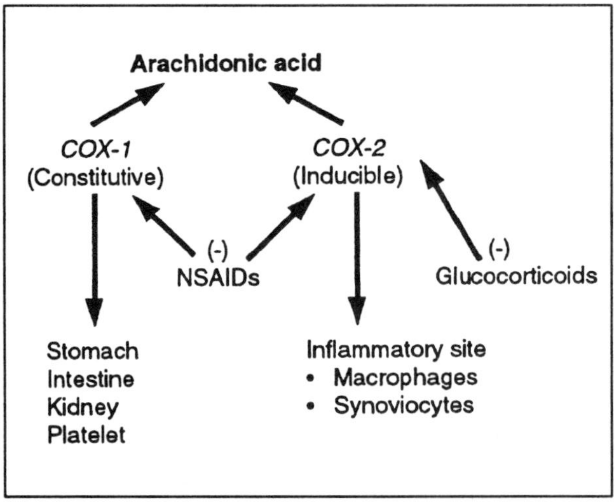


*Figure 18.10.* Pathway of arachidonic acid production. NSAIDs inhibit both COX-1 and COX-2 enzymes, while glucocorticoid steroids inhibit expression of COX-2. Reprinted from Needleman and Isakson (1997), pg. 7, Fig. 1, with permission of Oxford University Press.

and hypertension. A new class of COX-2 specific inhibitors has been developed in order to target the pain and inflammation pathway, while sparing effects on COX-1 in the platelets and gastric mucosa. COX2 inhibitors have no effect on thromboxane A2 but may decrease prostaglandin I2 and there is some concern about their effects on thrombosis which needs further investigation (Mukherjee et al., 2001).

The roles for NSAIDs and corticosteroids, which are potent anti-inflammatories, appear to be more limited than previously thought. As seen in the development of tendinopathy (see Section 18.3), the histology of overuse injuries to tendon does not involve a significant inflammatory component. There has been some suggestion that muscle fiber recovery and regeneration may be delayed (Mishra et al., 1995; Almekinders and Gilbert, 1986), and that bone healing may be prolonged (Friedman et al., 1992) when NSAIDs are used. Prostaglandin E2 production, which sensitizes afferent pain fibers and is inhibited by anti-inflammatories, also appears to be involved in the production of cytokines involved in tissue healing (Mishra et al., 1995). Corticosteroids, which have the strongest anti-inflammatory effect, can lead to cell death and inhibition of collagen synthesis, as well as, tendon atrophy and decreased load to failure (Khan et al., 1999). Results of steroid injections for several indications are controversial as they have shown little benefit in long-term outcomes (Newcomer et al., 2001). NSAIDs and corticosteroids may permit improved performance due to less pain, however, they may adversely affect the necessary process for healing. Further research is needed to determine the appropriate use of anti-inflammatories in chronic, overuse and acute soft-tissue injuries.

### 18.4.2 Concussion Management

Concussion is a serious injury in sports medicine that is being identified more commonly. Athletes in contact sports are bigger, stronger, and faster, resulting in higher impact collisions. In the last decade, the emergence of extreme sports such as skateboarding, snowboarding, and motorcross, that can involve high-risk maneuvers in the air increases the possibility of concussion. Fortunately, the awareness and the detection of concussions are improving in the medical community.

The definition of concussion is "a complex pathophysiological process affecting the brain, induced by traumatic biomechanical forces" (Concussion in Sport Group, 2002). Concussions were traditionally treated based on the experience of sports practitioners using guidelines that were based mainly on clinical symptoms, in particular loss of consciousness. These guidelines are controversial. Long-term neuropsychological effects of mild traumatic head injury on cognitive functioning are now felt to be more concerning than brief transient loss of consciousness.

The biochemical pathophysiology of concussion is becoming clearer with research. A large increase in extracellular potassium and excitatory amino acids immediately after the concussion appears to play a significant role. Potassium concentration can reach levels four to ten times normal or more, which may inhibit the action potential and consequently the function of neurons, which can result in loss of consciousness (Maroon et al., 2000) seconds to minutes after the injury. Hyperglycolysis, which begins 10 minutes after a concussion, may lead to a relative imbalance in glucose utilization following head injury (Maroon et al., 2000) and resulting in a state of metabolic depression. The roles of cerebrovascular auto-regulation and intracranial pressure in concussion have yet to be clarified.

Post-concussive effects on cognition are being investigated. Functional MRI suggests that individuals with minor traumatic brain injuries respond with increased processing loads demonstrated by increased or altered activation patterns in areas of the brain, compared to controls for variable tasks (McAllister et al., 1999). The differences were noted even though both groups showed similar task performance. The frontal lobes show different patterns of activation, with absence of activation of specific areas in an athlete with multiple concussions (Johnston et al., Clin J Sports Med 2001).

Neuropsychological testing has been tried for the last two decades. They focus on the athlete's performance on different tasks of memory, recognition, and reaction times, which are felt to indirectly measure brain function. The results are compared to baseline tests completed before the head injury, usually at the start of the season. Computer based tests have been developed and are becoming more commonly used at the professional, college and high-school levels. As further understanding of injury to the brain and recovery becomes clear, more clinically relevant guidelines for managing concussions can be developed.

### 18.4.3 Primary Prevention Issues

A continuing goal in sports medicine is to make sports safer and the best treatment for a sports injury is prevention. Analysis of the risk factors and injury surveillance studies enlighten our understanding of the epidemiology of sports injuries. Protective equipment, such as helmets, facial protection and mouthguards, has helped decrease head and face trauma. Rule changes, such as avoiding tackling with the head first in American football and avoiding cross-checking in hockey, have decreased the number of catastrophic head and neck injuries (Cantu and Mueller, 1999). Equipment changes make the playing environment safer, for example, break-away bases which help avoid lower extremity injuries in baseball by avoiding the force applied to the foot and ankle by stationary bases (Janda et al., 2001). Further funding and research should be directed to prevention efforts, as prevented injuries lead to reduced health care costs and days lost to injury.

An issue affecting performance and prevention is sports nutrition. As previously discussed, the body's systems must stay in balance. The amount of calories ingested should match the body's natural basal metabolic rate as well as the calories exerted from training and activities. The importance of calorie intake is paramount in the female athlete, particularly those participating in sports where thinness is desirable such as gymnastics, figure skating, track and dance. Over the last two decades, the female athlete triad has been identified consisting of: 1) eating disorder, 2) ammenorrhea (loss of menstrual periods for greater than 6 months), and 3) osteoporosis. In a study of female college (NCAA) athletes, 60 percent reported irregular periods, 3.4% had an eating disorder, 45.4% of athletes were not happy with present weight, and 12.6% were exercising excessive to lose weight (Evans et al., 2000) .

The calorie deficit decreases the energy available for metabolism which affects the hormonal function of the hypothalamus, which regulates reproductive hormone production, in particular estrogen. Low levels of estrogen can affect bone development, leading to inadequate mineralization resulting in osteopenia (low bone mineral density) and osteoporosis. As a result of training, women with these factors are at high risk for developing stress fractures. The treatment for the female athlete triad involves early recognition, identification of risk factors such as low bone density, counseling and correction of the calorie and hormonal imbalances. The female athlete triad is a serious condition, which requires a multidisciplinary team including physicians, a psychiatrist, and a dietician.

## 18.5 REHABILITATION

During the 20th century, scientists in the fields of physiology and medicine began to focus on kinesiology, the study of motion and biomechanics, "the science that examines forces acting upon and within a biological structure and effects produced by such forces" (Nigg, 1999). The field has advanced using high-speed cameras, digital video cameras, electromyography, radiographic marker systems and computers to analyze movement in three dimensions (Zheng and Barrentine, 2000). Accelerometers can track linear and angular velocities and acceleration forces, though there are presently limitations in detecting motion in all directions. Forces and torque applied are measured using force transducer platforms. Understanding human movement has implications in medicine, industry, and athletics.

The aims of rehabilitation are commonly to reduce pain, restore range of motion and strength, and to recover function. After injury, loss of strength, flexibility, and neuromuscular control can affect the athlete's performance. Stretching and strengthening protocols, modalities, manual therapies and sport-specific exercises are frequently used in the course of returning an athlete to competition.

## 18.5.1 Early Mobilization

Controlled early mobilization of muscle, tendon and ligament injuries is recommended for most musculoskeletal injuries, as opposed to traditional suggestion of rest. Sufficient time should be allowed for adequate granulation tissue to form in order to attain sufficient tensile strength to allow mobilization of the injured tissue. For muscle injuries, the period of immobilization should be short usually less than one week in order to limit the extent of connective tissue proliferation at the site of injury (Jarvinen and Lehto, 1993). Increased movement and isometric training can often begin after three days as symptoms allow, for a mild to moderate tissue injury. The muscle should be kept under tension to maximize healing response and limit contracture.

Immobilization has negative effects including degenerative changes at the myotendinous junction, as well as the muscle spindles (Józsa and Kannus, 1997). Muscle shortening and loss of sarcomeres leads to atrophy and weakening of the muscle. Also, tendons have decreased collagen synthesis and content of sulfated glycosaminoglycans. Noyes found a decrease of 39 percent in load to failure in ligaments that were immobilized (Noyes, 1977). As discussed in Section 18.2.8.1, articular cartilage receives its nutrients through the synovial fluid dynamics. Exercise and weight-bearing are essential to maintain cartilage health, improve bone mineralization, and prevent bone loss and osteoporosis.

## 18.5.2 Specific Rehabilitation Exercises

Appropriate rehabilitation exercise programs are designed to apply sufficient but not excessive stress to healing structures, based on the underlying pathology. For example, the biomechanics of the knee demonstrated that open kinetic chain exercises, where the distal segment (foot) is free to move without external resistance, applies tension to the anterior cruciate ligaments and high compressive forces to the patellofemoral joint (Escamilla et al., 1998). Consequently, many programs focus on closed chain exercises where the terminal segment is opposed by resistance, such as squats, which emphasize quadriceps strengthening and have more hamstring activity than leg press or leg extension exercises (Escamilla et al., 1998). Rehabilitation programs are best planned taking into account the injury and the expected course of healing, introducing specific exercises with gradually increasing difficulty with the functional goals of the athlete in mind.

In managing tendinopathy, eccentric loading seems to be the recommended trend. The concept is to stress the collagen along its longitudinal axis in order to align the collagen fibers and induce cross-link formation and remodeling. Alfredson et al. have been advocating heavy load eccentric calf muscle training for patients with chronic Achilles tendinosis (Alfredson et al., 1998).

The stabilization of the central core (trunk) in sports has benefits for injuries as well as performance. The trunk is often divided into the torso and the pelvis, and provides a stable foundation for movement of the extremities. The core is stabilized by strengthening the deep and superficial muscle groups of the spine. Dynamic lumbar muscular stabilization has become the most popular methods of improving proprioception and strengthening the low back, which facilitates control of intersegmental spinal motion and load on the spinal elements.

## 18.5.3 Proprioception

Proprioception is a somatic sense and is defined as "the afferent input of joint position sense or kinesthesia" (Lephart, 1994). Therefore, two components of proprioception include the "static" sense or joint position sense, which provides the individual with an appreciation of the orientation of different body parts with respect to one another, and the "dynamic" sense, which gives neuromuscular system feedback about the presence and rate of movement (Safran et al., 2001).

The afferent input is created by different mechanoreceptors found in the soft tissues that convert mechanical deformation into a frequency modulated neural signal. There are quick adapting mechanoreceptors, such as Pacinian corpuscle, that discharge within milliseconds of the onset of a continuous stimulus (Safran et al., 1999). There are also slow-adapting mechanoreceptors, for examples Ruffini endings and Golgi tendon organ that continue discharge in response to a continuous stimulus. Mechanoreceptors are found in ligaments, including the anterior cruciate ligament, muscles, tendon and other soft tissues. Proprioception may vary depending on the direction of motion and position of the joint as different mechanoreceptors may be recruited (Borsa et al., 1997).

When the mechanoreceptors are triggered, an involuntary protective reflex arc may occur. These proprioceptive short loops or spinal reflex afferent-efferent arcs (VanDuersen et al., 1998) are quicker (70-100 m/s) than the reflex arc for pain initiated by nociceptors (1 m/s). This suggests that proprioception may play a more significant role than pain in preventing injury (Safran et al., 1999). In the case of a sports injury, proprioception can be diminished due to damage to mechanoreceptors, loss of the tensile properties of static structures such as ligaments or the joint capsule, and latent response of muscles to provide reflex stabilization of a joint. Athletes with ACL deficient knees demonstrated a significant higher threshold to detect passive motion (TTDPM) particularly moving into extension at 45° more so than 15°, over angles in the range of approximately 0.77° to 1.1° (Borsa et al., 1997). Though they seem to be small measurable deficits, there were significant correlations with declining function on single leg hop testing (Borsa et al., 1997).

Proprioceptive training has become an important aspect of rehabilitation programs. Emphasis has been placed on retraining the proper firing patterns of muscles to provide dynamic stability. Prophylactic training programs are presently being investigated using balance exercises, muscle strengthening programs, and jumping protocols, to try to prevent injuries of ankles and the anterior cruciate ligaments.

## 18.6 SURGERY

### 18.6.1 The Arthroscopy Revolution

Arthroscopy has been one of the most significant advances in sports medicine. It enables minimally invasive surgery to be performed for both diagnostic and therapeutic purposes. Over the past three decades, the indications for arthroscopic surgery have expanded significantly as have the improvements in technology. Surgery time is reduced as well as surgical morbidity and complications. Procedures can be performed under general, regional or local anaesthesia, again reducing risks associated with having surgery.

The arthroscope is typically a rigid rod-lens system, transmitting light and images through an array of lenses to a video camera (Cautilli Jr., 1997) (Fig. 18.11). The field of vision is demonstrated on a video monitor in the operating room. Fiber-optic arthroscopes use fibrils that conduct light

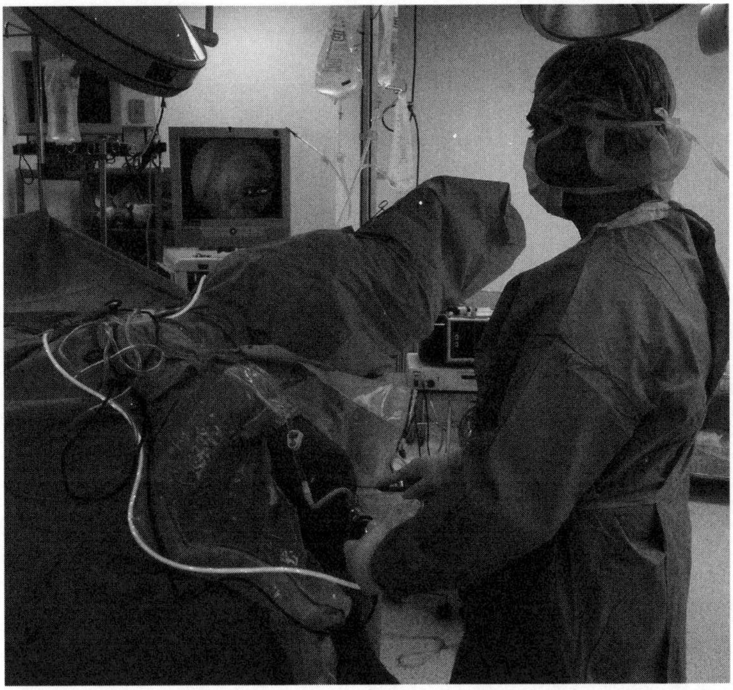

*Figure 18.11.* Arthroscopy of the left knee set up.

(Cautilli Jr., 1997). The scope diameters range from 2.7 mm to 7.5 mm and the angle of inclination can vary from 0° to 120°, enabling a larger viewing area (Fig. 18.12).

The arthroscope is introduced into the joint through a small carefully positioned incision, referred to as a portal. Subsequent portals are made to introduce an inflow cannula for fluid as well as other tools. Fluid irrigation via gravity or a pump system distends the joint and enables the surgeon to visualize through the arthroscope, as small debris and blood are cleared by the inflow and removed by the outflow. Monitoring intraarticular pressure and rate of flow control avoids unnecessary distension of the joint capsule. A pressure tourniquet may be used to reduce blood flow to some joints.

A number of instruments enable the surgeon to manipulate, cut, drill and repair structures. A blunt probe is used to feel, push and examine tissues. The armamentarium includes spinal needles, graspers, basket forceps, and surgical blades (Fig. 18.13). A motorized shaver, electrocautery, laser, thermal heat probe and motorized drill are other tools.

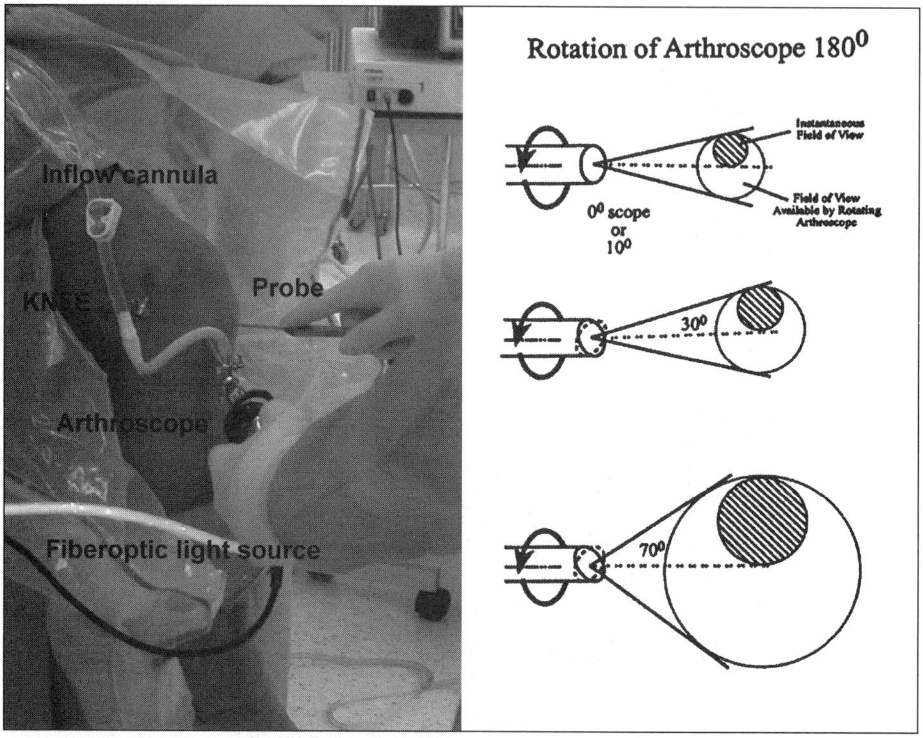


*Figure 18.12.* Left: Arthroscopy of left knee. Right: Arthroscope field of vision. Reprinted from Dukes (1997), pg. 8, Fig. 2-4, with permission of Elsevier Science.

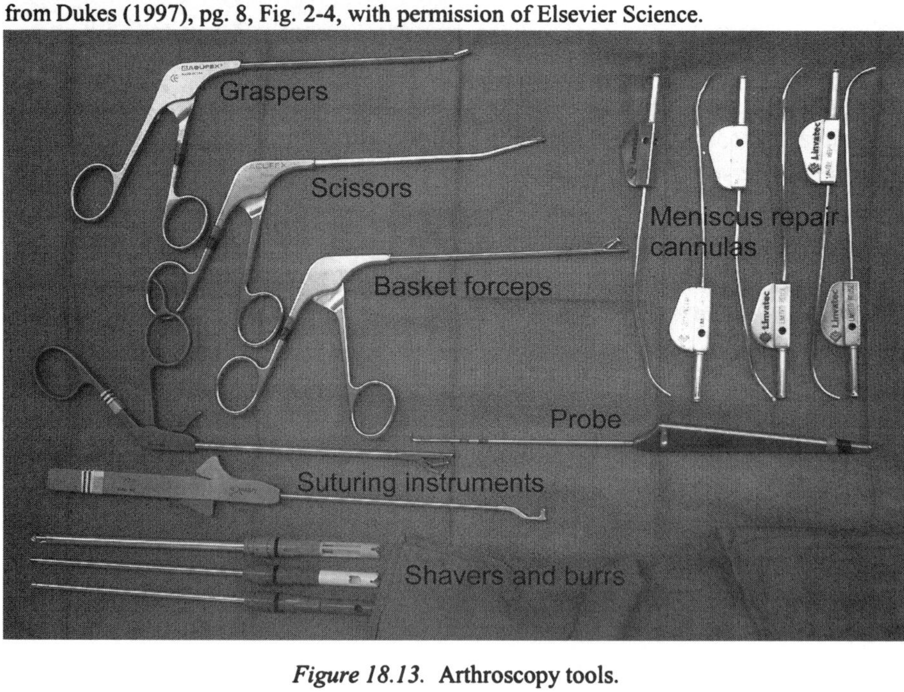


*Figure 18.13.* Arthroscopy tools.

## 18.6.2 Anterior Cruciate Ligament (ACL) reconstruction

### 18.6.2.1 Function of the ACL

An injury of particular importance is the rupture of the anterior cruciate ligament (ACL), which can cause significant dysfunction especially in athletes. The incidence of this injury appears to be increasing, particularly in females participating in cutting and twisting sports. Restoring the ligamentous properties of the ACL is necessary for successful recovery and demonstrates the importance of ligament function. The ACL functions under loads of 500 N to 700 N with normal activities and typically has an ultimate failure strength of 1700 N when loaded along the tibial axis and 2400 N when loaded along the axis of the ACL (Renstrom and Lynch, 1998).

Following an ACL tear, clinical instability is often detectable in the knee, causing giving way sensations for the athlete during cutting, pivoting and jumping. Without the stabilization of tibia by the ACL and the reflex, dynamic stabilization of the knee due to diminished proprioception (Borsa et al, 1997), the tibia can translate anteriorly in relation to the femur. When rotational forces are applied with the knee in flexion, a "pivot shift" effect can occur, where the knee joint subluxes out of position and reduces, causing a giving way sensation.

ACL injuries are often complicated by further injury to associated structures. Multiple ligament damage often suggests knee dislocation, which can also disrupt the neurovascular structures behind the knee. Historically, the ACL tears were seen in association with a collision type of cross blocking injury occurring in sports such as American football, soccer and rugby. The early combination of observed injuries was named the "O'Donaghue's Triad", consisting of injuries to the ACL, the medial collateral ligament (MCL), and the medial meniscus (cartilage). Since then, tearing of the lateral meniscus has been found to occur more frequently than the medial meniscus, and geographic bone bruise patterns seen on MRI over the femoral condyles and tibial plateaus demonstrate that the lateral side of the knee is most commonly affected. These findings support that the usual mechanism for an ACL injury involves valgus, rotational stress to the knee, for example when an athlete plants the foot and pivots or lands awkwardly on one leg from a jump, which may initially seem to the athlete to be a relatively minor injury.

### 18.6.2.2 How Can the ACL be Reconstructed ?

The challenge for the orthopaedic sports medicine surgeon is how to best restore the mechanical function of the ACL deficient knee and prevent future injury to knee structures when the athlete returns to activities. The history of the treatment of ACL tears demonstrates how the knowledge regarding the biomechanics of the ACL and biology of implanted tissues have advanced, enabling improved treatments. Initial attempts to address the problem have ranged from benevolent neglect to attempts to repair the injured ligaments directly either acutely or subacutely. Unfortunately, most of these attempts were doomed to failure, since the biology of the ACL and its location in the central aspect of the knee bathed by synovial fluid make it difficult for direct tissue healing of the ligament to occur (see Section 18.2.6.2). Another approach aimed to simulate the function of the anterior cruciate ligament using extra-articular repairs, particularly those that tightened the antero-medial or antero-lateral structures of the knee. An early intervention of this type was the pes transfer of Slocum (Slocum, 1968). These procedures typically gave initial short term improvement, but failed long term when the repairs would loosen and lose the ability to constrain the knee.

Surgeons discovered one way to replace a patient's ligament is to use the patient's own tissues that displayed similar characteristics, referred to as an autologous graft or "autograft". Giertz performed the first autologous graft in 1912 using a strip of tensor fascia lata. Ivar Palmer from Sweden published his thesis in 1938 on ligament injuries of the knee, which lays the foundations of current ACL reconstruction. In 1963, Jones used a patella tendon with bone block as a graft to reconstruct the ACL. With the introduction of arthroscopy, Clancy et al. has helped advance reconstruction methods and techniques using the patellar tendon (Clancy et al, 1982), which became the graft of choice by the late 1980's. Since that time, the hamstrings, namely the semitendinosis and gracilis tendons, and cadaver allografts have provided other options for reconstruction. Fulkerson has also used the quadriceps tendon.

Graft selection is important as it determines the tensile properties of the reconstruction, the type of fixation at the ends and some of the potential complications from surgery. The strength and tensile properties of the graft ideally must match that of the patient's inherent anterior cruciate ligament. Unfortunately, the quality of the collagen tissue in a graft is not equal to the original tissue. The collagen in a patellar tendon graft is only a quarter to half the normal size (Larson, 1993). The middle third of the patella has been a popular choice of graft for many surgeons (Fig. 18.15). The initial tensile strength of the ligament and good bone to bone healing due to the bone plugs at the ends of the graft provide good stability and fixation. However, many surgeons have found the hamstring tendons provide a successful graft for reconstruction. The hamstring tendons, typically the semitendinosis and gracilis tendons, taken from their medial insertion to the tibia, have more elasticity. Doubling the tendons by looping them over allows the hamstrings graft to match the strength of the patellar tendon graft (Fig. 18.14). Post-operatively, patients typically regain any hamstring strength deficits as the tendons heal and regrow (Pagnani et al., 1993).

Extrinsic grafts are alternatives for reconstruction. Allografts from cadaver specimens can be used to replace the ACL, typically using the bone-patella-bone tissue grafts or Achilles tendon, and sometimes tibialis posterior or tibialis anterior. They avoid the injury sustained in harvesting a graft from the patient and decrease the time of the operation. Though the allograft is tested and gamma irradiated to avoid transmission of infectious disease, immunogenicity and weakening of the graft are concerns of using these allografts. Noyes reported failure rates of 14% of allografts at late follow up at 5 to 9 years post-operatively (Noyes, 1996). Allograft medial collateral ligament complexes were also found to be weaker in comparison to autograft or native ligament tissues (Sabiston et al., 1990). Synthetic materials have been tried in the past to reconstruct or augment an autologous reconstruction. Ligament augmentation devices made from Dacron™ were tried to help support autologous grafts as they were maturing. Synthetic grafts made of Gortex™ were also used for reconstruction at one time. Unfortunately, the synthetic materials usually break down over a period of 2 to 5 years (Larson, 1993) and their use is not favored at this point in time.

Positioning the graft is important to restore the anatomy. The challenge is ideally to create an isometric graft, which maintains its length throughout the flexion and extension of the knee (Sapega et al., 1990). The position of the graft is determined most importantly by the placement of the tibial and femoral tunnels through which the graft is passed. The most appropriate biomechanical sites of insertion have been identified in cadaver experiments. Proper placement particularly of the femoral tunnel can affect the isometry of the graft and bending of the graft during movement. The tibial tunnel has moved from an eccentric anteromedial position to more posteriorly in relation to the intercondylar roof. The flexion angle of the knee and the force applied during tensioning of the distal end of the ACL graft during surgery affects the force in the graft and the magnitude of change in the anteroposterior position of the graft but did not affect these variables during knee flexion movements (Bylski-Austrow et al., 1990). The ideal amount of tension to be applied to the graft is still uncertain.

Fixation of the graft is suspected to be the most common cause for acute failure of the graft (Kurosaka et al., 1987). Fixation of the tibial side is more commonly a problem than the femoral end. Interference screws that are placed parallel to the tunnel into the bone plug of the graft, or fixation posts, which allow the graft to be tied down with suture, are commonly used. An endobutton technique is popular to fix the femoral end, which is performed by drilling the femoral tunnel through both cortexes of the femur and securing a button over the proximal opening, leaving the graft sutured to the button in place in the tunnel (Fig. 18.14). Methods of fixation of the tibial end include sutures tied over a post, staples, and interference screws (Fig. 18.15). Ultimately, the graft, whether there are bone plugs or it is purely tendinous, will mineralize within the tunnel (Rodeo et al., 1993). While the ends of the graft are maturing post-operatively, movement of the graft within the tunnels can lead to wear, particularly as the knee undergoes rehabilitation to regain range of motion. Therefore, the angles for drilling the tunnels may be important to minimize deformity of the graft, reducing the possibility of tunnel widening.

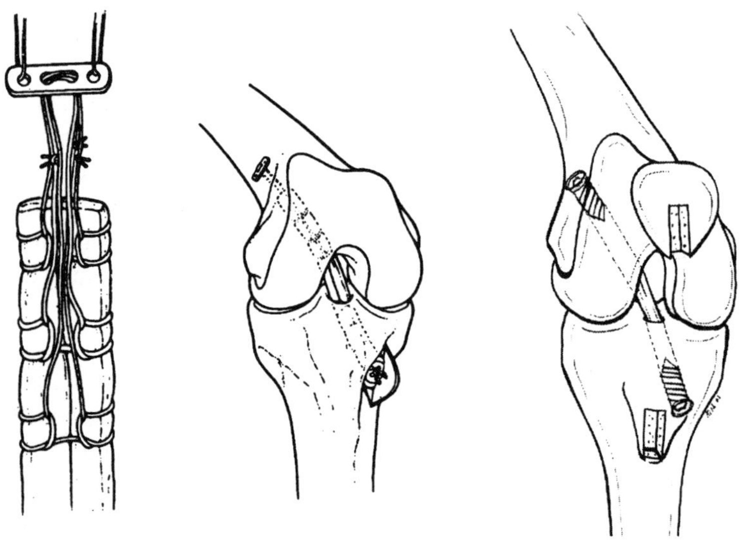

*Figure 18.14.* Left two figures: Preparation of hamstring tendon graft with endobutton and completed graft placement. Reprinted from Barrett et al. (1995), pg. 8, Fig. 2-4, with permission of Elsevier Science.

*Figure 18.15.* Right: Interference screw fixation of bone patella bone graft. (Figure drawn by S. Kish, 2003).

One of the major concerns regarding long-term results of ACL reconstruction is the development of osteoarthritis, which is discussed in detail by Clatworthy and Amendola (1999). Associated meniscal injuries, osteochondral injuries and ligamentous injury other than to the ACL can lead to chondral damage but their exact contribution to development of osteoarthritis still needs epidemiologic investigation. Malalignment from congenital causes or following meniscectomy can shift the load bearing axis and some experts recommend an osteotomy as well as ACL reconstruction for athletes with varus knee alignment and ACL deficiency. The effects of ACL reconstruction and biological factors need to be researched further.

Alternative treatments include primary repair and non-operative treatment. Kaplan et al. (1990) described results using multiple sutures to directly surgically repair the anterior cruciate ligament tear, however this led to unpredictable results with a failure rate of 17%. Another study showed that patients undergoing primary repair of the ACL had similar function to those treated without surgery (Odensten et al., 1985). Regarding non-operative treatment, while physical therapy is universally recommended for recovery, functional knee bracing is still a controversial topic. With appropriate physical therapy, which is considered as important to surgery in terms of recovery, the chances to return the athlete to a previous level of functioning is maximized.

At the present time, the great majority of athletes who sustain an anterior cruciate ligament injury can expect to have a satisfactory treatment of this condition. Similar to the experience with ACL injuries, the surgical treatment of posterior cruciate ligament tears is following the same stepwise progression of comprehension. Unfortunately, the results for PCL reconstruction at present are not nearly as successful as interventions for ACL tears. Nevertheless, as technology improves and the understanding of the biomechanics and healing is enhanced, the field of orthopaedic sports medicine will continue to advance to meet the challenge.

## 18.7 ON THE HORIZON

### 18.7.1 Gene Therapy

Major strides are being made in medicine in the fields of molecular genetics and immunology. The future of treatment for many diseases will involve targeting pathology at the cellular level in order to initiate the proper healing effect through gene therapy. Applications for gene therapy in sports medicine include enhancing bone healing, tissue repair and inflammation (Evans et al., 2000). Preliminary studies have demonstrated successful gene transfer to bone, articular and meniscal cartilage, ligament and tendon (Evans et al., 2000). Genetics may also help identify defects in collagen associated with pathological processes.

Genes are sequences of deoxyribonucleic acid (DNA), which are the genetic codes commonly for the synthesis of cell proteins. These sequences are located on specific areas on the chromosomes which are found in the nucleus of cells. The codes are transcribed into messenger RNA (mRNA) and transferred outside the nucleus into the cell. Ribosome proteins translate the mRNA into the desired proteins, which may perform specific cellular functions (Evans et al., 2000).

Vectors are used to transfer the gene to the specific target cells. Viral infections have served as an ideal vector as they infect cells and deliver their genes to the cell nucleus. Genetic manipulation of the viral structure has enabled the production of vectors that can efficiently integrate the genes into the host cells DNA, or simply deliver functional genes to nondividing cells (Evans et al., 2000). Non-viral vectors include plasmid DNA and DNA coated particles which are projected into cells using a "gene gun" (Evans et al., 2000). These methods are performed "ex-vivo" by altering cells genetically then reintroducing them into the body or "in-vivo" by delivering the vector directly to the site of interest.

A major area of possible gene therapy involves possible effects on stimulation and inhibition of cytokines. Cytokines are proteins that are released by cells to trigger biological activities of specific local target cells via receptor binding (Evans, 1999). They have been found to be involved in the regulation of the healing responses including angiogenesis, cell migration and division, matrix synthesis, differentiation of cells, and inflammation (see Table 18.3) (Evans, 1999). Gene therapy may be a means of delivering cytokines or help regulate their expression in order to enhance healing of ligaments, tendons and cartilage.

## 18.7.2 Transplant Technology

Similar to the concepts of ligament transplantation with cadaver allografts and patient autografts, surgical transplantations of cartilage are possible treatments for athletes with certain meniscal or cartilage injuries. Meniscal allografts from cadavers have been more successful than synthetic materials. The proteoglycan production from viable donor cells is important for healing of a meniscal allograft, which must be harvested from a recent casualty (Verdonk, 1997).

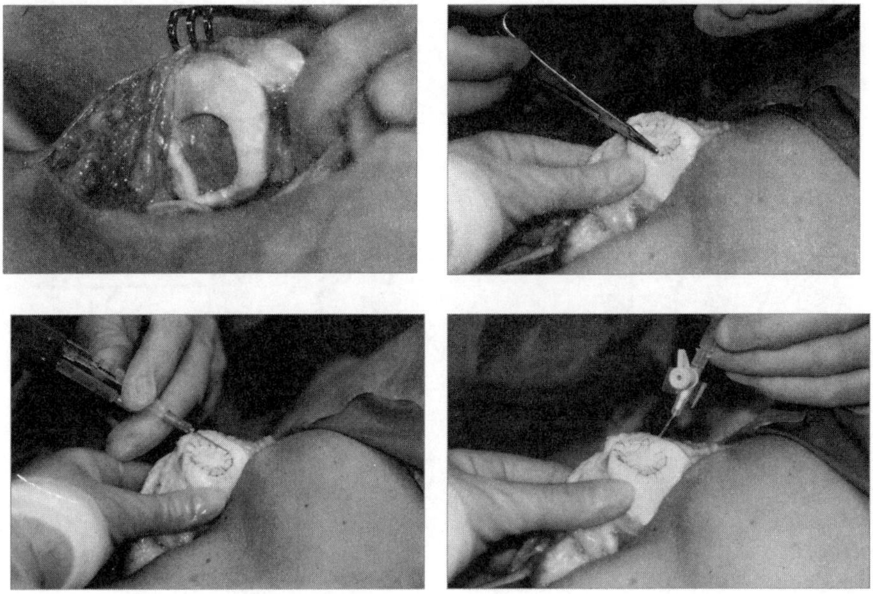

*Figure 18.16.* Pictures of Autologous chondrocyte implantation surgery. Reprinted from Peterson (2002) with permission from Genzyme.

Methods to transplant cells to focal areas of osteochondral injuries have been developed. Autologous chondrocyte implantation involves culturing a sample of chondrocytes taken from the femoral condyle then transplanting the cells 14 to 21 days afterwards into the area of a defect. The cells in the lesion are covered with a flap of periosteum, held in place by tiny 6-0 sutures and a fibrin sealant around the margins of the site (Fig. 18.16). In a multicenter, prospective trial, 84% of patients with autologous chondrocyte implantation (ACI) for repair of symptomatic, full, thickness, chondral lesions were improved after 36 months. Only 6% had graft failure (Micheli et al., 2001).

Transplanting osteochondral grafts has also been performed, with the advantage that the cartilage graft may match the mechanical characteristics of the original cartilage closely (see Section 18.2.8.1). The area of the defect can be filled by a large osteochondral graft from the patella, from a cadaver allograft, or by smaller osteochondral plugs from the superior femoral condyle, referred to as mosaicplasty (Fig. 18.17). Careful patient screening and irradiation of the tissues to be transplanted help address concerns of infection in donor allografts.

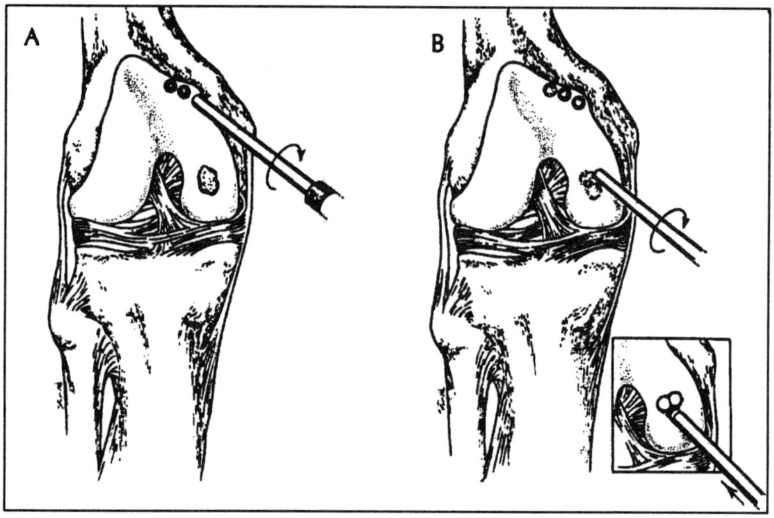


*Figure 18.17.* Osteochondral implantation or "Mosaicplasty". Reprinted from Menche et al. (1998), pg. 511, Fig. 5A, with permission from the American Academy of Orthoapedic Surgeons.

## 18.7.3 Extracorporeal Shockwave Therapy (ESWT)

During World War II, it was noted that submarine sailors sustained lung tissue damage from depth charge explosions, despite having no visible signs of external trauma (Krause, 1997). It appeared that tissues exposed to the shock waves dissipated and absorbed the energy. Since that time, clinical shock waves have been applied by creating a controlled explosion to produce a sonic pulse in order to obtain tissue effects (Ogden et al., 2001).

There have been three designs for musculoskeletal treatment devices based on different impulse generating principles (Fig. 18.12). The electrohydraulic principle uses a high voltage from a charged capacitor applied across electrode tips. The spark vaporizes the surrounding water creating a gas bubble filled with water vapor and plasma. The focal point can be directed at the area intended for therapy (Ogden et al., 2001). More commonly used, the electromagnetic principle passes an electrical current through a coil to produce a magnetic field. The strong magnetic field then causes an adjacent, highly conductive membrane to be forced rapidly away, compressing the surrounding fluid medium to produce a shock wave. A lens is used to focus the wave. Finally, the piezoelectric principle uses high voltage current applied to piezo crystals mounted on the inside of a sphere creating a rapid electrical discharge. The deformation of the crystals resulting from quick contraction and expansion produces a pressure pulse in the surrounding water steepening to a shock wave (Ogden et al., 2001).

The shock wave will be attenuated when reflected and refracted at boundaries between two different mediums, for examples muscle and bone (Ogden et al., 2001). Attenuation in water is 1000 times lower than air (Ogden et al., 2001). Part of the wave is reflected back and the other part is transmitted, resulting in energy dissipation. The energy can be focused on a small area and the kinetic energy delivered and absorbed by the target tissue.

Physical parameters of ESWT include the pressure distribution (in MPa), the energy flux density and the total energy. The energy flux density is a measure of the energy per square area describing the maximum amount of acoustical energy that is transmitted through an area of 1 $mm^2$ per pulse (Ogden et al., 2001). It reflects the energy released at a certain point (area below the squared pressure time curve) and is important when considering threshold values in generating certain biologic effects. The total energy applied to the tissue corresponds to the energy flux density as integrated over the entire region (represented by the number of pulses multiplied by the energy per pulse) (Ogden et al., 2001).

Shock waves generate high stresses that act on boundary interfaces and generate tensile forces that cause cavitation. On a cellular level, changes in membranes with cell destruction can occur. However, there is histologic evidence that osteoblasts may be activated (Haupt, 1997). ESWT has been used for nonunion and delayed union fractures, plantar fasciitis, lateral epicondylitis (tennis elbow), and calcific tendinopathy (Haupt, 1997). It is speculated that the shock waves stimulate or reactiviate the healing processes in tendon probably through microdisruption of avascular or minimally vascular tissues by encouraging revascularization, release of local growth factors and recruitment of appropriate stem cells (Ogden et al., 2001). Its full biologic effects on bone, cartilage, tendon and ligament are still unknown and are being investigated.

## 18.8 CONCLUSIONS

### 18.8.1 Evolution of the Athlete

In today's age, an athlete on the playing field can sustain the same injury as our caveman two millennia ago. For example, a football running back escapes a linebacker but plants his foot on the artificial turf and makes a cut, twisting his right knee. He is promptly attended to by the athletic trainer on the field. The physical examination of the knee suggests an anterior cruciate ligament injury, possibly with associated meniscal cartilage injury. He is helped of the field and given crutches to avoid weight on the leg and ice is applied immediately. He visits the team's sports medicine physician. Radiographs show no bony abnormality, however, his MRI of the knee confirms a complete midsubstance tear of the anterior cruciate ligament (ACL), with partial tearing of the medial collateral ligament, and a bucket-handle shaped tear of the medial meniscus. There is evidence of bone bruising over lateral aspect of the knee, consistent with the mechanism of an ACL injury. The athlete attends physical therapy in order to regain the range of motion and minimize muscular atrophy. He is able to weight bear without crutches within two weeks of his injury, but still has some pain due to residual injuries, and lacks function due to his loss of proprioception. The orthopaedic surgeon explains the importance of restoring stability to the leg. With ligament reconstruction, the mechanical properties of the knee to withstand rotational and perpendicular stresses will be more predictable. During the arthroscopic surgery, the medial meniscus is repaired in order to maintain alignment in the knee and the shock absorbing properties of the meniscus to protect the articular cartilage. The ligament is reconstructed using a hamstring autograft. Following day surgery, the athlete undergoes a strict six-month rehabilitation protocol based on previous research and experience of medical centers. As his muscle strength and proprioception is improved through closed chain exercise, the reconstructed ligament graft matures and remodels in accordance with the principles of tissue healing. Six-months after successful rehabilitation, the athlete gradually returns to sport-specific activities, looking forward to regaining his form for the next season.

## 18.8.2 Practically Speaking

From the athlete's perspective, medical attention should be sought early, especially if the injury affects mechanics and performance. The popular adage of modified activity, ice, compression, and elevation still applies in most acute injuries. However, it is important to have a proper diagnosis made so that appropriate management can be planned, whether physical therapy, modified activities or even surgery is required.

With the proper teamwork of health professionals, coaching staff, family and teammates, an athlete's injury can be managed efficiently to minimize time away from sport and risk of injury when they return. From the discussion of tissue healing, there are ranges of time expected for each specific injury to heal. For the individuals involved in sports, a general understanding of the physiology makes it easier to inform athletes of expectations and risk management. For the biomedical engineer, there are exciting areas of development in diagnostic and surgical technologies, as well as, in protective equipment design. For the clinician, further funding and research is needed to ensure treatments are evidence-based. In the meantime, the field of sports medicine will continue to advance with the hope that the lessons learned from the great laboratory of the sports arena can further benefit the health and fitness of the "athlete" in every individual.

## 18.9 ACKNOWLEDGMENT

Special thanks to M. Jenkins, S. Kish, L. Thain, M. Chisholm and M. Zeitoun for their assistance in producing this chapter.

## 18.10 REFERENCES

Ait-Haddou, R., and Herzog, W., 2000, A general mathematical framework for cross-bridge modeling, in: *Skeletal Muscle Mechanics: From Mechanisms to Function,* W. Herzog, ed., John Wiley & Sons, Ltd., Chechester, West Sussex, pg. 125.

Alexander, R. M., 1984, Elastic energy stores in running vertebrates, *Am. Zool.* **24**: 85-94.

Alfredson, H., Ljung, B. O., Thorsen, K., and Lorentzon, R., 2000, In vivo investigation of ECRB tendons with microdialysis technique – no signs of inflammation but high amounts of glutamate in tennis elbow, *Acta. Orthop. Scand.* **71**: 475-479.

Alfredson, H., Pietilä, T., Jonsson, P., and Lorentzon, R., 1998, Heavy-load eccentric calf muscle training for the treatment of chronic Achilles tendinosis, *Am. J. Sports. Med.* **26**: 360-366.

Alfredson, H., Thorsen, K., and Lorentzon, R., 1999, In situr microdialysis in tendon tissue: high levels of glutamate, but not prostaglandin $E_2$ in chronic Achilles tendon pain, *Knee Surg. Sports Traumatol. Arthrosc.* **7**:378-381.

Almekinders, L. C., and Deol, G., 1999, The effects of aging, anti-inflammatory drugs, and ultrasound on the in vitro response of tendon tissue, *Am. J. Sports Med.* **27**: 417-421.

Almekinders, L. C., and Gilbert, J.A., 1986, Healing of experimental muscle strains and the effects of nonsteroidal anti-inflammatory medication, *Am. J. Sports Med.* **14**: 303-308.

Amiel, D., Ishizue, K. K., Harwood, F. L., Kitabayashi, L., and Akeson, W. H., 1989, Injury of the anterior cruciate ligament: The role of collagenase in ligament degeneration, *J. Orthop. Res.* **7**: 486-493.

Arangio, G. A., 2001, Tendon problems, *Curr. Opin. Orthop.* **12**:112-119.

Åström, M., and Rausing, A.,1995, Chronic Achilles tendinopathy: A survey of surgical and histopathological findings, *Clin. Orthop.* **316**: 151-164.

Balduini, F. C., Shenton, D., O'Connor, K. H., and Heppenstall, R. B., 1993, Chronic exertional compartment syndrome: Correlation of compartment pressure and muscle ischemia utilizing $^{31}$P-NMR spectroscopy, *Clin. Sports Med.*, **12**: 151-165.

Baratz, M. E., Fu, F. H., and Mengato, R., 1986, Mensical tears: The effect of meniscectomy and of repair on intraarticular contact areas and stress in the human knee. A preliminary report, *Am. J. Sports Med.* **14**: 270-275.

Barrett, G. R., Papendick., L., and Miller, C., 1995, Endobutton button endoscopic fixation technique in anterior cruciate ligament reconstruction, *Arthroscopy.* **11**:3: 340-343.

Beck, B. R., 1998, Tibial stress injuries: An aetiological review for the purposes of guiding management, *Sports. Med.* **26**: 265-279.

Bennell, K., Matheson, G., Meeuwisse, W., and Brukner, P., 1999, Risk factors for stress fractures. *Sports. Med.* **28**: 91-122.

Best, T. M., and Hunter, K. D., 2000, Muscle injury and repair, *Phys. Med. Rehabil. Clin. N. Am.* **11**: 251-266.

Best, T. M., McElhaney, J. H., Garrett, W. E. Jr., and Myers, B.S., 1995, Axial strain measurements in skeletal muscle at various strain rates, *J. Biomech. Eng.* **117**:262-265.

Borsa, P. A., Lephart, S. M., Irrgang, J. J., Safran, M. R., and Fu, F. H., 1997, The effects of joint position and direction of joint motion on proprioceptive sensibility in Anterior Cruciate Ligament-deficient athletes, *Am. J. Sports. Med.* **25**: 336-340.

Buckwalter, J. A., and Mankin, H. J., 1998, Articular cartilage: Tissue design and chondrocyte-matrix interactions, *Instr. Course. Lect.* **47**: 477-486.

Bylski-Austrow, D. I., Grood, E. S., Hefzy, M. S., Holden, J. P., and Butler, D. L., 1990, Anterior cruciate ligament replacements: A mechanical study of femoral attachment location, flexion angle at tensioning, and initial tension, *J. Orthop. Res.* **8**: 522-531.

Cantu, R. C., and Mueller, F. O., 1999, Fatalities and catastrophic injuries in high school and college sports, 1982-1997: Lessons for improving safety, *Phys. Sportsmed.* **27**: 35-48.

Cardinal, E., Chhem, R. K., and Beauregard, C. G., 1998, Ultrasound-guided interventional procedures in the musculoskeletal system, *Radiol. Clin. North. Am.* **36**: 597-605.

Cautilli Jr., R., 1997, Introduction to the basics of arthroscopy of the knee, *Clin. Sports. Med.* **16**:1-15.

Clancy, W. G., Nelson, D. A., Reider, B., and Narecharia, R.G., 1982, Anterior cruciate ligament reconstruction using one-third of the patellar ligament augmented by extra-articular tendon transfers, *J. Bone Joint Surg. Am.* **64**: 352-359.

Clanton, T. O., and Coupe, K. J., 1998, Hamstring Strains in Athletes: Diagnosis and Treatment, *J. Am. Acad. Orthop. Surg.* **6**:237-248.

Clatworthy, M., and Amendola, A., 1999, The anterior cruciate ligament and arthritis, *Clin. Sports Med.* **18**: 173-198.

Committee on Sports Medicine, 2001, Medical Conditions Affecting Sports Participation, *Pediatrics*, **107**:5: 1205-1209.

Concussion in Sport Group: Aubry, M., Cantu, R., Dvorak, J., Graf-Baumann, T., Johnston, K. M., Kelly, J., Lovell, M., McCrory, P., Meeuwisse, W. H., and Schamasch, P., 2002, Summary and agreement statement of the 1st International Symposium on Concussion in Sport, Vienna 2001. *Clin. J. Sport Med.* **12**: 6-11.

Cotran, Kumar, and Robbins. *Robbins Pathologic Basis of Disease (4th Edition),* pp.1323-1324.

DeHaven, K. E., and Arnoczky, S. P., 1994, Meniscal repair, Part I: Basic Science. Indications for repair and open repair, *J. Bone Joint Surg. Am.* **76**: 140-152.

Dukes, K. M., 1997, Operating room environment, in: *Diagnostic and operative arthroscopy,* J. R. Andrews and L. A. Timmerman, eds., W.B. Saunders Co., Philadelphia, PA, U.S.A. pg. 8.

Escamilla, R. F., Fleisig, G. S., Zheng, N., Barrentine, S. W., Wilk, K. E., and Andrews, J. R., 1998, Biomechanics of the knee during closed kinetic chain and open kinetic chain exercises, *Med Sci Sports Exerc.* **30**: 556-569.

Evans, C. H., 1999, Cartilage and synovium, in: *Orthopedic Surgery: The Essentials, M. E. Baratz,A. D. Watson, and J. E. Imbriglia*, eds., Thieme Medical Publishers Inc, New York, U.S.A. pg. 33.

Evans, C. H., 1999, Cytokines and the role they play in the healing of ligaments and tendons, *Sports Med.* **28**: 71-76.

Evans, C. H., Ghivizzani, S. C., and Robbins, P. D., 2000, Potential applications of gene therapy in sports medicine, *Phys. Med. Rehab. Clin. North Am.* **11**: 405-417.

Evans, T. A., Putukian, M., Earl, J. E., et al., 2000, Frequency of specific entities of female athlete triad among female collegiate athletes at an NCAA Division I institution, *J. Ath. Training.* **35**: S88.

Fleisig, G. S., Andrews, J. R., Dillman, C. J., and Escamilla, R. F., 1995, Kinetics of baseball pitching with implications about injury mechanisms, *Am. J. Sports. Med.* **23**: 233-239.

Frank, C. B., 1996, Ligament healing: Current knowledge and clinical applications, *J. Am. Acad. Orthop. Surg.* **4**: 74-83.

Friedman, R. J., Acurio, M. T., Davis, R., and Classen, B. et al., 1992, Effects of growth factors and indomethacin on fracture healing, *Trans. Orthop, Res, Soc*, **17**:421.

Frost, H. M., 1989, The biology of fracture healing: an overview for clinicians: part 1, *Clin. Orthop.* **248**: 283-293.

Garrett, Jr., W. E., Safran, M. R., Seaber, A. V., Glisson, R. R., and Ribbeck, B. M., 1987, Biomechanical comparison of stimulated and nonstimulated skeletal muscle pulled to failure, *Am. J. Sports. Med.* **15**: 448-454.

Genant, H. K., Gordon, C., Jiang, Y., Lang, T. F., Link, T. M., and Majumdar, S, 1999, Advanced imaging of bone macro and micro structure, *Bone.* **25**: 149-152.

Ghivizzani, S. C., Oligino, T. J., Robbins, P. D., and Evans, C. H., 2000, Cartilage injury and repair, *Phys. Med. Rehab. Clin. North. Am.* **11**: 289-307.

Haupt, G., 1997, Use of extracorporeal shock waves in the treatment of pseudarthrosis, tendinopathy and other orthopedic diseases, *J. Urology.* **158**: 4-11.

Hildebrand, K. A., and Frank, C. B., 1998, Scar formation and ligament healing, *Can. J. Surg.* **41**: 425-429.

Huxley, A. F., 2000, Cross-bridge action: Present views, prospects, and unknowns, in: *Skeletal Muscle Mechanics: From Mechanisms to Function,* W. Herzog, ed., John Wiley & Sons, Ltd., West Sussex, U.K., pp. 7-31.

Hyman, J., and Rodeo, S. A., 2000, Injury and repair of tendons and ligaments, *Phys. Med. Rehabil. Clin. N. Am.* **11**: 267-288.

Janda, D. H., Bir, C., and Kedroske, B., 2001, A comparison of standard vs. breakaway bases: an analysis of a preventative intervention for softball and baseball foot and ankle injuries, *Foot Ankle Int.* **22**: 810-816.

Jarvinen, M. J., and Lehto, M. U. K., 1993, The effects of early mobilisation and immobilisation on the healing process following muscle injuries, *Sports Med.* **15**: 78-89.

Jarvinen, T. A. H., Kalimo, H., and Jarvinen, M., 1998, Healing of muscle injuries – biological, background and clinical treatment, in: *Controversies in Orthopedic Sports Medicine,* K. M. Chan,F. Fu, N. Maffulli,C. Rolf, M. Kurosaka, and S. Liu, eds., Williams & Wilkins Asia-Pacific Ltd., Hong Kong, pp. 475-485.

Johnston, K. M., Ptito, A., Chankowsky, J., and Chen, J. K., 2001, New Frontiers in diagnostic imaging in concussive head injury, *Clin. J. Sport Med.* **11**: 166-175.

Józsa, L. G., and Kannus, P., 1997, Effects of activity and inactivity on tendons, in: *Human Tendons, Anatomy, Physiology and Pathology*, Human Kinetics, Champaign, IL, U.S.A., pp. 98-160.

Kaplan, N., Wickiewicz, T. L., and Warren, R. F., 1990, Primary surgical treatment of anterior cruciate ligament ruptures: A long-term follow-up study, *Am. J. Sports. Med.,* **18**: 354-358.

Khan, K. M., Cook, J. L., Bonar, F., Harcourt, P., and Åström M., 1999, Histopathology of common tendinopathies: Update and implications for clinical management, *Sports. Med.* **27**: 393-408.

Kielty, C. M., Hopkinson, I., and Grant, M. E., 1993, Collagen: The collagen family: Structure, assembly, and organization in the extracellular matrix, in: *Connective Tissue and its Heritable Disorders,* P. M. Royce and B. Steinmann, eds., Wiley-Liss, New York, New York, U.S.A., pp. 103-148.

Kocher, M. S., DiCanzio, J., Zurakowski, D., and Micheli, L. J., 2001, Diagnostic performance of clinical examination and selective magnetic resonance imaging in the evaluation of intraarticular kknee disorders in children and adolescents, *Am. J. Sports Med.* **29**: 292-296.

Komi, P. V., Linnamo, V., Silventoinen, P., and Sillanpää, M., 2000, Force and EMG power spectrum during eccentric and concentric actions, *Med. Sci. Sport.* **32**:1757-1762.

Krause, H., 1997, Physik und Technik Medizinishcher Stoβwellemsysteme, in: *Extrakiorporate Stoβwellentherapie, Rompe, J.D.*, Chapman & Hall, GmbH, Weinheim, Germany, pp.15-34.

Kurosaka, M., Yoshiya, S., and Andrish, J. T., 1987, A biomechanical comparison of different surgical techniques of graft fixation in anterior cruciate ligament reconstruction, *Am. J. Sports Med.* **15**: 225-229.

Lackomy, H. K. A., 1998, Strength and power (Ch 1.3.2), in: *Oxford Textbook of Sports Medicine,* M. Harries, C. Williams, W. D. Stanish, and L J. Micheli, eds., Oxford University Press, Oxford, U.K., pp. 171-185.

Larson, R. L., 1993, Principles of replacement reinforcement, *Curr. Orthop.* **7**: 94-100.

Lephart, S., 1994, Reestablishing proprioception, kinesthesia, joint position sense, and neuromuscular control in rehabilitation, in: *Rehabilitation Techniques in Sports Medicine (2nd Ed.*), W. E. Prentice, ed., Mosby, St. Louis, Missouri, U.S.A. pp. 118-137.

Lieber, R. L., and Friden, J., 1993, Muscle damage is not a function of muscle force but active muscle strain, *J. Appl. Physiol.* **74**:520-526.

Lopes, V., Jones, J. A. H., Soan, P., and McWilliam, L., 1994, Temporomandibular ganglion or synovial cyst? A case report and literature review, *Oral. Surg. Oral. Med. Oral. Pathol.* **77**: 627-630.

MacIntosh, B. R., and Allen, D. G., 2000, Contractile changes and mechanisms of muscle fatigue, in: *Biomechanics and biology of movement,* B. Nigg, B. R. MacIntosh, and J. Mester, eds., Human Kinetics, Champaign, IL, U.S.A., pp. 365-383.

McAllister, T. W., Saykin, A. J., Flashman, L. A., and Sparling, M. B., Johnson, S. C., Guerin, S. J., Mamourian, A. C., Weaver, J. B., Yanofsky, N., 1999, Brain activation during working memory 1 month after mild traumatic brain injury: A functional MRI study, *Neurology.* **53**: 1300-1308.

Menche, D. S., Vangsness, C. T. Jr., Pitman, M., Gross, A. E., and Peterson, L., 1998, The treatment of isolated articular cartilage lesions in the young individual, in: *Instructional Course Lectures 47*, W. D. Cannon, ed., American Academy of Orthopaedic Surgeons, Rosemont, IL, U.S.A., pp. 505-515.

Merritt, C.R., 2001, Technology Update, *Radiol. Clin. North Am.* **39**:3:385-397.

Micheli, L. J., 1987, The traction apophysitises, *Clin. Sports Med.* **6**: 389-404.

Micheli, L. J., Browne, J. E., Erggelet, C., Fu, F., Mandelbaum, B., Moseley, J. B., and Zurakowski, D., 2001, Autologous chondrocyte implantation of the knee: multicenter experience and minimum 3-year follow-up, *Clin. J. Sport.Med.* **11**:4:223-8.

Micheli, L. J., Solomon, R., Solomon, J., Plasschaert, V. F., and Mitchell, R., 1999, Surgical treatment for chronic lower-leg compartment syndrome in young female athletes, *Am. J. Sports Med.* **27**:197-201.

Micheli, L. J., Stater, J. A., Woods, E., and Gerbino, P. G, 1986, Patella alta and the adolescent growth spurt, *Clin. Orthop.* **213**:159-163.

Mishra, D. K., Friden, J., Schmitz, M. C., et al., 1995, Anti-inflammatory medication after muscle injury. A treatment resulting in short-term improvement but subsequent loss of muscle function. *J. Bone. Joint. Surg. Am.* **77**: 1510-1519.

Mohler, L. R., Styf, J. R., Pedowitz, R. A., Hargens, A. R., and Gershuni, D. H., 1997, Intramuscular deoxygentation during exercise in patients who have chronic anterior compartment syndrome of the leg, *J. Bone Joint Surg. Am.* **79**: 844-849.

Mow, V. C., Homes, M. H., and Lai, W. M., 1984, Fluid transport and mechanical properties of articular cartilage: A review, *J. Biomech.* **17**: 377.

Mow, V. C., Kuei, S. C., Lai, W. M. et al., 1980, Biphasic creep and stress relaxation of articular cartilage in compression? Theory and experiments, *J. Biomech. Eng.* **102**: 73.

Mukherjee, D., Nissen, S. E., and Topol, E. J., 2001, Risk of Cardiovascular Events Associated with Selective COX-2 Inhibitors, *J.A.M.A.* **8**: 954-959.

Needleman, P., and Isakson, P. C., 1997, The discovery and function of COX-2, J. Rheumatol. **24** Suppl: 6-8.

Newcomer, K. L., Laskowski, E. R., Idank, D. M., McLean, T. J., and Egan, K. S., 2001, Corticosteroid injection in early treatment of lateral epicondylitis, *Clin. J. Sport Med.* **11**: 214-22.

Newman, A. P, 1998, Articular cartilage repair, *Am. J. Sports. Med.* **26**: 309-324.

Newman, J. S., Adler, R. S., Bude, R. O., and Rubin, J. M., 1994 Detection of soft-tissue hyperemia: value of power Doppler sonography, *Am. J. Roentgenol.* **163**: 385-389.

Nigg, B. M., 1998, Biomechanics as applied to sports (Ch 1.3.1), in: *Oxford Textbook of Sports Medicine,* M. Harries, C. Williams, W. D. Stanish, and L. J. Micheli, eds., Oxford University Press, Oxford, U.K., pp. 153-171.

Nigg, B. M., 1999, Selected Historical Highlights, in: *Biomechanics of the musculo-skeletal system (2nd edition),* B. M. Nigg and W. Herzog,, eds., John Wiley & Sons, Chichester, West Sussex, p. 7-8.

Nirschl, R. P., 1992, Elbow tendinosis/tennis elbow, *Clin. Sports. Med.* **11**: 851-70.

Noonan, T. J., Best, T. M., Seaber, A. V., and Garrett Jr., W. E., 1994, Identification of a threshhold for skeletal muscle injury, *Am. J. Sports. Med.* **22**: 257-261.

Noyes, F. R., and Barber-Westin, S. D., 1996, Reconstruction of the anterior cruciate ligament with human allograft: Comparison of early and later results, *J. Bone Joint Surg Am.* **78**: 524-537.

Noyes, F. R., 1977, Functional properties of knee ligaments and alterations induced by immobilization, *Clin. Orthop. Res.* **123**: 210-242.

Oakes, B. W., 1998, Tendon/ligament basic science (Ch 4.4.3), in: *Oxford Textbook of Sports Medicine*, M. Harries, C. Williams, W. D. Stanish, and L. J. Micheli, eds., Oxford University Press, Oxford, U.K., pp. 584-610.

Odensten, M., Hamberg, P., Nordin, M., Lysholm, J., and Gillquist, J., 1985, Surgical or conservative treatment of the acutely torn anterior cruciate ligament. A randomized study with short-term follow-up observations., *Clin. Orthop.* **198**:87-93.

Ogden, J. A., Toth-Kischkat, A., and Schultheiss, R., 2001, Principles of shock wave therapy, *Clin. Orthop.* **387**: 8-17.

Paessler, H, 1995, Minisymposium. Part 1: Anterior cruciate ligaments: the history of cruciate ligaments, *Sports. Exerc. Injury.* **1**: 166-171.

Pagnani, M. J., Warner, J. J. P., O'Brien, S. J., and Warren, R. F., 1993, Anatomic considerations in harvesting the semitendinosus and gracilis tendons and a technique of harvest, *Am. J. Sports Med.* **21**: 565-571.

Peterfy, C. G., Majumdar, S., Lang, P., Van Dijke, C. F., Sack, K., and Genant, H. K., 1994, MR imaging of the arthritic knee. Improved discrimination of cartilage, synovium and effusion with pulse saturation transfer and fat-suppressed T1-weighted sequences, *Radiology.* **191**: 413.

Peterfy, C. G., Roberts, T., and Genant, H. K., 1997, Dedicated extremity MR imaging: An emerging technology, *Radiol. Clin. North. Am.,* **35**: 1-20.

Rappeport, E. D., Mehta, S., Wieslander, S. B., Lausten, G. S., and Thomsen, H. S., 1996, MR imaging before arthroscopy in knee joint disorders?, *Acta. Radiologica.* **37**: 602-609.

Renstrom, P. A. F. H., and Lynch, S. A., 1998, An overview of anterior cruciate ligament management, in: *Controversies in Orthopedic Sports Medicine,* K. M. Chan, F. Fu, N. Maffulli,C. Rolf, M. Kurosaka, and S. Liu, eds., Williams & Wilkins Asia-Pacific Ltd., Hong Kong, pp. 5-22.

Rodeo, S. A., Arnoczky, S. P., Tozilli, P. A. et al., 1993, Tendon healing in a bone tunnel: a biomechanical and histological study in the dog, *J. Bone Joint Surg. Am.* **75**: 1795-1803.

Rorabeck, C. H., Bourne, R. B., Fowler, P. J., Finlay, J. B., and Nott, L., 1988, The role of tissue pressure measurement in diagnosing chronic exertional compartment syndrome, *Am. J. Sports Med.,* **16**: 143-146.

Sabiston, P., Frank, C., Lam, T., and Shrive, N., 1990, Transplantation of the rabbit medial collateral ligament. II. Biomechanical evaluation of frozen/thawed allografts, *J. Orthop. Res.* **8**:46-56.

Safran, M. R., Allen, A. A., Lephart, S. M., Borsa, P. A., Fu, F. H., and Harner, C. D., 1999, Proprioception in the posterior cruciate ligament deficient knee, *Knee. Surg. Sports. Traumatol. Arthrosc.* **7**: 310-317.

Safran, M. R., Borsa, P. A., Lephart, S. M., Fu, F. H., and Warner, J. J. P., 2001, Shoulder proprioception in baseball pitchers, *J. Shoulder. Elbow. Surg.* **10**: 438-444.

Sapega, A. A., Moyer, R. A., Schneck, C., and Komalahiranya, N., 1990, Testing for isometry during reconstruction of the anterior cruciate ligament, *J. Bone Joint Surg. Am.* **72**: 259-267.

Schaffler, M. B., Radin, E. L., and Burr, D. B., 1990, Long-term fatigue behaviour of compact bone at low strain magnitude and rate, *Bone.* **11**: 321-326.

Schepsis, A. A., Martini, D., and Corbett, M., 1993, Surgical management of exertional compartment syndrome of the lower leg: Long term follow up, *Am. J. Sports Med.*, **21**: 811-817.

Schumacher, Jr., H. R., Klippel, J. H., and Koopman, W. J., 1993, Structural molecules of connective tissues, in: *Primer on the Rheumatic Diseases Tenth Edition,*Arthritis Foundation, Atlanta, GA, U.S.A., pp. 16-27.

Seeger, L. L., 1998, Imaging sports injuires: The future, *Clin. J. Sports Med.* **8**: 47-51.

Slocum, D. B., and Larson, R. L., 1968, Pes anserinus transplantation. A surgical procedure for control of rotatory instability of the knee. *J. Bone Joint Surg. Am.* **50**: 226-42.

Stanish, W. D., and Wood, R. M., 1998, Overuse injuries of the knee (Ch 5.4), in: *Oxford Textbook of Sports Medicine*, M. Harries, C. Williams, W. D. Stanish, and L. J. Micheli, eds., Oxford University Press, Oxford, U.K., pp. 679-693.

Taaffe, D. R., Duret, C., Cooper, C. S., and Marcus, R., 1999, Comparison of calcaneal ultrasound and DXA in young women, *Med. Sci. Sports. Exerc.* **31**: 1484-1489.

Thain, L. M. F., Lee, S. L., and Downey, D. B., 2001, Ultrasonography of the lower leg: technique, anatomy and pathologic conditions, *Can. Assoc. Radiol. J.* **52**: 325-336.

Tung, G. A., and Brody, J. M., 1997, Contemporary imaging of athletic injuries, *Clin. Sports. Med.* **16**: 393-417.

Van der Wall, H., Storey, G.., Frater, C., and Murray, P., 2001, Importance of positioning and technical factors in anatomic localization of sporting injuries in scintigraphic imaging, *Semin. Nucl. Med.*, **31**: 17-27.

VanDuersen, R. W., Sanchez, M. M., Ulbrecht, J. S., et al., 1998, The role of muscle spindles in ankle movement perception in human subjects with diabetic neuropathy, *Exp. Brain. Res.* **120**: 1-8.

Verdonk, R., 1997, Alternative treatments for meniscal injuries, *J. Bone Joint Surg. Br.* **79**: 866-873.

Winter III, T. C., Teefey, S. A., and Middleton, W. D., 2001, Musculoskeletal ultrasound: An update, *Radiol. Clin. North. Am.* **39**:3: 465-483.

Wilson, A. M., and Goodship, A. E., 1994, Exercise-induced hyperthermia as a possible mechanism for tendon degeneration, *J. Biomechanics.* **27**: 899-905.

Woo, S. L. Y., Smith, B .A., Livesay, G. A., and Blomstrom, G. L., 1993, Why do ligaments fail?, *Curr. Orthop.* **7**: 73-84.

Zheng, N., and Barrentine, S. W., 2000, Biomechanics and motion analysis applied to sports. *Phys. Med. Rehabil. Clin. N. Am.* **11**: 309-322.

# SUBJECT INDEX